Techniques and Experiments in Organic Chemistry

Biological Perspectives and Sustainability

Techniques and Experiments in Organic Chemistry

Biological Perspectives and Sustainability

GREGORY K. FRIESTAD

THE UNIVERSITY OF IOWA

W. W. NORTON & COMPANY

Celebrating a Century of Independent Publishing

W. W. Norton & Company has been independent since its founding in 1923, when William Warder Norton and Mary D. Herter Norton first published lectures delivered at the People's Institute, the adult education division of New York City's Cooper Union. The firm soon expanded its program beyond the Institute, publishing books by celebrated academics from America and abroad. By midcentury, the two major pillars of Norton's publishing program—trade books and college texts—were firmly established. In the 1950s, the Norton family transferred control of the company to its employees, and today—with a staff of five hundred and hundreds of trade, college, and professional titles published each year—W. W. Norton & Company stands as the largest and oldest publishing house owned wholly by its employees.

Editor: Rob Bellinger
Senior Associate Managing Editor, College: Carla L. Talmadge
Developmental Editor: John Murdzek
Editorial Assistant: Aidan Windorf
Associate Director of Production, College: Benjamin Reynolds
Managing Editor, College: Marian Johnson
Media Editor: Marilyn Rayner
Chemistry Content Development Specialist: Dr. Richard L. Jew
Associate Media Editor: Liz Vogt
Media Project Editor: Jesse Newkirk
Media Assistant Editor: Manny Ruiz
Managing Editor, College Digital Media: Kim Yi
Ebook Producer: Sophia Purut
Marketing Director, Chemistry: Stacy Loyal
Design Director: Rubina Yeh
Designer: Anne-Michelle Gallero
Director of College Permissions: Megan Schindel
College Permissions Specialist: Josh Garvin
Photo Editor: Mike Cullen
Composition: GW, Inc./Project Manager: Gary Clark
Illustrations: Alicia Elliott, Spark Life Science Visuals
Manufacturing: Transcontinental Interglobe—Beauceville, Québec

Permission to use copyrighted material is included alongside the appropriate content.

ISBN 978-1-324-04576-2 (paperback)

W. W. Norton & Company, Inc., 500 Fifth Avenue, New York, NY 10110
wwnorton.com

W. W. Norton & Company Ltd., 15 Carlisle Street, London W1D 3BS

1 2 3 4 5 6 7 8 9 0

To my family, who always inspire me

To my students, who invariably teach me

Brief Contents

Contents

PART A
TECHNIQUES   1

Jeffrey B. Banke/Shutterstock.

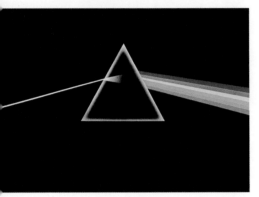
Records/Alamy Stock Photo.

PART B
EXPERIMENTS 143

Maria Uspenskaya/Shutterstock.

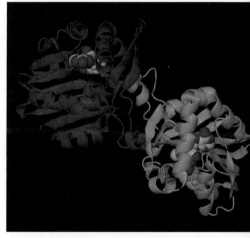

Courtesy of Dr. Angel Herráez, Universidad de Alcalá, Creative Commons cc-by 4.0.

Pictorial Press Ltd/Alamy Stock Photo.

Natalia Kuzmina/Alamy Stock Photo.

Design Pics Inc/Alamy Stock Photo.

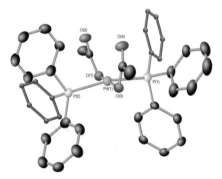

Republished with permission of The Royal Society of Chemistry. Scott, N. W. J.; Ford, M. J.; Schotes, C.; Parker, R. R.; Whitwood, A. C.; Fairlamb, I. J. S. The Ubiquitous Cross-Coupling Catalyst System 'Pd(OAc)₂'/2PPh₃ Forms a Unique Dinuclear Pd Complex: An Important Entry Point into Catalytically Competent Cyclic Pd₃ Clusters. *Chem. Sci.* **2019**, *10*, 7898–7906. DOI: 10.1039/C9SC01847F. ©2019; permission conveyed through Copyright Clearance Center, Inc.

Techniques Videos

Aqueous–Organic Extractions and Drying Organic Solutions

Boiling Point Measurement

Cleaning Glassware

Column Chromatography

Fractional Distillation

Gravity Filtration

Heating at Reflux

Melting Point Measurement

Neutralizing Acidic or Basic Solutions and Checking pH

Polarimetry

Preparing NMR Samples

Recrystallization

Rotary Evaporation

Simple Distillation

Thin-Layer Chromatography

Vacuum Filtration

Preface

A LAB CURRICULUM THAT EMPHASIZES CONNECTIONS AMONG CHEMISTRY, BIOLOGY, AND SUSTAINABILITY

Dear Instructor,

My mission in creating this textbook you're reading was to develop a green organic chemistry laboratory curriculum that inspires student engagement by emphasizing the connections among chemistry, biology, and sustainability. Incorporating concepts of green chemistry was a key aspect, not only to limit environmental impact and operational costs, but also to appeal to broader student and instructor interests. An important secondary consideration was to integrate the green chemistry with instruction in traditional techniques and glassware for the benefit of those students who need a practical foundation for careers in chemistry or other scientific disciplines and/or graduate study. Thirdly, the textbook needed to be flexible for use in one- or two-semester formats, whether the lab is integrated with the lecture or as a separate course.

How did this project start? At Iowa, we have a one-semester Organic Laboratory course for non-majors that meets twice a week plus lecture, with an enrollment of 150–200 per semester, including some chem majors who take it for scheduling reasons. Our non-majors course was taught without any major curriculum revision for many years. Over this time, many informal discussions about the need for revising some experiments, mainly for reasons of modernization and safety, prompted me to take action. Our graduate student teaching assistants and departmental lab staff recognized the need as well, and were willing and eager to test new experiments, so I began a significant curriculum revision, replacing a couple of experiments per semester. After a few semesters it became apparent that an emphasis on biological perspectives and sustainability had emerged as a coherent theme, and that it was an effective way to reach students with wide-ranging interests. It was then I realized that this emphasis, new for our course, could likely be of interest to the larger Chemistry Education community as well.

Why is a new book needed? Over the years, existing textbooks have tackled waste disposal, safety, and cost issues by downsizing the scale of reactions using specialized glassware that students won't likely see anywhere else. However, this approach is unsuitable for students who will need familiarity with standard glassware and realistic preparative scale reactions, whether it's in future employment or in graduate study. The microscale approach deals with the local classroom sustainability problem, but is less effective in teaching students to think about sustainability in real-world chemical processes where large scales are inevitable.

Compared with traditional lab texts—many of which contain hundreds of pages that are often not used in the typical undergraduate setting—a single text combining green chemistry principles and experiments with traditional organic techniques instruction offers cost and convenience advantages to both students and instructors. There is a clear need for this new textbook.

Goals

My motivations for this textbook likely mirror concerns that inspire curriculum revisions at other institutions. I wanted to

a. replace certain time-worn experiments that had grown somewhat stale,

b. enhance connections to biological chemistry,

c. devote increased attention to issues of sustainability, and

d. better coordinate topics between lecture and lab.

To address these issues, I began work on the following goals:

ENLIVEN EARLY SEMESTER INSTRUCTION ON TECHNIQUES

Teach techniques like melting point, recrystallization, and distillation as means to an end rather than ends in themselves, by placing them in the context of natural product isolation and synthetic reactions. Early semester work in the organic lab would provide students with opportunities to practice techniques in real-world context.

DRAW CONNECTIONS TO BIOLOGICAL CHEMISTRY

Use biology as a cross-disciplinary theme to enhance the appeal to organic lab course populations that often include a large proportion of biology, pre-med, pre-dental, and environmental science students. Examples include:

- Natural product isolation—gas chromatography measurement of the content of terpenes extracted from citrus peels (Chapter 10)
- Grignard reaction followed by a greener oxidation method—preparation of naturally occurring insect pheromones (Chapter 21)
- Plant material as a feedstock for an organic reaction—furfural from corn cobs (Chapter 22)
- Biomimetic thiamine-catalyzed furoin synthesis—a more reliable analog of the classic biocatalytic benzoin condensation (also Chapter 22)
- Enzyme-catalyzed organic reactions—perform biocatalytic reduction with natural enzymes from plant material (Chapter 23)

EMPHASIZE SUSTAINABILITY

Introduce green chemistry early, and emphasize sustainability in a greater proportion of the experiments, while still teaching fundamentals of practical organic chemistry. Students are more motivated when their study of organic chemistry is tied to real-life applications. Going forward in their careers, they will benefit from a foundation that emphasizes the responsibility and accountability of chemists in solving or preventing chemistry-related problems. In addition to introducing the 12 Principles of Green Chemistry in Chapter 1, the text's experiments:

- Use less hazardous reagents, e.g., green oxidation with hydrogen peroxide (Chapter 16)
- Use renewable resources, e.g., furfural from corn cobs (Chapter 22), light energy for promoting a cycloaddition (Chapter 18)
- Use solvent-free or aqueous reaction conditions, e.g., solventless aldol reaction (Chapter 24)
- Use catalytic transformations that generate less waste, e.g., enzyme-catalyzed reduction (Chapter 23), palladium-catalyzed coupling (Chapter 25)

ALIGN TOPICS WITH COVERAGE IN LECTURE COURSES

Use mechanisms and reactions typically covered in Organic I in the earlier experiments, moving mechanisms and reactions covered in Organic II to later in the book. Instructors of labs that are tied to the lecture course will find this organization useful.

Students who take the lab course later as a one-semester course (as we do here at Iowa) will also learn more effectively by reviewing and reinforcing topics in an organization that parallels lecture courses, building complexity in a logical way.

PROVIDE EXPERIENCE WITH MULTISTEP SYNTHESIS

Several experiments use the product from a previous step as the starting material. In this way, students gain a very personal perspective on reaction efficiency and why it is important. This is valuable for students who will go on in chemistry, of course, but it also benefits students moving on to health sciences careers, so that they can understand how pharmaceuticals are made and why they are so expensive.

BRIEF OVERVIEW OF THE BOOK

When this textbook began to emerge from my course redesign, the potential advantages to a broader range of instructors and students became clearer. As seen in the table of contents, the organization of Techniques instruction in Part A places general topics (safety, recordkeeping, and report-writing) into Chapters 1–4, followed by standard separations, purifications, physical properties, and spectroscopy in Chapters 5–8. Collecting all of this together in Part A allows for the instructor to reference specific chapters and sections as reading assignments, whether using experiments from this textbook or their own materials, or both. Part A is also available as a stand-alone Techniques book.

In Part B, experiments use a concise presentation that avoids redundancies in technique instruction; for example, if a recrystallization is needed, the experiment chapter refers the student to the appropriate section of Chapter 5 for instructions on that, rather than explaining it again. This promotes the idea that there is consistency and logic in lab procedure design, not just a collection of stepwise recipes.

Recognizing the need for some preliminary instruction before students are ready to carry out a complete synthetic reaction, Part B begins with a mix of "dry labs" with practical experience in exploring the literature of organic chemistry and interpreting ^{1}H NMR spectra (Chapters 9 and 11) and "wet labs" using basic techniques of extraction, recrystallization, and thin-layer chromatography (Chapters 10 and 12). Chapters 13–16 involve substitutions, electrophilic additions, and eliminations, reactions and mechanisms that are typically found in the first semester of the lecture course. This includes a vehicle for introducing polymer chemistry associated with synthesis of adipic acid. In Chapters 17–24, emphasis turns to reactions and mechanisms often covered in a second-semester course, focusing on cycloadditions, electrophilic aromatic substitution, and carbonyl additions, including nucleophilic acyl substitution.

This portion also carries the greatest emphasis on biological chemistry, including coverage of biopolymers (Chapter 22) and biocatalysis (Chapter 23). The introductory materials of Chapters 19–21 and 24 also are designed to connect organic chemistry topics with biology, such as the roles of aromatic halides in drug–protein interactions (Chapter 19); natural porphyrin-like compounds such as vitamin B_{12}, heme, and chlorophyll; and small molecules used in biological communication (Chapter 21). The aldol reactions of Chapter 24 offer lab activities in carbonyl addition and condensation chemistry that can help reinforce their connections with biosynthesis of fatty acids and polyketides.

The final three chapters of Part B cover more advanced or specialized topics, coverage of which is often somewhat instructor-dependent in the typical organic chemistry curriculum.

- Chapter 25 explores transition metal catalysis via an aqueous Suzuki reaction, offering coverage of a topic that increasingly demands attention in ACS-certified courses and preparation for chemistry careers.

- Chapter 26 serves up a unique opportunity to develop logical thinking and hypothesis generation through the practice of qualitative organic analysis. Although qualitative functional group tests and solid derivatives are justifiably viewed as obsolete in real-world practice, this chapter builds active engagement with the scientific method; students decide what experiment to do based on their evaluation of prior results. Early anxiety gives way to confidence as they learn to critically evaluate data that are sometimes contradictory, locate the key diagnostics in their spectra, and propose structural hypotheses. Their excited voices discussing their results as they pass through the hallways before and after lab tell me that they are actively learning and teaching each other—always a heartwarming moment for an organic chemistry instructor.
- Chapter 27 exposes students to molecular modeling and computational chemistry to examine aspects of structure and spectroscopy; this is a "dry lab" that fits nicely at the end of a semester when practical scheduling needs may preclude bench chemistry.

One of the components of any instruction in green chemistry is minimizing the production of waste materials. At Iowa, our Environmental Health and Safety officials track the amount of waste generated by the Chemistry Department, and gratifyingly, with the inception of this new curriculum, waste quantities were reduced to two-thirds of the prior amounts. Now, when I discuss the 12 Principles of Green Chemistry with my students, I can point to our own class as an example of the tangible impact we can all have upon environmental protection. My hope is that incorporating green chemistry experiments at your own institution will yield similar outcomes.

CONCLUDING REMARKS

During our development toward the goals outlined above, student engagement improved as judged by comments in student evaluations, and my own interest in teaching the course strengthened. What a great synergy!

Our non-majors' organic laboratory course at Iowa, with strong themes of biological chemistry and sustainability woven throughout, has proved very appealing to many hundreds of students who have already used the curriculum in this textbook. Their unsolicited positive comments and sincere engagement in the course have been very gratifying, and inspired my proposal to take the text to a wider audience. Quite a lot of critical thinking and adaptations based on instructor, student, teaching assistant, and reviewer feedback have been implemented to make this material more generalizable to other universities and colleges.

I thank you for taking the time to consider this textbook for your course, and I welcome your feedback so that I can continue to improve it in the years to come.

RESOURCES FOR INSTRUCTORS
Instructor's Guide

Written by the author of the text, the instructor's guide provides support resources for instructors incorporating these experiments into their course.

Based on Dr. Friestad's teaching experience of over 15 semesters of organic chemistry lab, the instructor's guide provides the support resources instructors need in a searchable digital format. The instructor's guide features instructions for all 19 experiments in the book that will make it easy to incorporate these experiments, including:

- Tips for success
- Predicted yields
- List of all glassware used in each experiment
- Prep sheets for the stockroom

RESOURCES FOR STUDENTS

Techniques Videos

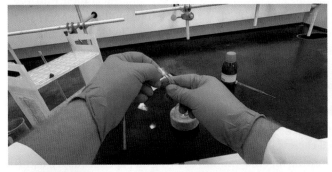

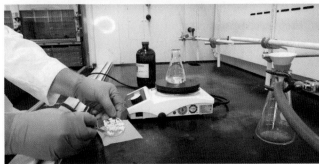
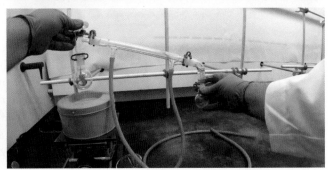

Courtesy of Gregory K. Friestad.

Developed by Dr. Friestad in collaboration with the Norton Chemistry team, 16 videos focus on the most common laboratory techniques used in the course. These videos are integrated into the ebook in both the Techniques chapters and the Experiments chapters where they are referenced, and are also included in the Smartwork pre-lab activities, ensuring students receive the support they need.

Smartwork Online Homework

Smartwork is an easy-to-use online homework system that helps students become better problem-solvers through a variety of interactive question types, book-specific hints, and extensive answer-specific feedback.

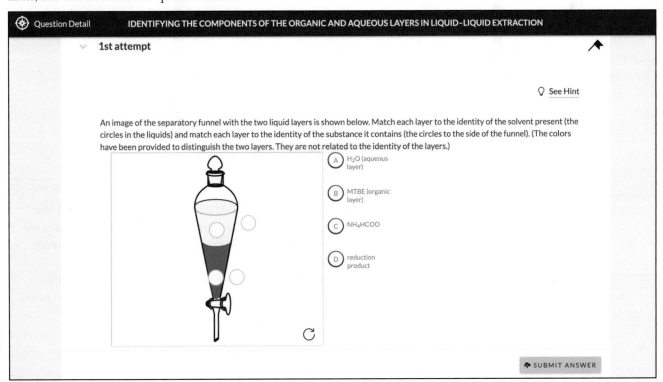

Get started quickly with our pre-made, pre-lab assignments that accompany each of the 19 experiments in the book, or take advantage of Smartwork's flexibility by customizing questions and adding your own content. Integration with your campus learning management system (LMS) saves you time by allowing Smartwork grades to report directly to your LMS gradebook, while individual and classwide performance reports help you see students' progress, identify challenging concepts, and intervene to help students who struggled with those concepts.

The Smartwork course features the following:

- **An expert author team.** The Smartwork course was authored by instructors who teach at a diverse group of schools.
- **An unparalleled chemical structure drawing tool.** Smartwork contains an upgraded 2-D drawing tool that mimics drawing on paper, reduces frustration, and helps students focus on the problem at hand. This intuitive tool supports both multistep mechanism and multistep synthesis problems and provides students with answer-specific feedback for every problem. The 2-D drawing tool has a variety of features that make drawing easy and efficient. Students are provided with templates, including a variety of common rings and a carbon chain drawing tool. In addition, Smartwork presents students with a toolbar of commonly used elements, the ability to add lone pairs with a single click, and easy-to-use features such as undo, redo, single-click erase, and zoom-in/zoom-out.

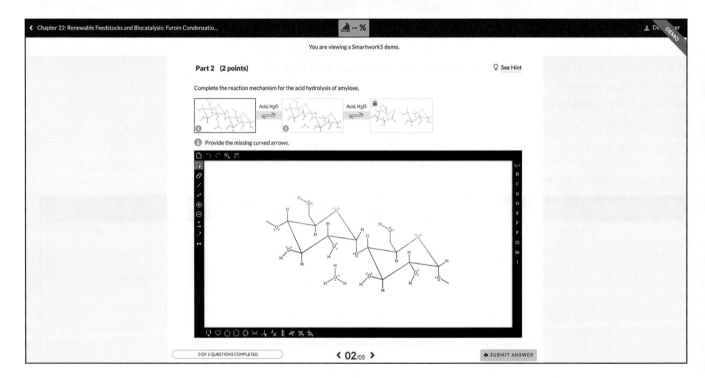

- **Pooled problem sets.** Every question has three versions that are "pooled" into one problem so different students receive different problems from the pools to encourage independent work. Instructors can choose our preset pools or create their own.
- **Video-based problems.** These questions use the Techniques Videos to ensure that students understand the techniques required and are prepared to succeed in the lab.

Ebook

Norton Ebooks offer an enhanced reading experience at a fraction of the cost of a print textbook. New videos are embedded directly in line with the text, helping students visualize the techniques as they are introduced. The ebook provides an active reading experience with the ability to take notes, bookmark, search, highlight, and read offline. Instructors can even embed their own videos or notes that students can see as they read the text. Norton Ebooks can be viewed on all computers and mobile devices.

The ebook is available at no extra cost with the purchase of a new print text, or it may be purchased as a stand-alone with Smartwork and the videos.

Preface for Students

A NOTE TO STUDENTS

Learning organic chemistry is an exciting endeavor because organic chemistry is the language of life. Indeed, organic chemistry is critical to communication, energy transfer and storage, nutrient uptake, growth, replication, and virtually everything else needed for life to exist on earth. Our understanding and manipulation of organic chemistry also impacts our daily lives in uncounted ways: treating our diseases (drugs), fueling our transportation (petroleum), maintaining our bodies (food), coloring our clothes (dyes), and constructing our homes (building materials). And don't forget, you wouldn't have mobile communication devices without organic chemistry. Understanding something so ubiquitous in our daily lives, and so critical to life itself, clearly enriches us.

Ask organic chemists what attracted them to the field, and most will recall their first opportunity to get into the lab, tinkering with the glassware, manipulating beautiful crystals or colorful liquids, and finding something new or unexpected. This laboratory course is intended to introduce you to those appealing joys, while also strengthening your understanding of the lecture material. The lab provides technical tools to be sure, but also builds practices of logical thinking that you'll find valuable in the future, wherever you are.

While many fantastic innovations have emerged from organic chemistry, we also know that hazards can exist throughout product life cycles, from design, development, production, and storage, then on through the use (or misuse) and disposal of the products. Lack of attention to potential hazards can create unintended consequences now or in the future. Can we anticipate these consequences, and minimize them, while we continue to creatively address the scientific questions and technological goals that are at the core of the human condition? I believe we can, and I want you and future students to learn organic chemistry while keeping this perspective in mind. That is part of the reason for this book.

Another important goal for me is to provide a cost-effective option for students in the organic chemistry lab. You likely have an organic chemistry textbook to accompany the lecture. Other lab textbooks can be just as large as that one, with hundreds of pages that go unused. I wanted to provide a more focused book, avoiding an "everything-under-the-sun" approach while delivering the instructional material you will need to succeed in the course, and beyond.

How to Use This Book

The Techniques part of the book is your resource for instructions on how to carry out all of the common procedures that are required in the Experiments part. Each experiment cross-references to the important Techniques sections so that you can study or revisit any background reading before performing the techniques. This will help you understand how each procedure works and why it is needed in the experiment.

Additionally, pay attention to the Learning Objectives provided in each chapter. These indicate key points that you should be able to explain, or key things you should be able to do, after reading the chapter and performing the experiment.

This textbook also comes with 16 Techniques Videos that you can use throughout the course—first, to familiarize yourself with each technique, then to examine again when preparing to conduct a specific experiment. These videos can help you build confidence before entering the lab, by showing that technical skills are accessible to you, even on the first try.

The process of learning organic chemistry will benefit you in intangible ways, with a fun and appealing combination of both logic and creativity. This combination stimulates the development of valuable critical thinking skills, which will be useful no matter what career path you eventually choose. In the lab, you will add hands-on experience that deepens your understanding of all the knowledge and skills you learn in the lecture classroom. I hope that you'll find that this engaged learning experience will reveal the lively and adventurous nature of organic chemistry as an endless source of innovation and a strong foundation for the health sciences.

Good luck, be safe, and have fun!
Greg Friestad

Acknowledgments

REVIEWERS

Carolina Andrade *(Lone Star College)*
Michael Ansell *(Las Positas College)*
Cosimo Antonacci *(Seton Hall University)*
Jesse Bergkamp *(California State University—Bakersfield)*
Michelle Boucher *(Utica College)*
Laura Brown *(Indiana University)*
Kathleen Brunke *(Christopher Newport University)*
Bobby Burkes *(Grambling State University)*
Christopher Callam *(The Ohio State University)*
Chad Cooley *(Indiana University)*
Cliff Coss *(Northern Arizona University)*
Sean Curtis *(Des Moines Area Community College)*
Nathan Duncan *(Maryville College)*
Jason Dunham *(Ball State University)*
Brendan Dutmer *(Highland Community College)*
Ola El-Rashiedy *(Penn State University–Abington)*
Douglas Flournoy *(Indian Hills Community College)*
Nicholas Greco *(Western Connecticut State University)*
Dustin Gross *(Sam Houston State University)*
Matthew Grote *(Otterbein University)*
Scott Hartley *(Miami University)*
Nicholas Hill *(University of Wisconsin—Madison)*
Daniel Holley *(Columbus State University)*
Kevin Jantzi *(Valparaiso University)*
Keneshia Johnson *(Alabama A&M University)*
Michael Justik *(Penn State Erie, The Behrend College)*
Renat Khatmullin *(Middle Georgia State University)*
Mike Koscho *(University of Oregon)*
Joseph Kremer *(Alvernia University)*
Shane Lamos *(Saint Michael's College)*
F. Andrew (Andy) Landis *(Penn State University–York)*
Rita Majerle *(Hamline University)*
Rock Mancini *(Washington State University)*
Sara Mata *(Indiana University)*
Vanessa McCaffrey *(Albion College)*
Sri Kamesh Narasimhan *(SUNY Corning Community College)*
Erik Olson *(Upper Iowa University)*
Steve Oster *(Middlebury College)*
Hasan Palandoken *(California Polytechnic State University–San Luis Obispo)*
Noel Paul *(The Ohio State University)*
Angela Perkins *(University of Minnesota)*
Joanna Petridou-Fischer *(Spokane Falls Community College)*
Brian Provencher *(Merrimack College)*
Matt Siebert *(Missouri State University)*

Chester (Chet) Simocko *(San José State University)*
Mackay Steffensen *(Southern Utah University)*
Anne Szklarski *(King's College)*
Gidget Tay *(Pasadena City College)*
Matthew Tracey *(University of Pittsburgh)*
Michael Wentzel *(Augsburg University)*
Laura Wysocki *(Wabash College)*
Kimo Yap *(California State University–Los Angeles)*
Hui Zhu *(Georgia Tech)*

About the Author

Courtesy of Gregory K. Friestad.

GREGORY K. FRIESTAD is associate professor in the Department of Chemistry at the University of Iowa, with research interests in asymmetric synthesis, free radical chemistry, organosilicon chemistry, and new synthetic methods with transition metal reagents and catalysts. Dr. Friestad earned a BS degree in chemistry from Bradley University and a PhD in organic chemistry from the University of Oregon, mentored by Bruce P. Branchaud.

Dr. Friestad teaches at the graduate and undergraduate levels, including more than 15 semesters teaching majors and non-majors organic laboratory courses. Since 2015, Friestad has been developing a new curriculum for the introductory organic laboratory, focusing on standard-scale experimental procedures to develop skills and experience that translate to research, while emphasizing sustainability and biological perspectives. Aside from practicing organic chemistry, he enjoys travel, live music, disc golf, and spending time outdoors with family.

Marchu Studio/Shutterstock.

PART A
TECHNIQUES

1

- Relate the importance of organic chemistry to humanity, using the example of antibiotics.

- Recognize how green chemistry principles can minimize undesirable impacts of organic chemistry.

- Apply specific metrics of green chemistry to measure the impacts of organic chemistry processes or procedures.

Organic Reactions and the Twelve Principles of Green Chemistry

ORGANIC CHEMISTRY

Organic chemistry happens all around us, and within us. We can use organic chemistry to create new substances with new properties; this creativity can coexist with all of the fascinating organic chemistry in nature.

Courtesy of Gregory K. Friestad.

hemistry has had a profound impact on human existence. The reactions of organic chemistry are how we convert readily available carbon-containing substances to new compounds and materials of greater value. This has fueled revolutionary innovations in all kinds of items in our daily experiences, from building materials to electronic devices, and from textiles to medicines such as the penicillins (**Figure 1.1**). Since organic reactions are involved in all these endeavors, learning the laboratory practice of organic chemistry can put you in a better position to understand these innovations, whether you intend to be on the front lines of discovery or a responsible end user of new inventions.

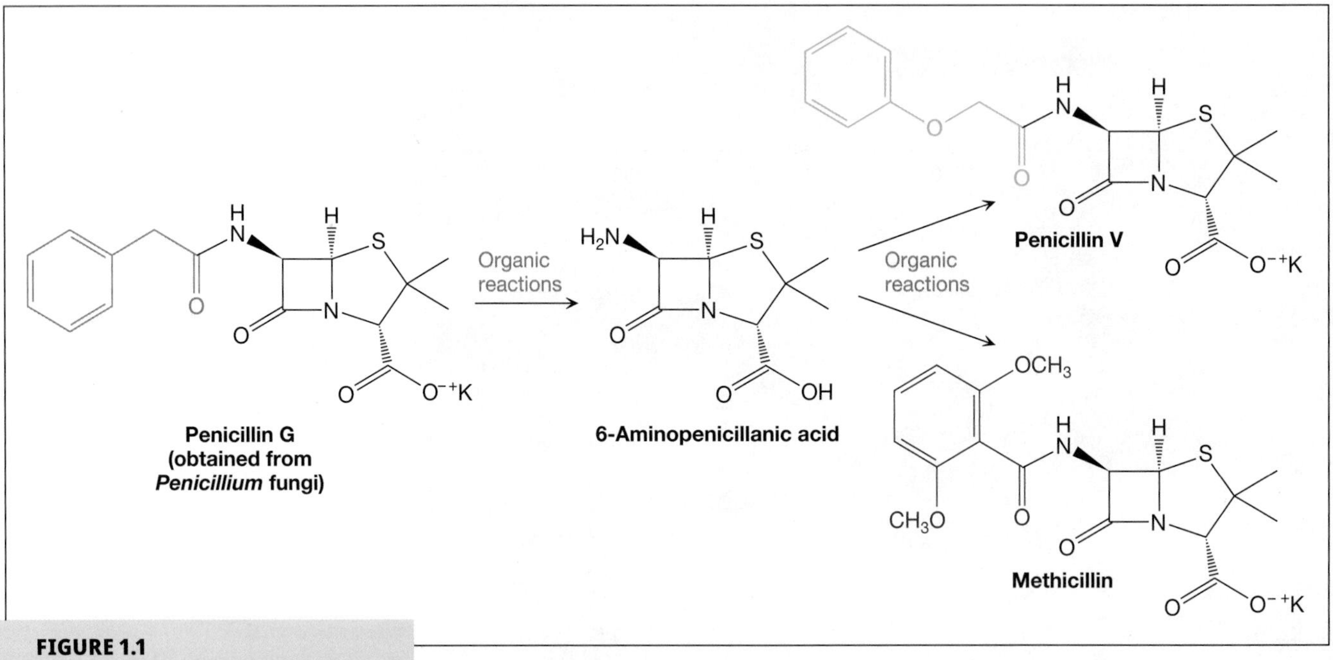

FIGURE 1.1

When bacteria developed resistance to penicillin G, transformations of the original compound to new penicillins, using organic reactions, led to the discovery and development of new antibiotics.

One of organic chemistry's most dramatic impacts has been on antibiotic drugs and the effects they have had on average life expectancy through the 20th century. Life expectancy worldwide increased dramatically from 1900 to 2010, and a comparison of the top 10 leading causes of death in 1900 versus 2010 indicates a connection to the widespread use of antibiotics (**Figure 1.2**).[1] The three top causes of death in 1900 were infectious diseases, for example, gastrointestinal infections. By 2010 these had almost disappeared from the top 10,[2] and during this same time period, there was a 30-year increase in life expectancy. In 1900, before antibiotics became a widespread tool for the treatment of infectious diseases, 30.4% of all deaths occurred among children ages 5 and below; in 1997, it was only 1.4%.[3]

A period of rapid increase in life expectancy came during 1936–1952, when antibiotics—first sulfa drugs, then penicillins—became widely introduced to the

[1]Arias, E. United States Life Tables, 2002. *Natl. Vital Stat. Rep.* **2004**, *53* (6), 1–40. https://www.cdc.gov/nchs/data/nvsr/nvsr53/nvsr53_06.pdf (accessed April 2022).

[2]Jones, D. S.; Podolsky, S. H.; Greene, J. A. The Burden of Disease and the Changing Task of Medicine. *N. Engl. J. Med.* **2012**, *366*, 2333–2338. DOI: 10.1056/NEJMp1113569

[3]U.S. Centers for Disease Control and Prevention. Achievements in Public Health, 1900–1999: Control of Infectious Diseases. *MMWR Morb. Mortal. Wkly. Rep.* **1999**, *48* (29), 621–629. https://cdc.gov/mmwr/preview/mmwrhtml/mm4829a1.htm (accessed April 2022).

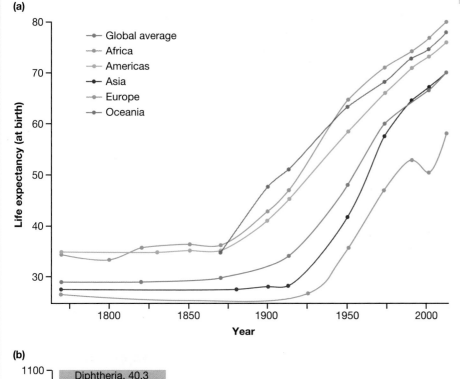

(a)

FIGURE 1.2

(a) Life expectancy change over time, and (b) the top 10 causes of death in 1900 and 2010.

Plotly Chart Studio. https://chart-studio.plotly.com/~amatelin/320 (accessed April 2022). Jones, D. S.; Podolsky, S. H.; Greene, J. A. The Burden of Disease and the Changing Task of Medicine. *N. Engl. J. Med.* **2012**, *366*, 2333–2338. DOI: 10.1056/NEJMp1113569

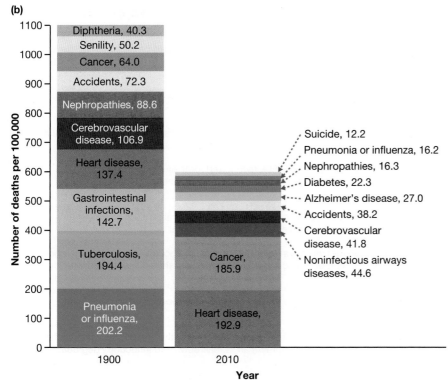

(b)

public. Unfortunately, the industrial scale production and use of these compounds has also had unintended consequences in the form of antibiotic resistance.[4] Because microbial organisms rapidly reproduce, they evolve new characteristics with alarming ease. A few in the population may have chance genetic mutations that allow them to escape the effects of the antibiotic. These few resistant bacteria

[4]King, A. Why Antibiotic Pollution Is a Global Threat. *Chemistry World*. The Royal Society of Chemistry, 2018. https://chemistryworld.com/news/why-antibiotic-pollution-is-a-global-threat/3009021.article (accessed April 2022).

survive, and their descendants, carrying the mutations, can multiply to cause an infectious disease that is more difficult to treat.

When it became clear that bacterial resistance was a problem, organic chemistry provided a solution! Chemists found that penicillin G could be converted to a simpler core structure, 6-aminopenicillanic acid (Figure 1.1). Then, by attaching different side chains to the amino group, many variations on the penicillin core structure were invented, and some were found to be active against resistant bacteria. For example, methicillin was introduced to clinical use in 1960. These new penicillins were not available from the natural source; they had to be prepared by chemists, through the use of organic reactions.

This brief case study of the impacts of antibiotics illustrates one way that organic chemistry contributes to incredible advances in quality of life and life expectancy. However, every organic reaction that contributes to such advances comes with a cost: Energy and natural resources are consumed, waste is generated, accidents can happen, and products require end-of-use disposal.

Fortunately, chemists are in a position to understand and anticipate the costs of the chemistry we do and the new compounds we invent! Therefore, chemists bear

FIGURE 1.3

The originally published 12 principles of green chemistry.

Anastas, P. T.; Warner, J. C. *Green Chemistry: Theory and Practice*; Oxford University Press: New York, 1998.

1. It is better to prevent waste than to treat or clean up waste after it is formed.

2. Synthetic methods should be designed to maximize the incorporation of all materials used in the process into the final product.

3. Wherever practicable, synthetic methodologies should be designed to use and generate substances that possess little or no toxicity to human health and the environment.

4. Chemical products should be designed to preserve efficacy of function while reducing toxicity.

5. The use of auxiliary substances (e.g., solvents, separation agents, etc.) should be made unnecessary wherever possible and innocuous when used.

6. Energy requirements should be recognized for their environmental and economic impacts and should be minimized. Synthetic methods should be conducted at ambient temperature and pressure.

7. A raw material or feedstock should be renewable rather than depleting wherever technically and economically practicable.

8. Unnecessary derivatization (blocking groups, protection/deprotection, temporary modification of physical/chemical processes) should be avoided whenever possible.

9. Catalytic reagents (as selective as possible) are superior to stoichiometric reagents.

10. Chemical products should be designed so that at the end of their function they do not persist in the environment and break down into innocuous degradation products.

11. Analytical methodologies need to be further developed to allow for real-time, in-process monitoring and control prior to the formation of hazardous substances.

12. Substances and the form of a substance used in a chemical process should be chosen so as to minimize the potential for chemical accidents, including releases, explosions, and fires.

a responsibility to take a leadership role not only in providing the innovations that society demands, but also in minimizing their costs. Throughout this text, you will learn the fundamentals of laboratory organic chemistry while considering various ways to minimize your impact on health, safety, and the environment. Let's start with some guiding principles.

1.1 THE TWELVE PRINCIPLES OF GREEN CHEMISTRY

In the late 1990s, synthetic organic chemists began to look at preventing negative impacts of their endeavors through a new and broader perspective, called **green chemistry**, which is summarized by a set of principles popularized by Paul Anastas and John Warner.[5] The purpose of green chemistry is to minimize the risk of negative outcomes while enabling innovations and creativity to flourish. It recognizes the value and power of chemistry, while calling on chemists to prevent negative impacts of their work in advance and by design. This is a long-term problem that requires consistent effort and new ideas, but significant progress has been made and will continue. There is reason for great optimism, especially with new generations of chemists approaching the field with exposure to green chemistry and sustainability always in mind.

The 12 principles of green chemistry are presented in their originally published form in **Figure 1.3**. The sections that follow present some examples of how these pertain to organic reactions and the organic chemistry laboratory studies upon which you are about to embark.

<< **green chemistry**
The design and implementation of products and processes that minimize or eliminate hazards of chemistry activities during all phases of product life cycles, including manufacture, use, and disposal.

| 1.1A | Prevent Waste: Avoid Producing Waste, so There Is No Need for Treatment or Cleanup |

Certain classes of reactions inherently produce more waste than others. Among the common reaction classifications (**Figure 1.4**), substitutions and eliminations by their very nature will produce one molar equivalent of waste for the amount of product formed. Even if the reaction can be improved to 100% efficiency, significant quantities of waste are formed. A useful metric for evaluating this is the environmental impact factor or **E-factor**,[6] which considers not only the yield of product, but also all other outputs of waste materials. It's worthwhile to do this calculation for experiments that you perform. For an aqueous solution of waste, consider only the mass of the waste solute; water is generally omitted from the E-factor calculations. In the companion Experiments manual, a number of reactions have been designed for aqueous conditions (or aqueous ethanol), thereby substituting innocuous and renewable solvents in order to minimize the output of flammable, petroleum-sourced, or halogenated solvents. An ideal synthesis would have an E-factor of zero. This calculation can use kilograms or grams, as long as the units match in the numerator and denominator.

<< **E-factor**
A measure of environmental impact of a chemical process, defined as the mass of waste produced divided by the mass of product obtained.

$$\text{E-factor} = \frac{\text{mass of waste (kg)}}{\text{mass of product (kg)}}$$

[5]Anastas, P. T.; Warner, J. C. *Green Chemistry: Theory and Practice*; Oxford University Press: New York, 1998.
[6]Sheldon, R. A. Organic Synthesis: Past, Present and Future. *Chem. Ind.* **1992**, *23*, 903–906.

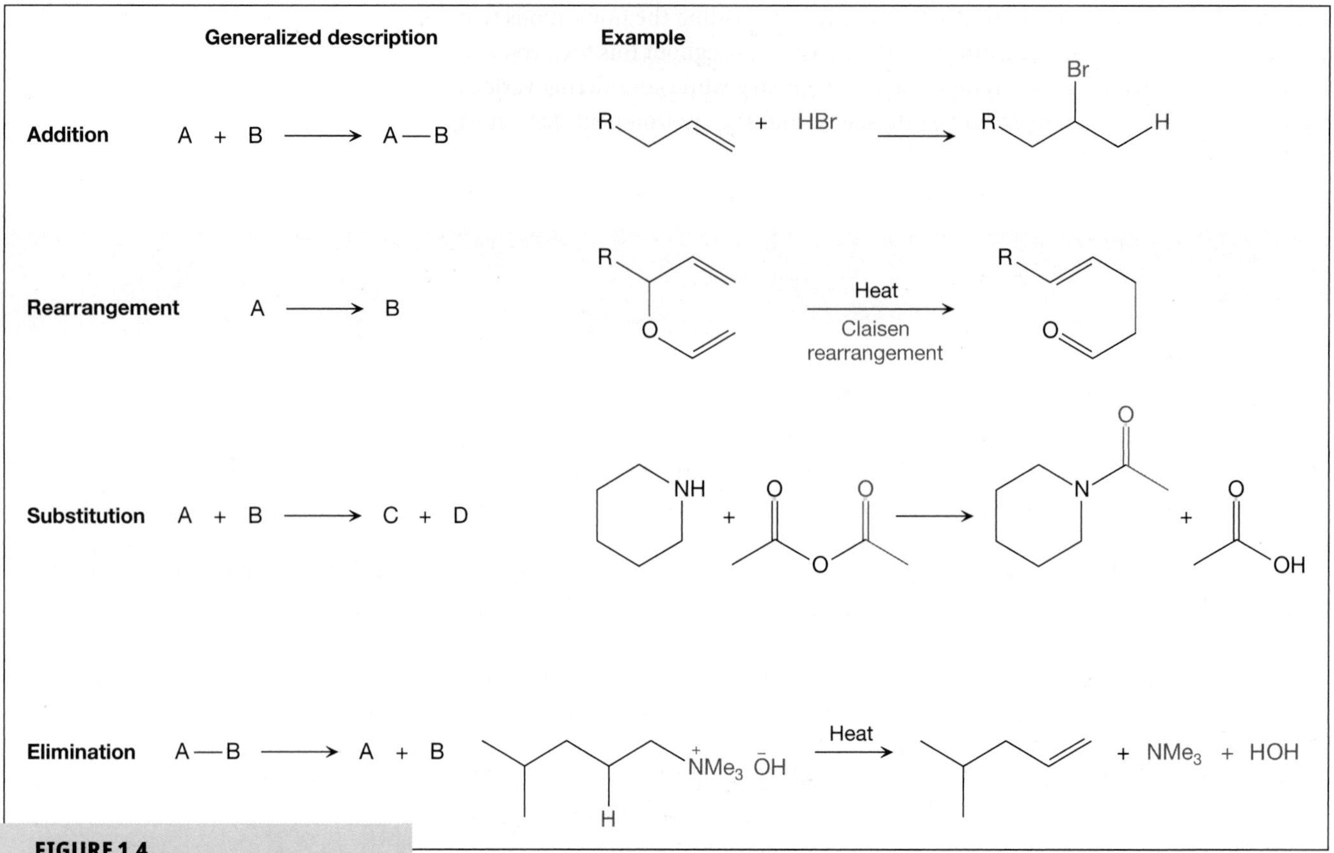
Generalized description
Example
Addition
A + B ⟶ A—B
R
+ HBr
Br
R
H
Rearrangement
A ⟶ B
R
O
Heat
Claisen rearrangement
R
O
Substitution
A + B ⟶ C + D
NH
+
O
O
O
O
N
+
O
OH
Elimination
A—B ⟶ A + B
H
NMe3 OH
Heat
+ NMe3 + HOH

FIGURE 1.4

Reaction types and examples illustrating side products and atom economy.

1.1B Atom Economy: Incorporate All Atoms from Starting Materials into Products

Atom economy is a measure of reaction efficiency that considers the outcome for all the atoms of the various reaction inputs. Certain types of reactions are very *high* on atom economy, such as addition reactions (Figure 1.4), because two reactants become one. If, on the other hand, most of the atoms from a very large reagent end up as by-product waste material, then the process has very *low* atom economy. In the companion Experiments manual, there are some cycloaddition reactions that incorporate all of the reactants into the products for excellent atom economy.

Atom economy is calculated in two ways, called intrinsic and experimental. Intrinsic atom economy can be calculated based on theory, from the balanced equation on paper, whereas experimental atom economy uses data from laboratory results.[7] Experimental atom economy is a somewhat confusingly named quantity because it uses **theoretical yield** as the numerator. The name comes from the denominator, where any excess amounts of reagents that are used in the lab are included, even if their amounts are beyond the stoichiometry required by the balanced reaction equation.

theoretical yield >> The maximum amount of product that could be obtained if all of the limiting reagent is converted to the product.

$$\text{Intrinsic atom economy (\%)} = \frac{\text{molar mass of desired product}}{\text{molar mass of all reactants}} \times 100$$

$$\text{Experimental atom economy (\%)} = \frac{\text{theoretical yield of desired product (g)}}{\text{actual quantity of all reactants used (g)}} \times 100$$

[7](a) Trost, B. M. The Atom Economy—A Search for Synthetic Efficiency. *Science* **1991**, *254*, 1471–1477. (b) For several examples of atom economy calculations, see Abhyankar, S. B. Introduction to Teaching Green Organic Chemistry. In *Green Organic Chemistry in Lecture and Laboratory*; Dicks, A.P., Ed.; CRC Press: Boca Raton, FL, 2012; pp 1–28.

Worked Example

Calculate the E-factor, intrinsic atom economy, and experimental atom economy for the S_N2-type Williamson ether synthesis from 2-naphthol and 1-iodobutane (**Figure 1.5**).

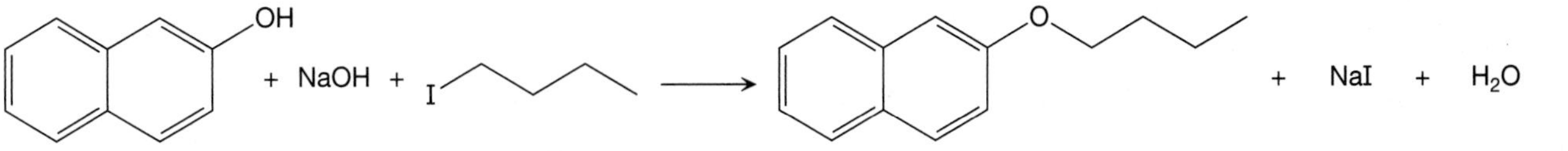

2-Naphthol   **1-Iodobutane**

1.00 g	0.560 g	1.60 g	1.29 g	(149.9 g/mol) (18.0 g/mol)
(144.2 g/mol)	(40.0 g/mol)	(184.0 g/mol)	(200.3 g/mol)	
Limiting reagent				

$$\text{Theoretical yield (mol)} = \frac{1.00 \text{ g}}{144.2 \text{ g/mol}} = 0.00693 \text{ mol}$$

$$\text{Experimental atom economy} = \frac{0.00693 \text{ mol} \times 200.3 \text{ g/mol}}{3.16 \text{ g}} \times 100\% = \boxed{44\%}$$

FIGURE 1.5

The S_N2 reaction between 2-naphthol and 1-iodobutane, with experimental amounts of the reactants and the main organic product.

To calculate the E-factor, we need to know the mass of waste material (NaI + H_2O) and the mass of desired product (the ether). The combined mass of NaI and H_2O can be calculated by subtracting the mass of desired product (1.29 g) from the sum of all the reactant masses (1.00 + 0.560 + 1.60 = 3.16 g). The remainder (3.16 – 1.29 = 1.87 g) is the amount of waste. The mass of waste divided by the mass of product is the E-factor.

$$\text{E-factor} = 1.87 \text{ g}/1.29 \text{ g} = 1.45$$

For the intrinsic atom economy, we first divide the molecular weight of the desired product (200.3 g/mol) by the sum of molecular weights of reactants, with each reactant molecular weight multiplied by the stoichiometry of the reactant in the balanced chemical equation. We then convert that value to a percentage by multiplying by 100%. In this case, the stoichiometry is 1:1:1, which simplifies the calculation:

$$\text{Intrinsic atom economy} = 200.3/(144.2 + 40.0 + 184.0) \times 100\% = 54\%$$

The experimental atom economy is the theoretical yield of the ether in grams, divided by the mass in grams of all of the reactants used, and which we calculated previously to be 3.16 g. As shown in Figure 1.5, the theoretical yield in moles is the amount of moles of the limiting reactant, 2-naphthol, which is obtained by dividing the amount used by its molecular weight. Multiplying this by the molecular weight of product gives a theoretical yield in grams of 1.39 g. We then divide this value by the sum of the reactant masses used and convert to a percentage:

$$\text{Experimental atom economy} = (1.39 \text{ g})(3.16 \text{ g}) \times 100\% = 44\%$$

Notice that this experimental atom economy is lower than the intrinsic atom economy. This is because excess amounts of NaOH and 1-iodobutane were used, beyond the 1:1:1 stoichiometry of the balanced reaction equation.

1.1C Low Toxicity: Design Alternative Processes/ Substances That Are Known to Have Lower Toxicity

The field of toxicology has advanced by leaps and bounds over the last few decades, and we now have a much better handle on the classes of compounds and regions of molecular structures that will likely yield toxicity problems, whether through workplace activity or environmental exposure. When designing a new organic compound to serve some purpose, toxicology expertise should be engaged early in the discussion. One example of this is modifying the lipophilicity of a compound, which is known to affect the degree to which it is transported across membrane barriers at the cellular level or the skin. If altering lipophilicity can minimize transport of a compound through the skin, then its exposure hazards can be reduced, and a less toxic material could be designed with the same intended function. Well-informed decisions up front may avoid wasting additional resources by reversing course later.

1.1D Maintain Function While Lowering Toxicity

Achieving a greater ratio of effectiveness to toxicity is important, whether it is by lowering the toxicity or by increasing the efficacy. Either approach is beneficial. A more highly efficacious compound may be used in lower quantity, so the by-products of its manufacture and use will also be minimized. The key point here is that risk is a function of both hazard and exposure: If a substance retains its function even when used in very low quantities, then its risk is lowered simply because the amount of the exposure is lower.

1.1E Avoid Solvents or Separation Agents Whenever Possible

column chromatography >>
A purification technique whereby a mixture of organic products is separated into its components due to differences in the rate of travel with a solvent through a column of insoluble powdered material, generally silica gel.

In an organic synthesis, a common purification method is **column chromatography**, which consumes large quantities of solvents and adsorbents. Sometimes these can be recycled, but sometimes they add to the waste stream without creating any new bonds or structural changes in the compound. On the other hand, if reactions can be designed so that the product can be moved on to the next reaction step without purification, all of this waste can be avoided.

1.1F Energy Efficiency: Minimize Impacts by Working at Ambient Temperature and Pressure

It takes a lot of energy to perform reactions outside of ambient conditions. Heating requires an input of energy, but cooling is energy intensive, too, because electricity is needed to run refrigerator devices and produce coolants like dry ice or liquid nitrogen. Pressurized reactions, whether high pressure or low pressure, require pumps that also use energy. Removing solvent requires energy input, too. Bristol Myers Squibb won a Presidential Green Chemistry Challenge Award for its redesign of the synthesis of the cancer drug Taxol. The redesigned route eliminated 10 solvents and improved energy efficiency by avoiding six drying steps.[8] On the much smaller academic lab scale, these factors may seem negligible. But, if reactions can be designed to avoid these energy inputs, then the energy efficiency improvements become very significant upon scale-up to industrial production.

1.1G Renewable Feedstocks: Avoid Depleting Natural Resources When Feasible

Organic chemistry makes extensive use of petroleum feedstocks, so organic chemists are often dismayed to see all these valuable materials simply burned away in combustion engines. We know that petroleum feedstocks are a finite resource, and gradually people are finding ways to access alternative feedstocks from agricultural products or waste materials. A well-publicized example is the use of corn and its by-products to generate ethanol. But beyond their use as fuel, a wider variety of chemical feedstocks can be made accessible as organic chemists develop ways to convert waste into replacements for petroleum products. One experiment in the companion Experiments manual shows you how to convert corncobs into furfural, which is the starting material for an addition reaction to form a carbon–carbon bond.

1.1H Fewer Synthetic Steps: Avoid Unnecessary Derivatization (e.g., Protecting Groups)

Synthesis sequences made up of multiple steps are prone to problems of functional group incompatibility along the way. In a hypothetical sequence of six steps, the functional group formed in step 3 may not be compatible with the reaction planned for step 6. To get around this, an extra reaction may be needed to protect the vulnerable functional group from an undesired reaction. Such extra steps can be detrimental to the overall efficiency of the synthesis sequence, and if possible, the sequence should be redesigned to avoid extra steps.

1.1I Catalytic Processes: Use Catalytic Processes of Superior Efficiency Relative to Stoichiometric Ones

Catalysis has remarkable potential to improve the efficiency of synthesis. Instead of using equal molar amounts of a reagent that ends up producing equivalent amounts of waste, a catalytic process recycles the reagent during the reaction. In this way, each molecule of the reagent is reused repeatedly (sometimes thousands of times). Nature uses enzymes in this way. Imagine if you had to consume equal amounts of reagents to digest all the food you eat. Instead, your digestive tract contains very small quantities of enzymes to do this work. Because they are catalytic, the enzymes are reused over and over, so they contribute almost nothing to the waste output. Furthermore, they lower the energy of activation of reactions, allowing them to occur at lower temperatures, minimizing energy inputs. In the companion Experiments manual, two experiments use a tungsten **catalyst** for oxidations, and

<< **catalyst**
A compound that lowers the energy of activation for a chemical reaction but is not consumed by the reaction.

---

[8]U.S. Environmental Protection Agency, Office of Pollution Prevention and Toxics. Presidential Green Chemistry Challenge Award Recipients 1996–2016, 2016. https://epa.gov/greenchemistry/document -green-chemistry-challenge-award-recipients-1996-2016 (accessed April 2022).

in two other experiments, biocatalysis (vitamin B$_1$ and a ketoreductase enzyme) is used for the nucleophilic addition to a carbonyl and the **enantioselective** reduction of a ketone, respectively.

1.1J Innocuous After Use: Avoid Products That Persist After Use and Are Unsafe After Degradation

There are numerous heartbreaking stories of unanticipated outcomes from industrial wastes that were not disposed of by today's standards, and continue to persist in the environment or accumulate in organisms. In Times Beach, Missouri, for example, which is a town outside of St. Louis, an industrial waste oil containing dioxin was sprayed on roads in an effort to keep dust down. It persisted in the environment and reached a level from 300 to 700 times greater than that deemed safe by the U.S. Centers for Disease Control and Prevention. Eventually the entire town was purchased by the federal government and the residents resettled elsewhere to avoid further exposures. Awareness raised in response to events such as these has led to much better standards for waste handling, but accidents still happen. If an alternative to dioxin could have been designed that degrades on exposure to light, this problem may have been avoided. Situations such as these present opportunities and responsibilities for organic chemistry, and designing compounds that degrade in a harmless fashion is an active area of research.

1.1K In-Process Monitoring: Control Processes in Real Time, Prior to the Formation of Hazardous Materials

A variety of negative impacts occur when a reaction gets out of control. For example, pressure may build up, reaction vessels may exceed their capacity, and accidents can result. Many of these situations can be addressed easily in an academic lab. On the industrial scale, however, these breakdowns can be much more significant. If the quality of the product is low because of a problem with process control, it may be unsuitable for the market and will become part of the waste stream. Or, a problem with temperature in a reactor may cause a nontoxic reaction to begin producing an unanticipated toxic by-product. These situations call for analysis of the reactions and processes throughout, so that any loss of control can be detected immediately and corrected.

1.1L Avoid Accident-Prone Materials: Avoid the Possibility of Release, Explosion, and Fire

Accidents in the lab are an important topic for everyone to discuss, so that they can be prevented when possible and so that their impact can be minimized when the unanticipated does occur. One way to do this is to avoid a reaction if you know that it comes with a high risk of release, explosion, or fire. The severity of the problems is magnified on a larger scale, so organic chemists should implement safer alternatives early in the development phase, before scaling up the synthesis for production.

Two different syntheses of oxirane (ethylene oxide) are shown in **Figure 1.6**. The first uses stoichiometric amounts of reagents to reach oxirane in two steps. The second is a one-step process using a catalyst (recall that a catalyst is recycled, not consumed, in a reaction). Refer to these syntheses to address the questions that follow.

1. (a) Prior to scaling up for industrial production, chemists developing the lab-scale synthesis of oxirane in Figure 1.6a found that a reaction producing 0.264 kg oxirane also produced 0.99 kg $CaCl_2$ and 0.33 kg HCl. Calculate the E-factor for this reaction under these conditions. (b) These chemists found that the synthesis in Figure 1.6b yielded 2.40 kg oxirane, with 0.72 kg waste material. Calculate the E-factor under these conditions.

2. Calculate the intrinsic atom economy for each of the syntheses in Figure 1.6.

3. Which of the syntheses in Figure 1.6 would be preferred from the green chemistry perspective? Use both E-factor and intrinsic atom economy to justify your answer.

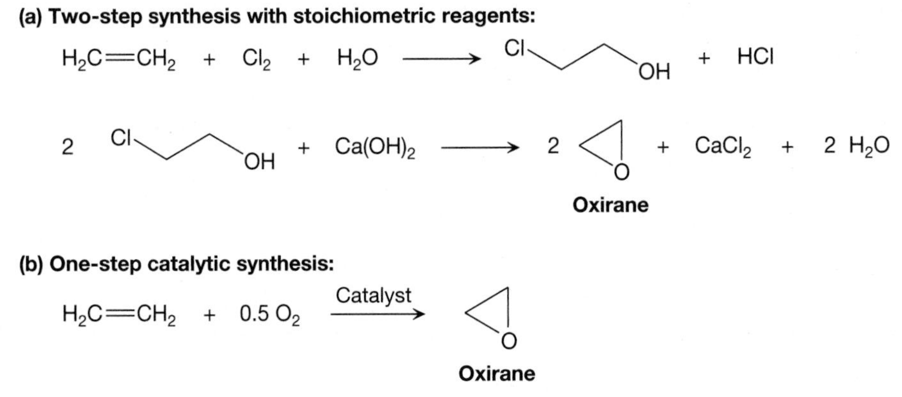

FIGURE 1.6

(a) Two-step synthesis with stoichiometric reagents:

$$H_2C{=}CH_2 \; + \; Cl_2 \; + \; H_2O \longrightarrow Cl{\sim}\!\!\diagdown_{OH} \; + \; HCl$$

$$2 \; Cl{\sim}\!\!\diagdown_{OH} \; + \; Ca(OH)_2 \longrightarrow 2 \; \triangleleft\!\!|_O \; + \; CaCl_2 \; + \; 2\,H_2O$$

Oxirane

(b) One-step catalytic synthesis:

$$H_2C{=}CH_2 \; + \; 0.5\,O_2 \xrightarrow{\text{Catalyst}} \triangleleft\!\!|_O$$

Oxirane

Two different routes for the industrial production of oxirane.

2

LEARNING OBJECTIVES

- Identify and implement safe practices in the organic chemistry laboratory.

- Locate and interpret hazard data using SDS information and standardized labeling symbols.

- Use appropriate personal protective equipment and approved waste disposal methods.

- Demonstrate knowledge of your location's safety procedures and equipment by taking a quiz.

Working Safely in the Organic Chemistry Laboratory

LAB SCENE

Workers have a responsibility to themselves and to others to work safely in the organic chemistry laboratory.

Thomas Barwick/Getty Images.

Instructional laboratories generally involve experiments that are chosen because they are relatively safe. Still, you need to learn the tools and procedures that chemists use to operate safely in the lab. This is a key component of any chemistry lab course, to keep everyone as safe as possible not only during the course, but also in future lab work. The skills you learn in this course will be useful in a variety of related fields.

Safety in the chemistry instructional laboratory is a responsibility that must be shared among all who will be present in the lab. Students, instructors, and laboratory staff all have the same interest in avoiding accidents. The topic of safety should be an ongoing discussion involving all of these participants, with updates to the lab procedures and guidelines as best practices continue to evolve. Some basic safety guidelines are a great starting point. Additional site-specific safety protocols may be slightly different than the general guidelines here, so be sure to follow the protocols provided by your lab instructor.

2.1A Preparation

Come to lab prepared. General instructions for preparation are covered in Chapter 3 of this text. Most of the risks associated with laboratory work can be minimized by thorough preparation prior to arrival. Any chemistry laboratory can be a dangerous place if procedures are not followed according to plan. You have to know the plan in order to follow it, so prepare in advance.

2.1B Supervision

Depending on your institution, your lab may be supervised by the professor or instructor in charge of the course, or your lab may be supervised by a teaching assistant (TA), associate instructor (AI), graduate teaching fellow (GTF), or other staff member. We will use the term "lab instructor" to refer to any one of these persons who may be supervising your lab.

Students are allowed in the laboratory only during their assigned times and with proper supervision. Do *not* enter the lab if one of these experienced authorities is not present. If you must leave the lab for any reason, inform your lab instructor.

2.1C Unauthorized Experiments

All experiments must be approved by the lab instructor. Instructional labs are not the time or place to invent new lab procedures or try out unauthorized experiments.

2.1D Laboratory Attire

Dress appropriately in the lab to minimize the risk of injury. Follow the attire guidelines in place at your school, which may be slightly different from these general guidelines:

- Wear shoes that cover the entire foot. Open-toed shoes, sandals, flip-flops, canvas shoes, or shoes with perforations are unacceptable.

- Cover your legs. Shorts, short skirts, and short dresses are unacceptable.
- Wear a shirt that covers you completely. Muscle shirts, tank tops, or anything that leaves your arms or torso exposed are unacceptable.
- Do not wear loose clothing.
- Tie back long hair.

2.1E | Hazard Data and SDS Access

Specific hazard data on millions of different compounds can be found in **Safety Data Sheets (SDSs)**. These are readily available online. For example, if you will be handling dichloromethane, you should first access the SDS to find out what hazards are listed for that compound. To find it, simply search "SDS dichloromethane" in your web browser. Chemical suppliers are required to provide the SDS for any product they sell. Keep in mind that the information you find there may be addressed to people who handle much larger quantities than we would use in an instructional lab.

<< **Safety Data Sheets (SDSs)** Safety advisory resources generally provided by chemical suppliers, providing hazard data for any chemical compound that is sold commercially. Also known as Material Safety Data Sheets (MSDSs).

2.1F | Hazard Labeling Pictograms

The Occupational Safety and Health Administration (OSHA) has standardized pictograms to alert users of various hazards presented by compounds, as shown in **Figure 2.1**. Always check the label of a chemical container to review these hazards before handling the chemical.

2.1G | Contact Lenses

Many schools and workplaces do not allow use of contact lenses in the laboratory without special approval due to medical reasons. If contact lenses are allowed at your location, they must be accompanied by goggles that completely contact the skin around the eyes to protect from splashes. Contact lenses are *not* eye protection devices.

2.1H | Food

Eating, drinking, or the use of any tobacco product (including e-cigarettes or vaping devices) is prohibited in the laboratory. This includes chewing gum, cough drops, throat lozenges, and the like.

2.1I | Medical Conditions

People with conditions that could be adversely impacted by exposure to organic chemicals should consult with their health care provider. Some organic chemicals are potential hazards specifically to the fetus or to young children. Those who are pregnant, nursing, or who suspect they may be pregnant are strongly advised to consider the advice of their health care provider, and may wish to take this course at a later time.

2.1J | Service Animals

Those with service animals should work in consultation with the lab instructor to determine how to safely comply with institutional policy guidance regarding service animals in labs.

FIGURE 2.1

OSHA.

Hazard Communication Standard Pictogram

The Hazard Communication Standard (HCS) requires pictograms on labels to alert users of the chemical hazards to which they may be exposed. Each pictogram consists of a symbol on a white background framed within a red border and represents a distinct hazard(s). The pictogram on the label is determined by the chemical hazard classification.

HCS Pictograms and Hazards

Health Hazard	Flame	Exclamation Mark
• Carcinogen • Mutagenicity • Reproductive Toxicity • Respiratory Sensitizer • Target Organ Toxicity • Aspiration Toxicity	• Flammables • Pyrophorics • Self-Heating • Emits Flammable Gas • Self-Reactives • Organic Peroxides	• Irritant (skin and eye) • Skin Sensitizer • Acute Toxicity (harmful) • Narcotic Effects • Respiratory Tract Irritant • Hazardous to Ozone Layer (Non-Mandatory)
Gas Cylinder	**Corrosion**	**Exploding Bomb**
• Gases Under Pressure	• Skin Corrosion/ Burns • Eye Damage • Corrosive to Metals	• Explosives • Self-Reactives • Organic Peroxides
Flame Over Circle	**Environment** (Non-Mandatory)	**Skull and Crossbones**
• Oxidizers	• Aquatic Toxicity	• Acute Toxicity (fatal or toxic)

For more information:

Occupational Safety and Health Administration

U.S. Department of Labor   www.osha.gov   (800) 321-OSHA (6742)

OSHA 3491-01R 2016

2.2A Fume Hood

Most organic chemistry operations should be conducted in a fume hood (**Figure 2.2a**). Fume hoods are cabinets with an exhaust flow designed to pull vapors out of the room from behind the work area, minimizing chemical exposure to people during lab activities. They generally have a hood sash with shatter-resistant glass that can be raised and lowered, and some also have panels that slide from side to side. As much as possible, chemists should work with the hood sash closed. Items that are not actively in use should not be stored in the hood; anything that inhibits the designed airflow can result in unnecessary hazard exposure. Hoods are not storage areas.

(a)

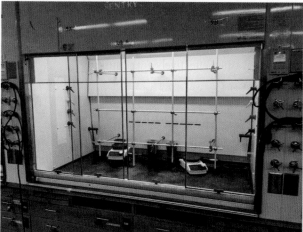

(b)

(c)

FIGURE 2.2

(a) A fume hood with shatter-resistant glass sash in the closed position. (b) An eyewash station. (c) A safety shower.

Courtesy of Gregory K. Friestad.

2.2B Personal Protective Equipment (PPE)

A variety of measures can be taken for personal protection during chemistry laboratory work. The type of PPE a chemist chooses will depend on the quantities of materials being handled and the types of hazards that will be encountered. In the undergraduate organic chemistry laboratory, safety goggles and gloves are generally suitable. If further measures are warranted, your lab instructor will notify you.

SAFETY GLASSES

You must wear eye protection at all times in the lab! Goggles that contact the skin in a continuous loop around the eye area are preferred, because they provide better protection from these hazards. The minimum eye protection should be safety glasses with side-shields to protect against splashes or objects that may approach from various directions.

GLOVES

Gloves may or may not be required for all experiments at your school. Make sure you know which experiments require gloves before you begin. Gloves may be reuseable or disposable. Reuseable gloves are generally thicker and more durable, and may be suitable for longer exposures. Disposable gloves, such as nitrile examination gloves, protect your hands from brief chemical exposures. They are not generally suitable protection for full immersion into chemicals other than water. If disposable gloves have holes, tears, discoloration, or swelling of the glove material after contact with a chemical, remove the gloves, wash your hands, and get a new pair. Be cautious about what you touch with gloves; if you touch pencils, pens, or cell phones, they'll be contaminated with whatever is on the outside of your gloves. *Gloves should not be worn outside the lab.* Always remove gloves before touching a doorknob or any other surface that others generally touch without gloves. If you see someone wearing gloves in the hall, how do you know which doors or other surfaces they may have contaminated?

2.2C | Eyewash Station and Safety Shower

These are emergency tools that can prevent serious injuries in case of an accident. Know where eyewash stations and safety showers (**Figure 2.2b** and **Figure 2.2c**) are located in advance of the need to use them. If a chemical is splashed into the eyes, you do not want to waste any time trying to locate the eyewash. The safety shower, for emergency use only, will rapidly deposit a large amount of water, and is effective in the event clothing catches fire, or in case of a larger chemical spill affecting enough body area that it cannot be washed effectively at the sink.

2.2D | Fire Extinguishers

Know the location of every fire extinguisher in the lab. Most organic solvents are flammable and ignite readily when exposed to a source of ignition. Thus, open flames are *not* permitted in the laboratory, unless specifically directed by your lab instructor. Smoking is strictly forbidden. In case of fire, it is also important to know the locations of safety showers and the nearest building exit.

2.3 WASTE DISPOSAL

Laboratory waste (solvents, solids, sharps, etc.) must be disposed of properly. **Table 2.1** lists general guidelines for waste disposal. **Figure 2.3** shows examples of waste containers that are properly labeled and stored. The labeling of containers and guidelines for disposal at your school may be somewhat different, and should be reviewed by your lab instructor before you begin the experiment. Make sure you know what type of waste you have before you dispose of anything. If you are unsure how to dispose of something, ask your lab instructor. No waste should ever go into the sink without permission.

TABLE 2.1

Waste Disposal Guidelines

CATEGORY	EXAMPLES	METHOD OF DISPOSAL
Organic waste	Waste solvents, acetone, reaction intermediates, products	"Organic Waste" bottles in hood, separate bottles for non-halogenated or halogenated (e.g., dichloromethane)
Aqueous waste, hazardous	Dilute acids and bases, aqueous washes from extractions	"Aqueous Waste" bottle in hood
Aqueous waste, nonhazardous	Water baths, ice baths, etc., that have *not* been contaminated with hazardous materials	Sink
Sharps	Broken glassware, test tubes, TLC plates, thermometers, etc., but *not* mercury thermometers!	Plastic "Sharps" bucket
Solid chemical waste, hazardous	Used filter papers, used drying agents, insoluble organic and inorganic solids	"Solid Waste" container in hood
Other solid waste, hazardous	Paper towels used for lab cleanup, used gloves, empty vials, other laboratory waste	Plastic solid waste bucket
Nonhazardous waste	Paper towels used for drying clean water from hands or glassware, other garbage not associated with laboratory activities	Trash container
Mercury	Broken thermometers or manometers	For any type of mercury spill, special cleanup procedures are required. Do *not* put mercury in the trash or sink. Contact laboratory staff immediately.

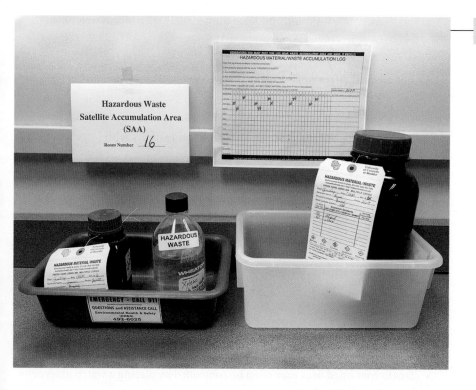

FIGURE 2.3

Waste accumulation site, with containers properly labeled. Note that liquid waste is kept within secondary containment (the plastic bins).

Courtesy of Derek J. Hayes, University of Colorado Boulder.

If an accident occurs, always inform your lab instructor and other laboratory staff as soon as it is safe to do so. If any accident involves a hazard that is beyond your capability to respond, remove yourself and others from the area immediately and notify your lab instructor and other laboratory staff.

2.4A | Chemical Spills

Any spill should be cleaned up immediately using agents found in spill kits that are supplied in the lab room. These contain solid neutralizing agents that are designed for specific types of spills. Do not use the spill kit reagents for spills on your body (see Chemical Exposures, below).

- Acids are treated with a magnesium oxide–containing agent (e.g., Spill-X-A).
- Base spills are treated with a citric acid–containing agent (e.g., Spill-X-C).
- Solvent spills are treated with an absorbent solid agent (e.g., Spill-X-S).
- Use caution in responding to any liquid spill; many liquids look like water. Even if it is confirmed to be water, clean it up to avoid the hazards of a slip or fall.
- Report the use of spill cleanup reagents to your lab instructor, so the reagents can be replaced.

2.4B | Chemical Exposures

Eyewash stations and safety showers are standard in organic chemistry labs. Your instructor is responsible for showing you the locations of these, and how to operate them.

If you get chemicals in your eyes,

1. go to an eyewash station and flush immediately with lots of water;

2. report the accident; and

3. get medical attention!

If you get chemicals on your skin,

1. go to the sink or safety shower and wash immediately with large amounts of water;

2. remove contaminated clothing;

3. continue to wash the area with water;

4. report the accident; and

5. get medical attention!

2.4C | Fires

If there is a chemical fire in the hood or bench area:

- Small, contained fires may often be controlled by placing a watch glass or large beaker over the vessel.

- Larger fires may require the use of a fire extinguisher.
- Uncontrolled fires require using the alarm to contact the fire department.

If there is a fire on you or your neighbor,

- don't panic;
- shout for help;
- roll on the floor to smother the flames; and
- *walk* to the nearest safety shower.

2.4D Broken Glassware

Cuts sustained from breaking glassware are among the most common lab injuries. Handle broken glassware with caution. All broken, cracked, or chipped glass should be disposed of in a properly labeled glass waste container.

If you break something, keep in mind what chemicals are in or on it. Alert your lab instructor and don't try to pick up small pieces with your fingers. Use a broom, dustpan, or other necessary cleanup supplies that are available, and avoid incidental skin contact.

2.4E Injury

Report any injury to your lab instructor, no matter how small. For treatment of cuts, it is best to be on the safe side and visit a health care provider in case small bits of glass remain in the cut. Take note of what chemicals may have been present, in case medical personnel ask for this information.

Most schools require an incident report to be filed for any and all injuries, regardless of the severity. Use the reporting protocols as advised by your lab instructor.

2.5 SAFETY QUIZ INFORMATION

All students enrolled in chemistry lab courses should complete a safety quiz at the beginning of each course. The details of how this is administered may vary depending on your lab instructor, so check your syllabus or any other online course materials before the course begins. Many questions about safety can be answered using simple common sense, but you will also need to prepare by reading this chapter and any other safety materials assigned by your lab instructor. You should be able to answer some of the questions based on information discussed in your previous chemistry courses. Other questions may require some knowledge of the layout of the laboratory and building.

2.5A Some Reminders to Assist You with the Quiz

- Safety goggles are the prime protection for your eyes. Contact lenses are *not* eye protection. In fact, contact lenses can trap liquids against your eye, making it more difficult to flush them out.
- Safety equipment, such as showers, eyewash stations, and fire extinguishers, are found in every lab. Be sure you know where they are in your lab. Have your lab instructor explain how to operate them.

- If you spill anything on your skin, wash it off immediately. Don't experiment by trying to run reactions on yourself or another student.
- If you or another student have an accident in lab, tell your lab instructor as soon as possible. Don't try to wait out the entire lab session.
- If you break glassware, don't just throw the broken pieces into any waste container. Check in your lab for a place where broken glass can be safely disposed of.
- Eating, drinking, and chewing gum are all prohibited in all chemistry labs, as is the use of all tobacco and vaping products.
- Closed-toed shoes (not sandals), long pants or skirts (not shorts or mini skirts), and shirts (not tank tops or muscle shirts) must be worn in the lab.
- Keep books, backpacks, and coats out of the aisles and off the bench tops. Ask your lab instructor where these items can be safely stored during lab periods.
- Check the location of the closest exits from your lab and from the building so that you can exit the building quickly in case of a fire.
- Never work alone or without supervision in the laboratory. Your lab instructor will be present throughout the entire lab period.
- Come to the lab prepared. Always be sure that you understand what you are doing before you do it. If you have questions, *ask*!

LEARNING OBJECTIVES

- Identify how records of daily lab activities affect interpretation, attribution, and reproducibility of scientific discoveries.

- Implement an orderly procedure for recording information and observations before, during, and after lab activities.

- Locate physical properties and hazard data for organic compounds.

- Create a flowchart to organize lab activities in advance.

Pre-Lab Preparation: The Laboratory Notebook

LIBRARY AND INTERNET SOURCES

Organic laboratory work requires advance preparation and information gathering, including searches of handbooks, databases, and the like.

Ammentorp Photography/Alamy Stock Photo.

INTRODUCTION

The laboratory notebook serves as a permanent chronological record of experimental work, whether it is prepared in hardcopy format or with digital notebook software. When there is a legal dispute over who was the first to find a lucrative new discovery or invention, the laboratory notebook is a key piece of evidence. So, keeping a proper lab notebook could be worth a lot of money and prestige! Also, sometimes you need to reproduce an experiment exactly as it was done the first time, and taking good notes will make this possible. These notes are also needed to write laboratory reports explaining your results. For all of these reasons, your notebook needs to be thorough and accurate. As a general rule, a good notebook is one from which someone else can repeat your experimental work in the same way that you have done it.

3.1 GENERAL GUIDELINES

1. The laboratory notebook must be composed of pages attached within a binding. The pages must be numbered, and if your instructor requires it, may have a pressure-activated copy page ("carbon copy" or carbonless copy).

2. Name, course, and lab section number must be written on the cover or front page of your lab notebook.

3. Make sure your handwriting is legible.

4. Always use permanent ink, not pencil.

5. Notes must be written at the time the observations are made. Write it down NOW!

6. Write all data in your notebook—weights, temperatures, everything. When recording experimental data, always include units.

7. Use complete sentences, or at least enough words so that someone else can follow the train of thought.

8. Do not erase. If you make an error, draw a single line through it, add initials and date, and continue. The original statement should still be legible.

9. If you are using a lab notebook with pressure-activated copy ("carbon copy"), follow the direction of your instructor on which pages (original or duplicate) can be removed when submitting notebook pages for grading. Never remove both the original and copy pages from your notebook. You should always keep a backup record of your work.

10. Date every page as you use it.

There is considerable variation among lab instructors in how they address the general guidelines listed above. Naturally, you will need to incorporate specific instructions as given by the lab instructor of your course. While the lab notebook information described here is geared toward the use of a hardcopy notebook, most of these general guidelines also apply if you are using an electronic laboratory notebook. The key is to have a record that is informative, contemporaneous, and permanent.

The first several pages of your notebook should be reserved for a Table of Contents. From there, each experiment recorded in your notebook should contain sections A–G, which are outlined below. Sections A–G contain information that should be entered and completed prior to the laboratory period in which you begin the experiment.

3.2A | Title

Start each experimental write-up with an accurate, descriptive title.

3.2B | Purpose

Discuss the general purpose of the experiment in at most two or three sentences. This should include a statement of the hypothesis to be tested, or a question that will be answered if the experiment is successful. If the experiment is a synthesis (as opposed to a technique), write the chemical equation, including reagents and expected product(s). For multistep syntheses, write one equation for each transformation, including the preparation of reagents.

3.2C | References

Cite the reference(s) you use in preparing your notebook. This includes the sources of physical properties and hazards of any compounds you'll use, as well as the source upon which your experimental procedure is based. In most cases this will be the lab textbook and/or your lab instructor's in-house course materials. While the sources of information may seem obvious while you're in this course, it's good to build a regular practice of including the references in your notebook.

3.2D | Chemical Properties

Make a table that lists the chemical properties of all reactants and reagents that you will be using in the experiment. This table should include the name of each compound, its molecular weight (MW), density (d), melting point (mp), and boiling point (bp). Boiling point should include the pressure at which it was measured, if it wasn't under standard atmospheric pressure. Add the source or sources of these data to the References section of your notebook. Along with the properties, you should add a column where you list the amounts of each component you will use. The amounts should be expressed in terms of the units you measure (g or mL). For the reactants, include the moles, so that you can identify the **limiting reagent**. The molar equivalents of any reagents should also be listed; this will help you predict what unreacted materials may remain after the reaction is complete, so that you can handle and dispose of them properly. A **molar equivalent** is simply the ratio of moles of reagent divided by moles of the limiting reactant.

<< **limiting reagent**
The reaction component that is consumed first when other reactants are present in larger amounts. Many organic reactions have 1:1 stoichiometry, where 1 mol of reactant theoretically gives 1 mol of product; in these cases the reactant of lowest molar quantity is the limiting reagent.

<< **molar equivalent**
The ratio of moles of a reactant or reagent relative to the moles of limiting reagent, without adjusting for the stoichiometry of the reaction.

3.2E Safety Guidelines

Make a table listing the safety hazards of the compounds that you will use, including solvents. For each compound, list the toxicity (if known), the flash point (in °C), and any other important safety information (e.g., flammable, corrosive, irritant, etc.).

The chemical properties (section 3.2D) and safety guidelines (section 3.2E) may be combined into one table.

3.2F Equipment

Sketch any equipment setups or apparatus that you will use for the first time. Include in your drawing the positions of any clamps that are used. If you have already drawn the apparatus for an earlier experiment, you need only indicate the page in your notebook where the drawing can be found.

3.2G Pre-Lab Flowchart

The pre-lab flowchart is a visual depiction of the order of steps you'll do in the lab, and the connections between the steps. It will help keep you organized and efficient in the lab, and will also help with identifying desired products versus waste materials. For more details, see section 3.4.

3.2H Experimental (during Lab)

This section of your notebook is written during the course of a laboratory period. It is a record of what you do as you do it, and it must be completed before you leave the lab for the day. Some instructors may require portions of these records to be submitted at the end of the lab period, so be sure to keep the following records during your lab work:

1. As you move through the experiment, note any deviations from the procedure and flowchart. Such deviations might be intentional, in response to directions from the lab instructor, or an unintentional mistake that you will need to note in your records as an aid in interpreting the results.

2. Keep a log of both your actions and your observations. Any reader should be able to repeat the experiment as you ran it based on what you have written. Include any thoughts you have about what may be going on, or how the experiment might be changed in the future.

3. Make sure to record any melting points, boiling points, weights, etc., before you leave the lab whether you think you need them or not. Chances are that you will. Drawings of all thin-layer chromatography plates should also be included here. Alternatively, a hard copy print from a photograph may be permanently attached to the notebook page in place of a drawing.

4. Record your progress and observations completely and accurately. The information included here may help you understand later if your experiment was successful or what went wrong.

5. At the end of each day initial and date what you have written.

6. Follow the directions from your lab instructor about whether to submit these notebook pages before leaving the lab.

3.3 RESOURCES FOR CHEMICAL PROPERTIES AND SAFETY GUIDELINES

The following references (and many others) are available in either the laboratory or the library. Some are also available online. You should familiarize yourself with them, because you will use them frequently throughout the semester.

3.3A | Chemical Properties (See Section 3.2D)

1. *Aldrich Chemistry: Handbook of Fine Chemicals*; Sigma-Aldrich, 2007–2008 (or later edition). An online product search on the MilliporeSigma website (https://www.sigmaaldrich.com/US/en) provides access to the same information previously available in the handbook.

2. *CRC Handbook of Chemistry and Physics*, 99th ed. (internet version 2018); Rumble, J. R., Ed.; CRC Press/Taylor & Francis. Online at https://hbcp .chemnetbase.com.

3. *The Merck Index*, 12th ed.; Budavari, S., Ed.; Merck: Whitehouse Station, NJ, 1996 (or later edition). Online at https://www.rsc.org/merck-index.

3.3B | Safety Guidelines (See Section 3.2E)

1. *Safety in Academic Chemistry Laboratories: Best Practices for First- and Second-Year University Students*, 8th ed.; Finster, D. C., Ed.; American Chemical Society: Washington, DC, 2017.

2. Lewis, R. J. *Sax's Dangerous Properties of Industrial Materials*, 11th ed.; Wiley: Hoboken, NJ, 2004. Online at https://onlinelibrary.wiley.com/.

3. Safety Data Sheets are provided by chemical suppliers; for example, MilliporeSigma (https://www.sigmaaldrich.com/US/en).

3.4 PRE-LAB PREPARATION: THE PRE-LAB FLOWCHART

Working in the organic chemistry lab requires some advance preparation to formulate a "game plan" to organize your activities in the lab. This makes the activities safer and more efficient.

3.4A | What Goes in the Pre-Lab Flowchart?

An organized plan is important for each experiment that involves hands-on work with chemicals. While the pre-lab flowchart described here is one way to organize your experimental plan, you should check with your lab instructor to see if they have an alternative way to organize the experimental plan.

The flowchart is a visual representation of the inputs and outputs from the various operational procedures during an experiment, and it provides an abbreviated road map to help you know what to do next in the lab. Preparing the flowchart will help you to think more deeply about the experiment you are doing—not only about the reaction itself, but also the practical reasons behind each of the procedures you perform. The flowchart, then, will help to maximize what you can learn from each experiment. If you are unsure about the reasons behind a specific step or action that is required in an experiment, ask questions!

3.4B Some General Instructions

1. Write a balanced equation for the reaction you will perform, if appropriate. A balanced equation will include any by-products of the reaction that will need to be separated. It is much easier to figure out what each step accomplishes if you know all of the reactants and products.

2. Provide a couple of words or a brief phrase to denote each operation and input/output, and don't go overboard with excessive detail. In most cases, a single page will be sufficient for the entire flowchart.

3. Write the flowchart in your notebook and submit a copy of it, using the submission method your instructor requires.

3.4C Example of a Flowchart Created from an Experimental Procedure

A representative experimental procedure, "Synthesis of (R)-(+)-3-Methyladipic Acid," is described below in text format.[1] The reaction is shown in **Figure 3.1**, and a sample flowchart is provided (**Figure 3.2**), generated from the text instructions. As you read the text, examine the flowchart to see how it illustrates the same information in a graphical format. There is a pathway showing how to move through the various steps of the experimental procedure, and branch points that help track different materials that are generated. Every experiment is different, though, so the number of steps and branch points will vary. The flowchart you submit may not look exactly like the one in the sample.

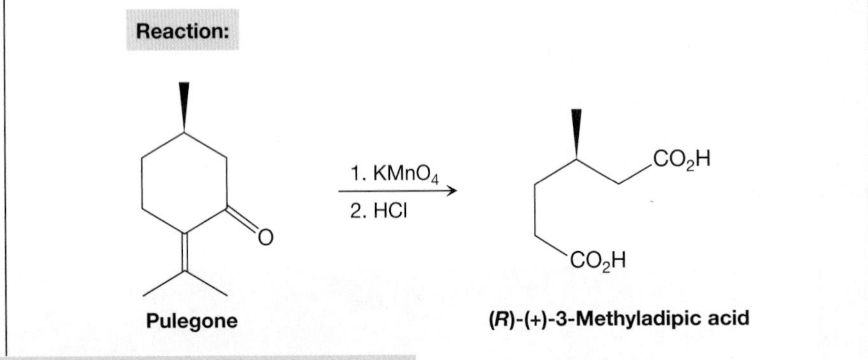

It's worth noting that this example synthesis is an older procedure that uses a lot of $KMnO_4$. Pay attention to the types and amounts of wastes that are generated, and how they are handled. The flowchart will help you determine whether or not this procedure does a good job of satisfying green chemistry principles (see Chapter 1). As you proceed through your own work in the lab, you can use a flowchart for a similar review of how well your own experiments address green chemistry principles.

Figure 3.3 shows a balanced equation of the reaction steps.

To a 250-mL Erlenmeyer flask containing 40 mL of distilled water, add 5 mL of pulegone. Swirl to mix the components (a two-phase mixture should result), then add 5 g of $KMnO_4$. Continue swirling for 10 minutes, and then allow the mixture to stand for approximately 2 hours, swirling occasionally. After this time, heat the mixture in a boiling water bath for 10 minutes. *CAUTION: The oxidation of pulegone is*

Reaction:

1. $KMnO_4$
2. HCl

Pulegone

(R)-(+)-3-Methyladipic acid

FIGURE 3.1

[1]Scott, W. J.; Hammond, G. B.; Becicka, B. T.; Wiemer, D. F. Oxidation of (R)-(+)-Pulegone to (R)-(+)-3-Methyladipic Acid. *J. Chem. Educ.* **1993**, *70*, 951–952.

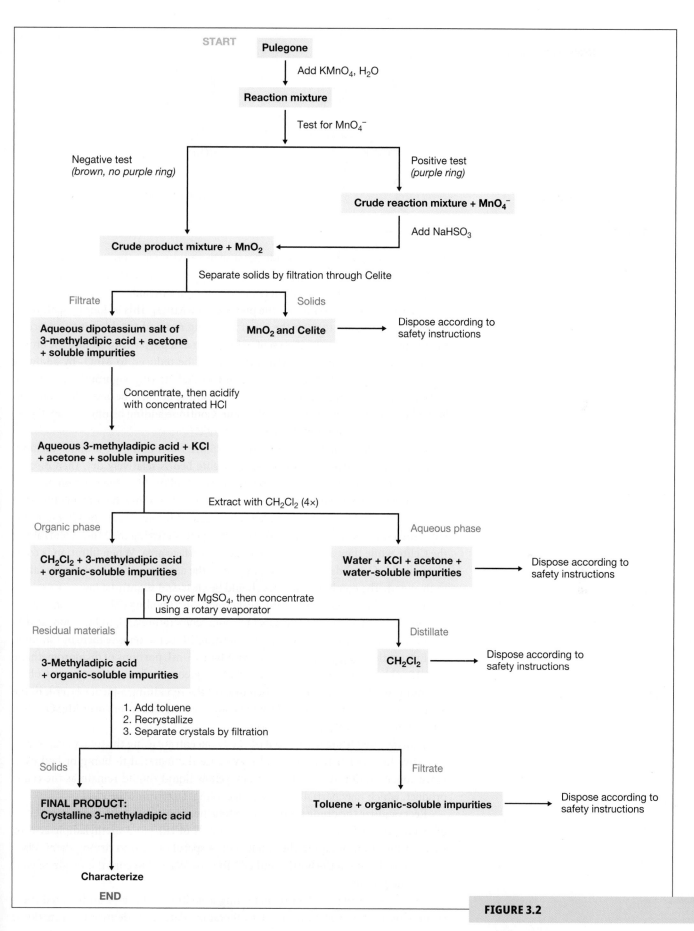

FIGURE 3.2

Representative pre-lab flowchart: synthesis of (*R*)-(+)-3-methyladipic acid.

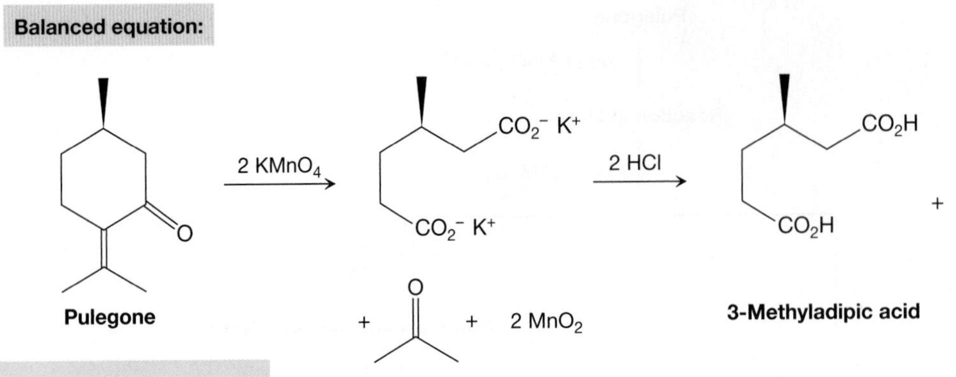

Balanced equation:

Pulegone →(2 KMnO₄)→ 3-Methyladipic acid

FIGURE 3.3

exothermic. Keep an ice bath handy because the mixture may boil over if it proceeds too quickly. After the heating period, allow the reaction to cool to room temperature, stopper the flask, and store it until the next laboratory period.

Test the reaction mixture for the presence of $KMnO_4$. This is done by withdrawing a drop of the mixture on the tip of a stirring rod and touching it to a piece of filter paper. Permanganate, if present, will appear as a purple ring around the brown MnO_2 solids. Remaining permanganate should be reduced to MnO_2 by adding a small portion of solid sodium bisulfite (0.1 g) and stirring vigorously. Do this in the hood, though, and don't add too much bisulfite at once, because the reduction is exothermic and some foaming will occur. Continue adding bisulfite in small portions until the test for permanganate is negative.

Prepare a slurry of 2–3 mL of Celite in 25 mL water and filter it under vacuum. Continue to apply the vacuum until the Celite bed is relatively dry, then discard the water. To remove the fine precipitate of MnO_2 solids, filter the reaction mixture through the Celite bed in parts. The filtration will proceed much more quickly if the solids are allowed to dry out completely between additions. Wash the filtered solids with three successive 10-mL portions of water, gently stirring the MnO_2 solids above the Celite while they are wet to maximize the surface area. When filtration is complete, the MnO_2 solids should be discarded in the appropriate solid waste container in the hood. The combined filtrates should be clear and slightly yellow.

Transfer the filtrate to a 250-mL beaker, add 2 or 3 boiling chips, and concentrate it to about 15 mL on a hot plate. Cool to room temperature and, in the hood, acidify with 10 mL of concentrated HCl. If solids form at this point, remove them by filtration.

Extract the cool aqueous solution with four 15-mL portions of dichloromethane (4 × 15 mL). Be sure you know which layer is aqueous and which is organic. Collect the four organic layers in a flask, then discard the remaining aqueous layer into the appropriate bottle in the hood. Dry the combined organic layers over $MgSO_4$, then remove the solids by gravity filtration.

Transfer the filtrate to a tared flask, and concentrate using the rotary evaporator. Excessive heat is not required, and may cause the material to bump or boil over. Once concentration is complete, a thick yellow liquid should remain as the crude product, which may partially solidify on cooling. Determine the yield.

Recrystallize the crude product from toluene, using 5 mL toluene for every gram of crude product. Collect the resulting crystals of (R)-(+)-3-methyladipic acid by vacuum filtration, scraping the solids with a spatula to aid in drying them. Place the toluene filtrate in the bottle marked "Toluene Waste." Record the weight of your crystalline product.

Characterize the product by obtaining a melting point and an infrared spectrum. Compare the melting point to literature data and identify the functional groups in your infrared spectrum.

Calculate your percent yield based on the amount of pulegone you used.

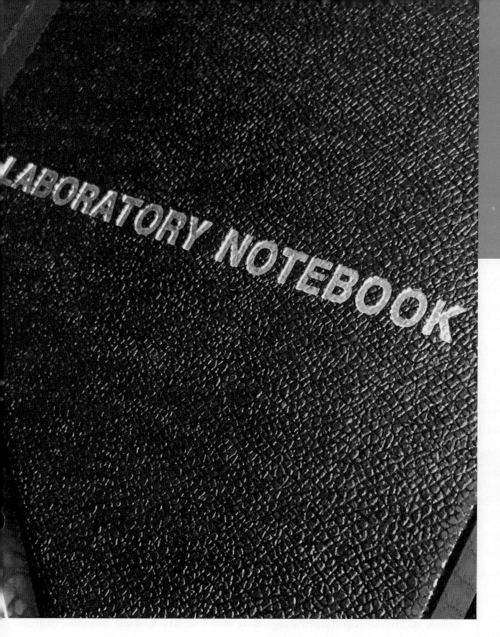

4

LEARNING OBJECTIVES

- Demonstrate your understanding of the importance of technical writing for various audiences.

- Report results and discussion in a format that clearly communicates them to your audience.

- Practice technical writing, using the components typically found in chemistry research journal articles.

The Laboratory Report

LABORATORY NOTEBOOK

Keeping an informative, contemporaneous, and permanent notebook that records your laboratory observations and data provides the foundation for an excellent laboratory report.
Courtesy of Gregory K. Friestad.

INTRODUCTION

After you have completed an experiment, including data analysis and interpretation, you will need to share your work with others. An essential component of doing science is communicating your experimental results to those outside the lab via written reports. Different audiences may require different writing styles. The audience could be scientists with a background in some other discipline, or members of the general public who happen to be interested in the field. Or, they could be experts in the same type of science, working on similar projects in a competing lab somewhere else in the world. Your audience could be your boss, too, who is about to make a case to the board of directors that your salary is due for a raise. Regardless of the audience, you need to write clearly and effectively about the work you have done in the lab.

If only an informal report is needed, then simply copying notebook pages and attaching them to a brief summary of the findings may suffice. In many cases, however, a more professional, technical report is needed, so a formal lab report must be submitted. Formal scientific writing takes experience and practice, and fortunately for you, you will have a number of opportunities to write formal lab reports for this course.

4.1 FORMAT

Technical writing is usually expected to follow certain formatting requirements that are imposed by journal publishers or by convention within the workplace. Similarly, when your instructor assigns a formal lab report, there is a specific format that is *required*. General formatting guidelines provided in this chapter should be followed closely unless your lab instructor gives separate directions. Your instructor may include additional requirements regarding font size, line spacing, margins, pagination, etc.; follow your instructor's requirements.

A typical report will be three or four pages long. These reports should be typed using the format outlined below. It should be complete, but concise. In the chemical literature, journals often have page limits for reports, although appendices such as data tables, spectra, copies of lab notebook pages, and other forms of data are not included in the page limit. Your lab instructor will notify you of any page limits for your reports.

Some rules for writing are commonly accepted for scientific literature and should be followed when generating your report. They are as follows:

1. Write in the past tense when describing what you did.

2. Some instructors prefer that you write in an active voice: "Recrystallization gave the pure product." Others prefer that you write in passive voice: "The product was purified by recrystallization." Use your lab instructor's advice on this point. However, you should *not* write in the first person: "I purified the product by recrystallization."

3. Incorporate data into the text of the report when explaining things, even if the data are already in a table. This helps to clarify the discussion.

4. Write with the three C's in mind: clear, concise, and complete.

Lab reports (and research journal articles) contain specific sections. These are explained here.

4.2A Title Page

a. Title

b. Experiment number

c. Identifying information such as your name, section number, TA, and report submission date.

4.2B Purpose

Discuss the general purpose of the experiment in at most two or three sentences (e.g., "The purpose of this lab was to investigate and compare various methods of distillation."). This should be more than a simple restatement of the title. If you are performing a synthesis, include the **balanced chemical equation**. The chemical equation provided in a textbook is not always balanced. Organic chemists often omit inorganic or other by-products such as HCl or CH_3OH from an organic reaction equation involving more complicated organic structures. In the lab, though, for safety reasons you must plan for how to contain or handle these by-products, so the balanced equation is essential.

<< **balanced chemical equation**
An accounting of all the atoms of reactants, defining how the atoms are distributed in products. The mole ratios of components in such an equation is called the stoichiometry.

4.2C Experimental

During the course of the experiment, you should have recorded in your notebook a clear written account of the procedures that you followed. That will become your experimental section. It will be a concise description of what you did and how you obtained the data you present, sticking to the facts only (interpretations and conclusions come later). In some cases, if you have not deviated from a procedure from the textbook, your instructor may suggest that you cite that as a source rather than merely copying the procedure. You should indicate where the source information can be found in your notebook pages. Attach these pages to your report as an appendix.

4.2D Results and Discussion

This is the most important section of the report. Here you should interpret your own experimental data and discuss what those data mean. You will need to explain clearly how you interpreted the results. It usually works best to divide this section into parts that correspond to the individual components of the experiment (e.g., each synthetic step). Do not include a detailed experimental account—this should already be written in your notebook pages. Instead, concentrate on your results. In a sentence or two, tell what you did and the outcome (e.g., "Vacuum filtration provided the product, benzoic acid."), then give the supporting evidence of that outcome (e.g., "The infrared spectrum showed a strong peak at 1689 cm^{-1}, indicating that the product may contain a functional group with a conjugated $C=O$."). Give all the evidence you have that a particular step was successful (or was not!). Not all experiments will use the same types of supporting evidence. If you have done a synthesis type of experiment, the following are some examples of supporting evidence that should be included in your report:

- Amount (in grams) and percent yield.
- Physical appearance: color, state, etc.
- Physical properties: melting point or boiling point, with literature value (cite your source) for comparison. Melting points should be reported as a range, as this provides evidence of purity.
- Thin-layer (TLC) or gas (GC) chromatographic data. For TLC, report the R_f value(s) and the solvent system used (drawings of TLC plates should be in your notebook). For GC, report the retention times and relative amounts of each component, and comment on the number of other peaks present, if relevant (the GC output, clearly labeled with experimental conditions such as column type, flow rate, and temperature, should be attached in an appendix).
- Spectroscopic data: infrared (IR) and/or nuclear magnetic resonance (NMR) spectra. List these peaks in the order they appear from left to right in the spectrum. For IR data, report key diagnostic absorbances only (in cm^{-1}), along with an assignment of the functional group of each absorbance. For NMR data, report the position in ppm, the integration (number of protons), and the multiplicity (e.g., 3.6 ppm, ^3H, triplet), and assign each signal to a proton (or protons) in the compound. Coupling constants for multiplets, if required, would also be included here (e.g., $J = 6.7$ Hz). Include a drawing of your compound (number the carbon atoms) in this section to facilitate identification and interpretation of specific resonances. If more than two or three peaks are reported, present them in a table. Examples of such tables are given in the corresponding spectroscopy chapters, Chapters 6–8. Explain the significance of the observed peaks, and note expected peaks that are absent, too, where relevant. For example, if an alcohol is being oxidized to a ketone, and no C=O peak is observed in the IR spectrum, this should be reported because it is important evidence that the reaction did not work as expected. In an appendix, attach a hard copy of each original spectrum, clearly labeled so that a reader knows at which point each spectrum was obtained (i.e., before recrystallization or after).

Once you have presented the results and supporting data, draw specific conclusions based on the data that *you* obtained. For a synthesis type of experiment, these will focus on identity and purity of the product: Did you really make the compound or not? Did the distillation work or not? Is your product pure? If not, does the evidence suggest what might have happened? The answers to these kinds of questions *must* be consistent with *your* data, and *must* be supported with experimental evidence *from your own work* as compared to literature-based expectations. Clearly explain how you drew your conclusions. Such evidence should include a comparison of the physical and spectral properties of your product with those of the starting material, and with literature values (with citation) expected for the product whenever possible. Discuss the yield of the product you obtained. If it is unreasonably low, then suggest possible explanations.

4.2E Conclusion

Summarize the main points in the Results and Discussion section, and indicate whether the Purpose was accomplished. Was your experiment successful? Where appropriate, discuss what your specific outcome tells you about a general theory or class of reactions. Did any of your results agree or disagree with your expectations? If a new technique was utilized, comment on its effectiveness relative to what you were trying to accomplish.

4.2F References

Provide citations to all of the references you used in preparing, carrying out, and analyzing the experiment. This includes the sources of procedures, the references you consulted in analyzing the data, and primary or secondary literature you used for comparisons of melting and boiling points, IR spectra, NMR spectra, and the like. Chapter 9, on the chemical literature, provides extensive examples of how to properly cite your sources.

4.2G Appendices

You may not always have material for each of the appendices listed here, but the material you do have must be labeled and appear in the following order:

1. Appendix A: Calculations (percent yields, R_f values, etc.). For multistep synthesis experiments, include a percent yield calculation for each synthetic step. In other cases, where multiple calculations of the same type are done, a sample calculation is sufficient. Include the equation you utilized, complete with units.

2. Appendix B: Spectra (IR, NMR), GC printouts, etc.

3. Appendix C: Experimental records (notebook pages), including any pictures or drawings you have of your experimental setup.

If your lab instructor has assigned any post-lab questions, be sure to attach the questions and your answers to them.

4.3 EVALUATION OF LAB REPORTS

The quality of a lab report is based on (a) following the required format, (b) including the relevant data, and (c) interpreting the appropriate data to draw conclusions of relevance to the purpose of the experiment. All of these are independent of whether the experiment worked well.

Still, this is a laboratory course. So, experimental results are typically a component of the grade. It is likely, then, that your instructor may consider your experimental success, such as the quality and amount of the product, in the grading evaluation. Some instructors may require a sample of the product to be submitted, in addition to the report; follow your instructor's instructions on this.

Regardless of how successful or unsuccessful the experiment seems to have been, a well-written report will still show significant gains in knowledge of practical organic chemistry. This is true even in cases where poor yield or purity was observed. In such cases, you should still put in the effort to write a strong report, because even unsuccessful experiments can be useful tools to meet the learning objectives. Most instructors place more value in achieving the learning objectives rather than focusing solely on the yield.

5

LEARNING OBJECTIVES

- Explain how separation techniques purify compounds.
  - » *Explain how filtrations separate solids from liquids.*
  - » *Explain how extractions separate compounds of differing solubilities.*
  - » *Explain how distillations separate liquids with different boiling points.*
  - » *Explain how recrystallizations purify organic solids.*
  - » *Explain how chromatography separates compounds for analytical and preparative purposes.*
- Select the appropriate purification methods for different types of mixtures.
- Explain how boiling points, melting points, densities, and optical rotations are measured.
- Use measurements of physical properties as evidence of identity and purity.

Purifications of Organic Compounds and Determination of Their Physical Properties

PANNING FOR GOLD

Separation techniques for mixtures in the organic lab depend on the physical properties of the components. Here, a specialized pan aids in using water to separate heavier gold particles from lighter particles of dirt or gravel.

Jeffrey B. Banke/Shutterstock.

INTRODUCTION

The organic chemistry laboratory is where new materials are made—materials with biological or physical properties that are both interesting and practical. Preparing these materials, purifying them, providing evidence of their identity, and measuring their properties are all activities you will experience in this course. There are a variety of techniques to synthesize, purify, characterize, and measure the compounds, which depend on a compound's characteristics. This chapter introduces you to some of the fundamental tools needed in the lab, and should be consulted regularly as a general resource for your work throughout the term.

5.1 SEPARATION OF LIQUIDS AND SOLIDS BY FILTRATION

Filtration is the physical separation of solid and liquid phases—familiar to anyone who has made a cup of coffee. It can be carried out in a number of ways, depending on the nature of the compounds. Generally, the mixture of solids and liquid is poured into a funnel, where it passes through a porous material (cloth, filter paper, or porous glass disc) that retains the solid phase, allowing the liquid phase to pass through. Different techniques and glassware are used, depending on whether the flow of liquid is aided by gravity or vacuum.

5.1A Gravity Filtration

Gravity filtration tends to be used when the liquid phase contains the material of interest, and the solid is an undesired material. Examples include removing the **drying agent** $MgSO_4$ from an organic solution, removing activated charcoal from a decolorized solution, or removing an insoluble side product from a reaction mixture.

drying agent >>
A salt that rapidly adsorbs water to form a hydrate. In its dehydrated form, a drying agent is added to organic solutions in order to remove water.

The key to carrying out this process efficiently is to use a fluted filter paper in a simple conical funnel with an open stem (**Figure 5.1a**). A filter paper is "fluted" by folding in half, then folding in small wedge-shaped segments in alternating directions with each fold passing through a point at the center of the original circle shape (**Figure 5.1b**). Upon opening, all the creases radiate outward from the center of the filter paper, providing increased surface area for the free flow of solvent through all parts of the paper when it is placed into a conical funnel (**Figure 5.1c**). If the filter paper is made into a cone shape without the fluting, it will lie flat against the inner walls of the funnel and filtration will occur mostly at the tip of the filter paper; this is inefficient, particularly when filtering fine particulates, which can clog the filter paper.

diatomaceous earth >>
Sedimentary deposits from the fossilized remains of diatoms. In powdered form, this material is insoluble in organic solvents and is used to aid in filtration, preventing fine particulates from clogging the filter. Also known as Celite or Filter-Aid.

Occasionally the solid is so finely divided that it even clogs the pores of a properly fluted filter paper, causing the flow to be unacceptably slow. In such cases, **diatomaceous earth** ("Celite" or "Filter-Aid") may be employed; it helps to distribute fine particles so that they don't accumulate in the filter paper. It is usually wetted to form a slurry, using the same solvent present in the mixture, and then filtered first. This leaves a firm pad of Celite on the filter paper prior to adding the mixture. Some Celite may also be added directly to the mixture to be filtered; as the resulting slurry is poured onto the pad of Celite in the funnel,

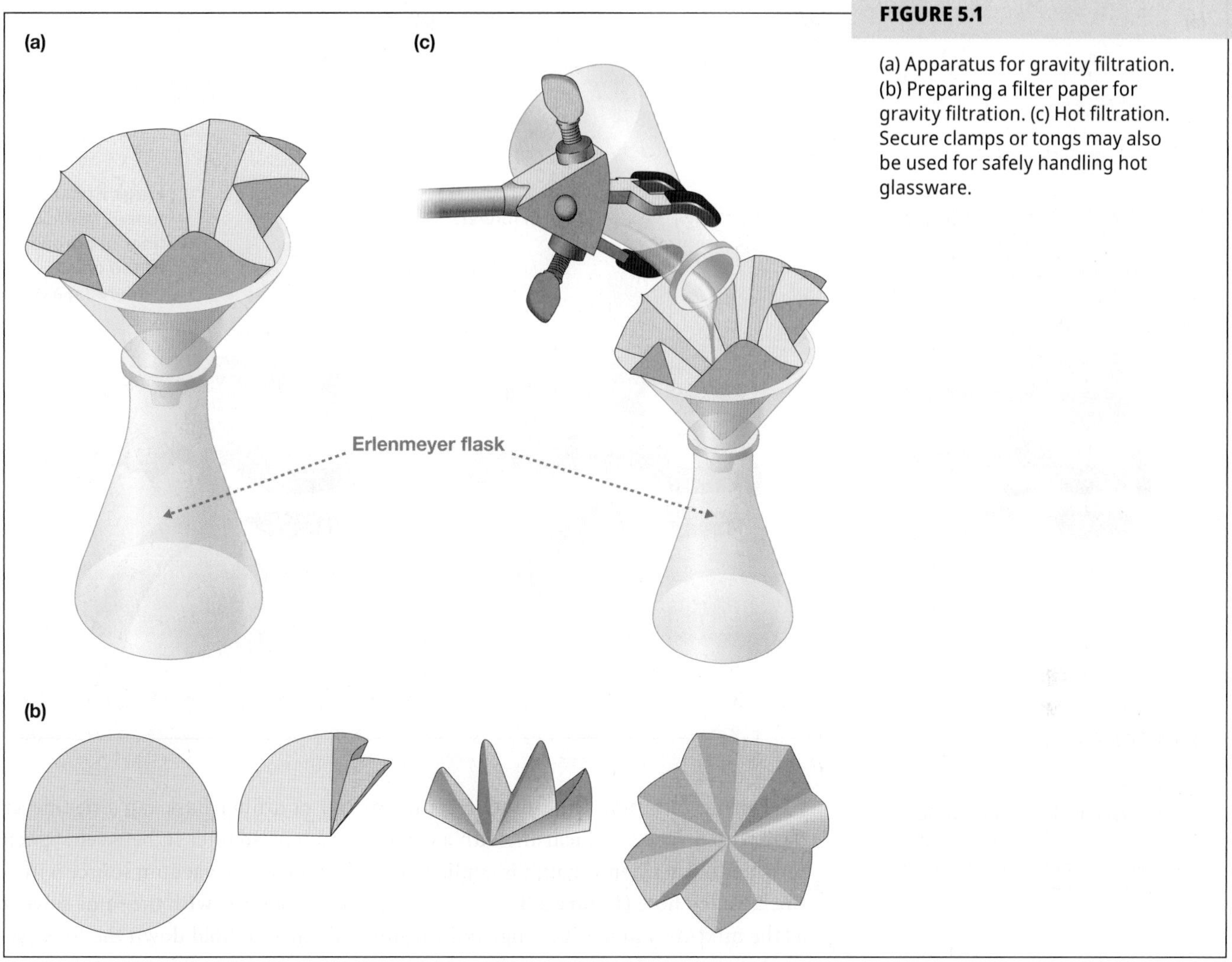

FIGURE 5.1

(a)

(c)

Erlenmeyer flask

(b)

FIGURE 5.1

(a) Apparatus for gravity filtration. (b) Preparing a filter paper for gravity filtration. (c) Hot filtration. Secure clamps or tongs may also be used for safely handling hot glassware.

particulates will be dispersed throughout the Celite, helping to avoid clogging the filtration.

Recrystallization (section 5.4) is a purification process in which the desired product forms crystals from solution. Gravity filtration is often used during recrystallization, before the crystals form, to remove undesired insoluble materials. This procedure is often called "hot filtration" (Figure 5.1c). In this case, the conical funnel is placed into an Erlenmeyer flask containing a small amount of solvent, and the apparatus is placed on a hotplate and warmed by heating to a gentle boil. An initial rinse of the filter using a small amount of the hot solvent mixture can help bring the apparatus up to temperature. Keeping the entire apparatus warm as the filtration proceeds allows the desired compound to stay in solution so that it does not crystallize in the filter paper along with the unwanted solids.

 Gravity Filtration

5.1B Vacuum Filtration (Suction Filtration)

Vacuum filtration should be used when the material of interest is the solid, not the liquid. It is used during a recrystallization to recover the desired compound after it has cooled and crystallized. A filter paper is chosen with a proper size to lie flat upon the porous surface inside a Büchner or Hirsch funnel (**Figure 5.2**), trimming excess paper if necessary. For small Hirsch funnels, this means a 1-cm filter paper (about the size of a thumbnail).

Vacuum Filtration

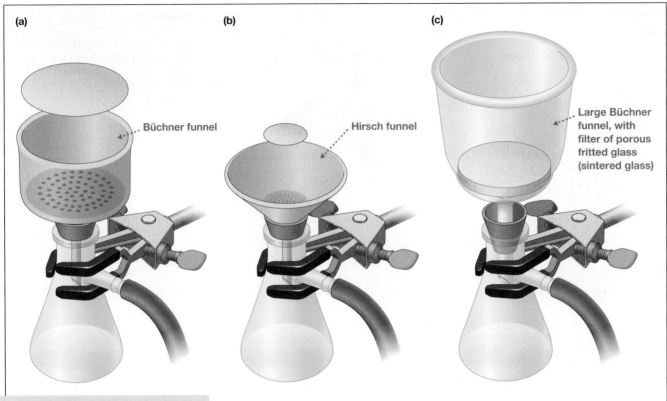

Büchner funnel

Hirsch funnel

Large Büchner funnel, with filter of porous fritted glass (sintered glass)

FIGURE 5.2

(a–c) Büchner and Hirsch funnels used in vacuum filtration. Smaller Hirsch funnels, which hold a 1-cm filter paper, are preferred for small amounts of material.

During filtration, the funnel rests in the top of a filter flask (an Erlenmeyer flask with a vacuum sidearm), with a conical neoprene adapter in between, which provides a seal when vacuum is applied by connecting to the vacuum source with a thick-walled hose (**Figure 5.3**). The filter paper is wetted first with the same solvent as the mixture you are filtering, and vacuum is applied to hold down the filter paper before the mixture is added to the funnel; this prevents solids from bypassing the filter paper. After the filtration is complete, the solid can be washed with an additional portion of cold solvent, and then dried by using the vacuum to pull air through the solid.

FIGURE 5.3

(a) The equipment for a vacuum filtration, showing a Büchner funnel, black neoprene adapter, filter flask, and filter paper. The filter flask should be secured with a clamp before use. (b) The wet filter paper is sealed to the surface of the Büchner funnel by applying vacuum before pouring in the solid/liquid mixture.

Courtesy of the University of Iowa.

(a)

(b)

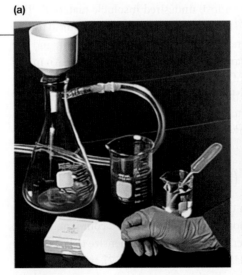

Liquid–liquid extraction is designed to partition compounds between two liquid phases. Organic reactions are often followed by a **workup** that includes an extraction to separate water-soluble by-products from organic products. Extraction is particularly useful in the separation of water-soluble acidic and/or basic components from an organic mixture.

The success of liquid–liquid extraction depends on two factors. First, the two solvents must be *immiscible*, or not mutually soluble, so that they form two separate layers. Second, solutes must have different solubilities in the two liquid phases. Thus, an organic compound in the presence of two immiscible solvents, such as water and diethyl ether, will distribute (partition) itself between the two phases until equilibrium is reached. Typically, organic compounds are mostly in the organic (diethyl ether) phase, and ionic compounds (salts) are in the aqueous (water) phase.

At equilibrium, the ratio of concentrations of the solute in each layer is constant, and may be defined as the **partition coefficient** or the *distribution coefficient, K*:

$$\text{Distribution coefficient } (K) = \frac{[X]_B}{[X]_A} = \frac{\text{solubility of X in solvent B}}{\text{solubility of X in solvent A}}$$

where $[X]_B$ is the concentration of solute in solvent B (generally the organic phase) and $[X]_A$ is the concentration of solute in solvent A (generally the aqueous phase). This relationship is independent of the total concentration of the solute and the actual volumes of the two solvents. The distribution coefficient has a constant value for each combination of solute and solvent. In pharmaceutical chemistry, this coefficient is frequently cited as the "logP" of a drug, and is measured using octanol and water as the two-phase system. The logP reflects the **lipophilicity** of a drug, which has an impact on how it is absorbed in the body, where it is concentrated, and how it is eliminated.

Imagine that you have a solution of some solute (S) in solvent A, and it is mixed with a second solvent B, which is immiscible with A. The solute can be transferred between phases in an equilibrium process (**Figure 5.4**). Eventually the solute will reach equilibrium at *K*. The layers are separated. If the distribution coefficient *K* is very large ($K > 100$), the solute is virtually all found in solvent B. More commonly, though, not all the solute will be transferred in a single extraction. However, one or two additional extractions with a small amount of solvent B will recover whatever solute may have remained in solvent A after the first extraction. Extracting with two or three smaller portions of solvent is more efficient than a single extraction with a larger amount of solvent. Generally speaking, any organic compound with $K > 1$ can be efficiently extracted from aqueous solution.

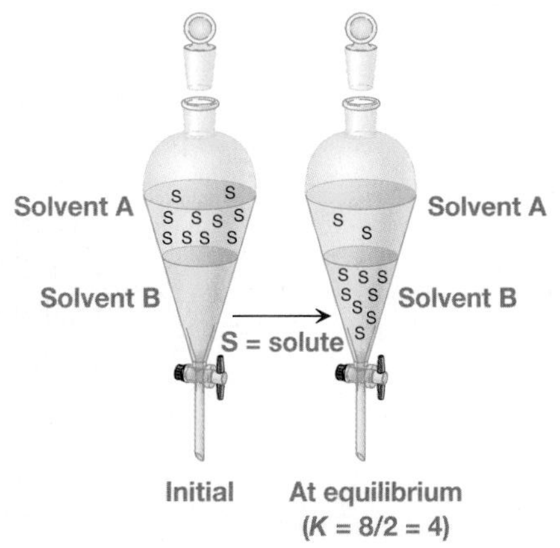

<< workup
A procedure for separation of an organic product from other materials after a reaction is complete, often involving deactivation or quenching of excess reagents, liquid–liquid extraction, and solvent removal. A two-phase aqueous–organic extraction in a separatory funnel is commonly used to separate organic products from water-soluble by-products.

<< partition coefficient
A ratio that describes the portion of a compound dissolved in an organic phase versus an aqueous phase when the compound is mixed with both phases and allowed to reach a solubility equilibrium. Also known as the distribution coefficient.

<< lipophilicity
A term describing the degree of solubility in an oil-like solvent.

FIGURE 5.4

Consider a 10-g sample of a compound with $K = 4.0$, partitioned between organic and aqueous phases. At equilibrium distribution, four parts (10 g $\times$ 4/5 = 8.0 g) of the compound will be in the organic phase and one part (10 $\times$ 1/5 = 2.0 g) of it will be in the aqueous phase. With a single extraction of 60 mL of organic solvent, 8.0 g of the compound will be obtained by evaporating the organic phase, only an 80% recovery. However, by dividing the same 60 mL of solvent into three 20-mL portions, the total amount of compound recovered may be increased to $>$ 99%:

$$\text{First extraction:} \quad 10 \text{ g} \times 4/5 = 8.0 \text{ g}$$
$$\text{Second extraction:} \quad (10 \text{ g} - 8.0 \text{ g}) \times 4/5 = 1.6 \text{ g}$$
$$\text{Third extraction:} \quad (10 \text{ g} - 8.0 \text{ g} - 1.6 \text{ g}) \times 4/5 = 0.32 \text{ g}$$
$$\text{Total:} \quad 9.92 \text{ g} \left(99.2\% \text{ recovery}\right)$$

In cases where $K < 1$, a simple extraction process will not give a satisfactory recovery of organic solute from an aqueous solution. In this case, however, the distribution coefficient can be altered by adding an inorganic salt such as sodium chloride to the aqueous layer. Organic compounds are generally less soluble in a saturated salt solution than in water, so the addition of NaCl shifts the equilibrium of solute between the two phases toward the organic layer, thereby increasing the distribution coefficient, and increasing the efficiency of extraction. This process is called *salting out*. Conversely, water tends to move into the saturated salt (brine) layer to help solvate the inorganic ions. Thus, saturated aqueous salt solutions are frequently used as preliminary drying agents to extract water molecules from the organic layer into the aqueous layer.

5.2B | Extraction to Obtain Organic Products from Reaction Mixtures

Extraction is an important tool for the preliminary purification of a reaction product, where inorganic by-products or organic salts can be washed away by extraction into an aqueous phase. In this way, the desired product of a reaction may be separated from unreacted starting materials, unwanted by-products, and impurities. In a typical extraction sequence, a two-phase system of water and an organic solvent (commonly diethyl ether or *tert*-butyl methyl ether) is utilized to remove unwanted water-soluble impurities (inorganic salts, low-molecular-weight polar organics, etc.) from the organic reaction medium. Usually the organic product is a neutral compound that remains in the ether phase.

Acidic or basic impurities that might otherwise remain in the organic layer can also be removed by formation of their corresponding water-soluble salts (**Figure 5.5**). For example, an acidic impurity (RCO_2H), in the presence of aqueous NaOH, forms a charged species, a *salt* ($RCO_2^-Na^+$), thereby making it soluble in the aqueous layer, and effectively separating it from the uncharged components in the organic layer. An analogous procedure can be applied for the removal of basic impurities upon treatment with dilute acid.

5.2C | Extraction for the Separation of Organic Compounds

In the same way that acidic and basic impurities can be removed from a neutral organic compound, the acidic and basic properties of organic compounds can be utilized to separate the components of a mixture. Organic acids (carboxylic acids and

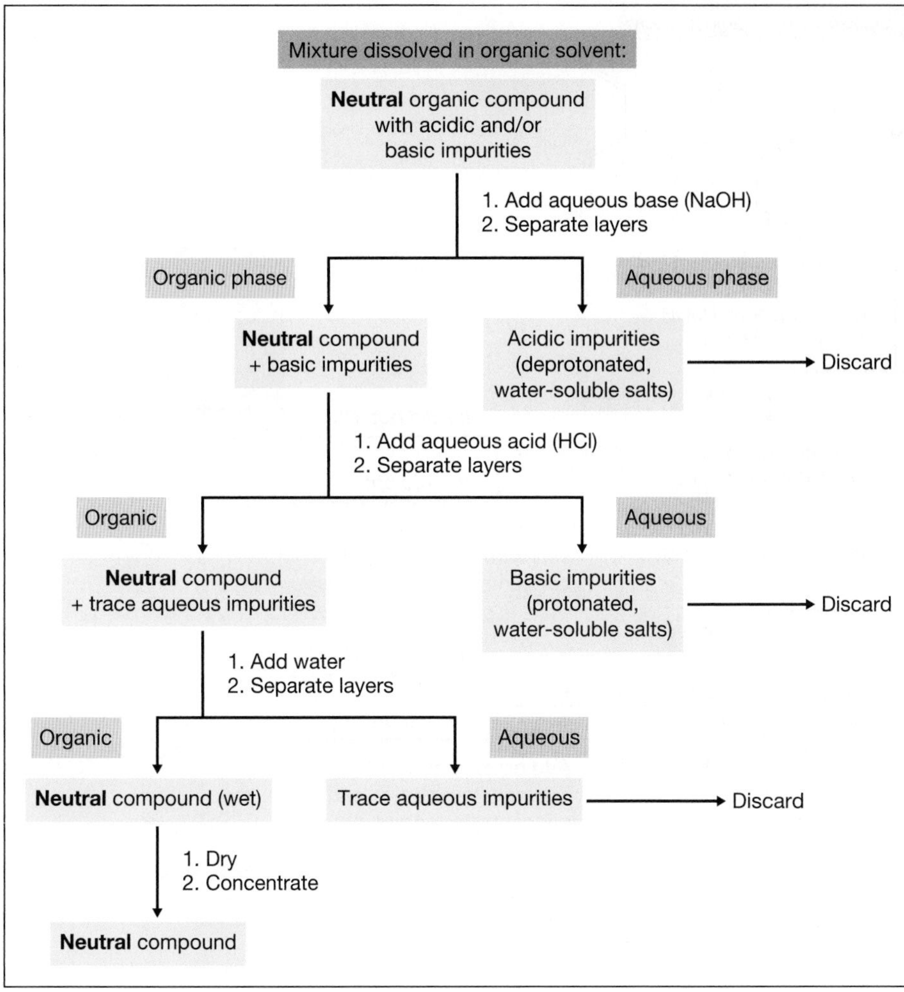

FIGURE 5.5

A general extraction sequence for the removal of acidic and basic impurities from a neutral organic compound. If one type of impurity (acid or base) is known to be absent, that portion of the sequence may be skipped.

phenols) and organic bases (amines) can be readily separated from each other and from neutral compounds by the extraction protocol outlined in **Figure 5.6**. Thus, stronger organic acids such as carboxylic acids ($pK_a \approx 5$) are easily converted into their sodium salts by reaction with sodium bicarbonate (for bicarbonate, H_2CO_3, $pK_a = 6.4$). Weaker organic acids such as phenols ($pK_a \approx 10$) require a stronger base such as sodium hydroxide. If both a phenol and a carboxylic acid are present, their differences in acidity allow their selective separation by extraction with the appropriate base. Aqueous sodium bicarbonate converts only the carboxylic acid to a salt, so only the carboxylic acid component is then drawn into the aqueous layer. If a phenol is present, it would be left behind in the organic phase because it is not acidic enough to be converted to a salt by $NaHCO_3$, and it could be later extracted into a more strongly basic NaOH solution.

Conversely, organic bases such as amines are converted into water-soluble hydrochloride salts by reaction with hydrochloric acid, and they may be separated from neutral and acidic substances by extraction with aqueous HCl.

Once the organic and aqueous phases have been separated, the various components can be isolated—by removing the solvent from the organic layer, and by neutralizing and subsequently filtering or extracting the aqueous layer. Using these principles it is possible to separate the various components from rather complex organic mixtures.

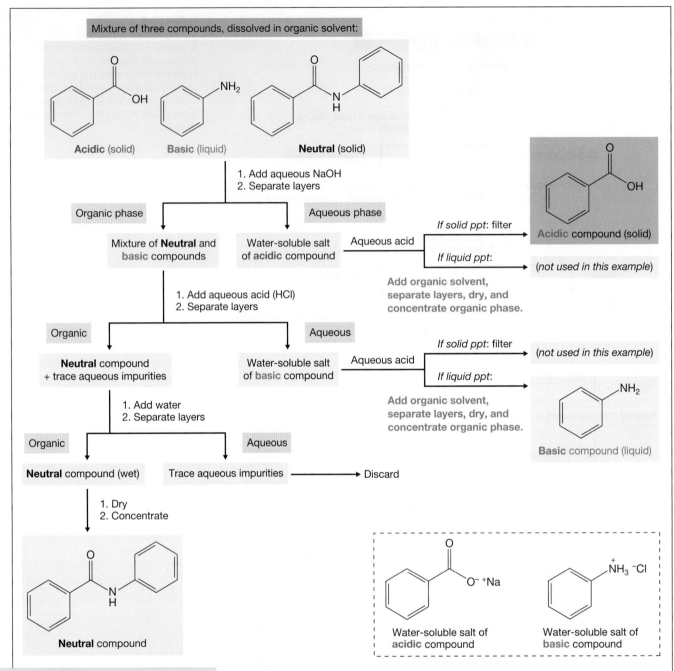

FIGURE 5.6

General approach to the separation of neutral, acidic, and basic compounds from a mixture. This is the same as Figure 5.5, except that the water-soluble salts are neutralized with acid or base so that the acidic and basic compounds can be recovered from the aqueous fractions. *Note:* ppt = precipitate; can be solid or liquid (oily droplets).

5.2D Practical Aspects of Extraction

Extraction is carried out in a separatory funnel (**Figure 5.7**), which is a cone-shaped enclosed funnel with a stopper at the wider end and a valve or stopcock at the narrower end. The separatory funnel rests in an iron ring with the narrow end down. The two liquid phases are placed in the funnel, which is closed, inverted with the narrower end up, swirled or shaken gently. Pressure may build up, especially with solvents having low boiling points, or if gas is produced by a reaction, such as that of $NaHCO_3$ (sodium bicarbonate) to make a salt from an acid, producing CO_2. Gas pressure is vented from the separatory funnel by opening the stopcock briefly while it is pointed upward, into a hood, and away from people. The separatory funnel is then placed back in the iron ring, and the stopper is removed from the top. The layers will separate, and the more dense layer can be drained off into an Erlenmeyer flask through the stopcock at the narrow end. As the phase boundary approaches the stopcock, begin to close the stopcock to slow the flow rate; this will avoid overshooting.

IDENTIFYING THE PHASES

An extremely important aspect of separations and extractions using a separatory funnel is the correct identification of the organic and aqueous layers. The easiest and most accurate way to predict which layer will be on top and which layer will be on bottom is to compare the densities of the two solvents being used. The denser layer will be the bottom layer, and the less dense layer will sit on top of it. Organic solvents such as hexane, diethyl ether, and ethyl acetate are less dense than water, while halogenated organic solvents such as chloroform and dichloromethane are more dense than water. However, the presence of other solutes in the aqueous phase, such as NaOH or NaCl, changes its density, so it is wise to confirm the identity of the organic and aqueous layers by adding a few drops of water to each; the water will dissolve only in the aqueous fraction.

If a layer is discarded before identifying it properly, an important component may be lost, and an experiment may have to be repeated from the start! Therefore, *save all the layers of any extraction until the product is isolated.* This way, if layers were incorrectly identified, they are all still available, and the product can be retrieved.

EMULSIONS

Sometimes a mixture that you know should separate into organic and aqueous phases fails to separate in the separatory funnel. There may appear to be a third, milky layer, or the entire mixture may appear to be one cloudy phase. This combined phase is called an **emulsion**, and it interferes with the extraction, making it hard to drain just one phase from the separatory funnel. To address this situation, add brine (saturated aqueous sodium chloride) to the mixture, mix it gently (shaking vigorously tends to make emulsions worse), and wait patiently. If partial separation occurs, drain out all the lower clear phase and repeat the separation with a fresh portion of the solvent in the lower phase.

Examples of liquid–liquid extractions. Worked Examples 1 and 2 represent common cases where compounds are separated based on their acid and base solubility properties.

Worked Example 1

A student has a mixture of *p*-toluic acid (pK_a = 4.36), *p-tert*-butylphenol (pK_a = 10.16), and acetanilide (pK_a = 22); see **Figure 5.8**. How can these components be separated from one another?

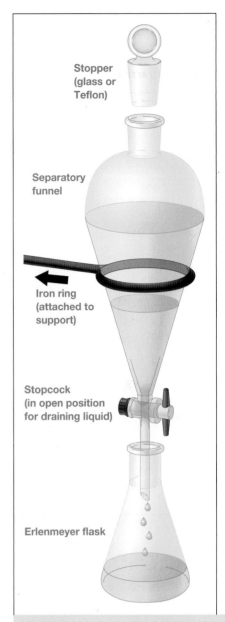

FIGURE 5.7

Separatory funnel.

<< emulsion
A nonhomogeneous mixture of two or more phases, such as organic and aqueous, that resists separation into two clearly defined layers.

FIGURE 5.8

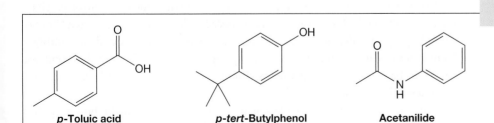

***p*-Toluic acid**	***p-tert*-Butylphenol**	**Acetanilide**

Solution

First, dissolve the mixture in organic solvent such as *tert*-butyl methyl ether or ethyl acetate. In this mixture, only *p*-toluic acid is acidic enough to be converted into a salt by HCO_3^-, so selectively extract it into aqueous $NaHCO_3$ solution. Next, extract the *p-tert*-butylphenol left behind in the organic phase using a more strongly basic aqueous NaOH solution. Separately neutralize each of the extracts by adding aqueous HCl, which will precipitate the water-insoluble *p*-toluic acid and *p-tert*-butylphenol. Then recover these two compounds by vacuum filtration. The third compound, acetanilide, is a neutral compound—not acidic enough to react with either NaOH or $NaHCO_3$ in aqueous solution—so it will remain dissolved in the nonpolar organic solvent throughout the sequence of extractions. Recover it by drying the organic layer with a drying agent such as anhydrous magnesium sulfate ($MgSO_4$) or sodium sulfate (Na_2SO_4), separating the drying agent by decanting or filtering (see Figure 5.1), and removing the organic solvent by concentrating it on the rotary evaporator. After all three of the components have been separated in this fashion, they can be recrystallized (section 5.4) to obtain pure substances.

Worked Example 2

A student is preparing an amide from 4-methylaniline and pentanoic acid via the acid chloride (**Figure 5.9**). Usually there are leftover carboxylic acid and amine that must be separated from the amide product. How should the student remove any leftover reactants from the desired amide product?

FIGURE 5.9

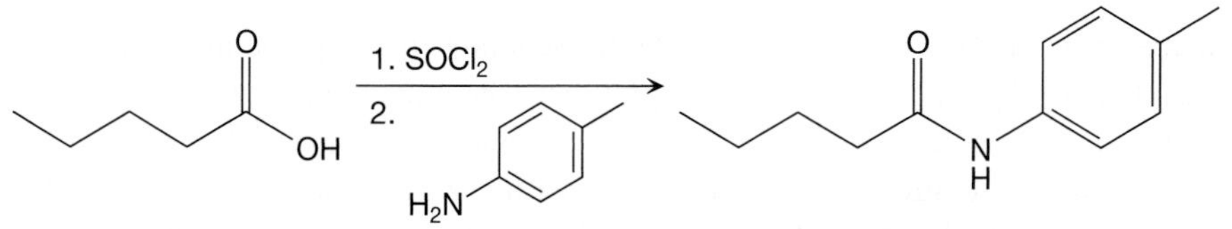

Solution

The amide product is neutral, whereas the leftover reactants are acidic (pentanoic acid) and basic (4-methylaniline). To remove any leftover reactants, dissolve the product mixture in an organic solvent such as ethyl acetate, then extract sequentially with aqueous HCl, aqueous NaOH, and water. Dry the organic phase over sodium sulfate, concentrate it on the rotary evaporator, and recrystallize (section 5.4) it to obtain the pure amide product.

5.3 PURIFICATION BY DISTILLATION

Distillation is a common method for purifying organic liquids. It involves phase changes between the liquid and gas phases because the mixture is subjected to cycles of boiling and condensing. Heating vaporizes a volatile compound, separating it from its less volatile contaminants. The vapor phase can then be cooled to condense the vapor back to the liquid phase (**distillate**), which is collected. Ideally the distillate is a single component of high purity.

The particular type of distillation used often depends on the boiling points of the compounds to be isolated, and the specifics of the desired separation. For example, *simple distillation* can easily separate two mutually soluble substances which differ in boiling points by 80°C or more. Simple distillation can sometimes

distillate >>
The liquid output collected from a condenser during distillation, a process of boiling and condensing that separates a liquid from other components of a mixture.

be successful when boiling points are 40–80°C. When the differences in boiling point are less than 40°C, *fractional distillation* may be required.

5.3A | Simple Distillation

In a simple distillation apparatus (**Figure 5.10**), a distillation flask (or boiling flask) is connected to a distillation head, which is simply a vertical tube with a thermometer port and a downward-angled sidearm for distillate to drain out. Complete condensation from vapor back to liquid (the *distillate*) is ensured by attaching a jacketed condenser to the sidearm; compressed air or cold water is passed through the jacket to keep the condenser cool. A curved distillation adapter connects the condenser to a receiving flask, where the distillate collects, and also provides a vacuum or inert gas inlet.

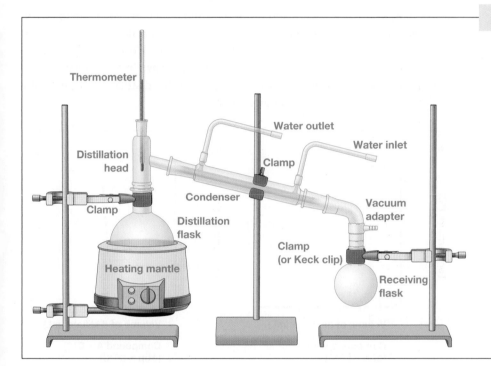

FIGURE 5.10

Simple distillation apparatus. The thermometer is inserted so that its reservoir bulb is below the sidearm of the distillation head. Heating may be supplied by sand bath, oil bath, or heating mantle (as shown). A magnetic stirrer may be added below the heat source.

It's important to clamp the apparatus securely. Two clamps should be used: The first at the neck of the distillation flask and the second on the condenser. Both clamps should be securely connected to a ring stand or built-in support bars in a fume hood. Be cautious in applying second and third clamps; tighten them only enough to support weight, and avoid twisting and pressure on other parts of the clamped apparatus, which can cause breakage. After the apparatus is set up and clamped, check carefully that all the ground glass connections remain tightly sealed. Sometimes it is necessary to switch the receiving flask during distillation; this is facilitated if a plastic clip ("Keck clip") is used in place of the clamp on the neck of the receiving flask.

The sample mixture is placed in the distillation flask by disconnecting the distillation head and pouring the liquid through the neck of the flask with the aid of a funnel. Alternatively, the liquid can be poured in through the opening created by removing the thermometer adapter; use a long-stem funnel to ensure that this mixture travels toward the distillation flask, not the receiving flask. Insert a boiling stick or boiling stone in the distillation flask, or use a magnetic stirrer, to facilitate smooth boiling with smaller bubbles. This prevents sudden eruptions of large volumes of vapors, known as "bumping," which can cause undistilled material to splash into the condenser, contaminating the distillate.

 Simple Distillation

A caution before heating: Never heat a sealed system—dangerous pressure can build up and cause an accident. Before you begin applying heat, make sure the vacuum adapter hose connector is unblocked so that the pressure can equalize with the external atmospheric pressure.

Heat is applied to the distillation flask, raising the temperature within, and increasing the vapor pressures of the components. When the sum of those vapor pressures reaches atmospheric pressure (or the pressure within the apparatus), the material in the distillation flask begins to boil. Vapors are carried upward to the distillation head, and their temperature is monitored. A cooled condenser slopes downward from the distillation head, ensuring that liquid condensing from cooled vapors will drain toward the receiving flask. As the distillation progresses, the receiving flask can be periodically changed, so that different fractions of distillate are collected in different flasks. Each fraction can be analyzed by gas chromatography to determine the ratio of components it contains.

The temperature of a distillation process is monitored by the thermometer in the distillation head. How the temperature varies over time depends on the vapor pressure of the liquid being distilled, and this depends, in turn, on the liquid's composition. When distilling a pure liquid, there is no change in the composition as condensate is removed from the system. As a result, the distillation proceeds at a relatively constant temperature (**Figure 5.11a**).

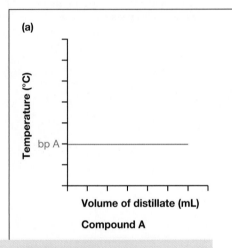

(a)

Compound A

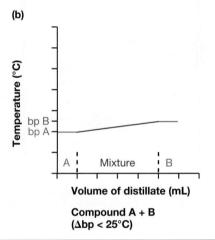

(b)

Compound A + B
(Δbp < 25°C)

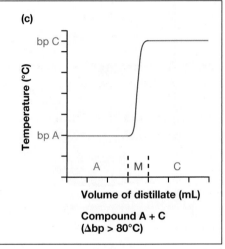

(c)

Compound A + C
(Δbp > 80°C)

FIGURE 5.11

Temperature readings at the distillation head during a simple distillation of (a) a relatively pure compound, (b) a mixture of two components having boiling points (bp) differing by < 25°C, and (c) a mixture of two components with boiling points differing by > 80°C.

For any liquid mixture, though, the composition and temperatures change during the distillation. Dalton's law and Raoult's law describe this behavior.

Dalton's law: The vapor pressure of a liquid is the sum of the partial pressures of the individual components:

$$P = P_A + P_B$$

Raoult's law: At a given temperature and pressure, compound A has a partial pressure (P_A) in a mixture that is equal to the vapor pressure of the pure compound (P_A^{pure}) multiplied by its mole fraction (X_A) in the mixture:

$$P_A = (P_A^{pure})(X_A)$$

As the distillation proceeds, the composition of both the liquid and the vapor change; early in the distillation, the component of lower boiling point (greater partial pressure, P_A) is more rapidly removed from the system. As a result, the temperature increases throughout the distillation (**Figure 5.11b** and **Figure 5.11c**). A phase

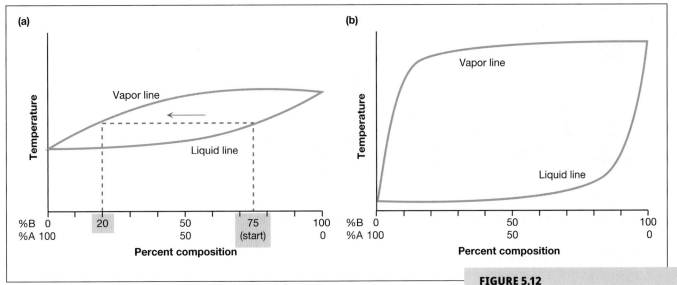

FIGURE 5.12

Phase diagrams for a mixture of two components (A and B) having (a) similar (< 80°C) boiling points and (b) widely differing (> 80°C) boiling points. In both cases, the boiling point of A is less than the boiling point of B.

diagram (**Figure 5.12**) is a plot of composition (mol %) versus temperature (T), where the lower curve is the liquid line and the upper curve is the vapor line. These diagrams help to explain the temperature-versus-volume behavior (Figure 5.11) of liquid mixtures, and additionally can be used to determine the composition of both the liquid and the vapor phase at any temperature throughout the distillation.

When examining a liquid–vapor phase diagram, keep in mind that the phase change from liquid to vapor is *not* accompanied by a change in temperature; the vapor and liquid are in equilibrium at the same temperature. However, the compositions of the vapor and liquid are different (Figure 5.12), which explains how distillation can change the ratio of the two components. The horizontal line represents the phase change at constant temperature from liquid to gas phase. This horizontal line intersects the liquid and vapor lines at two different compositions. The dotted lines in Figure 5.12 show that a boiling liquid starting at a ratio of 75% B and 25% A is in equilibrium with a vapor consisting of 20% B and 80% A. Condensing that vapor and cooling it gives a liquid which is enriched in A, the compound with the lower boiling point.

5.3B Fractional Distillation

In fractional distillation, an extra vertical column called a *fractionating column* is added to the apparatus (**Figure 5.13**). The physical principles that describe a simple distillation hold true for a fractional distillation as well, but the fractionating column offers a much larger surface area. The vapor undergoes a continuous cycle of condensation and revaporization as it passes up through the fractionating column, and each sequential revaporization is equivalent to another simple distillation. Thus, the composition of the vapor is progressively enriched as it moves up the column. As a result, the temperature behavior of a fractional distillation over time resembles that shown in Figure 5.11c, even when two components have similar boiling points. With an apparatus like the one shown in Figure 5.13, it is possible to cleanly separate components of liquid mixtures in which the boiling points differ by as little as 25°C. Compounds with boiling points closer than this can be separated, but it may require a more specialized apparatus, such as a longer fractionating column. It should be noted that the fractional distillation technique generally takes more time, and can result in greater material losses because of the extra glassware surface area, so it is used only when the boiling points are too close to separate by simple distillation.

(a)

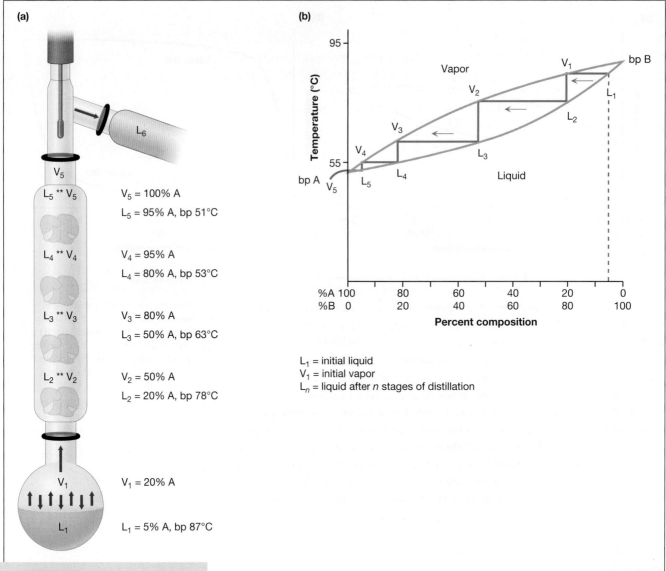

$V_5 = 100\%$ A
$L_5 = 95\%$ A, bp 51°C

$V_4 = 95\%$ A
$L_4 = 80\%$ A, bp 53°C

$V_3 = 80\%$ A
$L_3 = 50\%$ A, bp 63°C

$V_2 = 50\%$ A
$L_2 = 20\%$ A, bp 78°C

$V_1 = 20\%$ A

$L_1 = 5\%$ A, bp 87°C

FIGURE 5.13

(a) Fractional distillation of a simple two-component liquid mixture of initial composition 5% A and 95% B, where the boiling point of A is less than the boiling point of B. (b) The effect of fractional distillation is illustrated on the phase diagram.

 Fractional Distillation

5.3C | Vacuum Distillation

Vacuum distillation is particularly appropriate for compounds with very high boiling points, or for those that decompose at high temperature. The boiling point of a liquid is the temperature at which the vapor pressure of that liquid is equal to the applied pressure. So, the vacuum reduces the applied pressure within the simple or fractional distillation apparatus, thereby reducing the temperature at which the liquid boils. To accomplish this, the distillation apparatus is sealed from the atmosphere and connected to a vacuum pump. The distillation is carried out in the same way, except some care is needed to change flasks and collect fractions without disrupting the vacuum. A stopcock placed between the vacuum adapter and the receiving flask allows the distillation to proceed under vacuum while the receiving flask is removed and replaced.

The joints between the components of the apparatus need to be tightly sealed before vacuum is applied. The ground glass surfaces usually seal well enough on their own, but occasionally two joints will not match well enough for a good seal. Grease can be applied sparingly to help seal ground glass joint connections, but the grease can contaminate the desired products, it makes a big mess of the glassware, and it is difficult to clean up. Grease should only be used upon instruction from the TA or instructor. A thin strip of Teflon tape between the ground glass surfaces of the joint accomplishes the same objective with much less mess.

Boiling can be quite erratic in a vacuum distillation; occasionally the material in the boiling flask will "bump," or suddenly burst upward within the apparatus. Bumping can send the unpurified liquid mixture into the condenser and then into the collection flask. Use a boiling stick, boiling stone in the distillation flask, and/or rapid magnetic stirring to avoid this.

5.3D | Rotary Evaporation

Organic solvents are often used to dissolve reactants before carrying out reactions or separations in the organic lab. The solvents must subsequently be removed to recover the organic compounds that are in solution. Solvents having a low boiling point (usually 80°C or below) are commonly used to facilitate separation by vacuum distillation. There is a specialized device for this type of process called a rotary evaporator, or "rotovap" (**Figure 5.14**), that rotates a round-bottom flask during distillation, increasing the surface area of the solvent for more rapid solvent evaporation. The solvent evaporation causes the flask to get cold, so often a water bath is used to keep the flask at a constant temperature.

 Rotary Evaporation

By reducing the pressure, the boiling points of common organic solvents can be lowered to room temperature or below. This means that not much heat is required to rapidly boil away the solvent, leaving behind any solutes. Reducing the pressure and thus the amount of heat required helps to avoid damaging heat-sensitive organic compounds.

CAUTION: Before operating the rotary evaporator, check it carefully to make sure its glass parts have no cracks. Your sample should be in a round-bottom flask that is also free of any cracks (replace if necessary). Cracked glassware can break under vacuum, causing glass pieces to fly. This is one reason why you should always wear eye protection in the lab.

To operate the rotary evaporator, the flask containing organic solvent is affixed to the ground glass joint that points down toward the water bath. Hold the flask with your

FIGURE 5.14

Rotary evaporator.
Courtesy of Gregory K. Friestad.

hand under it until you have securely clamped the ground glass connection together. Turn on the rotation of the flask, turn on the vacuum, and close the venting valve. Finally, lower the flask so that it just touches the water bath. Solvent will be removed. After the sample reaches a constant volume, reverse the order of steps to remove your flask: Open the venting valve, turn off the vacuum, raise the flask out of the water, stop the rotation, and place your hand beneath the flask while unclamping it.

5.3E | Steam Distillation

Steam distillation is used to purify organic compounds that are immiscible with water by distilling them along with water. Two immiscible liquids both contribute to the vapor pressure in the distilling flask, and as the water distills, the vapor phase (steam) carries some of the organic compound along with it. After the condensate is collected, the water can be separated from the organic compound because the two are immiscible.

A main reason for using this approach is to lower the temperature at which the organic compound distills. This not only makes it a more energy-efficient process, but also is more likely to avoid degradation of heat-sensitive organic functional groups.

The water may provide other beneficial effects. For example, boiling plant material along with water may disrupt cell walls, helping to release an organic compound from the matrix and allow it to distill more efficiently. This is a procedure commonly used in separating essential oils from various plants.

5.4 RECRYSTALLIZATION OF ORGANIC SOLIDS

If impurities are present in organic solids, the organic solid may be purified by recrystallization from a solution. When the amount of solute in solution exceeds the solubility, the solution is "supersaturated," and the solute will come out of solution. If you are fortunate, the solute will form crystals as it comes out of solution. As crystals grow, the molecules assemble in a tightly packed repeating lattice structure, which usually accommodates no impurities. Therefore, as a crystalline solid forms from a solution, impurities are excluded from the crystals. A rapid precipitation of powdery or amorphous solids may not exclude impurities, so a slow crystallization is preferred in most cases. Higher purity is usually observed when the solid forms crystals of highly regular geometric shapes or "crystal habits" that are characteristic to the compound. These may appear as needles, cubes, columns, or other geometries.

In the standard technique, crystallization is induced by lowering the temperature of a saturated solution. The compound is mixed with the minimum amount of hot solvent, just enough to dissolve the compound. Then the solution is allowed to cool, so that it becomes supersaturated. The crystals that form are recovered by vacuum filtration. The steps are summarized in more detail below.

 Recrystallization

5.4A | Summary of the Basic Steps of Recrystallization

For most routine work, the following sequence of steps will generally be effective. Further details and more specialized techniques are described later.

1. Choose a solvent. The key is to identify one in which the solute has high solubility when hot, and low solubility when cold.

2. Dissolve the substance. First, heat the solvent in a separate Erlenmeyer flask. To an Erlenmeyer flask containing the substance, slowly add the minimum amount of hot solvent needed to dissolve the desired substance, and no more.

3. Optional hot filtration. If the desired substance is dissolved, but the impurities are not, remove the impurities via gravity filtration (see Figure 5.1) while keeping the solution and filtrate hot.

4. Cool slowly. Remove the solution from heat and allow it to stand undisturbed while crystallization proceeds. Here, patience is a virtue. After crystallization occurs, cool in an ice bath to ensure completion.

5. Recover the crystals. Use vacuum filtration (see Figure 5.2) to separate the crystals from the liquid filtrate (the **mother liquor**).

6. Optional repeat. Some of the desired substance may remain in the mother liquor. Evaporate some or all of the solvent and repeat the sequence to obtain a second batch (crop) of crystals.

<< **mother liquor**
The liquid phase that remains after removal of crystalline solid, e.g., by filtration.

5.4B Choosing a Recrystallization Solvent

The fundamental requirement is that the compound is soluble at high temperature, and mostly insoluble at low temperature. In that scenario, a hot saturated solution can be cooled, forcing the solute to come out of solution. There is no single ideal recrystallization solvent to apply for all situations, and sometimes trial and error is required.

Several solvents are listed in **Table 5.1**, with boiling points and hazard notes. Except for dichloromethane and water, all of them present some flammability hazards, and should be kept away from flames. Diethyl ether, dichloromethane, and benzene introduce additional risks, and should generally be avoided for recrystallizations in the instructional lab except with special instructor permission.

TABLE 5.1

Some Solvents Suitable for Recrystallization

SOLVENT	bp (°C)	HAZARD NOTES
Diethyl ether	35	Extremely flammable
Dichloromethane (CH_2Cl_2)	40	Possible carcinogen
tert-Butyl methyl ether (MTBE)	55	
Acetone	56	
Hexane	69	
Ethyl acetate	77	
Ethanol	78	
Benzene	80	Carcinogen
Water	100	

To choose a good solvent, place 5–10 mg of the solid to be recrystallized into each of several vials, and add a few drops of various solvent candidates. An ideal solvent will not dissolve the solid when cold, but will dissolve most or all of it when heated. If the solid dissolves immediately in the cold solvent, reject it. If the solid doesn't dissolve at all, even at the boiling point of the solvent, reject it as well.

5.4C Recrystallization from Two-Solvent Systems

Two-solvent systems can give you much better control over solubility. The organic solid should have a high solubility in one of the solvents (the "good" solvent), and little or no solubility in the other (the "poor" solvent). Poor solvents can be either too polar (water) or not polar enough (hexane) to dissolve the organic solid. Once a pair is chosen, the polarity of the two-solvent system can be modified simply by varying the ratios of the two solvents. To enable this, the good and poor solvents must be miscible (mutually soluble with each other at all proportions). It is often helpful to choose a good solvent that has a lower boiling point than the poor solvent, so that the good solvent can be evaporated if too much has been added. Although the labels of "good" and "poor" may depend on the type of compound being recrystallized, **Table 5.2** lists some common two-solvent systems that are often useful for recrystallization of typical organic compounds.

In a typical two-solvent procedure, a small amount of the poor solvent is added to the compound, and the mixture is heated on a hot plate. Occasionally the compound dissolves in the poor solvent; if this occurs, simply cool and recrystallize as described previously (see section 5.4A) for the single-solvent method. If the compound does *not* dissolve in the poor solvent, then slowly add the good solvent, while keeping the mixture warm, until the compound is just dissolved. (If most of the material dissolves readily, leaving a persistently insoluble precipitate or cloudiness, remove these insoluble impurities via hot gravity filtration.) Remove the homogeneous solution from the heat and allow it to cool, so that it becomes supersaturated. If crystallization does not occur, cool the flask further in an ice/water bath. (Additional advice on inducing crystallization is given below.) Recover the crystals by vacuum filtration. A cold mixture of the two solvents may be used to wash the crystals, using a lower ratio of good solvent to avoid dissolving the crystals.

TABLE 5.2

Two-Solvent Systems for Recrystallization

GOOD SOLVENT	POOR SOLVENT
Ethanol (bp 78°C)	Water (bp 100°C)
Acetone (bp 56°C)	Water (bp 100°C)
tert-Butyl methyl ether (bp 55°C)	Hexane (bp 69°C)
Dichloromethane (bp 40°C)	Hexane (bp 69°C)
Ethyl acetate (bp 77°C)	Hexane (bp 69°C)

5.4D Inducing Crystallization

Crystals don't always form spontaneously; you may need to intervene to induce crystallization. This process is a combination of skill, art, and luck. There are various ways to induce crystallization, but the basic process is to prepare a saturated solution of the organic solid, then change the conditions in some way that lowers the solubility. This is usually accomplished by lowering the temperature (because the solubilities of solids decrease with decreasing temperature), causing the solution to be supersaturated—that is, the amount of material in solution is greater than its solubility will allow—so that the solid must come out of solution. If crystals don't grow spontaneously, crystallization can be encouraged by introducing **nucleation sites** where crystal growth can initiate. The best way to do this is by "seeding" the supersaturated solution with a trace amount (< 1 mg—not enough to affect the yield calculation) of the same compound that has been previously crystallized. Sometimes, the seed crystals initiate a rapid crystal growth, which can be strikingly beautiful to

nucleation site >>
The origin of crystal growth from solution, where molecules begin to organize to form a crystalline solid.

watch. If seed crystals are unavailable, crystallization can sometimes be induced by scratching the inside of the Erlenmeyer flask with a glass rod or pipet, which creates nucleation sites on the surface of the glass.

The best purification occurs when crystals are allowed to grow slowly from a homogeneous solution over a period of time, usually from a few minutes to a few hours. Forcing a compound out of solution by changing the solubility too rapidly leads to a liquid or amorphous solid precipitate rather than crystals. A precipitate is likely to trap more impurities within the solid, thus giving an unsatisfactory purification.

5.4E Recovering the Crystals

After crystallization has occurred, most of the impurities are generally left in the liquid phase. The two-phase mixture can then be separated using vacuum filtration to recover the crystals. The liquid filtrate, called the mother liquor, contains soluble impurities, along with additional desired compound that had yet to be crystallized. A second or third batch (or crop) of crystals can sometimes be obtained by evaporating some of the solvent from the mother liquor and then cooling it again. Thin-layer chromatography (section 5.5) can be used to assess whether the mother liquor still contains enough of the desired compound to make it worth pursuing an additional crop of crystals.

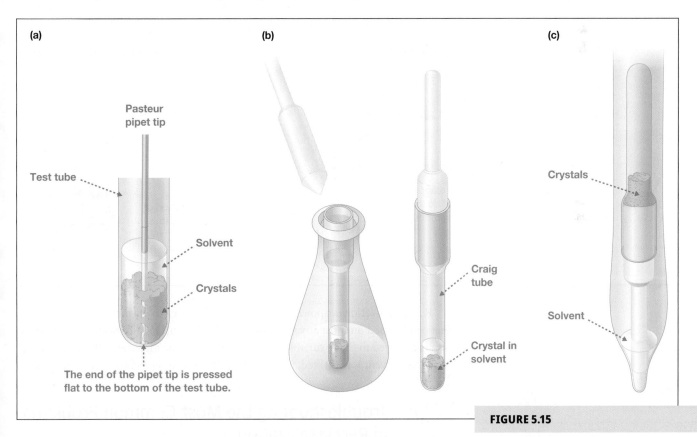

(a) Pasteur pipet tip

Test tube

Solvent

Crystals

The end of the pipet tip is pressed flat to the bottom of the test tube.

(b)

(c) Crystals

Craig tube

Crystal in solvent

Solvent

FIGURE 5.15

(a) Pipet tip pressed to bottom of test tube for microscale filtration. (b) Craig tube for separating crystals from solvent with (c) the aid of a centrifuge.

On small scales (e.g., < 50 mg), recrystallizations are sometimes more conveniently performed in a test tube, and the solvent may be removed with a Pasteur pipet. Holding the tip of the pipet firmly against the bottom of the tube creates a very small gap that will not let larger crystals pass as the solvent is drawn up into the pipet (**Figure 5.15a**).

A more elegant test tube recrystallization uses a Craig tube (**Figure 5.15b**). This tube has a smaller diameter restriction above the crystallization mixture, and a glass insert that nests into it. The glass pieces are machined so that they fit tightly together and only solvent can pass. Once assembled with the crystallized mixture inside, the

Craig tube is inverted inside a larger test tube (**Figure 5.15c**) that is placed in a centrifuge. Centrifugal force pushes the liquid past the insert and the liquid collects in the larger tube, leaving the crystals behind in the Craig tube.

5.4F | Hot Filtration of Less Soluble Materials

Depending on how the desired compound is formed or isolated, there may also be impurities that are less soluble than the desired compound. When recovering the compound by vacuum filtration, these less soluble impurities will not be separated. Less soluble impurities may be detected as a persistent cloudiness that does not clear up upon adding more solvent. These insoluble materials may be removed by hot filtration or gravity filtration prior to crystallization (**Figure 5.16**). After the hot filtration, the filtrate contains the desired compound. Crystallization may then be induced, and finally, the desired compound may be isolated by vacuum filtration.

FIGURE 5.16

Different types of filtration and their purposes during a recrystallization.

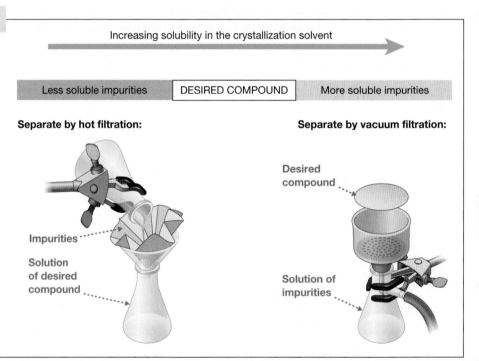

Increasing solubility in the crystallization solvent

| Less soluble impurities | DESIRED COMPOUND | More soluble impurities |

Separate by hot filtration:

Impurities

Solution of desired compound

Separate by vacuum filtration:

Desired compound

Solution of impurities

Recrystallization is a powerful, versatile, and cost-effective method of purification that can be used on most any scale, from milligrams up to kilograms. Experimentation is required to find a suitable solvent from which the material will crystallize, but once that solvent is identified, it is usually quite easy to reproduce on a larger scale.

5.4G | Troubleshooting the Most Common Problems in Recrystallization

PRECIPITATION

If the solid comes out of solution quickly, this is referred to as precipitation, and some of the impurities are often trapped within a solid precipitate. Such solids may resemble a crust or powder with no visible crystallinity, and are said to be amorphous. The material should be re-dissolved by heating, or by adding more of the good solvent, if a two-solvent system is being used, and then allowed to cool more slowly.

OILING

If the material comes out of solution at a temperature above its melting point, it will precipitate as an oil. Cooling this two-phase mixture may result in solidification rather than a proper crystallization, and the solid that results likely contains significant amounts of impurities. A small additional portion of solvent should be added, and the mixture reheated to form a homogeneous solution. With more solvent present, the hot solution may cool below the compound's melting point before it becomes supersaturated, so that it comes out of solution as a solid rather than as an oil.

NOTHING COMES OUT OF SOLUTION

If neither crystals nor oil are forming, there's probably too much solvent, or if a two-solvent system is being used, there may be too much of the good solvent. Excess solvent can be removed by using a rotary evaporator, or by adding a boiling chip and placing the Erlenmeyer flask back on the hot plate. If a two-solvent system is being used, and if the good solvent has a *lower* boiling point than the poor solvent, then boiling will help because the good solvent will be more rapidly evaporated. If the good solvent has a *higher* boiling point, then it will make the problem worse. In this case, add more of the poor solvent while heating, then allow the solution to cool again.

5.4H | Specialized Techniques for Recrystallization

Occasionally a compound can be best crystallized by very slowly changing the composition of the solvent. This can be achieved by slow diffusion of the poor solvent into the good solvent, which may take hours or days. The methods described here are not commonly used in the introductory organic experiments, but they can be very handy in the event that typical recrystallization techniques fail.

SOLVENT LAYERING

The solvent layering technique requires two miscible solvents, where the good solvent is more dense than the poor solvent. A solution of the compound in the good solvent is placed in a test tube, and the poor solvent is slowly added by pipet at the surface, avoiding mixing, so that the less dense poor solvent rests on top as a separate layer. The test tube is then stoppered and allowed to stand undisturbed. Over time (e.g., waiting until the next lab period), the solvents will diffuse across the interface of the layers. As more of the poor solvent mixes into the lower layer of good solvent, the solubility of the compound slowly decreases, leading to crystallization.

SOLVENT VAPOR DIFFUSION

Figure 5.17 shows how a poor solvent can be made to diffuse via its vapor phase into a solution of the compound in the good solvent. A small, open container of the compound dissolved in a minimum amount of the good solvent is placed inside a larger container of the poor solvent. The larger container must have an air-tight seal to prevent the vapors from escaping. The two solvents will both be in equilibrium between their liquid and gas phases, and over a few hours or days, will diffuse into one another, reducing the solubility of the compound and causing it to crystallize.

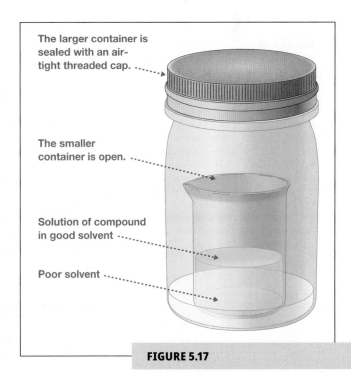

The larger container is sealed with an air-tight threaded cap.

The smaller container is open.

Solution of compound in good solvent

Poor solvent

FIGURE 5.17

Simple apparatus for solvent vapor diffusion during recrystallization.

5.5A Background

Chromatography is one of the most ubiquitous methods of analyzing and purifying organic compounds from mixtures. This technique, originally used to separate plant pigments, encompasses a variety of sophisticated methodologies that allow for the separation, isolation, and identification of the components of a mixture. While there are many types of chromatography, the fundamental basis for this technique is the distribution of the individual components of a mixture between two phases: the **stationary phase and the mobile phase**. For any given compound (A), there is a rapid equilibrium between phases; the compound spends some time adsorbed on the stationary phase and some time dissolved in the mobile phase.

**stationary phase
and mobile phase >>**
In chromatography, the stationary phase is a motionless material that accompanies another phase that is moving (the mobile phase). Compounds differ in how strongly they associate with the stationary phase, causing them to be separated as they travel at different rates along with the mobile phase.

$$A_{(mobile)} \xrightleftharpoons{K} A_{(stationary)} \qquad K = \frac{[A_{(stationary)}]}{[A_{(mobile)}]}$$

The equilibrium constant K depends upon the intermolecular attractions that the compound experiences with both the stationary and mobile phases. If there is a weak intermolecular attraction to the stationary phase, the compound will be mostly in the mobile phase ($K < 1$), and will travel rapidly along with the mobile phase. Conversely, if the compound has strong intermolecular attractions with the stationary phase, the compound will travel much more slowly because it spends most of its time in the stationary phase ($K > 1$).

Figure 5.18 depicts a cross section of a chromatographic separation at three different time points, with mobile and stationary phases. The mobile phase is moving left to right. Compound A has a weaker association with the stationary phase than compound B ($K_A < K_B$). As a result, less of A is adsorbed on the stationary phase at

FIGURE 5.18

Cross section of a chromatography column, showing chromatographic separation of two components A and B from a mixture, where the mobile phase is moving from left to right. (a) The initial mixture of A and B shows that both A and B are together in the mobile phase. The progress of the separation can then be seen at (b), an intermediate time point, and (c), an even later time point, where it becomes clear that A is moving faster than B because A spends more time in the mobile phase ($K_A < K_B$).

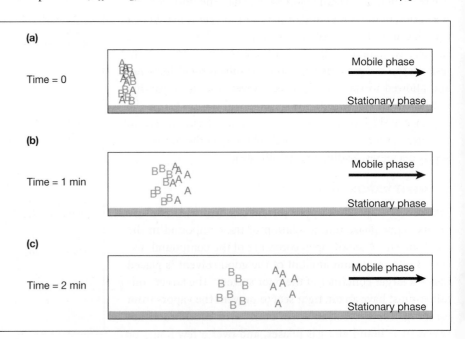

any given time. As the mobile phase moves from left to right, B moves more slowly than A because a greater proportion of B is in the stationary phase. Alternatively, A moves faster than B because it "spends more time" in the mobile phase.

5.5B Gas Chromatography

Gas chromatography (GC) is a common analytical technique used to separate volatile organic compounds. It may be applied to identify the compounds within an unknown mixture (qualitative analysis) or to determine their relative amounts (quantitative analysis).

In a typical gas chromatography instrument (**Figure 5.19**), a long tube called a GC column is placed in a temperature-controlled oven. The oven includes both a heating element and a cooling fan, so that temperature can be accurately adjusted either up or down. The GC column is where the separation takes place. It is a stainless steel or glass capillary tube, 2 m or more in length, coated on its interior surface with a stationary phase. The mobile phase is an inert gas (usually helium), also called a carrier gas, which is passed through the column at a controlled **flow rate**. A small amount (e.g., 1 μL) of a liquid or gaseous sample is injected into the column. The compounds in the sample are carried along by the mobile phase and detected as they emerge from the outlet.

A flame ionization detector (FID) is commonly used with GC. The outflow from the GC column passes through a hydrogen–air flame, and when an organic compound is present, ions are produced and attracted by a voltage in the detector. The resulting signal is plotted versus time to produce a chromatogram, with peaks appearing at specific times that are characteristic of the compounds present.

As in other types of chromatography, compounds travel through the column at different rates because they exist in equilibrium between the stationary and mobile phases, with different equilibrium constants for association with the two phases. They spend some time adsorbed ("stuck") on the stationary phase as a liquid, and some time moving with the carrier gas as a vapor. Compounds that associate more strongly with the stationary phase take longer to pass through the column. Compounds with higher vapor pressure are more associated with the mobile phase. Consequently, *boiling point* is the most important property for separation via GC.

Polarity, however, may also affect GC behavior. If two compounds have similar boiling points but very different polarities, separation can still occur. For many GC columns the *less* polar one will emerge from the column first. Solid samples can be run on the GC by first dissolving them in an appropriate solvent. Care should be taken to inject only volatile solids; if the solid cannot be vaporized at the temperatures of the injection port and oven, it may clog the system or damage the column.

<< **flow rate**
In chromatography, the rate of input and output of a mobile phase.

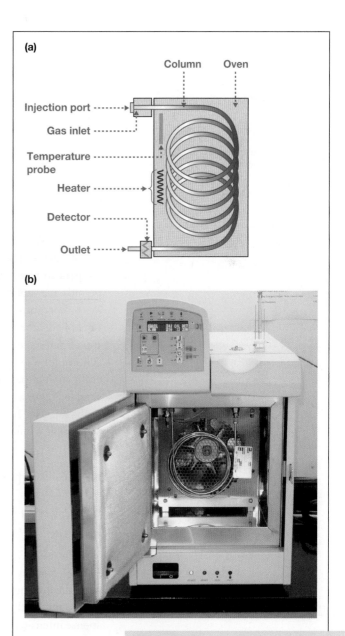

FIGURE 5.19

(a) Schematic diagram of a gas chromatograph. (b) A typical gas chromatograph, with the oven door open to reveal the column and cooling fan. The oven door is closed during operation.

(b) Courtesy of Gregory K. Friestad.

A plot of GC detector response versus time is called a gas chromatogram (**Figure 5.20**). Each peak in the chromatogram has a specific **retention time**, which is the amount of time between when the sample was injected and when it emerged from the column.

The detector response is proportional to the amount of compound passing through it, so the area under a peak is proportional to the total amount of compound in the sample. The ratio of peak areas in a single chromatogram is equal to the ratio of compounds in the mixture, as long as we assume the compounds give the same intensity of response to the detector. In the introductory organic lab, this is usually a reasonable assumption. For more precise measurements, however, the peak areas of different components must be compared with standards of defined concentrations in order to account for the different detector responses.

retention time >>
In gas chromatography, the amount of time from sample injection to emergence of an analyte from the column.

FIGURE 5.20

GC chromatogram of a three-component mixture. The table shows that retention times often increase as boiling points increase.

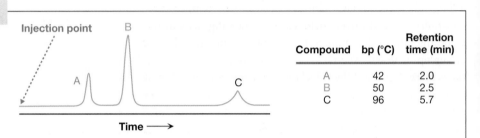

Compound	bp (°C)	Retention time (min)
A	42	2.0
B	50	2.5
C	96	5.7

In gas chromatography, the oven temperature affects retention times—high temperature leads to a short retention time and little separation because all compounds are vaporized and thus move at the same rate as the mobile phase. A very low temperature leads to impractically long retention times because the compounds remain adsorbed on the stationary phase. In addition, diffusion causes the peaks to spread out as the retention time increases, so compounds that are retained in the column for a long time may appear as broad, ill-defined peaks. The injector is generally maintained at a much higher temperature than the column to ensure that the sample is completely vaporized before it reaches the column and does not condense in the injector. The temperature of the detector is also set to prevent condensation of the sample components.

5.5C | Interpreting GC Data

USING GC TO QUANTIFY THE RATIOS OF COMPONENTS OF A SAMPLE MIXTURE

We can use GC to quantitatively evaluate a mixed sample of known components. The detector response is proportional to the amount of compound passing through it, so the area under a peak (the integration) is proportional to the total amount of compound in the sample. Assuming two compounds A and B give the same response intensities at the detector, the ratio of peak areas in a single chromatogram is equal to the ratio of compounds A and B in the mixture. Peak areas for A and B are calculated using the following equation:

$$\text{Area} = (\text{peak height})(\text{peak width at 1/2 height})$$

Using this method, the ratio in a two-component mixture can be found by simply dividing the area of the larger peak by the area of the smaller one. If there are more components, the smallest of the peaks is identified, and the relative amount of each component is calculated by dividing the area of each larger peak by the smallest peak.

Worked Example

Compounds A and B have equal detector responses. For the sample chromatogram shown in **Figure 5.21**, area$_B$/area$_A$ ≈ 2.5, so the ratio of component B to component A in this sample is 2.5 to 1.

This area calculation is an effective but "low tech" way of integrating the areas under the peaks. Modern GCs can integrate peaks very accurately, and typically the integration data for each peak are included in a table along with the printed chromatogram.

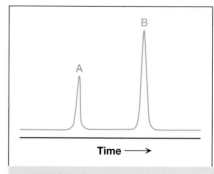

FIGURE 5.21

USING GC TO DETERMINE PURITY

When evaluating the purity of a compound by GC, the observation of a single, large peak suggests there is only one component in the mixture. However, more than one component may have the same retention time. For example, in **Figure 5.22**, the peak on the left shows a sample that appears to be pure. The chromatogram on the right shows a sample that is impure. To confirm that a peak consists of just one component, additional evidence may be needed, such as using a different GC column, or a variety of conditions (different temperatures, flow rates, etc.). If a single peak is still seen in all these analyses, this strengthens the evidence that the sample is pure.

FIGURE 5.22

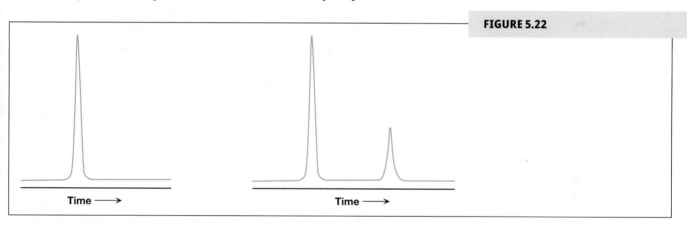

USING GC TO IDENTIFY AN UNKNOWN

Retention times in GC can help to identify unknown compounds. One method is to compare the retention time of the unknown with those of authentic samples of known compounds. The presence of a peak does not conclusively identify the structure of the compound, so usually further information about an unknown is needed to establish its identity.

If there are only a couple of possibilities for the identity of the unknown, and authentic samples are available, then you can add a known compound to the sample (this technique is called "spiking"). If the "spiked" known compound matches the unknown, then one peak will increase in size. If a new peak appears, or if the one peak develops a shoulder, then the "spiked" compound doesn't match the unknown.

Additional information about each peak may be obtained with an instrument that routes the GC output directly into a mass spectrometer (GC-MS). As the components come out of the end of the GC along with the mobile phase, they are routed into the mass spectrometer, and a mass spectrum is obtained from the components of each separate peak in the chromatogram. Mass spectrometry is discussed in detail in Chapter 8.

5.5D | Thin-Layer Chromatography

Thin-layer chromatography (TLC) is a separation technique that is used to determine the purity of a compound, the status of an ongoing reaction, or as a preliminary means of identification. The basic principles discussed for GC chromatography apply here as well—namely, there is a stationary phase and a mobile phase, and the analytes travel with the mobile phase at different rates depending on how strongly they associate with the stationary phase. In TLC, the stationary phase is the TLC plate, a thin layer of finely powdered silica gel (SiO_2) or alumina (Al_2O_3) that is affixed to a glass slide or to a thin sheet of aluminum or plastic. The SiO_2 or Al_2O_3 of the stationary phase is a polar solid to which the components of a mixture may adsorb with different affinities depending on their polarities. The mobile phase may be a single organic solvent or a mixture of a nonpolar organic solvent, such as hexane or petroleum ether, plus varying amounts of a more polar solvent, such as ethyl acetate, to adjust the polarity of the solvent mixture. As the mobile phase travels through the stationary phase, the components of a mixture are carried along at different rates because of their different affinities for the stationary phase, resulting in separation.

Like gas chromatography, TLC is used primarily as an analytical technique, because the amount of material loaded onto the TLC plate is generally very small and often not worth recovering afterward. When the materials must be recovered in quantities that are useful for subsequent experiments, the closely related technique of column chromatography (section 5.5F) is preferred. TLC can be very helpful in choosing appropriate conditions for the column chromatography.

To carry out a TLC analysis, a capillary tube (open on both ends) is dipped into a solution of the analyte, causing the solution to be drawn up into the capillary tube. A spot of the sample is deposited from the capillary tube onto the TLC plate by briefly touching the capillary tube to the surface along a lightly-drawn pencil line (the origin, or baseline) near the bottom of the TLC plate (**Figure 5.23**). One or more standards may also be spotted along this line, so that their TLC properties may be compared with the sample. The plate is developed by placing it in a covered beaker or jar that contains a small amount of the appropriate solvent. The level of the solvent in the beaker must be below the level of the origin line bearing the initial sample and standard spots. To facilitate consistent development of the plate, the atmosphere in the jar should be saturated with solvent vapors. A filter paper or a small section of a paper towel is used to help keep the atmosphere in the container saturated, but the TLC plate should not touch the filter paper.

Capillary action draws the solvent up the plate, and the leading edge of the solvent can usually be observed visually as it travels. This is called the **solvent front**. When the solvent front is near the top of the plate, the development is complete, so the plate is removed from the beaker and the solvent front is immediately marked with a pencil before the mobile phase evaporates.

The location of each spot is then noted. Plates that contain a UV fluorescent material facilitate this. If the spots lack color for visualization, they can be visualized using an ultraviolet lamp, if the compound absorbs UV light. For some compounds, staining the plate with a chemical stain is preferred. A simple chemical stain entails placing the TLC plate into a closed jar containing a few crystals of iodine; the iodine vapor reacts to give a color at the location of spots on the plate. More widely effective is dipping the plate into a dilute aqueous $KMnO_4$ solution, then heating it on a hot plate or with a heat gun. The plate will appear purple, except for yellow spots where compounds are located. Other dip-and-heat stains include anisaldehyde, 2,4-dinitrophenylhydrazine, and phosphomolybdic acid.

After developing the plate, an initial spot containing more than one component should now show multiple spots that traveled different distances from the origin.

solvent front >>
In thin-layer chromatography, the distance traveled by the solvent during development of a plate.

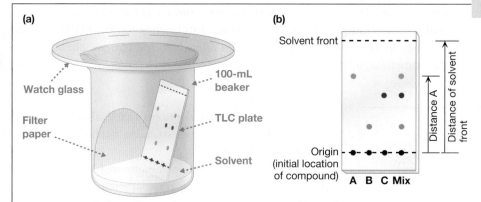

FIGURE 5.23

Typical TLC experimental setup and results. (a) A TLC plate in the developing chamber. The origin (the initial spot location) is above the level of solvent. (b) A developed plate, which compares the separated three-component mixture with three standards A, B, and C.

The components can be analyzed by determining the distance each spot traveled and comparing it with standards. For each spot on the TLC plate, a characteristic value R_f or "ratio to the front" (also sometimes called "retention factor") can be calculated. For example, for the yellow spot A (Figure 5.23b), the R_f is calculated as follows:

$$R_f = \frac{\text{distance A from origin}}{\text{distance of solvent front from origin}}$$

The R_f value is defined as the ratio of the distance traveled by a spot (measured from the center of the spot) to the distance traveled by the solvent (Figure 5.23). The R_f value is characteristic for a given compound as long as the polarities of the stationary phase and mobile phase are carefully controlled. This level of control is difficult to achieve, however, so there are no tables of R_f values in the chemical literature. Instead, standards of the compounds known to be in the mixture are included as separate spots on the plate (Figure 5.23), so that each standard R_f value can be matched with the components of a mixture on the same plate, ensuring the same conditions for the analysis.

The difference in R_f values between two compounds will also vary with the solvent, although generally the R_f values appear in the same order (i.e., the higher R_f compound will still be the highest). The choice of developing solvent is crucial. With a solvent that is too polar, all of the spots will run to the top of the plate, and there will be no difference in R_f. With a very nonpolar solvent, on the other hand, the spots will not move from the baseline, and again there will be no difference in R_f. Generally, the best separations are achieved by selecting a solvent that moves the spots to the middle areas of the plate—namely, $R_f = 0.3$–0.7. To simplify the selection of solvent, mixtures of polar and nonpolar solvents can be used in various ratios to adapt the solvent polarity to the polarities of the compounds in the mixture. A commonly used solvent pair is ethyl acetate and hexane.

 Thin-Layer Chromatography

5.5E Interpreting TLC Data

USING TLC TO EVALUATE THE PROGRESS OF A REACTION

When evaluating a reaction mixture, the disappearance of the spot representing starting material and the appearance of a new spot over time indicate that the original compound has been converted to something else. This means that the reaction is proceeding or has gone to completion. Generally speaking, the more polar a compound, the lower its R_f value, and vice versa. Thus, for a given set of conditions, the R_f values of two spots on a TLC plate may provide some evidence of the identity of a compound, and the success (or failure) of a reaction.

TLC is convenient to carry out directly on reaction mixtures because volatile solvents like CH_3OH evaporate from the plate and are not usually visible in TLC analysis. Also, inorganic by-products and reagents (such as NaCl or KOH) are often too polar to move, so they usually don't interfere with the analysis, although they may sometimes be visible as a spot at the origin.

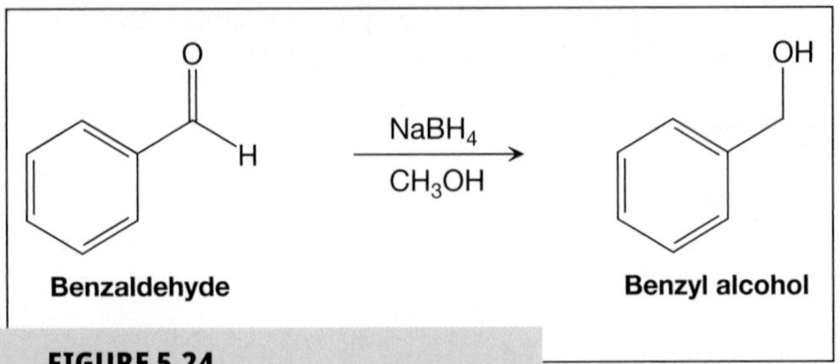

Worked Example

Upon treatment with sodium borohydride ($NaBH_4$), benzaldehyde is reduced to give the corresponding alcohol, benzyl alcohol, as shown in **Figure 5.24**.

The progress of the reaction is monitored by using a glass capillary tube to withdraw a trace amount of the reaction mixture for analysis by TLC. The reaction mixture is sampled before the reaction begins, after 10 minutes, and again after 1 hour, and the analysis is carried out on three different TLC plates (**Figure 5.25**). In each of the three TLC plates, authentic standards are placed on the plate for comparison; spots A and C are authentic samples of benzaldehyde and benzyl alcohol, respectively.

As the plates are developed, the spots move vertically from the origin in a lane. The spots in lane B (the reaction mixture) are compared with lanes A and C (the standards) to determine if the reaction is complete. After 30 minutes, the reaction is incomplete, because lane B contains both benzaldehyde and benzyl alcohol. It is successfully completed after 1 hour, however, because the higher R_f benzaldehyde spot is no longer present in lane B. The relative R_f values in this case are as expected—that is, the aldehyde has a higher R_f value than the alcohol, because the alcohol is more polar (it can serve as both a hydrogen bond donor and a hydrogen bond acceptor).

Benzaldehyde | NaBH₄ / CH₃OH → | **Benzyl alcohol**

FIGURE 5.24

FIGURE 5.25

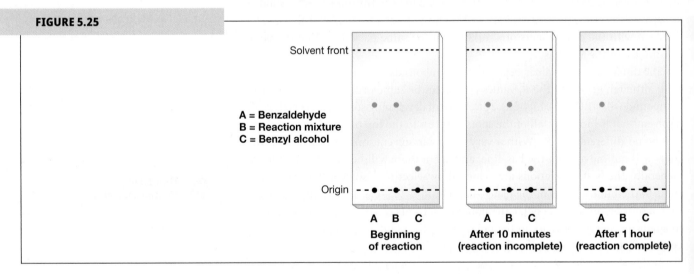

A = Benzaldehyde
B = Reaction mixture
C = Benzyl alcohol

Solvent front

Origin

A B C
Beginning of reaction

A B C
After 10 minutes (reaction incomplete)

A B C
After 1 hour (reaction complete)

USING TLC TO IDENTIFY COMPOUNDS IN MORE COMPLEX MIXTURES

More complex mixtures can also be analyzed by TLC, because the procedure used for the two-component mixture of benzaldehyde and benzyl alcohol can be extended to several components. That is, the analyte is placed in one lane, and an extra lane is added for each of the authentic standards to be compared with the analyte, using a wider TLC plate if necessary to fit additional lanes. The presence or absence of the various components can be determined by the presence or absence of a spot in the analyte lane at the R_f of the standard.

Worked Example

Nutritional supplements may contain components that may or may not be beneficial, so it is important to determine which ingredients are present or absent from a particular commercial product. The nutritional supplement is placed in one lane, and each authentic standard is placed in its own lane. After developing, the absence of a spot in the analyte lane at the R_f of an authentic standard shows that that component is absent (or below the level of detection). If that spot is present at the matching R_f, however, it indicates the presence of that component. The presence of a spot is weaker evidence than the absence of a spot, though, because TLC alone is insufficient to prove the identity of the component. Many compounds may have the same R_f, so the spot could be from something else. Further evidence, such as IR and NMR spectra, may be needed in order to confirm its identity.

USING TLC TO INDICATE PURITY

A pure compound should produce a single spot in TLC. Two (or more) spots in a single lane indicate that the compound is impure. However, while TLC can often show clearly that a substance is impure, TLC may fail to detect an impurity sometimes, too. Impurities that are low in concentration, or unresponsive to UV light or chemical stain, may be present but not visible. Different compounds (and impurities) may exhibit very similar R_f values on TLC and thus may appear as a single spot, making it impossible to distinguish them using this method. As such, other experimental techniques should be used to confirm the purity of a substance, even if it appears to be a single compound by TLC.

Worked Example

Analysis of a sample by TLC shows the presence of two components, the desired compound, B, and a higher R_f impurity, A. After purification by column chromatography (section 5.5F), a series of fractions are obtained, and to find out which fractions contain A and which contain B, they are analyzed by TLC. A spot for each fraction is placed in separate locations along the baseline on the TLC plate (Figure 5.25). After developing the plate, the R_f values of spots from each fraction are compared with the standards to see if they match with A or B (or contain both). The desired compound can then be recovered from the fractions in which it has been observed by TLC.

5.5F | Column Chromatography

Column chromatography involves a mobile phase of organic solvent that passes through a column of finely powdered solid material (often SiO_2 or Al_2O_3). The mobile phase carries the components of the mixture through the column at different rates, thus allowing them to separate. This separation is analogous to TLC, but can be used for purification of larger amounts of material. The method is especially important in cases where crystallization and distillation are unsuccessful. It is easiest to apply when the desired compound and its impurities have different R_f values in thin-layer chromatography (TLC), because it is governed by the same principles as TLC, except that the adsorbent or stationary phase is packed in a glass tube or *column*, rather than spread on a thin plate (**Figure 5.26**).

In column chromatography, the stationary phase of SiO_2 (silica gel) or Al_2O_3 (alumina) is a polar solid to which the components of a mixture may adsorb with different affinities, depending on their polarities. The mobile phase is commonly a nonpolar organic solvent, such as hexane or petroleum ether, plus varying amounts of a more polar solvent, such as ethyl acetate, to adjust its polarity.

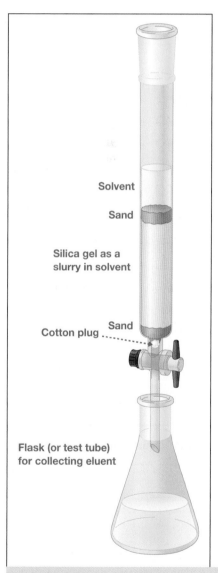

Solvent

Sand

Silica gel as a slurry in solvent

Sand

Cotton plug

Flask (or test tube) for collecting eluent

FIGURE 5.26

Setup for column chromatography, with elution in progress. Columns for this purpose are available in a variety of diameters and lengths, depending on the quantity of material to be separated.

More polar materials associate more strongly with the polar stationary phase, whereas nonpolar compounds associate weakly with the polar stationary phase. Because it is mostly in the mobile phase, a nonpolar compound is carried along more rapidly through the column. A polar compound spends most of the time bound to the stationary phase, so it moves more slowly. If the polarity of the mobile phase is increased, both components travel more rapidly through the column.

The sample mixture starts at the top of the column, and the mobile phase is passed through the column, moving the different components of the sample through the column at different rates. The components exit the column at different times, and can be collected in a number of fractions corresponding to materials that are less polar (early fractions) and more polar (later fractions).

5.5G | Typical Procedures for Column Chromatography

CHOOSING A SOLVENT

As with TLC, the choice of solvent system is crucial for good separation, and the best separation is often achieved by using solvent mixtures. As a rule of thumb, a good solvent system to start with is one in which the least polar component of the mixture has an R_f value of 0.3 in TLC. A single solvent mixture may be used to develop the column, or a solvent system that gradually increases in polarity (a polarity *gradient*) may be utilized. For example, a column may be developed starting with a low-polarity solvent, such as 10:1 hexane/ethyl acetate, and as fractions are collected, the developing solvent is changed from 10:1 to 5:1, then 3:1 hexane/ethyl acetate.

PACKING THE COLUMN

A cotton plug is loosely tamped into the constricted outlet tube at the bottom of the column. Some columns are made with a sintered glass frit to retain the column packing, and the cotton plug is unnecessary for columns of this type. A layer of sand is then added, and the top of the sand is made flat by gently tilting and tapping the side of the tube. The stationary phase is packed into the column, avoiding any gaps, bubbles, or cracks. The column may be packed "wet" by adding some solvent first and then pouring a solvent–adsorbent slurry into the tube, or "dry" by filling it with dry adsorbent and then adding the solvent. *CAUTION: Handle silica gel in the hood—fine particles of silica dust are harmful if inhaled.* It is important to keep the top and bottom surfaces of the adsorbent column as flat as possible—otherwise the separation will not be as effective. A protective layer of sand is placed on top to protect the top surface during the further addition of solvents.

 Column Chromatography

LOADING THE COLUMN

In rare cases, a solid mixture may be added directly. More commonly, the mixture to be separated is dissolved in a small amount of the chosen solvent [about three times the amount of the mixture, plus a few drops of dichloromethane (CH_2Cl_2), if needed to dissolve the mixture] and added carefully at the top of the column, so as not to disturb the packing. After this is allowed to drain down to the surface of the adsorbent, flow is stopped and an additional small portion of solvent is used to rinse the source flask, and this is also added at the top of the column. At this point there will be a visible band in the silica gel, usually white or yellowish, just below the sand. For good separation, this band must be as narrow as possible. Thus, avoid using excessive solvent to load the column because that will make this band wider.

DEVELOPING THE COLUMN

The column is developed, or *eluted*, by adding solvent to the top and collecting fractions of the *eluate* that comes out of the bottom. Gravity may be enough to elute the column at a reasonable rate. Often more efficient elution is obtained using "flash" chromatography, in which a slight air or nitrogen pressure is added to push the solvent through the column more rapidly. As the column is eluted, fractions may be collected in test tubes or small flasks, depending on the volume and the number of fractions. Add more solvent as needed—do not allow the solvent level to drain below the top surface of the column packing.

RECOVERING THE PURIFIED COMPOUND(S)

After the fractions are collected, they are analyzed by TLC to determine which fractions contain the compounds of interest (**Figure 5.27**). Ideally the compounds will be in pure form, as confirmed by a single spot on TLC. Fractions that contain the same single component may be combined, and the solvents can be removed by rotary evaporation to obtain the pure compound. Fractions with no detectable component at all may be discarded. Some fractions may contain mixtures, and they can be subjected to a repeat of the column chromatography if it is critical to recover every last milligram of material.

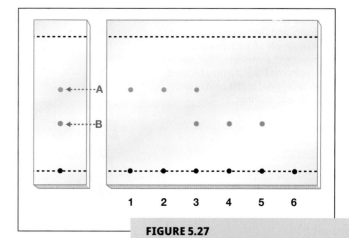

FIGURE 5.27

TLC plates of a mixture prior to column chromatography (*left*) and six fractions collected from the column (*right*). Fractions 1, 2, 4, and 5 show clean separation of the two pure components, while fraction 3 contains a mixture.

5.6 DETERMINATION OF PHYSICAL PROPERTIES

5.6A | Boiling Point Determination

The boiling point (bp) of a liquid is the temperature at which the vapor pressure of the liquid equals the atmospheric pressure. It is a characteristic physical property of a compound, so experimental boiling point data can be compared with literature values to provide evidence about the identity of a compound.

The best way to determine the boiling point of a liquid is to distill it and record the temperature at which it comes over in the distillation. When determining the bp of an unknown liquid, however, where only a small amount of sample is available, a full distillation is out of the question. So, in order to determine the boiling point in such circumstances, a microscale technique has been developed to obtain boiling point data on liquid samples as small as 5–10 drops.

In a microscale boiling point determination, 0.5–1.0 mL of sample is placed in a small (75 × 150 mm) test tube that is attached to a thermometer via a rubber band (**Figure 5.28**). A capillary tube, with one end closed, is placed, open-end down, into the sample. The test tube/thermometer is then submerged in an oil bath. The oil bath is warmed slowly, until the open end of the capillary has a steady and rapid stream of bubbles emerging from it. The temperature is then gradually lowered while watching the stream of bubbles slow to a stop. When the last bubble emerges, the vapor pressure of the liquid equals the atmospheric pressure, and at that moment the temperature is recorded as the boiling point.

 Boiling Point Measurement

FIGURE 5.28

Experimental setup for microscale boiling point determination. (a) The sample tube, capillary tube placed in the sample with the open end down, and thermometer assembly. (b) A simple oil bath in a beaker for heating the immersed sample tube. An electric hotplate may be used to apply heat to the beaker with magnetic stirring.

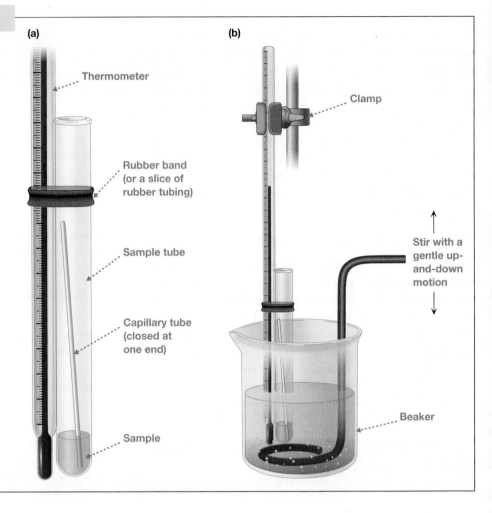

decomposition >>
An undesired reaction that diminishes the amount of a desired compound.

Decomposition may occur when heating some organic compounds to their boiling point, especially while they are exposed to air. If the analysis is to be repeated, use a fresh sample, assuming sufficient material is available.

5.6B Melting Point Determination

Melting points of crystalline solids are simple to obtain and are useful for identifying organic compounds and assessing their purity. A sample of the compound is placed in the closed end of a capillary tube by pressing the open end onto the solid (**Figure 5.29**), inverting it, and tapping it gently on the benchtop to move the crystals to the closed end. The tube should contain about 2–3 mm of solid. If the solid is difficult to get to the bottom, the tube can be dropped down a long tube held vertically on the benchtop.

The capillary containing the compound is slowly heated in an oil bath along with a thermometer, as in Figure 5.28, or in a heated metal block device such as a Mel-Temp (Figure 5.29). While observing the sample, the temperature is monitored with a thermometer or thermocouple. Heating at a rate of less than 5°C per minute, especially near the expected melting point, gives the best results. Two temperatures are recorded, one when melting first begins, and another at the point when melting is complete and the sample is a homogeneous liquid. Experimental melting point data are reported as a range of these two temperatures in the following format: 112–113°C.

Melting Point Measurement

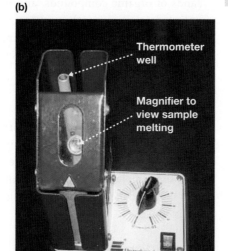

(a)

Melting point sampling:
1. Press the open end of the capillary tube into the solid.
2. Invert the capillary tube.
3. Tap the closed end of the capillary tube on the bench top to move the solid down into the closed end.

(b)

Thermometer well

Magnifier to view sample melting

MEL-TEMP®

FIGURE 5.29

(a) A solid sample is placed in a capillary tube for melting point determination. (b) A heated block apparatus heats the sample capillary tube, which can be viewed through a magnifier while monitoring the temperature via thermometer.

Courtesy of Gregory K. Friestad.

More sophisticated instrumental methods are usually not needed, but in some circumstances a technique called differential scanning calorimetry (DSC) can be used for a more rigorous quantitative analysis of phase transitions. In DSC, a sample along with a reference are heated slowly, and the amount of energy (heat) that each absorbs is measured over time. While the sample is undergoing a phase transition (i.e., at the melting point), it absorbs more energy than the reference. The energy input, or heat flow, is plotted versus temperature, and the melting point appears as a peak in the plot. DSC is used in the analysis of liquid crystals, polymers, and pharmaceuticals, and is also a tool for monitoring protein denaturation in biochemistry.

USING THE MELTING POINT TO ASSESS PURITY

Dissolved solutes (including impurities) lead to melting point depression. If the experimental melting point is more than 2–3°C less than the literature value for that compound, it may be considered impure. Usually, this melting point depression is also accompanied by a broadening of the melting range. Greater amounts of impurities lead to more melting point depression, perhaps by 20°C or more for a very impure sample, along with a more dramatic broadening of the range. Pure substances generally melt over a 1–2°C range. It is impractical to use melting point to quantify impurities in most situations, because there may be more than one impurity and their identities may be unknown. In some cases the impurity may not cause a problem for subsequent use of the compound, but if increased purity is needed, recrystallization is recommended.

USING THE MELTING POINT FOR IDENTIFICATION

To confirm the identity of a compound synthesized in the lab, or to identify an unknown compound, the experimental melting point range is simply compared with literature data. Impurities cause melting point depression, so for an impure compound, the experimental melting point may appear lower than the literature value. If the experimental melting point range is broader than two degrees, then melting point depression may also be occurring, and the melting point may be an unreliable means of identification. Melting point data are available for many thousands of organic compounds, and may be found in journal articles, handbooks, online databases, or other sources. Tables of melting point data for compounds (and their crystalline derivatives) are especially useful for identifying unknowns. These are usually organized by functional group, so the functional group must be known in order to use such tables.

MIXED MELTING POINT

If the identity of an unknown is narrowed to a couple of possibilities with similar melting points, and standards of high purity are available, then a mixed melting point determination can identify the compound. Each standard is separately mixed with the sample, and a melting point is determined on each mixture. The mixture that shows no melting point depression is the one in which the sample and standard are identical.

5.6C | Density Measurement

Occasionally, the density of an organic liquid can be used to distinguish between alternative structures. For example, the boiling points of 1-bromo-4-chlorobenzene and 1-fluoro-4-iodobenzene are close enough that measuring them might be inconclusive (**Table 5.3**).

Their densities, though, are significantly different. Densities can be very quickly and conveniently measured, with accuracy to the second decimal place, simply by transferring liquid from a plastic 1.00-mL graduated syringe into a tared vial or small flask. Take care to read the start and end points on the syringe graduations to two significant figures (it doesn't have to be the full 1.00 mL, as long as the exact volume is known). Weigh the flask immediately to avoid loss to evaporation, and then calculate density by dividing the mass by the volume (g/mL). More precise measurements are possible using larger volumetric pipets, but higher precision is usually unnecessary for common organic lab operations.

TABLE 5.3

Properties of Two Halogenated Benzenes

COMPOUND	bp (°C)	DENSITY (g/mL)
1-Bromo-4-chlorobenzene	196	1.576
1-Fluoro-4-iodobenzene	189	1.925

5.7 POLARIMETRY

Most light is unpolarized, meaning that its photons have electric field oscillations in all possible directions. Polarized lenses restrict the light that can pass through the lens so that only the light oscillating in one plane can pass (**Figure 5.30**). The light that passes through is then called *plane-polarized light*. For a beam of light to pass through two such lenses, you would need to rotate one of the lenses until the angle of the polarization matches. This simple operation can be viewed using two lenses from a pair of polarizing sunglasses.

When plane-polarized light is passed through a sample of water or hexane (or any other achiral material), the plane on which it's polarized does not change. However, when plane-polarized light is passed through a solution of naturally occurring glucose (or any other chiral material), the plane-polarization is rotated

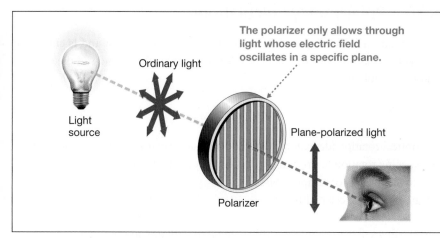

FIGURE 5.30

A polarizing lens restricts light so that only the light oscillating in a certain plane can pass.

Karty, J. *Organic Chemistry: Principles and Mechanisms*, 3rd ed.; W. W. Norton: New York, 2022; p 245. Person's eyes: BLACKDAY/Shutterstock.

to a new angle (**Figure 5.31**). This is called **optical rotation**. The magnitude of that rotation can be measured on an instrument called a *polarimeter*. When the measurement is taken under standardized conditions of concentration, temperature, and wavelength, the optical rotation can be used to calculate **specific rotation**, a characteristic property of glucose.

This property is called *optical activity*, and a compound with optical activity is said to be *optically active*. Optical activity is closely associated with the stereochemistry of organic compounds. Two compounds that are related as enantiomers (nonsuperimposable mirror images) rotate plane-polarized light with equal magnitudes but in opposite directions. That is, one enantiomer rotates plane-polarized light clockwise (+), whereas the other enantiomer rotates it counterclockwise (−). When the specific rotation is reported in the literature, it is reported with both a sign and a magnitude. Thus, data from a laboratory sample can be correlated with literature data. Such a correlation can provide evidence of the identity of the compound, and enantiomeric purity can be determined using calculations described in the next section.

<< **optical rotation**
In polarimetry, the number of degrees by which a sample rotates plane-polarized light.

<< **specific rotation**
In polarimetry, an optical rotation value that has been corrected for wavelength, concentration, and path length, allowing comparison of optical rotation data despite differences in these quantities.

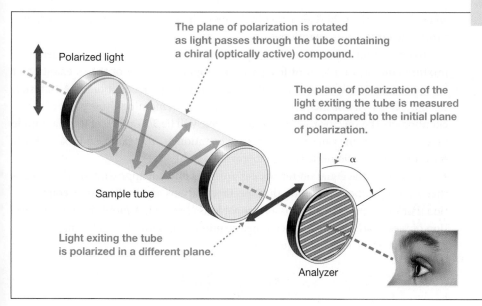

FIGURE 5.31

Plane-polarized light passes through an optically active sample solution, and a polarizing lens is rotated until it is aligned to allow light through.

Karty, J. *Organic Chemistry: Principles and Mechanisms*, 3rd ed.; W. W. Norton: New York, 2022; p 246. Person's eyes: BLACKDAY/Shutterstock.

5.7A | Polarimetry Calculations

Optical rotation (α; see Figure 5.31) is read from the instrument and is converted into specific rotation ($[\alpha]$) by correcting for cell pathlength (l) and sample concentration (c), as follows:

$$[\alpha] = \frac{\alpha}{c \cdot l}$$

α = optical rotation (degrees of rotation of the sample; readout from instrument)
$[\alpha]$ = specific rotation (corrected for concentration and pathlength)
c = concentration of sample in g/mL
l = pathlength of cell (1 dm)

This allows data to be more readily compared with previous reports in the literature. Specific rotation also depends on temperature and the wavelength of light (the "sodium D-line" at 589 nm is typically used). In the literature, data are reported as $[\alpha]_D^{25}$, where the subscript is the wavelength (here the sodium D-line), and the superscripted number is the temperature (in this case, 25°C).

Measuring the optical rotation can be used to determine the enantiomeric ratio of a sample that is a mixture of enantiomers. A racemic mixture (equal parts of both enantiomers) is optically inactive, because the two opposing optical rotations cancel each other out. If there is an excess of one enantiomer, then the sample is optically active. A single enantiomer of 100% enantiomeric purity will have a specific rotation that matches the literature value for the enantiomer, and the sign of that rotation can be used to determine which enantiomer it is.

When a mixture is nonracemic, however, the *enantiomeric excess* (%ee) describes its enantiomeric purity, which can be approximated by comparing the observed calculated specific rotation with that found in the literature for the known pure enantiomer,

$$\%ee \approx \% \text{ optical purity} = ([\alpha]_{calc}/[\alpha]_{lit}) \times 100$$

where $[\alpha]_{calc}$ is the value calculated in the preceding equation and $[\alpha]_{lit}$ is the literature value for the pure enantiomer.

To better understand %ee, imagine a 4:1 mixture of + and − enantiomers. The mixture consists of a total of five parts. The optical activity of the − enantiomer is canceled out by the optical activity of one part of the + enantiomer. As a result, the optical activity of the mixture comes from the remaining three parts of the + enantiomer. The specific rotation of this sample will be 3/5 or 60% of the magnitude of what would be measured for the pure + enantiomer. This sample is 60% optically pure, or 60% enantiomeric excess (abbreviated %ee).

The reason the equation for optical purity is only an approximation for %ee is that it requires the assumption that specific rotations are unaffected by concentration. For many purposes, that is a reasonable assumption. A more accurate calculation of %ee is achieved by directly measuring the amount of R and S enantiomers, then applying the following equation:

$$\%ee = \frac{(R - S)}{(R + S)}(100)$$

The accurate measurement of R and S enantiomer ratios is generally achieved by use of chiral chromatography with a stationary phase that has chiral structure. This can be achieved with gas chromatography (GC) or high-performance liquid chromatography (HPLC). If the stationary phase is chiral, then the two enantiomers will associate with the stationary phase with different affinities, in which case they will pass through the chromatographic column at different rates. A detector can integrate the peaks corresponding to the two enantiomers and determine their ratio.

5.7B | Sample Preparation for Polarimetry

 Polarimetry

Polarimetry is usually measured on dilute solutions—namely, concentrations near 1 mg/mL. Typical organic solvents for polarimetry are methanol (CH_3OH) and chloroform ($CHCl_3$), which should be handled in the fume hood, although very polar substances such as sugars and amino acids are frequently measured in aqueous solution. If you want to compare the optical rotation value you obtain experimentally with a literature value, choose your solvent and concentration to match the literature report, insofar as possible.

The concentration is important, so it must be accurately determined, usually to three significant figures. This means that both mass and volume must be measured to three significant figures. The best way to do this is to transfer the sample to a tared volumetric flask, measuring the mass of sample on an analytical balance. The volumetric flask or tube should be chosen so that its volume is slightly larger than the volume of the polarimetry cell. Dilute the sample with the chosen solvent, mix thoroughly, and fill the volumetric flask up to the marked volume line. Invert the stoppered flask several times to ensure that the solution is homogeneous. Alternatively, the mass of the sample may be determined in a vial or flask, then transferred quantitatively to the volumetric flask using at least three small portions of solvent.

Clean the polarimetry cell carefully before and after use, and be sure that any cleaning solvents are completely removed from the cell. When transferring the sample from the volumetric flask to the polarimetry cell, no further solvent can be used, or the concentration data will become inaccurate. If there is sufficient sample available, a small portion of the sample can be used to rinse the polarimetry cell before filling it. Before measuring optical rotation, inspect the cell visually to make sure the light path is not obstructed by bubbles.

Polarimetry is a nondestructive measurement, so a small precious sample may be recovered, if necessary, and used for other analytical or synthetic purposes.

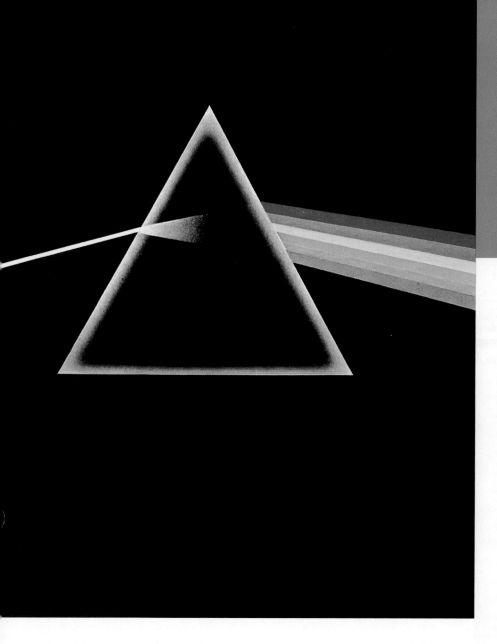

LEARNING OBJECTIVES

- Apply mathematical relationships that connect wavelength, frequency, energy, absorbance, and concentration.

- Relate infrared and ultraviolet radiation absorbance to structural features of organic compounds.

- Differentiate and identify organic compounds, using IR and UV data.

- Report IR and UV data in the appropriate formats.

Absorption Spectroscopy in Organic Chemistry

DIFFERENT COLORS, DIFFERENT ENERGIES

Visible white light contains different colors of light at different wavelengths, as part of a broader electromagnetic spectrum.

Records/Alamy Stock Photo.

INTRODUCTION: THEORY OF ABSORPTION SPECTROSCOPY

molecular bond vibrations >>
Changes in the lengths and angles of bonds, stimulated by absorbance of light in the infrared region of the electromagnetic spectrum.

A bsorption spectroscopy is a valuable tool for structure determination in organic chemistry. Absorptions of radiation from various parts of the electromagnetic spectrum are associated with certain structural characteristics in organic compounds (**Figure 6.1**). For example, ultraviolet (UV) and infrared (IR) light have different frequencies, so irradiating organic compounds with these forms of light impacts the structure in different ways. The absorption of ultraviolet/visible (UV/vis) light promotes an electron from one molecular orbital to a higher-energy orbital, whereas the absorption of infrared light causes changes in **molecular bond vibrations** such as stretching and bending motions.

	Higher frequency (ν) Shorter wavelength (λ) Higher energy (E)				Lower frequency (ν) Longer wavelength (λ) Lower energy (E)
ν (cm^{-1})	10^6	10^4	10^2	10^0	10^{-1}
λ	10 nm	200–700 nm	1–100 μm	10 mm	100 mm
Name	X-rays	Ultraviolet–visible	Infrared	Microwave	Radiofrequency (RF)
Molecular transition process	Excitation of core atomic electrons	Excitation of molecular valence electrons	Stretching and bending of molecular bonds	Rotational transitions of bonds	Transitions of nuclear spin magnetic dipoles
ΔE (kJ mol^{-1})	1200	300	10	10^{-4}	10^{-6}
		UV–vis spectroscopy	**IR spectroscopy**		**NMR spectroscopy**

FIGURE 6.1

The electromagnetic spectrum and its relationship to the energies associated with molecular transition processes.

These structural features respond to light from different parts of the electromagnetic spectrum. To understand why, we first review the basics of the wave and energy properties of light. Frequency (ν = the number of waves that pass a certain point during a given time period) is related to wavelength (λ = the distance between one point on a wave to the same point on an adjacent wave) by the equation,

$$\nu = c/\lambda$$

where c = speed of light = 3×10^8 m s^{-1}. Frequency is related to energy E by the equation,

$$E = h\nu$$

where h = Planck's constant = 6.6×10^{-34} J·s. Combining these two equations, we get

$$E = hc/\lambda$$

Therefore, the energy associated with light is *directly* proportional to its *frequency*, and *inversely* proportional to its *wavelength*. In other words, higher frequencies (shorter wavelengths) of light have higher energies, whereas lower frequencies (longer wavelengths) have lower energies.

According to the electromagnetic spectrum in Figure 6.1, UV light has a shorter wavelength than IR light. Promoting an electron to an unfilled molecular orbital is matched with the amount of energy imparted by one photon of UV light, so that is the transition that happens when the photon is absorbed. Meanwhile, the energy associated with vibration of specific bonds is matched with specific amounts of energy of photons in the IR region of the electromagnetic spectrum. When an IR photon is absorbed, vibrations are excited to a higher-energy state; in other words, the energy of the photon is converted to vibrational energy within the molecular structure. Specific energies in the IR region correspond to various stretching and bending vibrations of bonds within an organic compound. Excitations of bond rotations, on the other hand, require less energy than bond stretching and bending; these are associated with absorption of energy in the microwave region of the spectrum.

Lower still on the energy scale of the electromagnetic spectrum is radiofrequency radiation. This is the frequency range you use when you listen to your local FM radio stations (about 100 MHz). Energy associated with this radiation is matched with transitions of magnetic dipoles associated with nuclear spins, and is used for nuclear magnetic resonance (NMR) spectroscopy, which we will discuss further in Chapter 7.

In the sections that follow, we explain how to use molecular transitions associated with ultraviolet, visible, and infrared light for the analysis and characterization of samples prepared in the organic laboratory. Specifically, this chapter describes how to prepare samples, acquire spectra, interpret results, and report the data in a scientific report.

6.1 ULTRAVIOLET-VISIBLE SPECTROSCOPY

The absorption of light in the ultraviolet (UV) and visible (vis) region ($\lambda = 200$–700 nm) is associated with an **electronic transition** from an occupied lower-energy molecular orbital to an unoccupied higher-energy one in a molecule. When light's specific wavelength and energy match the difference in energy between these two orbitals, a photon can be absorbed by the molecule and an electron is promoted to the higher-energy orbital.

Instruments that detect the absorbance of UV–vis light work by passing the light through a cell (called a cuvette) containing a solution of the sample in a solvent that is transparent in the 200–700 nm range. Typical solvents include water, methanol, hexane, and acetonitrile. The sample is prepared by dissolving a known mass of the sample in such a solvent, then diluting it to a known volume, to achieve a concentration of about 10^{-4} M or lower. To achieve precision at very low concentrations, volumetric flasks and volumetric pipets may be used, with quantitative **serial dilutions** as needed. Once prepared, the sample is placed in a cuvette, and the spectrum is acquired. Interference from the solvent may be removed by acquiring a reference spectrum of a blank cuvette containing only the solvent, either prior to acquiring the sample spectrum (within a single beam instrument) or simultaneously (with a double beam instrument). The spectrum of the blank solvent is electronically subtracted from the sample spectrum.

The UV–vis absorbance can be detected at a single chosen wavelength, or by using a photodiode array detector, at all wavelengths in the region simultaneously. The latter results in a spectrum plotted with wavelength on the x axis and absorbance on the y axis (**Figure 6.2**). Notice that the absorbance is broad with respect to the wavelength. When data are presented in tables or databases, the wavelength is reported at its maximum absorbance (λ_{max}).

<< **electronic transition**
Movement of an electron from one orbital to another one of higher energy, stimulated by the absorbance of light in the ultraviolet–visible region of the electromagnetic spectrum.

<< **serial dilutions**
Using a portion of a diluted sample as the concentrate in a subsequent dilution, multiplying the dilution effect.

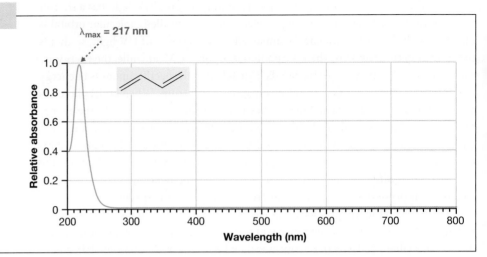

6.1A   Correlating λ_{max} with Structural Features

Organic compounds that have useful UV–vis spectra in the 200–700 nm region generally have functional groups with a π bond. This is because π bonding and antibonding orbitals are close enough in energy to allow light in this range to be absorbed. Exciting an electron in ethylene from π to π^* orbitals corresponds to absorption at $\lambda_{max} = 171$ nm, just outside the useful range. Conjugation of the π bond, though, causes this energy gap ΔE to diminish (**Figure 6.3**), so 1,3-butadiene absorbs at $\lambda_{max} = 217$ nm. Each successive π bond in conjugation further diminishes the gap between π and π^*, thereby moving the absorbance to longer wavelengths.

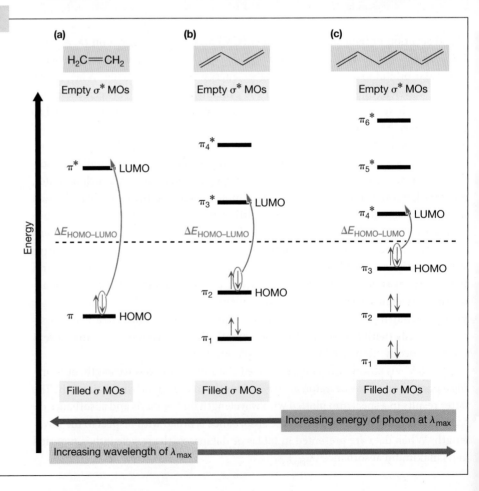

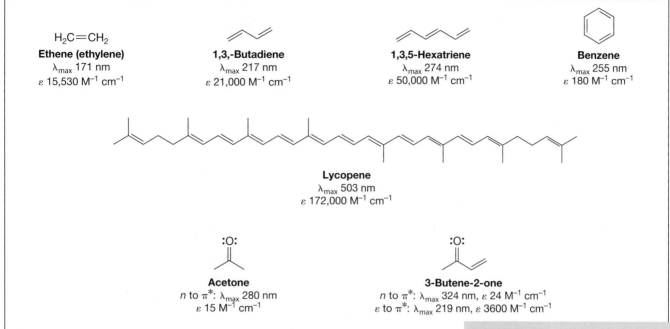

Ethene (ethylene)
λ_{max} 171 nm
ε 15,530 M^{-1} cm^{-1}

1,3,-Butadiene
λ_{max} 217 nm
ε 21,000 M^{-1} cm^{-1}

1,3,5-Hexatriene
λ_{max} 274 nm
ε 50,000 M^{-1} cm^{-1}

Benzene
λ_{max} 255 nm
ε 180 M^{-1} cm^{-1}

Lycopene
λ_{max} 503 nm
ε 172,000 M^{-1} cm^{-1}

Acetone
n to π^*: λ_{max} 280 nm
ε 15 M^{-1} cm^{-1}

3-Butene-2-one
n to π^*: λ_{max} 324 nm, ε 24 M^{-1} cm^{-1}
ε to π^*: λ_{max} 219 nm, ε 3600 M^{-1} cm^{-1}

FIGURE 6.4

Selected UV–vis data for a variety of organic compounds.

For example, *trans*-1,3,5-hexatriene absorbs at $\lambda_{max} = 274$ nm and lycopene at $\lambda_{max} = 503$ nm (**Figure 6.4**). Lycopene is found in ripe tomatoes and contributes to their intensely red color. The π bonds at the ends of lycopene are not conjugated with the rest, so they do not affect its λ_{max}.

Lone pairs of electrons associated with an atom involved in a π bond populate non-bonding orbitals, or *n* orbitals, and these can also be promoted to higher-energy **π^* orbitals** by absorption in the UV–vis region. This is referred to as an *n* to π^* transition. As with alkenes, conjugation also causes these transitions to occur with longer wavelengths. This can be seen by comparing the λ_{max} for acetone and 3-butene-2-one in Figure 6.4.

<< **π^* orbital**
Antibonding orbital generated from combination of the *p* orbitals of two atoms.

6.1B Using UV–Vis Spectroscopy for Quantitative Analysis

Another feature of the UV–vis spectrum is the intensity of each peak, or its absorbance (*A*), measured on the *y* axis (see Figure 6.2). Absorbance is related to the **molar absorptivity or molar extinction coefficient** (ε), of the compound, which is a constant that is characteristic of the specific compound. A compound that absorbs in the visible light region with a high ε, such as lycopene, will appear strongly colored even at lower concentrations. Absorbance is also dependent on the pathlength of the sample cell or cuvette, in centimeters (*b*), and the concentration of the sample (*c*). These three terms relate to absorbance according to Beer's law:

$$A = \varepsilon \cdot b \cdot c$$

When acquiring data for known compounds, ε and *b* are usually known, and *A* is measured by the instrument, leaving *c* as the only variable. Thus, Beer's law can be used to calculate *c*, the concentration of an analyte in solution, from a UV–vis spectrum.

<< **molar absorptivity or molar extinction coefficient**
A quantity that describes how strongly a compound absorbs light; it varies among compounds and is a characteristic physical property of compounds that absorb light. It appears as ε, a constant in Beer's law $A = \varepsilon \cdot b \cdot c$. This constant enables measurements of absorbance (*A*) to be used to calculate sample concentration (*c*).

Worked Example

The extraction of lycopene from vegetable material yielded a solution of lycopene. After diluting to 1/1000th of the original concentration, UV–vis spectroscopy of the diluted solution showed $A = 0.200$ at 503 nm ($b = 1.0$ cm). What is the original concentration of lycopene prior to dilution?

Solution

According to Figure 6.4, $\varepsilon = 172{,}000$ M^{-1} cm^{-1} for lycopene. Rearranging Beer's law to $c = A/\varepsilon b$, then substituting the known values of A, ε, and b, we get

$$c = \frac{A}{\varepsilon b} = \frac{0.200}{\left(172{,}000 \ M^{-1} \ cm^{-1}\right)} = 1.16 \times 10^{-6} \ M$$

for the diluted solution. The original concentration was 1000 times that, or 1.16×10^{-3} M.

6.2 INFRARED SPECTROSCOPY

Functional groups absorb infrared (IR) radiation at specific frequencies that match the energies associated with different stretching and bending vibrations of their bonds (**Figure 6.5**). The wavelength range (micrometers) is slightly longer than that of visible light. Expressed in wavenumbers (v) with units of waves per centimeter (cm^{-1}), the frequency range of infrared spectroscopy is approximately 500–4000 cm^{-1}. Absorbances at higher wavenumbers are connected to vibrations of higher energy, and this in turn correlates to structural features of specific bonds. Various types of functional groups have vibrations at frequencies that are broadly distributed across this range. Infrared spectroscopy is therefore an excellent method for identifying organic functional groups in experimental samples.

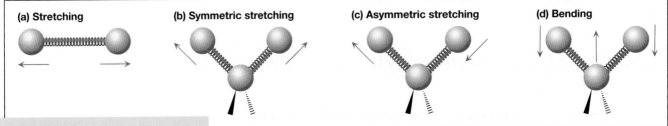

(a) Stretching   **(b) Symmetric stretching**   **(c) Asymmetric stretching**   **(d) Bending**

FIGURE 6.5

Various vibrations of organic compounds.

Gilbert, T.; Kirss, R.; Bretz, S.; Foster, N. *Chemistry*, 6th ed.; W. W. Norton: New York, 2020; p 409.

6.2A Infrared Absorption and Molecular Transitions

The bonds of organic molecules stretch and bend at specific vibrational frequencies, and the physics of this can be related to that of springs with weights attached. Hooke's law describes the vibration of a spring in terms of the masses of the weights and the stiffness of the spring (**Figure 6.6**). The frequency of vibrations is proportional to the stiffness of the spring, and inversely proportional to the masses of the weights. Hooke's law can be applied to covalent bonds as shown in Figure 6.6.

Hooke's law predicts, all other things being equal, that bonds between lighter atoms will vibrate at higher frequency than those between heavy atoms. And, all other things being equal, stronger bonds will have higher frequency vibrations than weaker bonds. These predictions are confirmed by measurements of the stretching frequencies of a C—H bond (3000 cm^{-1}) versus a C—D bond (2200 cm^{-1}), and the stretching frequencies of a C=C bond (1650 cm^{-1}) and a C≡C bond (2200 cm^{-1}). A wide range of organic functional groups and bonds are correlated to various vibrational frequency ranges, as observed in IR spectra (**Figure 6.7**). Tables of such data can be useful for specifying the functional group and its neighboring structural features in more detail (**Table 6.1**). Intensities of the peaks are noted as strong (s), medium (m), or weak (w) in the table. These intensities are related to bond polarization; for example, peaks related to C=O bond vibrations have strong intensities while C=C bonds are weaker.

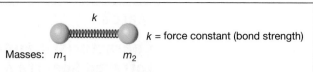

Vibrational frequency ν is given by Hooke's law:

$$\nu = \frac{1}{2\pi c}\left(\frac{k}{\mu}\right)^{1/2}$$

where c = speed of light, k = force constant, μ = reduced mass =

$$\frac{m_1 m_2}{m_1 + m_2}$$

FIGURE 6.6

Relationships of atomic masses and bond strengths to vibrational frequency.

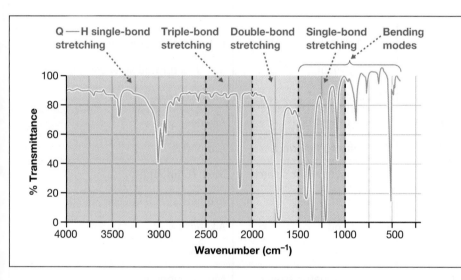

FIGURE 6.7

Representative IR spectrum showing the parts of the spectrum where various types of bond vibrations absorb in the infrared frequency range. Q = C, N, or O.

Karty, J. *Organic Chemistry: Principles and Mechanisms*, 3rd ed.; W. W. Norton: New York, 2022; p 777.

TABLE 6.1

Characteristic Functional Group Absorbances in Infrared Spectra

FREQUENCY (cm^{-1})	BOND	FUNCTIONAL GROUP
3640–3610 (s, sh)	O—H stretch, free hydroxyl	Alcohols, phenols
3500–3200 (s, br)	O—H stretch, H–bonded	Alcohols, phenols
3400–3250 (m)	N—H stretch	Primary, secondary amines, amides
3300–2500 (m, br)	O—H stretch	Carboxylic acids
3330–3270 (m, s)	—C≡C—H: C—H stretch	Alkynes (terminal)
3100–3000 (m)	C—H stretch	Aromatics
3100–3000 (m)	=C—H stretch	Alkenes
3000–2850 (m)	C—H stretch	Alkanes

(continued)

TABLE 6.1

Characteristic Functional Group Absorbances in Infrared Spectra *(continued)*

FREQUENCY (cm⁻¹)	BOND	FUNCTIONAL GROUP
2830–2695 (m)	H—C=O: C—H stretch	Aldehydes
2260–2210 (v)	C≡N stretch	Nitriles
2260–2100 (w)	—C≡C— stretch	Alkynes, non-symmetrical
1760–1665 (s)	C=O stretch	Carbonyls (general)^a
1760–1690 (s)	C=O stretch	Carboxylic acids
1750–1735 (s)	C=O stretch	Esters, saturated aliphatic
1740–1720 (s)	C=O stretch	Aldehydes, saturated aliphatic
1730–1715 (s)	C=O stretch	α,β-Unsaturated esters
1715 (s)	C=O stretch	Ketones, saturated aliphatic
1710–1665 (s)	C=O stretch	α,β-Unsaturated aldehydes, ketones
1680–1630 (s)	C=O stretch	Amides
1680–1640 (m)	—C=C— stretch	Alkenes
1600–1585 (m)	C—C stretch (in-ring)	Aromatics
1550–1475 (s)	N—O asymmetric stretch	Nitro compounds
1500–1400 (m)	C—C stretch (in-ring)	Aromatics
1470–1450 (m)	C—H bend	Alkanes
1370–1350 (m)	C—H rock	Alkanes
1360–1290 (m)	N—O symmetric stretch	Nitro compounds
1335–1250 (s)	C—N stretch	Aromatic amines
1320–1000 (s)	C—O stretch	Alcohols, carboxylic acids, esters, ethers
1300–1150 (m)	C—H wag (—CH$_2$X)	Alkyl halides
1250–1020 (m)	C—N stretch	Aliphatic amines
1000–650 (s)	=C—H bend	Alkenes
950–910 (m)	O—H bend	Carboxylic acids
910–665 (s, br)	N—H wag	Primary, secondary amines
900–675 (s)	C—H "oop"	Aromatics
850–550 (m)	C—Cl stretch	Alkyl halides
725–720 (m)	C—H rock	Alkanes
700–610 (br, s)	—C≡C—H: C—H bend	Alkynes
690–515 (m)	C—Br stretch	Alkyl halides

(s) = strong, (m) = medium, (w) = weak, (br) = broad, (sh) = sharp, (v) = variable
^{a}Conjugation of other unsaturated groups with the carbonyl will lower the C=O frequency by 20–40 cm⁻¹.

6.2B | Acquiring an Infrared Spectrum

The spectrum is obtained by passing IR radiation through a sample of the compound and measuring the light that is transmitted through the sample. The result is plotted in percent **transmittance** (%T) versus frequency (**Figure 6.8**), with the frequency expressed in units of wavenumbers (waves per centimeter, cm^{-1}). It is useful to understand how transmittance relates to absorbance. Transmittance and percent transmittance are calculated as follows:

$$T = I/I_0$$
$$\%T = (I/I_0)100$$

where I_0 is the initial light intensity and I is the intensity of light after it has passed through the sample. Transmittance is inversely related to absorbance through a logarithmic function:

$$A = \log_{10}(1/T)$$
$$A = 2 - \log_{10}(1/\%T)$$
$$\%T = (1/10^A)100$$

From these equations, we can see that when transmittance is 100%, the absorbance is 0, and working in the other direction, an absorbance of 2 corresponds to 1% transmittance.

<< transmittance
A fraction I/I_0 of light intensity remaining after it has passed through a sample (I), relative to its initial intensity (I_0). This fraction, expressed as a percentage (%T), is the unit commonly found on the y axis of an infrared spectrum.

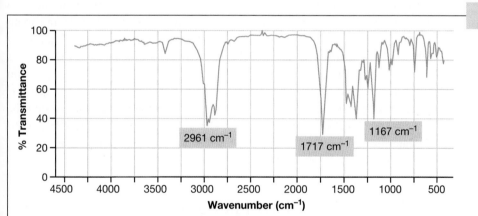

FIGURE 6.8

A typical infrared spectrum.

Karty, J. *Organic Chemistry: Principles and Mechanisms*, 3rd ed.; W. W. Norton: New York, 2022; p 774.

Usually it is desirable to subtract background peaks that mainly result from humidity and/or CO_2 in the air. A background spectrum is recorded and the instrument can subtract it from the sample spectrum to obtain a cleaner result. Using the instrument software, the peaks in the resulting spectrum can be labeled in units of cm^{-1}, and a spectrum can be printed so that it can be attached to a lab report.

The best quality spectrum is one in which the largest peak is in the range of 10–50%T. If the largest peak is above 50%T, then it is likely that smaller peaks may be lost in instrumental noise. If there are peaks at 0% transmittance, then there is too much sample and the spectrum gives little indication of the relative intensity of the peaks. In either case, adjust the amount of sample and acquire another spectrum.

SAMPLE PREPARATION FOR LIQUIDS

The simplest liquid sample preparation procedure is to place a drop of a liquid sample between two salt (NaCl) plates, forming a thin film of the sample, and then place the sandwiched sample in the path of the IR beam. If there is too much sample (peaks bottom out at 0%T), take the plates apart, wipe one off with a tissue to remove some sample, and acquire the spectrum again. To avoid scratching, do not contact the plate with a pipet. *CAUTION: Water will dissolve salt plates or damage them. Avoid exposing salt plates to water, whether it is present in the sample itself, or moisture from your fingers. Avoid contact with the faces of the salt plates.*

SAMPLE PREPARATION FOR SOLIDS

For solids, the same procedure may be used by placing a couple of milligrams of solid (the tip of a spatula) on the salt plate, adding a drop of $CHCl_3$ to make a solution, then sandwiching the solution between two salt plates. For solids that are sparingly soluble in $CHCl_3$, a spectrum may sometimes be obtained by heating the solid briefly with $CHCl_3$ in a test tube and placing a drop of the supernatant solution on the salt plate. After the $CHCl_3$ evaporates, the sample remains as a thin film. If this is still unsuccessful, a nujol mull can be prepared by grinding the solid into a fine powder and mixing it with nujol. Nujol is a mineral oil, with strong absorbances in the C—H region that can interfere with the spectrum. If $CHCl_3$ is used, a background spectrum should be obtained with just $CHCl_3$ so that it can be subtracted from the spectrum.

A procedure called *attenuated total reflectance (ATR)* provides another option that is particularly useful for insoluble solids. In ATR, which requires a special attachment for the instrument, the solid is pressed against a small window made of a germanium or zinc selenide crystal. The IR beam passes through that window and bounces off the surface of the solid. Small amounts of IR radiation are absorbed at the surface of the solid sample, resulting in an infrared spectrum from solids that otherwise would be difficult to analyze. The ATR method also works well for liquid samples.

Another technique for solid samples that are insoluble in organic solvents is to prepare a *KBr pellet* under high pressure. A mixture of about 2 mg solid and about 200 mg of anhydrous KBr is ground together in a mortar and pestle. Then, using a mechanical press, pressure is applied to squeeze the KBr into a pellet (various types of presses are available; check with your lab instructor for more specific operation instructions). Sufficient force produces a pellet that looks transparent to the eye. The KBr is also transparent in the IR region, so when the pellet is placed into the IR beam, any absorbance is due to the analyte. Note that anhydrous KBr can absorb water from the air or from fingertips, which will obscure analyte peaks in the O—H region (3200–3600 cm^{-1}).

6.2C | Interpretation of IR Data

Within the IR spectrum, there are two main ranges that are used for different purposes. The range from 1650–4000 cm^{-1} provides the most useful diagnostic information for identifying functional groups present in the sample. The range from 500–1650 cm^{-1} is often called the **fingerprint region**, and is only useful in certain circumstances. The fingerprint range often contains too many peaks to interpret each one individually for an unknown sample. Instead it is most useful when comparing the IR spectrum of a sample of high purity with a known literature spectrum to confirm that they match. As its name suggests, it is rare to have two compounds with an IR spectrum showing the same "fingerprint" of peaks in the fingerprint region.

fingerprint region >>
The lower-energy portion, or smaller wavenumber portion, of an infrared spectrum, typically 700–1600 cm^{-1}. This region of the spectrum contains many peaks that are sometimes difficult to assign, yet are useful in identification of an unknown by direct comparison of its fingerprint region to that of a known standard.

CHARACTERIZATION OF A KNOWN COMPOUND BY IR

To characterize the features of a known compound using IR, perform the following steps.

1. Know the structure of the compound you are attempting to characterize.

2. Examine the structure and identify the functional groups you would expect to see in the IR.

3. Determine the position (cm^{-1}) at which you would expect to find each peak for a given functional group. For some functional groups (e.g., carboxylic acid, nitro, etc.), you should expect to see several diagnostic peaks. If the compound is a reaction product, focus most attention on those peaks that correspond to the functional group that changed during the reaction; these peaks will be diagnostic for the success or failure of the reaction.

4. Obtain a good clear IR of your sample. If the compound is a solid, attenuated total reflectance (ATR) can be used. If ATR is not available, the spectrum can be obtained using a thin film of a solution of the solid in CHCl$_3$, placed between salt plates. A spectrum may also be obtained from a "nujol mull"—a slurry of the solid in mineral oil. The largest peak should have a nonzero value for % transmittance (%T).

5. Considering your preliminary expectations, identify those peaks in the spectrum that are diagnostic for your compound.

6. If unexpected peaks appear in your IR spectrum, you will need to explain them. Common impurities include unreacted starting materials, reagents, water, and solvents.

7. The IR spectra for millions of compounds can be found in the primary chemical literature (journal articles) or in online databases, either in graphic form or as a listing of peak frequencies. These data may be compared with your spectrum in order to strengthen your confidence in structure assignments and the identification of impurities. For best comparison, choose a literature spectrum obtained using the same method (e.g., ATR, KBr pellet, solution, film, etc.).

8. If you know that your sample was a single compound of high purity, you may compare the fingerprint region of your IR spectrum with one from the literature to see if they match. For an impure compound there can be a lot of extraneous peaks in the fingerprint region, so use caution in interpreting this part of the spectrum.

EVALUATION OF AN UNKNOWN SAMPLE BY IR

Preliminary evaluation: strongly diagnostic peaks for functional groups.

1. Determine what you would expect to see (peak position, number of peaks) for each of the possible major organic functional groups in your sample. Consider also any special circumstances that could affect what you see (e.g., hydrogen bonding, amine substitution, etc.).

2. Obtain a good clear IR of your sample. If the compound is a solid, attenuated total reflectance (ATR) can be used. If ATR is not available, the spectrum can be obtained using a thin film of a solution of the solid in CHCl$_3$, placed between salt plates. A spectrum may also be obtained from a "nujol mull"—a slurry of the solid in mineral oil. The largest peak should have a nonzero value for % transmittance (%T).

3. Based on your IR spectrum, determine which functional groups may be present in your sample. You will probably be unable to narrow it down to a single group, but you should be able to eliminate a number of options based on peaks that are *not* present (e.g., if no C=O stretch is present, then esters, carboxylic acids, aldehydes, and ketones can be eliminated from consideration). Often, peaks in the range 3100–3600 cm^{-1} have characteristic shapes that are useful: U-shape for alcohol O—H, W-shape for NH$_2$, V-shape for NH, and a sharp spike for alkyne C—H.

Subsequent evaluation: peaks with more ambiguous interpretations.

1. Using the information you have obtained from functional group tests and other sources, reinspect your IR spectrum to gain additional support for your findings. Consider additional functional groups, including double and triple bonds, halides, nitro groups, ethers, etc. IR spectra are usually very complex. Do not try to read more into the spectrum than is actually there.

2. If the number of possible compounds is small, locate IR spectra in the chemical literature for each of the possibilities and compare all the peaks, including the fingerprint region. If your unknown sample is pure, its spectrum should closely match the known spectrum found in a book or online database. If there are more than a handful of possible compounds, it can be tedious to find a fingerprint match. First try to narrow the possibilities using other data, such as physical properties (mp, bp) or number of carbons.

6.2D | Reporting IR Data

Report all information from your IR spectrum that is relevant to the identification and purity of the compound(s) that you have prepared. IR data should be reported in table format as shown in the example that follows. Peak positions should be reported in cm^{-1} with an assignment to a type of bond (e.g., O—H, C=O, etc.) and a specific functional group (ester, alcohol, etc.). When the absence of a peak is significant, mention in the report that this peak is absent and interpret what that absence implies.

Worked Example: IR Spectra of Ethyl Levulinate

When a student esterified levulinic acid (a carboxylic acid), the final product obtained was ethyl levulinate (**Figure 6.9**).

FIGURE 6.9

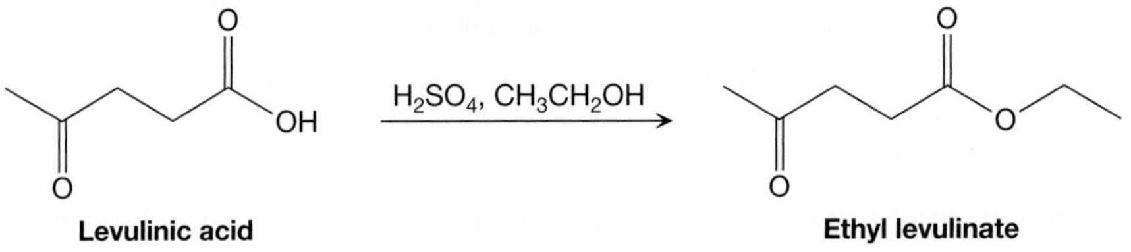

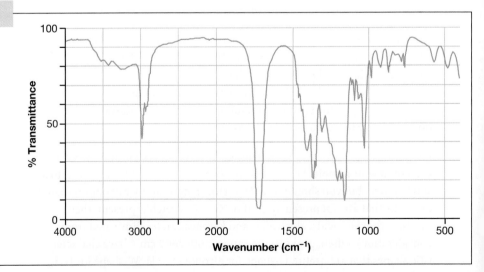

Levulinic acid   Ethyl levulinate

The IR spectrum in **Figure 6.10** was obtained of the reaction product.

FIGURE 6.10

Spectral Database for Organic Compounds, SDBSWeb. National Institute of Advanced Industrial Science and Technology. https://sdbs.db.aist.go.jp/ (accessed July 2022). Reprinted by permission.

The student interpreted the spectrum and reported the data in the Results and Discussion section of the lab report, using data from **Table 6.2**.

TABLE 6.2

IR Data for Ethyl Levulinate

FREQUENCY (cm^{-1})	BOND	FUNCTIONAL GROUP
2950	C—H	Alkane
1740	C=O	Ester
1720	C=O	Ketone
1160	C—O	Ester

Note that the two carbonyl peaks appear to partially overlap. In the body of the report, the student discussed what was meant by the presence of each of the peaks indicated in the table, and also discussed the absence of a broad O—H peak from 3300–2500 cm^{-1}, which would be expected from the CO$_2$H functional group in the starting material. The student noted that the absence of this O—H absorbance indicated that the reaction was successful and had gone to completion. The small O—H absorbances in the range of 3200–3500 cm^{-1} were attributed to small amounts of water or ethanol contaminating the sample.

6.3 SAMPLE PROBLEMS

1. (a) Match the compounds in **Figure 6.11** with the spectra in **Figure 6.12**.
   (b) For each of the spectra in Figure 6.12, identify at least two peaks that are diagnostic for the structure you matched to it. Write the wavenumber of the peak and assign it to a specific bond vibration (e.g., C—H stretch).

FIGURE 6.11

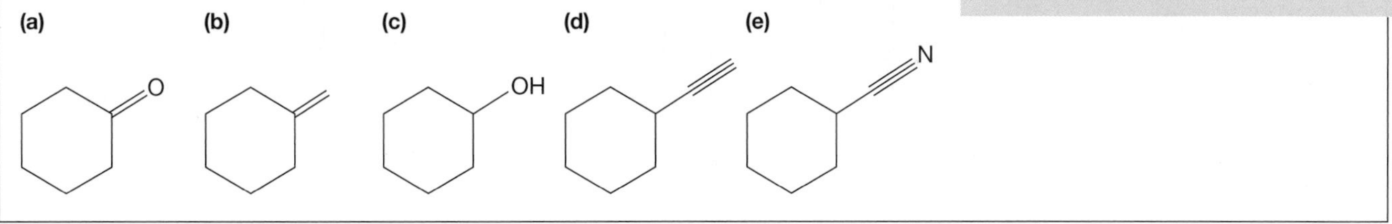

FIGURE 6.12

Spectral Database for Organic Compounds, SDBSWeb. National Institute of Advanced Industrial Science and Technology. https://sdbs.db.aist.go.jp/ (accessed July 2022). Reprinted by permission.

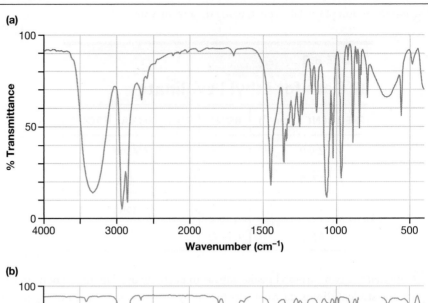

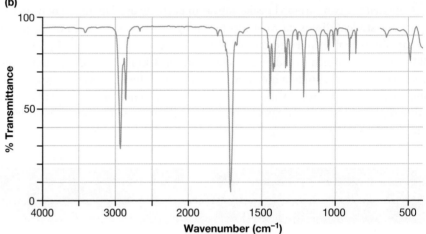

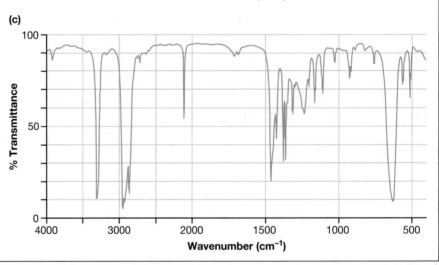

(continued)

FIGURE 6.12 (continued)

(d)

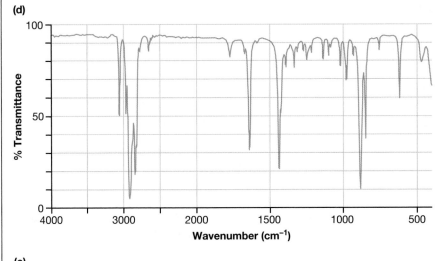

(e)

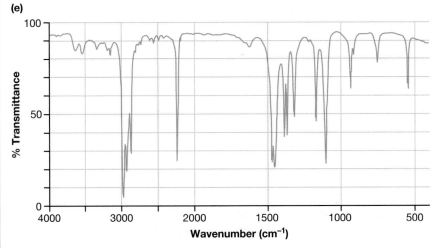

2. Propose a structure for each of the spectra in **Figure 6.13**, using the formula provided.

   a. C_7H_8O
   b. C_4H_9NO
   c. $C_4H_8O_2$
   d. C_7H_6O

FIGURE 6.13

(a)

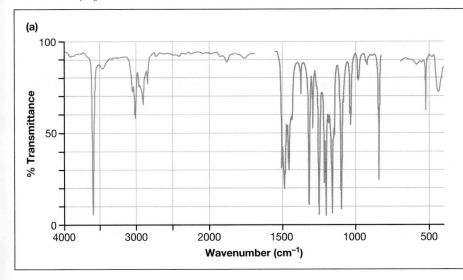

When acquiring the IR spectrum of this compound under other conditions, the peak in the 3200–3600 cm^{-1} region appeared more broad and U-shaped.

Spectral Database for Organic Compounds, SDBSWeb. National Institute of Advanced Industrial Science and Technology. https://sdbs.db.aist.go.jp/ (accessed July 2022). Reprinted by permission.

(continued)

FIGURE 6.13 *(continued)*

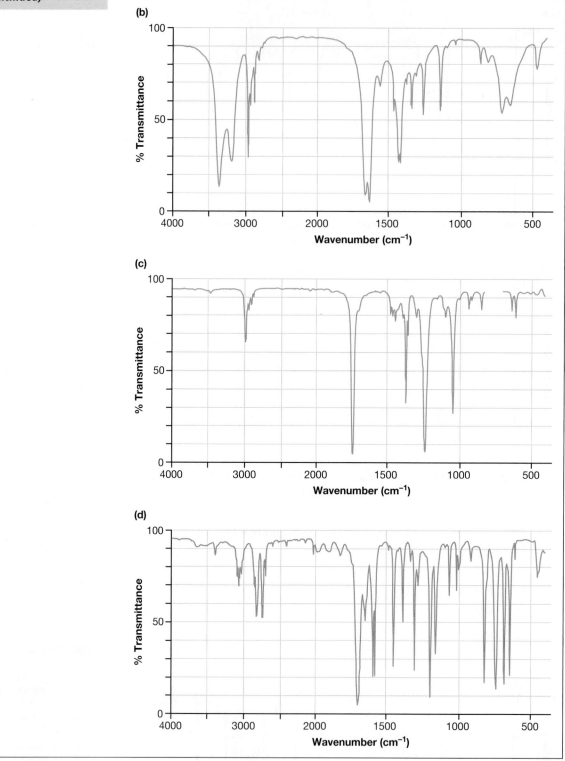

LEARNING OBJECTIVES

- Relate radiofrequency radiation absorbance to structural features of organic compounds.

- Identify four types of information associated with organic compound structures found in ¹H NMR spectra: Interpret or predict
  » *the number of signals,*
  » *the integrations of peaks,*
  » *the chemical shifts of peaks,*
  » *and the multiplicities of peaks.*

- Distinguish impurities from the analyte of interest, using ¹H NMR spectra.

- Identify organic compounds from ¹H and ¹³C NMR spectra.

- Report NMR data in appropriate formats.

NMR Spectroscopy

CAN YOU PICTURE A BELL FROM ITS RING?

The ringing of bells produces characteristic sounds that can be related to their shapes and sizes. We can "listen" to information about shapes and sizes of molecules through NMR spectroscopy.

Top: Alexander Blinov/Alamy Stock Photo. Left: Stefan Rotter/ Shutterstock. Right: Daan Kloeg/Shutterstock.

Nuclear magnetic resonance (NMR) spectroscopy is one of the most important diagnostic tools available to an organic chemist. It provides a spectrum of peaks containing information on the number of protons (^1H) and carbons (^{13}C), as well as their proximity to electronegative atoms. The NMR spectrum also indicates how structural features are connected, even at locations distant from the functional groups. The infrared (IR) spectrum is a very sensitive indicator of the types of functional groups, but provides less information about connectivity. As a result, NMR and IR work synergistically to determine the structure of an organic compound. In many cases, NMR can be used to determine a complete chemical structure in a very short period of time. NMR can also define product ratios, measure purity, and identify impurities.

7.1 ¹H NMR SPECTROSCOPY

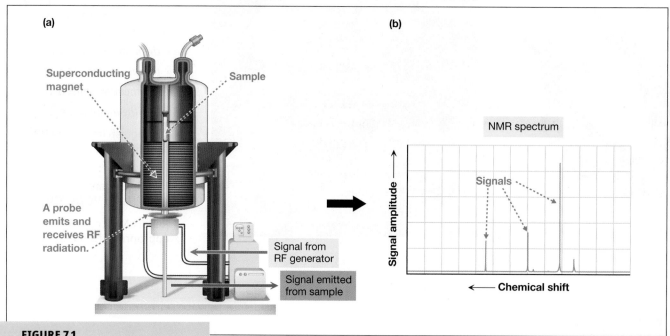

FIGURE 7.1

(a) A typical high-field nuclear magnetic resonance (NMR) spectrometer and (b) the output NMR spectrum.

Karty, J. *Organic Chemistry: Principles and Mechanisms*, 3rd ed.; W. W. Norton: New York, 2022; p 820.

magnetic dipole transitions >>
A change in the spin state of a magnetically active nucleus in response to absorption or emission of radiofrequency energy.

7.1A Theory of ¹H NMR Spectroscopy

Radiofrequency radiation is the part of the electromagnetic spectrum used for NMR spectroscopy (Figure 6.1). When a sample is placed into the magnetic field of a large superconducting magnet (**Figure 7.1**), and radiofrequency energy is used to monitor the **magnetic dipole transitions** of the nuclei, a spectrum is obtained that can be used to determine the structures of the compounds in the sample. Many labs use smaller, inexpensive, low-field instruments that operate with permanent magnets; although these furnish spectra with lower resolution, they are suitable for basic applications.

The spinning charge of a **spin-active nucleus** creates a small magnetic dipole, called μ (**Figure 7.2**). When the nucleus is under the influence of a much larger external magnetic field H_0, then μ has a tendency to align with H_0 in one of two **spin states**. The magnetic dipole μ of the nuclear spin will be either aligned in the same orientation as H_0 (the lower energy α state) or aligned in the opposite orientation (the higher β energy state).

The energy difference ΔE between these α and β states corresponds to the amount of energy imparted by absorbing a photon in the radiofrequency range of the spectrum. When the frequency of irradiation is matched, or in *resonance* with ΔE, it can be absorbed, and this absorption causes an excitation from the α state to the higher energy β state (**Figure 7.3**). The spin is said to "flip" from α to β. Over a short period of time (usually within a couple of seconds), the excited state β undergoes relaxation, back to the lower energy α state, and energy is emitted.

<< **spin-active (or magnetically active) nuclei**
Nuclei that exhibit nuclear magnetic resonance, and can give signals in NMR spectroscopy. Generally this refers to nuclei with a nonzero nuclear spin quantum number (I). Nuclei of ^1H and ^{13}C have $I = 1/2$, are magnetically active, and give NMR signals, while the ^{12}C nucleus has $I = 0$ and is not spin-active.

<< **spin states**
Different spin energy levels in which the magnetic dipole of a nucleus is aligned with or against an external magnetic field.

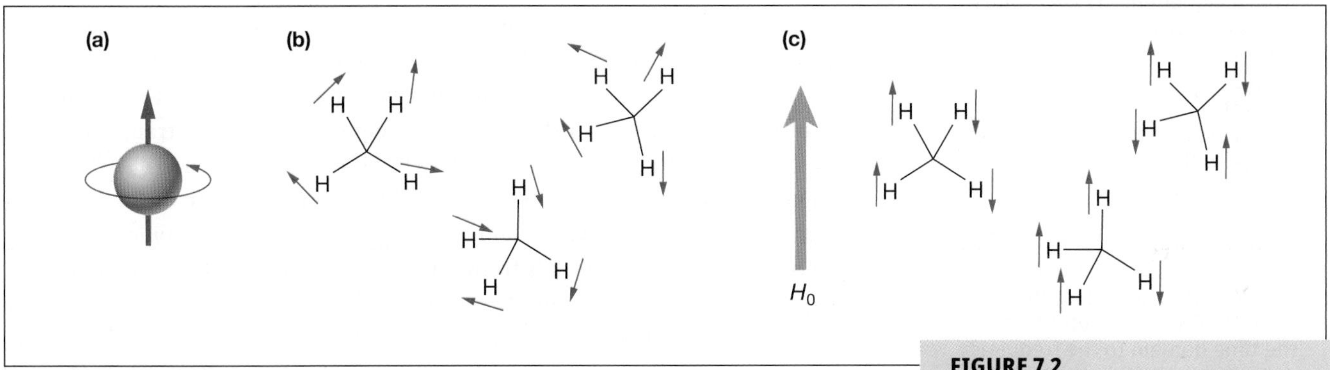

FIGURE 7.2

(a) A spinning nucleus creates a magnetic dipole μ, indicated by the solid arrow. (b) A sample of methane (CH_4) in the absence of any external magnetic field; the magnetic dipole of each hydrogen nucleus is randomly oriented. (c) A sample of methane in a strong external magnetic field (H_0); the magnetic dipoles of each hydrogen are aligned with (α) or opposing (β) the external field, with a slight excess in the lower energy α orientation.

The sample contains a population of nuclei distributed between the α and β spin states, with the lower energy state slightly in excess (**Figure 7.4**). Absorption of radiofrequency (RF) energy causes an increase in the population of the β state. The net magnetization associated with the population of dipoles changes as a result. As the population relaxes back to its initial distribution, and the net magnetization returns to its prior equilibrium condition, small fluctuations in the magnetic field in the sample create a current in the radiofrequency receiver, generating the NMR signal. The signal strength is low, so the NMR spectrum is obtained using a series of RF pulses, with a short delay after each pulse. With numerous repetitions, the signals become magnified and random noise is eliminated, improving the quality of the spectrum.

Nuclear magnetic resonance is extremely sensitive to small differences in the environment around the nucleus, including other nearby nuclei and their associated electrons. When several nuclei are present in different parts of a molecule, their emissions occur at slightly different frequencies. The emitted signals are all

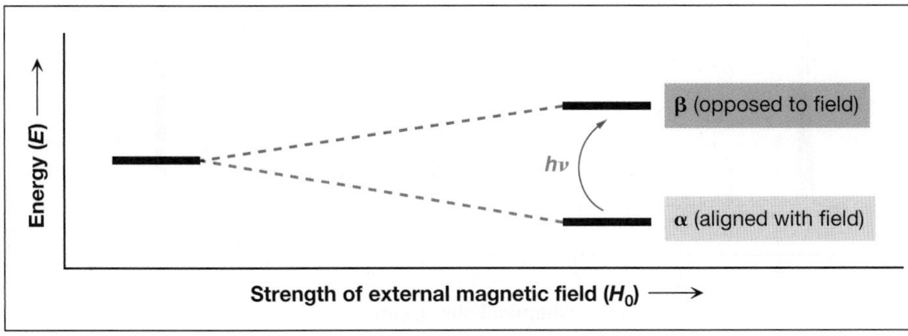

FIGURE 7.3

The difference in energy ($\Delta E = h\nu$) between α and β states is proportional to the applied external magnetic field (H_0). Absorption of a radiofrequency photon of energy $h\nu$ causes the magnetic dipole μ to transition from the α energy level to the β energy level.

FIGURE 7.4

Changes in population of nuclear spins in α and β states upon irradiation. The frequency v is in the radiofrequency (RF) region of the electromagnetic spectrum, associated with a change of the net magnetization in the sample. The relaxation back to the original population distribution results in emission of energy.

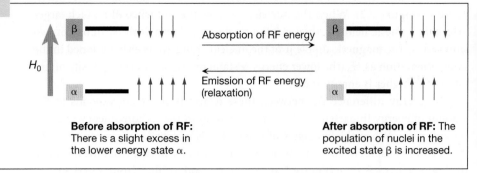

Before absorption of RF: There is a slight excess in the lower energy state α.

After absorption of RF: The population of nuclei in the excited state β is increased.

free induction decay (FID) >>
Raw form of radiofrequency emission signals from nuclei relaxing from a higher energy spin state to a lower energy spin state, plotted with respect to time. This output from a Fourier transform NMR instrument is mathematically converted into the NMR spectrum, plotted with respect to frequency (Hz).

Fourier transform >>
Mathematical process to convert the free induction decay output of an NMR spectrometer from the time domain to the frequency domain, resulting in the NMR spectrum.

contained together in a data set called a **free induction decay (FID)**, and to make use of this information, it is processed by a computer using **Fourier transform** operations to translate the emissions into a spectrum (**Figure 7.5**). The spectrum contains peaks in different positions that correspond to different nuclei in the molecular structure of the sample, and is plotted with frequency on the x axis, in units of parts per million (ppm), and intensity of absorbance on the y axis. The units of ppm are relative to the operating frequency of the RF transmitter and receiver; for example, 1 ppm at a 300 MHz (1 MHz = 1 megahertz = 10^6 Hz) operating frequency corresponds to 300 Hz—that is, one millionth of the operating frequency. The frequency values along the x axis, called the *chemical shift*, describe how far away each peak is from the tetramethylsilane (TMS) standard that is assigned a value of 0.00 ppm. By expressing the chemical shifts in ppm rather than Hz, the spectrum for a given compound will have the same numerical values on the x axis regardless of the operating frequency of the instrument. Each peak offers information about the molecular structure surrounding each of the nuclei, as judged by the chemical shift and other information contained in the peaks.

Nuclei with nonzero nuclear spin can be examined by NMR. When there is an even number of protons and also an even number of neutrons in the nucleus, the nuclear spin $I = 0$. These nuclei give no NMR signal; ^{12}C (6 protons, 6 neutrons) and ^{16}O (8 protons, 8 neutrons) are common examples. ^1H (1 proton, 0 neutrons) and ^{13}C (6 protons, 7 neutrons) are the most common nuclei that organic chemists subject

FIGURE 7.5

Example of a ^1H NMR spectrum from a research-grade high-field instrument with a 300 MHz operating frequency. Each vertical rise of the blue line measures the area, or integration, of the corresponding peak.

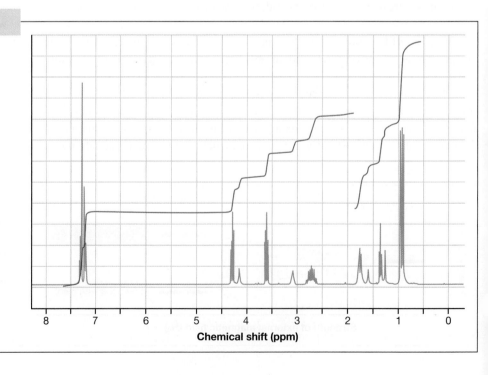

Chemical shift (ppm)

to NMR spectroscopy. The combination of protons and neutrons is an odd number for these nuclei, and each has $I = 1/2$. These nuclei are present throughout organic structures, not just in the functional groups. For that reason, NMR makes it possible to glean detailed information about portions of molecular structures independent of functional groups—information that is usually unavailable from UV–vis or IR spectroscopies.

Sample Problem

Two peaks appear 0.20 ppm apart in a spectrum obtained with an instrument operating at a frequency of 400 MHz. How far apart are they in Hz?

Solution

At an operating frequency of 400 MHz, one millionth of that, or one ppm, is 400 Hz. Use this relationship of 400 Hz per ppm to convert ppm to Hz:

$$\Delta\delta \ (\text{Hz}) = 0.20 \text{ ppm} \times \frac{400 \text{ Hz}}{1.00 \text{ ppm}}$$
$$= 80 \text{ Hz}$$

7.1B Interpreting Data in the NMR Spectrum

There are four main types of information available from a ^{1}H NMR spectrum. Each of these correlates to specific structural details, as follows:

1. Number of signals: The number of distinct signals corresponds to the number of different types of H, which is related to molecular symmetry.

2. Chemical shift: The frequencies of peaks (i.e., their location on the x axis of the spectrum) indicate electronic effects of nearby atoms and functional groups.

3. Integration: The area under each peak indicates how many H's give rise to that peak.

4. Multiplicity: A peak may have finer details because the signal is split into more than one part; the splitting indicates how many H's are present on neighboring atoms and may also reveal information about how they are spatially arranged relative to each other.

In ^{13}C NMR spectroscopy, the number of ^{13}C signals and their chemical shifts are the key elements to evaluate. The spectrum is acquired and processed in such a way that integration and multiplicity are not generally useful. We return to ^{13}C NMR in more detail near the end of this chapter.

In the next four sections, we explain in detail how to find and use all four types of information from ^{1}H NMR spectra. For a careful analysis of the NMR spectrum, especially when the identity of the compound is unknown, all four can be evaluated. However, not all of these types of information may be necessary to solve every problem. If two possible structures have a different predicted number of signals, for example, then one of them may be ruled out simply by counting the signals in the observed spectrum.

7.1C Number of Signals: Equivalent and Nonequivalent Hydrogens

The number of signals, or peaks, in a spectrum corresponds to the number of nonequivalent hydrogens. If hydrogens are nonequivalent, they will have different chemical and magnetic environments that depend on nearby structural features.

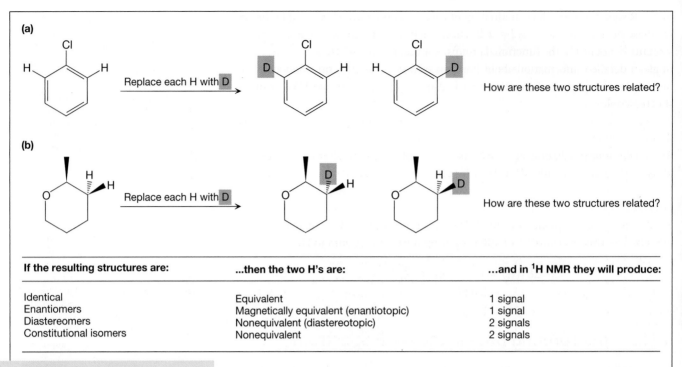

(a)

Replace each H with D → How are these two structures related?

(b)

Replace each H with D → How are these two structures related?

If the resulting structures are:	...then the two H's are:	...and in ¹H NMR they will produce:
Identical	Equivalent	1 signal
Enantiomers	Magnetically equivalent (enantiotopic)	1 signal
Diastereomers	Nonequivalent (diastereotopic)	2 signals
Constitutional isomers	Nonequivalent	2 signals

FIGURE 7.6

Depiction of how to determine whether two hydrogens will appear in the same signal or in different signals. Sequentially replace each by another atom, then determine the relationship between the resulting structures. *Solutions:* (a) Identical, 1 signal. (b) Nonequivalent (diastereotopic), 2 signals.

This means they will have different resonance energies, and this will give a peak in the spectrum at different frequencies. Equivalent hydrogens, on the other hand, all appear in the same signal. This arises when there is symmetry in a structure, such as an internal plane of symmetry that makes the left and right sides of a structure identical, or a symmetry axis such as the rotation about a C—C bond that makes all the H's of a methyl group equivalent.

Sometimes a symmetry plane or symmetry axis may not be obvious, so there is a simple test that can help determine if any two hydrogens in a structure are equivalent or nonequivalent (**Figure 7.6**). First, draw two copies of the structure, including the two hydrogens you want to test. On one structure, replace one of the two hydrogens with another atom (such as D, deuterium), and on the second structure make the same replacement of the other hydrogen. Now, determine how these two structures are related. Are they identical or different? Are they constitutional isomers or stereoisomers? If they are stereoisomers, are they enantiomers or diastereomers?

If structures A and B are

- *identical,* then the original two hydrogens before replacement will be equivalent, and these H's will appear in the same signal in the NMR spectrum.
- *constitutional isomers,* then the original two hydrogens before replacement will be nonequivalent, and these H's will have different peaks in the NMR.
- *enantiomers,* then the original two hydrogens before replacement will be enantiotopic, and these H's will be equivalent in the NMR.
- *diastereomers,* then the original two hydrogens before replacement will be diastereotopic, and these H's will be nonequivalent in the NMR.

To remember these last two, keep in mind that enantiomers have the same physical and spectroscopic properties (except optical rotation), but diastereomers have different properties—including nuclear resonance energy and other properties related to spectroscopy.

Some examples of structures that offer one or more of these scenarios are presented in Figure 7.6. Predict how many signals should appear in each ¹H NMR spectrum.

Certain types of hydrogens that are involved in hydrogen bonding, specifically O—H, N—H, and S—H, may not be detected reliably in all conditions. Hydrogen bonding interactions between molecules containing these bonds can lead to exchange of the hydrogens between molecules in a process known as **chemical exchange of hydrogen**. When the exchange is fast, the H's will be moving around among many different magnetic environments, and the signal in an NMR spectrum will be an average of all those different environments. This can lead to a signal that is broad, and may appear at chemical shifts that vary with concentration and other sample conditions. With carboxylic acids, the hydrogen of the CO_2H group can be so broad that you might not see it clearly as a peak, though it may be obvious in the integration.

Chemical exchange can help identify which peaks in a spectrum belong to exchangeable hydrogens—that is, those that are attached to electronegative atoms. **Deuterated NMR solvents** that are hydrogen-bond donors, such as CD_3OD or D_2O, can exchange their deuterium atoms for the exchangeable hydrogen atoms in the sample. The ^1H NMR spectra acquired in these solvents, therefore, will *not* show any signals for H's that are in O—H, N—H, or S—H bonds. To facilitate comparison with a reference spectrum acquired in $CDCl_3$, a few drops of D_2O can be shaken with the $CDCl_3$ solution in the NMR tube, and a new spectrum acquired. Comparing the spectra acquired before and after deuterium exchange makes it possible to assign the peaks that disappeared to the exchangeable hydrogens in the structure.

<< **chemical exchange of hydrogen**
Rapid proton transfer reactions between molecules containing hydrogens attached to electronegative atoms. This is commonly observed with hydrogen bond donor groups such as OH, NH, and SH.

<< **deuterated NMR solvent**
A solvent that has had its ^1H atoms replaced by ^2H, or deuterium (D) atoms that do not absorb radio-frequency energy in the same frequency range as ^1H. The solvent properties remain the same, but the deuterated solvent does not appear in a ^1H NMR spectrum. Trace amounts of the ^1H solvent are generally still present, resulting in a very small solvent peak; this can be used for calibration of the spectrum.

7.1D Chemical Shifts

The frequencies of the peaks (in units of ppm) in the NMR spectrum can be correlated to structural features, and this allows us to distinguish between hydrogen nuclei that are in different environments. There is a significant difference in the frequency of the peaks associated with the methyl groups of dimethyl ether (CH_3OCH_3, 3.2 ppm), and dimethyl sulfide (CH_3SCH_3, 2.1 ppm). Once we know the frequency of absorption for these two compounds, we can use the NMR spectrum of a laboratory sample to identify whether one or both of these compounds may be present. Frequencies of ^1H NMR peaks are known for millions of organic compounds, and when a new compound is made, the frequencies are measured and recorded in journals and databases so that other chemists can use the data for comparison.

Rarely does a spectrum have only one peak at one frequency, however. A more complicated molecule—one with lots of different types of hydrogens—affords peaks at a variety of frequencies. As first mentioned when we introduced the theory of ^1H NMR spectroscopy, NMR data are generally presented in a plot where the y axis is intensity of absorbance and the x axis is chemical shift (δ), measured in units of ppm (parts per million). ΔE (or ν) increases from right to left. These chemical shifts vary slightly depending on the environment around the hydrogens (i.e., what functional groups are nearby). Certain chemical shift ranges in the spectrum correlate very well to specific types of functional group environments (**Figure 7.7**).

The electronegativity of neighboring atoms dramatically affects the chemical shifts. The electrons around the nucleus cause magnetic **shielding**, which counteracts the external field H_0. Nearby electronegative atoms have an electron-withdrawing effect that lessens the shielding; this effect is called **deshielding**. Shielded hydrogens appear to the right, or *upfield*, while deshielded hydrogens appear at the left of the spectrum, or *downfield*. Figure 7.7 shows that a hydrogen in the absence of any functional group, electronegative atom Z, or π bond appears at about 1 ppm. One electronegative atom, such as a halogen or oxygen, shifts the resonance frequency downfield by about 2 ppm. These electronegative Z groups have an additive effect on the deshielding. For example, the hydrogens of CH_2Cl_2 give a peak at 5.28 ppm, and the hydrogen of $CHCl_3$ appears at 7.26 ppm. The chlorine atoms are diminishing the electron density around

<< **shielding/deshielding**
The effect on a nucleus from nearby electron density that changes the frequency of radio-frequency absorption. A nearby electronegative atom reduces electron density around a nucleus, resulting in deshielding.

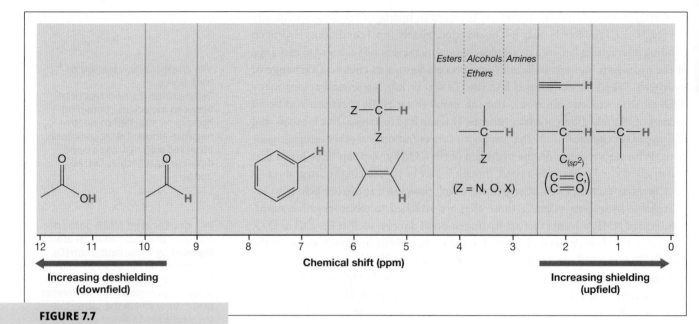

FIGURE 7.7

How chemical shift correlates with different functional group environments.

the H's, so there is less of a shielding effect. Deshielding increases the energy differ-ence between the α and β spin states (**Figure 7.8**), causing the emission to be of higher energy and farther downfield.

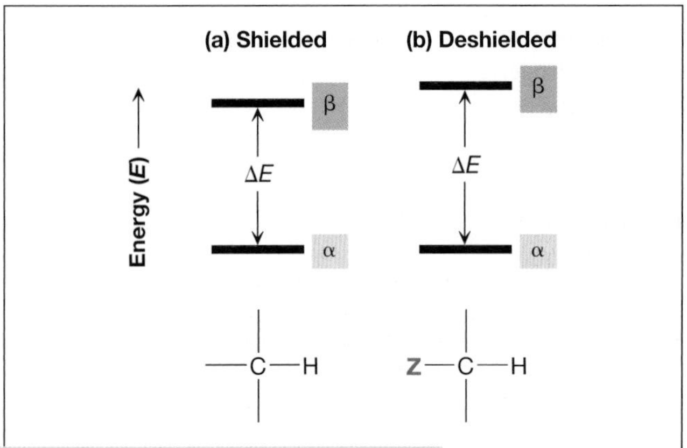

It is not only electronegativity that can cause deshield-ing. The benzene ring causes hydrogens to appear down-field, even though it contains no electronegative atoms. This is due to a phenomenon called **magnetic anisotropy**, which is caused by circulating electrons in π systems (**Figure 7.9**). The circulating π electrons create a current, inducing an-other magnetic field that reinforces the applied magnetic field, H_0, in the location of the attached hydrogens. Due to anisotropy, protons near double bonds or aromatics are shifted downfield (to the left in the spectrum) in comparison with an alkane H (**Figure 7.10**). Protons directly attached to aromatics or double bonds are shifted the most. The down-field shift of terminal alkyne H is less dramatic, although triple bonds do exhibit anisotropy.

(a) Shielded   **(b) Deshielded**

FIGURE 7.8

Effects on the relative energy levels of the α and β states for a hydrogen nucleus that is either (a) shielded or (b) deshielded by an electronegative atom Z.

magnetic anisotropy >>
An induced magnetic field that can be additive to or subtractive from the applied field experienced by a nucleus, depending on where the nucleus is located relative to the source of the induced field. This is commonly observed with ¹H nuclei located near aromatic rings and other π systems.

7.1E   Integration

A tool from calculus, the integral, gives us another piece of information about the peaks in our NMR spectrum. Integration is a measurement of the area under a curve, and this can be used to measure the area under a peak in an NMR spectrum. In NMR spectroscopy, the integral of a peak is proportional to the number of hydro-gens giving rise to that peak. The units of integration don't matter, because they are relative values used to compare two peaks. As long as the units are the same for both peaks, the ratio of the integrals for each peak will be useful. So, whether the integral ratio is 120:40 or 7.5:2.5, the ratio is the same (3:1), and it corresponds to the ratio of the number of hydrogens giving rise to the two peaks. Depending on how the op-erator formats the printed spectrum, the integral measurements may be illustrated graphically, with a line inscribed over the spectrum (the blue line in **Figure 7.11**) or the numerical values may be printed on the spectrum.

FIGURE 7.9

(a)

Circular movement
of π electrons
= Ring current

H_0

(b)

Ring current creates a
local magnetic field (B_{loc}).

The H atoms feel an additional magnetic field.
The induced field adds to the applied field, H_0.

Anisotropy effects upon chemical
shift of hydrogens located near the
π system of benzene.

Karty, J. *Organic Chemistry: Principles and Mechanisms*, 3rd ed.; W. W. Norton: New York, 2022; p 833.

FIGURE 7.10

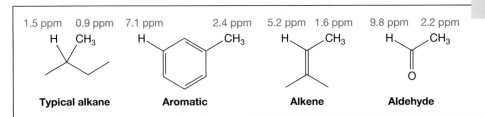

| 1.5 ppm | 0.9 ppm | 7.1 ppm | 2.4 ppm | 5.2 ppm | 1.6 ppm | 9.8 ppm | 2.2 ppm |

Typical alkane — **Aromatic** — **Alkene** — **Aldehyde**

Typical chemical shifts of H and
CH_3 attached to the sp^2-hybridized
carbons of an aromatic ring, alkene,
and aldehyde.

FIGURE 7.11

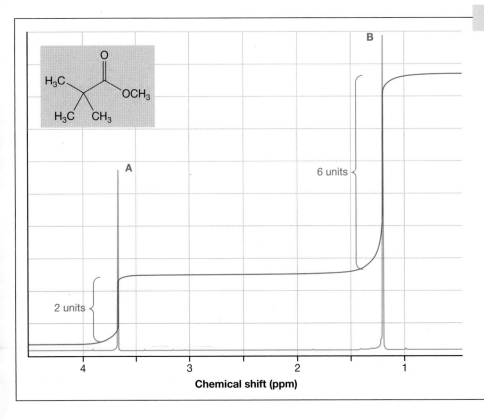

Chemical shift (ppm)

^{1}H NMR spectrum of methyl
2,2-dimethylpropanoate (300 MHz,
$CDCl_3$). The integration of the
spectrum is shown with the blue line.
Gridlines are added to the spectrum;
in this example, each horizontal
gridline is one unit to measure the
rise in the integral line. Any unit
(mm, cm, etc.) may be used, as long
as it is consistent for all peaks within
the spectrum.

© Sigma-Aldrich Co. LLC. Reproduced
with permission from Merck KGaA,
Darmstadt, Germany and/or its affiliates.

Example 1

Using the information we have discussed so far, let's interpret the ^{1}H NMR spectrum of methyl 2,2-dimethylpropanoate (Figure 7.11). How do the number of signals, their chemical shifts, and their integrals correlate with the structure of the compound?

Number of Signals

Two peaks are present, labeled **A** and **B**. This means that the compound has two different types of hydrogens. There are more than two hydrogens in the formula, however, so symmetry must be making some of them equivalent. In a methyl group, for example, all three hydrogens are equivalent.

Chemical Shift

Peak **A** appears at 3.7 ppm, quite a bit farther downfield than peak **B** at 1.2 ppm. The hydrogens giving rise to peak **A**, then, are deshielded, so there must be an electron-withdrawing atom nearby. In the structure, the H's of the *tert*-butyl group are four bonds away from the nearest electronegative atom (either oxygen atom), whereas the H's of the methoxy group are only two bonds away, and the methoxy group would therefore experience deshielding.

Integration

There is a horizontal line inscribed upon the spectrum, and it rises each time it passes a peak. The amount of the rise is a measurement of the integral, or area under the peak. In this case, the ratio of the integrals **A** and **B** is 2:6, and this ratio is proportional to the ratio of hydrogens giving rise to peaks **A** and **B**. The number of hydrogens must be an integer, so dividing the larger number by the smaller one, we get a ratio of 1:3. According to the structure (Figure 7.11), there are 12 hydrogens—three equivalent hydrogens in the OCH_3 group and nine equivalent hydrogens in the *tert*-butyl group, which accounts for the ratio of 1:3.

Example 2

After extracting a compound, C_3H_4BrClO, from a marine organism, you need to figure out its structure. You have previously deduced that you have one of the two isomeric compounds **A** and **B** shown in **Figure 7.12**. Match the ^{1}H NMR spectra in Figure 7.12 with these two structures.

FIGURE 7.12

Two predicted ^{1}H NMR spectra for two isomers of C_3H_4BrClO. The integral ratios have been calculated, and are shown at the top of each peak.

nmrdb.org: Tools for NMR Spectroscopists [Online]. https://www.nmrdb.org/ (accessed April 2022). CC-BY-4.0 https://creativecommons.org/licenses/by/4.0/

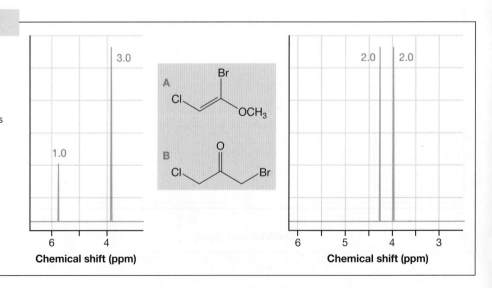

First narrow down the possible structures using the molecular formula along with functional group data from infrared or NMR (for this problem, the structures are already provided in Figure 7.12). Then, for each structural hypothesis, predict the number of signals and their relative integrals. Finally, compare these predictions with the observed spectrum, ruling out any structures that are inconsistent with observations.

Each of the structures has two different types of hydrogen, so each would be expected to have two signals in the NMR spectrum. Each of the spectra in Figure 7.12 has two signals, so we cannot differentiate them by number of signals. Structure **A** has an alkene, and a hydrogen attached to an alkene generally appears at 4.5–6.5 ppm (see Figure 7.7). Only the spectrum on the left has a peak that matches this expected chemical shift, so we might conclude that that spectrum corresponds to structure **A**. Can we strengthen this conclusion by analyzing the integration? Structure **A** would be expected to have two peaks, one for a single alkene hydrogen and the other for a methyl group (3 hydrogens); the two peaks would have an integral ratio of 1:3. This also matches the spectrum on the left, and strengthens our conclusion because two independent pieces of evidence both lead to the same conclusion.

The spectrum on the right in Figure 7.12 has two peaks with an integral ratio of 2:2, the ratio we would expect for structure **B**. How do we know which hydrogens belong to which signal? Recall that nearby electronegative atoms deshield the nearby H's, shifting their signals to the left (downfield). Because Cl is more electronegative than Br, the hydrogens nearer the Cl should be more deshielded, and farther downfield.

7.1F | Multiplicity (Signal Splitting)

In the NMR spectra we have seen so far, there has been only one line, or one maximum on the y axis, within each signal. Each of these peaks, called a *singlet*, is what we observe for hydrogens that are several bonds away from any other nonequivalent hydrogens. Commonly, there are nonequivalent hydrogens located in a **vicinal** (3 bonds away) or **geminal** (2 bonds away) relationship, and in these cases there will be **signal splitting**. Instead of a singlet, the peak will appear to have more than one maximum within it—it has been split into a *multiplet*. The number of maxima within a multiplet is called the **multiplicity**, and this can be correlated with the number of nonequivalent H's that are nearby. Multiplicity is therefore a powerful tool to unravel how the atoms are connected within the structure.

Signal splitting occurs when nonequivalent hydrogens are within two or three bonds of each other because their spins are coupled. They are close enough that the magnetic dipole of one H can add to or subtract from the overall magnetic field of the other. If the hydrogens are four bonds apart or more, then their magnetic dipoles have very little impact on each other, and usually no splitting is observed. The strong external field H_0 is accompanied by the small magnetic dipoles associated with nearby hydrogen nuclei. If a nearby dipole is aligned with H_0, it adds to H_0, whereas if it is aligned opposite H_0, it subtracts from H_0. This, in turn, affects the frequency of the absorbance. Thus, one nearby hydrogen can split a signal into two signals, because its dipole has two possible alignments (i.e., with or against H_0).

The presence or absence of signal splitting is associated with particular structural features (**Figure 7.13**). Most commonly, signal splitting is observed with vicinal hydrogens (neighboring hydrogens, on two carbons that are attached to each other) and in order to observe it, the vicinal hydrogens must be nonequivalent. Normally geminal hydrogens (hydrogens attached to the same carbon) are equivalent, though

<< **vicinal**
The relationship between atoms that are separated by three bonds.

<< **geminal**
The relationship between atoms that are separated by two bonds.

<< **signal splitting**
Creation of two or more lines from one NMR signal caused by induced magnetic fields from nearby nuclear spins (usually ^{1}H), which can be additive to or subtractive from the applied field, depending on whether the nearby nuclear spin is aligned with or against the applied field. The effect is most common for ^{1}H nuclei that are 2 or 3 bonds apart, but longer-range splitting is sometimes observed.

<< **multiplicity**
The number of lines within a signal when signal splitting occurs due to neighboring nuclei, usually [1]H.

FIGURE 7.13

Nearby hydrogens may or may not give rise to signal splitting.

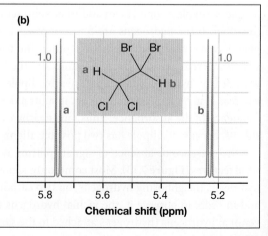

Vicinal: Splitting is observed (if H_a and H_b are nonequivalent).

Geminal: Splitting is observed (if H_a and H_b are nonequivalent).

Separated by 4 or more bonds: Usually no splitting is observed.

they can show splitting if they are nonequivalent. There is no splitting of signals in Figure 7.12, where the geminal H's in structure **B** are equivalent. Geminal splitting can sometimes be seen in cases where cis–trans isomerism is possible, such as a CH_2 of a substituted cyclohexane. One H is cis to a substituent, and the other is trans, so they are diastereotopic, and nonequivalent.

Coupling and signal splitting can be described by two simple rules. First, to observe coupling, the vicinal hydrogens must be in different environments (nonequivalent). Second, the splitting occurs according to the $N + 1$ rule. When there are N equivalent hydrogens nearby (vicinal, usually), they will split the signal into $N + 1$ peaks:

- No vicinal hydrogens: Signal appears as a single peak, called a *singlet* (**Figure 7.14a**; $N = 0$; $N + 1 = 1$).
- One vicinal hydrogen: Signal appears as two peaks, called a *doublet* (**Figure 7.14b**; $N = 1$; $N + 1 = 2$).
- Two vicinal hydrogens: Signal appears as three peaks, called a *triplet* (**Figure 7.14c**; $N = 2$; $N + 1 = 3$).

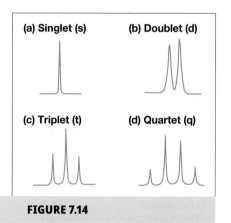

(a) Singlet (s)   (b) Doublet (d)

(c) Triplet (t)   (d) Quartet (q)

FIGURE 7.14

The five spectra in **Figure 7.15** illustrate these features. In spectrum (a), the two vicinal hydrogens are equivalent, so no coupling is observed. In spectrum (b), the two vicinal hydrogens are nonequivalent because of the different environments around Cl and Br. This gives rise to two different signals, and each of the two hydrogens splits the other into a doublet.

FIGURE 7.15

Examples of vicinal coupling that leads to signal splitting.

nmrdb.org: Tools for NMR Spectroscopists [Online]. https://www.nmrdb.org/ (accessed April 2022). CC-BY-4.0 https://creativecommons.org/licenses /by/4.0/

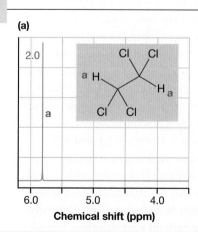

(continued)

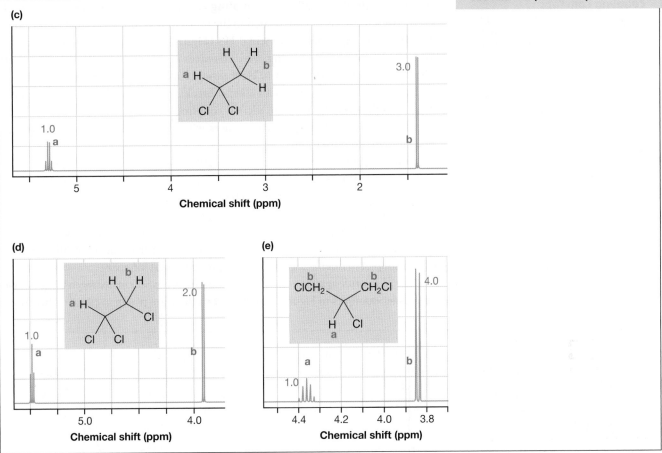

FIGURE 7.15 *(continued)*

In spectrum (c), H_a is split into a *quartet* (**Figure 7.14d**), because there are three vicinal hydrogens in the neighboring methyl group ($N = 3$, $N + 1 = 4$); its signal has four peaks within it, and they are of unequal intensity. Integrating the area under each part of the peak reveals that the ratio of the four peaks within the quartet is 1:3:3:1, which is typical for a quartet. Similarly, the triplet in spectrum (d) consists of three peaks of unequal intensity in a ratio of 1:2:1. This is typical of a triplet. In both spectra (c) and (d), H_a splits the signal for the H_b hydrogens into a doublet.

In spectrum (e), there are two sets of hydrogens labeled H_b on different carbons, but because of the symmetry in the structure, all four of them are equivalent. Thus, H_a is split into a signal of multiplicity 5 ($N = 4$, $N + 1 = 5$), or a quintet. The signal for H_b is split only by one neighbor (H_a) and is observed as a doublet. What will be the area ratios of these peaks within the quintet? You can predict the ratios by using Pascal's triangle (**Figure 7.16**). Simply put, Pascal's triangle gives you the area ratios for all multiplets, from singlets to doublets to triplets to quartets to quintets to sextets to septets and so on. In the chemical literature, once you get beyond quartets, the higher multiplets are often just referred to as multiplets.

COUPLING TO MORE THAN ONE TYPE OF HYDROGEN

If two different types of vicinal hydrogens split a signal, then more complicated splitting patterns can result, because the magnitude of the splitting is not always equal. The spacing between the peaks in multiplets is not necessarily the same

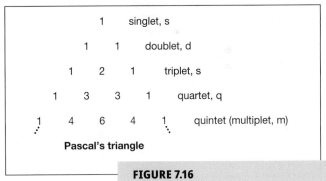

FIGURE 7.16

Pascal's triangle predicts the relative peak intensity in multiplets.

from one molecule to the next, or in different parts of a molecule. This spacing within a multiplet is called the **coupling constant** (J), and is measured in hertz (Hz). The variations in J are related to the molecular geometry—most importantly the **dihedral angle** between vicinal H's (**Figure 7.17**). When bond rotations are restricted by a π bond or a ring, the differences in coupling constants become very significant, ranging from 0 to 10 Hz for those linked by C—C σ bonds, and up to 18 Hz for trans vicinal H's on an alkene. Freely rotating single bonds in saturated acyclic molecules often lead to vicinal coupling constants around 7 Hz, because the contributions of different conformers with various dihedral angles are averaged by the rapid rotations.

Some examples are shown in **Figure 7.18** for a hydrogen H_a that is split by two different vicinal neighbors H_b and H_c, resulting in a *doublet of doublets*. The signal is split by H_b into a doublet, then each of those lines is split again by H_c. If the two coupling constants are almost equal, the doublet of doublets appears to be a triplet, and the overlap of two lines in the middle causes the middle of the peak to have a higher intensity. When one coupling constant is larger than the other, all four peaks of the doublet of doublets are visible with nearly equal intensity.

The $N + 1$ rule applies to cases when there are N *equivalent* hydrogens. The doublet of doublet examples in Figure 7.18 occur when there are two *nonequivalent* hydrogens

FIGURE 7.17

The Karplus curve, depicting the relationship of dihedral angle φ to magnitude of coupling constant J for vicinal couplings (three-bond coupling to hydrogens attached on adjacent carbons).

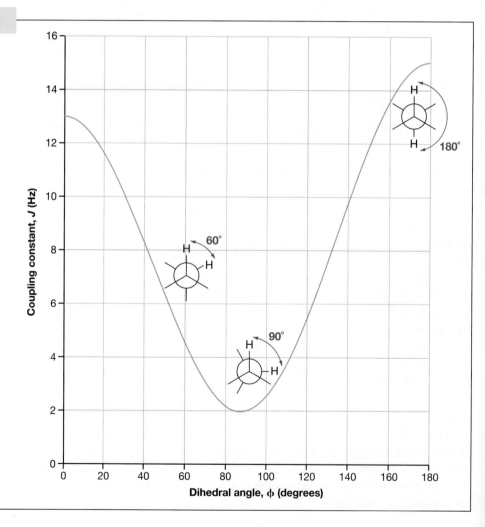

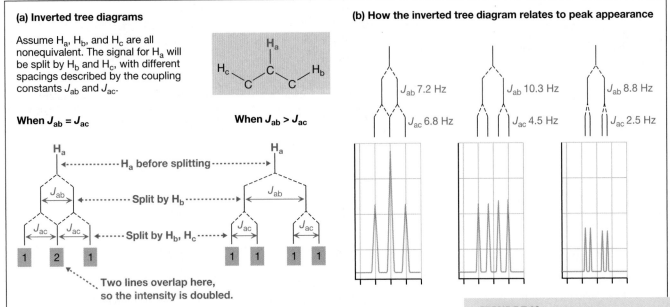

(a) Inverted tree diagrams

Assume H_a, H_b, and H_c are all nonequivalent. The signal for H_a will be split by H_b and H_c, with different spacings described by the coupling constants J_{ab} and J_{ac}.

When $J_{ab} = J_{ac}$

H_a

H_a before splitting

J_{ab} — Split by H_b

J_{ac} J_{ac} — Split by H_b, H_c

1 2 1

Two lines overlap here, so the intensity is doubled.

When $J_{ab} > J_{ac}$

H_a

J_{ab}

J_{ac} J_{ac}

1 1 1 1

(b) How the inverted tree diagram relates to peak appearance

J_{ab} 7.2 Hz

J_{ac} 6.8 Hz

J_{ab} 10.3 Hz

J_{ac} 4.5 Hz

J_{ab} 8.8 Hz

J_{ac} 2.5 Hz

FIGURE 7.18

Coupling of nonequivalent hydrogens H_b and H_c to H_a, when the coupling constants J are different, leads to a *doublet of doublets*.
(a) An inverted tree diagram helps to describe how the signal of H_a is split, first by H_b, then by H_c.
(b) The appearance of the doublet of doublets varies depending on the relative values of the coupling constants.

splitting the signal. In this latter case, instead of the $N + 1$ rule, there is an expanded form of the $N + 1$ rule that treats the two different hydrogens separately:

$$\text{Multiplicity} = (N + 1)(M + 1)$$

The effect of one of the hydrogens N is described by the $N + 1$ rule, and we introduce a label M for the different set of hydrogens and handle its effect similarly; then we multiply the two effects together. With two different hydrogens splitting the signal, both N and M are 1, and we have

$$\text{Multiplicity} = (1 + 1)(1 + 1) = 4$$

The signal in such a case will have four separate lines, and we can see this in the doublet of doublet cases (Figure 7.18). Further examples of multiplicity that give rise to doublets of triplets or even more complicated patterns are shown in **Figure 7.19**. You'll rarely encounter these more complex situations in introductory organic laboratory courses, but they are commonly seen in more advanced courses and in research labs.

COUPLING IN ALKENES

For alkenes, there is no rotation of the π bond under typical conditions, and this means that the dihedral angles for vicinal H's attached to the planar alkene are fixed, varying only slightly from the values of either 0° or 180°. The cis H's have a dihedral angle of 0° and coupling constants around 8–10 Hz, whereas trans H's have a dihedral angle of 180° and coupling constants around 15–17 Hz. As a result, it is possible to assign cis or trans configuration using the coupling constants, especially if the spectrum of both isomers is available for comparison. Also, geminal coupling may be observed in alkenes when there are two nonequivalent H's attached at one end of an alkene. Alkene geminal coupling is typically quite small, 0–2 Hz, so it may not clearly show up in the spectrum without an expansion plot of the peak in question.

For *n*-butyl vinyl ether (**Figure 7.20**), the alkene region of the spectrum shows coupling constants of 14.3 Hz, 6.8 Hz, and 1.8 Hz. The peaks at 4.2 and 3.9 ppm, assigned to H_b and H_c, respectively, are not simple doublets, as would be expected for vicinal coupling; instead, they show a second smaller coupling constant that is attributed to geminal coupling.

(a)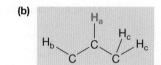

(b)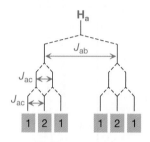

Splitting of H$_a$, by two **nonequivalent** sets of H's ($N = 1$ and $M = 2$). Coupling constants (J_{ab} and J_{ac}) are different:

(c)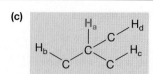

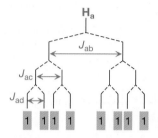

Splitting of H$_a$, by three **nonequivalent** sets of H's ($N = 1$, $M = 1$, $P = 1$). Coupling constants (J_{ab}, J_{ac}, and J_{ad}) are different:

Multiplicity = $(N + 1) = 3$

Peak splitting pattern: triplet (t)

Multiplicity = $(N + 1)(M + 1) = 6$

Peak splitting pattern: doublet of triplets (dt)

Multiplicity = $(N + 1)(M + 1)(P + 1) = 8$

Peak splitting pattern: doublet of doublet of doublets (ddd)

FIGURE 7.19

Inverted tree diagrams for H$_a$ in signals with various types of multiplicities, compared to (a) a standard triplet that follows the $N + 1$ rule. Relative intensities of each line within the multiplet are shown in red. When two nonequivalent sets of H's couple with H$_a$, the $N + 1$ rule is multiplied by $M + 1$, giving rise to more complex patterns. (b) When $N = 1$ and $M = 2$, a doublet of triplets (dt) arises. When a third different type of H is splitting, another multiplication with $P + 1$ is included. (c) When N, M, and P are all 1, the result is a multiplicity of 8, all with equal intensities, also known as a doublet of doublet of doublets (ddd).

RECIPROCITY OF COUPLING

An interesting feature of the spectrum of *n*-butyl vinyl ether (Figure 7.20) is that the same geminal coupling constant of 1.8 Hz is observed at both of the peaks at 4.2 and 3.9 ppm. This is a normal occurrence, because signal splitting between two H's is mutual, and the coupling constant J_{ab} observed at H$_a$ will also be observed at H$_b$ in the same magnitude. This feature can be useful when assigning a peak to a specific hydrogen in a structure. For example, knowing that H$_b$ and H$_c$ will have small geminal couplings allows us to assign them to the peaks at 4.2 and 3.9 ppm. However, which one is which? The third alkene peak at 6.4 ppm has coupling constants of 14.3 and 6.8 Hz that correspond to trans and cis relationships, respectively. The hydrogen that is trans to H$_a$ should have the larger coupling constant 14.3 as seen at 4.2 ppm, so we assign this peak to H$_b$.

FIGURE 7.20

^{1}H NMR spectrum of *n*-butyl vinyl ether. H$_a$ is a doublet of doublets because of coupling to cis (H$_c$) and trans (H$_b$) protons; H$_c$ and H$_b$ are nonequivalent and have different vicinal coupling constants with H$_a$. Expansion plots of the peaks at 4.2 and 3.9 ppm show evidence of geminal coupling ($J_{bc} = 1.8$ Hz).

© Sigma-Aldrich Co. LLC. Reproduced with permission from Merck KGaA, Darmstadt, Germany and/or its affiliates.

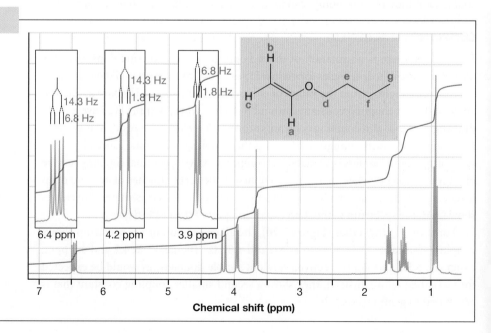

Impact of Rapid C—C Rotation on Multiplicity

Based on the structure of *n*-butyl vinyl ether (Figure 7.20), what would you predict the multiplicities of H_e and H_f to be? H_e should be coupled to both H_d and H_f, and H_d and H_f are nonequivalent, so you might say that H_e should have a total multiplicity of

$$\text{Multiplicity} = (N + 1)(M + 1) = (2 + 1)(2 + 1) = 9$$

The single C—C bonds in the butyl group are rapidly rotating, however, which causes the coupling constants to be almost the same, at 7 Hz. This is because the coupling constant results from an average dihedral angle from a composite of all the rotamers. When the coupling constants are all indistinguishable, which is typical for *n*-alkyl groups, then the appearances of the multiplets can be better predicted by grouping N and M together. In this case, $N = 2$ (for H_d) and $M = 2$ (for H_f), and thus for H_e, the multiplicity is

$$\text{Multiplicity} = (N + M + 1) = (2 + 2 + 1) = 5$$

The peak appears as a quintet. Although it is not a true quintet, because the hydrogens represented by N and M are not equivalent, the appearance of the peak suggests that it seems to be following the $N + 1$ rule with $N = 4$. This is typical behavior for acyclic alkyl groups like the *n*-butyl group, and this greatly simplifies multiplicities, both in prediction and interpretation.

7.1H Coupling and Chemical Exchange

Polarized bonds to hydrogen, such as N—H, O—H, and S—H, serve as hydrogen-bond donors, and can participate in *chemical exchange* of hydrogen as discussed previously in relation to analyzing the number of signals. If the exchange process is slow, it may not have any impact on vicinal coupling. In such circumstances, the O—H hydrogen in CH_3OH, for example, would be coupled to the methyl group, giving rise to a signal that is split into a quartet. However, the rate of chemical exchange can be faster at higher concentrations or in the presence of acid. When the exchange is rapid, the molecular environment around the hydrogen nucleus is rapidly changing, and the signal is formed from an average of all the various environments in which the hydrogen exists. This causes the signal to appear broadened, and also leads to the loss of coupling information. So, when interpreting spectra of such compounds, it's important to be aware that we don't necessarily know whether the spectrum was acquired under conditions that promote chemical exchange. Be prepared to see these signals broad or narrow, at variable chemical shift, and with or without signal splitting.

7.1I Recap of ¹H NMR Data Interpretation

To summarize, there are four main types of information that can be obtained from the ^1H NMR spectrum:

1. Number of different kinds of hydrogens (number of signals).

2. Functional group and environments around these hydrogens (chemical shift).

3. Number of hydrogens in each environment (integration).

4. Number of hydrogens attached at the neighboring atom (signal splitting or multiplicity).

These are four very powerful pieces of information to help you determine the molecular structure of a compound. Given a formula of an unknown compound, which can be determined by other types of experimental techniques, NMR data can complete the picture in many cases, telling us how those atoms are connected and reveal the structure.

7.1J | Additional Real-Life Practical Situations with ^{1}H NMR Spectra

OVERLAPPING OR UNRESOLVED SIGNALS

The NMR spectra obtained from experimental samples are not always as perfectly behaved as those presented in a textbook. There may be hydrogens that are non-equivalent on the theoretical level, but in practice their environments and resonance energies are so similar that routine-use instruments cannot *resolve* the signals (i.e., cannot see separate peaks). In 1-heptanol (**Figure 7.21**), for example, the CH_2 units that are farther away from the functional group are not all resolved; instead, they produce a large peak of unclear multiplicity at 1.3 ppm that can be simply labeled "m" for multiplet. The integration, which is much greater than expected for a single CH_2 unit, is what indicates that several different signals are contained in this peak. In a simple structure like this, it may be unnecessary to resolve all of those CH_2 units. Spectra produced by higher field NMR instruments with more powerful magnets may be able to resolve all the signals of a very complicated structure, but for simpler compounds like 1-heptanol, the information gained may not be worth the expense of such an instrument.

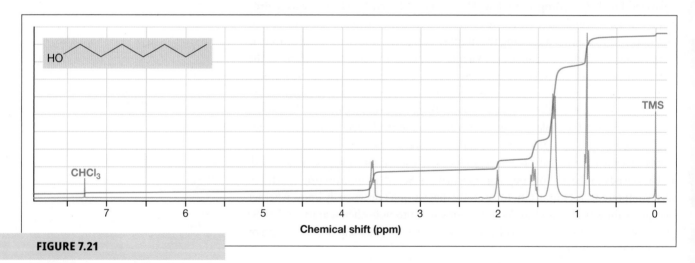

FIGURE 7.21

^{1}H NMR spectrum of 1-heptanol (300 MHz, CDCl$_3$). At first glance, it appears there are only five signals, and that seems inconsistent with the structure. However, the peak at 1.3 ppm actually contains four signals that are unresolved because their environments are very similar. The large integration at 1.3 ppm is the clue that should help alert you to situations like this.

© Sigma-Aldrich Co. LLC. Reproduced with permission from Merck KGaA, Darmstadt, Germany and/or its affiliates.

SOLVENT PEAKS AND CALIBRATION

Solvents used to prepare solutions for ^{1}H NMR spectroscopy are usually the same kinds of solvents normally found in a typical organic lab, except the hydrogens have been substituted with deuterium (^{2}H, or D). Deuterium is the isotope of hydrogen that contains one proton and one neutron in the nucleus. The deuterium nucleus absorbs radiofrequency energy outside the range of the ^{1}H spectrum, so these solvents do not interfere with ^{1}H signals from the sample. Even with great care to replace all the hydrogens with deuterium, there will be traces of hydrogen left in the solvents, so a small peak from the solvent usually appears in the spectrum. The chemical shifts of these peaks are listed in **Table 7.1**, and they are often used to calibrate the frequency (x axis, in ppm) of the spectrum (notice the peak for CHCl$_3$ in Figure 7.21). Alternatively, a trace of $(CH_3)_4$Si (tetramethylsilane,

TMS) can be added to the solvent. TMS produces a singlet at 0.00 ppm (see Figure 7.21), out of the range of peaks observed in typical organic compounds, and this also serves as a standard for frequency calibration.

Trace amounts of water are generally seen in experimental ^{1}H NMR spectra. This may arise from incomplete drying of a sample before preparing the NMR tube, incomplete drying of the NMR tube after cleaning, or an impurity in the source bottle of the NMR solvent. Deuterated dimethylsulfoxide (DMSO-d$_6$) is notoriously hygroscopic, meaning that it absorbs moisture from humid air, so it often gives a very large water peak in the ^{1}H NMR spectrum. More nonpolar solvents such as CDCl$_3$ or benzene-d$_6$ (deuterated benzene) may be kept fairly dry, but even transferring the solvent to the NMR tube can lead to traces of moisture due to evaporative cooling and condensation at the pipet tip. As a result, you should always expect to see a water peak, and you should neglect it in the interpretation of the spectrum. To help you identify it, the chemical shifts for the water peak in various deuterated solvents are also listed in Table 7.1.

TABLE 7.1

Common NMR Solvents and Their Peaks in ^{1}H NMR Spectra

DEUTERATED NMR SOLVENT	FORMULA	δ (ppm)	BOILING POINT (°C)
Acetone	$(CD_3)_2C{=}O$	2.05 2.84 (H$_2$O)	57
Benzene	C_6D_6	7.16 0.40 (H$_2$O)	80
Chloroform	$CDCl_3$	7.26 1.56 (H$_2$O)	61
Dimethyl sulfoxide	$(CD_3)_2S{=}O$	2.50 3.33 (H$_2$O)	189
Methanol	CD_3OD	3.31 4.87 (H$_2$O)	65
Water	D_2O	4.79	101

SECOND-ORDER NMR SPECTRA

Multiplet peak intensities in the sample spectra so far have been mostly very orderly, appearing in patterns that follow Pascal's triangle. As two sets of coupled hydrogens get closer in chemical shift, however, the individual peaks become distorted, so that it appears they are "leaning" toward each other; the inside edges of the set of two multiplets have a higher intensity. The series of spectra in **Figure 7.22** illustrates this effect for compounds with the general structure XCH$_2$CH$_2$Y, which would be expected to provide two triplets near 3 ppm. On the far left, the two triplets are about 0.6 ppm apart, and appear almost undistorted. From left to right, the chemical shifts of the two signals get closer together, leading to progressively more distorted multiplets. At the extreme case where X and Y are identical, there would be only one singlet because all four H's would be equivalent. This is an example of second-order effects, and the distortion caused by such effects can make it harder to determine multiplicity. On the other hand, the distortion is indicative of peaks that are coupled, so it can sometimes help in assigning which peaks belong to neighboring protons.

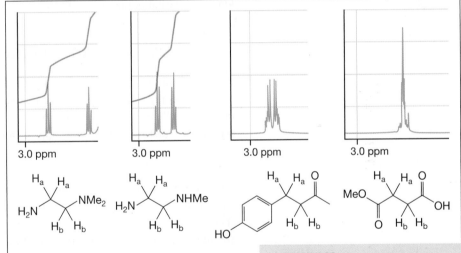

FIGURE 7.22

Distortion of multiplets in second order NMR spectra (300 MHz). From left to right, the chemical shifts of H$_a$ and H$_b$ become closer together, leading to a progressively more distorted appearance.

© Sigma-Aldrich Co. LLC. Reproduced with permission from Merck KGaA, Darmstadt, Germany and/or its affiliates.

LONGER-RANGE COUPLING

In certain cases, coupling may be observed between hydrogens that are more than three bonds apart. The most common examples are seen with aromatic hydrogens, where small four-bond coupling constants can sometimes be observed in the range

FIGURE 7.23

(a) Hydrogens in aromatic rings and a comparison of their longer-range four-bond coupling constants with the standard vicinal 3-bond coupling of ortho hydrogens.
(b) Representative examples of doublet of doublet peaks arising from aromatic hydrogens with both 3- and 4-bond couplings.

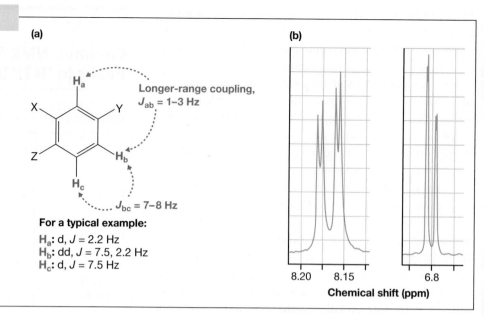

For a typical example:

H_a: d, J = 2.2 Hz
H_b: dd, J = 7.5, 2.2 Hz
H_c: d, J = 7.5 Hz

of J = 1–3 Hz (**Figure 7.23**). These small couplings are not always resolved in a spectrum. A peak that may look like a doublet with J = 7–8 Hz, when viewed more closely by "zooming in," may actually be a doublet of doublets with one very small coupling constant. Peak assignments can often be made without considering these longer-range couplings, although they can occasionally be useful.

7.1K Dealing with Spectra of Mixture Samples

It's very common to obtain NMR spectra of mixtures. In experimental samples, even of purified compounds, there may be small amounts of the solvents used in processing the sample by recrystallization or chromatography. If a reaction is incomplete, the leftover starting material may appear in the NMR spectrum.

IDENTIFYING EXTRA PEAKS

You can usually identify the small peaks corresponding to residual ^{1}H in the deuterated solvent used to prepare the sample (see Figure 7.21). These have well-known chemical shifts (Table 7.1) that allow for these peaks to be assigned. Solvent peaks are often very small relative to the rest of the peaks in the spectrum, and their integration may appear to be much less than one H. It would be absurd to have a structure with one-tenth of a hydrogen in it, so when you have attempted to reduce all the integration data to integers, and a peak is left as a noninteger, this is a clue that it is not part of the same structure. It is likely an impurity. To identify it, gather NMR data (from literature or prediction) about all the components you used in the experiment: starting materials, reagents, solvents, etc. Do the same for any by-products expected. For example, if you are conducting an E1 elimination reaction, you may be trying to make an alkene but there could be products from the leaving group, or from S_N1 as a side reaction.

QUANTIFYING RATIOS IN MIXTURE SAMPLES

Integration helps to identify which peaks may be assigned to different components of a mixture. It also can be used to calculate the ratio of two components. The integration of two peaks within the same compound allows us to determine the mole ratio of hydrogens involved in those peaks. Within the same structure, all peak integrals should be equal on a per hydrogen basis. Similarly, for two peaks from two different compounds, the integration will be equal to the mole ratio of the hydrogens involved. First, the integrations are converted to a per hydrogen basis by dividing each by the number of equivalent hydrogens giving rise to the peak. Then, the ratio of these equals the mole ratio of the compounds in the mixture.

Example

In an esterification reaction to make ethyl acetate from ethanol and acetic acid, the ^{1}H NMR spectrum of the product (**Figure 7.24**) shows two peaks in the region expected for a CH_3CH_2—O group at 4.1 ppm and 3.7 ppm, suggesting the product is actually a mixture of ethyl acetate and ethanol. Calculate the ratios of the components.

To solve this problem, locate a peak from each of the components, and divide the integral values for those peaks by the number of H's assigned to those peaks. This gives you the integrations adjusted to a per hydrogen basis. The mole ratio is then given by the ratio of these two numbers. If desired, the mole ratio can be converted to a weight percentage by converting each of the mole amounts to masses by dividing each by the molecular weight. The calculations are shown in **Figure 7.25**.

Using this technique to measure ratios of components requires that the two components are both known. If their structures are unknown, you will be missing key information on (a) how many hydrogens give rise to the peak, and (b) molecular weight. These cases present much more advanced problems, requiring further separation and experimentation to identify the unknown components.

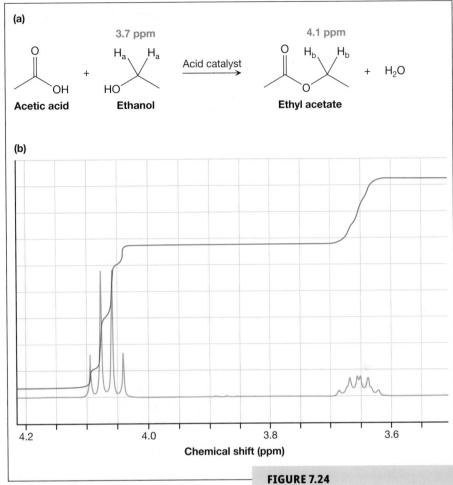

FIGURE 7.24

(a) The esterification reaction of acetic acid with ethanol. (b) A ^{1}H NMR spectrum showing a mixture of the product ethyl acetate and leftover ethanol. The integral values using arbitrary units are 260 units (4.07 ppm) and 120 units (3.65 ppm).

FIGURE 7.25

NMR integration data can be used to calculate weight percentage of components within a mixture.

Mole ratio = $\dfrac{\text{(integral b)/number of H}_b\text{'s}}{\text{(integral a)/number of H}_a\text{'s}}$ = $\dfrac{260}{120}$ = 2.17:1

This sample contains 2.17 moles of ethyl acetate for each mole of ethanol.

Because both integrals are measured on 2H peaks, the "number of H" terms cancel.

Mass ratio = $\dfrac{2.17 \text{ mol EtOAc} \cdot 88 \text{ g/mol}}{1 \text{ mol EtOH} \cdot 46 \text{ g/mol}}$ = $\dfrac{191 \text{ g}}{46 \text{ g}}$ = 4.15:1

This sample contains 4.15 g of ethyl acetate for each gram of ethanol.

Weight % ethyl acetate = $\dfrac{\text{mass EtOAc}}{\text{mass EtOAc + mass EtOH}}$ × 100 = $\dfrac{4.15 \text{ g}}{5.15 \text{ g}}$ × 100 = 80.6%

This sample is 80.6% ethyl acetate by weight.

SOLVENT CHOICE

Deuterated solvents are generally chosen because deuterium (^{2}H, or D) does not give peaks of its own in the ^{1}H NMR spectrum. A wide variety of deuterated solvents are commercially available, including deuterated dimethylsulfoxide (DMSO-d_6, CD_3SOCD_3), deuteroacetone (CD_3COCD_3), deuterobenzene (C_6D_6) and deutero-chloroform ($CDCl_3$). $CDCl_3$ is one of the most widely used solvents in NMR because it dissolves a wide range of compounds. A small peak is always visible from the solvent because not all solvent molecules have had all the hydrogens replaced with deuterium. For example, very small amounts of $CHCl_3$ in the $CDCl_3$ will always give a small peak at 7.26 ppm.

Needing to recover your sample from the NMR tube may influence the solvent choice, especially for very small-scale work, where the entire sample may be used for NMR acquisition. Since ^{1}H NMR is nondestructive, the sample can generally be recovered and used for further analysis or reactions. In such cases, choose your solvent carefully! Usually removing the solvent is as simple as transferring the sample to a round-bottomed flask and swirling it while applying vacuum. A rotary evaporator may also be used. Evaporation works very well for solvents with boiling points around 80°C or below (see the boiling point data in Table 7.1). Note that DMSO has a high boiling point (189°C), so it may be very difficult to recover your compound if it has been dissolved in DMSO. Instead of using evaporation, a more complicated liquid–liquid extraction may be needed. When possible, avoid DMSO if you anticipate having to recover the sample.

CALIBRATION

Chemical shifts of protons are reported in parts per million (ppm) relative to the position of a sharp peak given by protons in tetramethylsilane (TMS), which is set at 0.00 ppm when the instrument is calibrated. Stock solutions of NMR solvents sometimes contain 0.05–3% TMS, so if a spectrum shows a peak at 0.00 ppm, it may be TMS. The spectrum can also be calibrated to the solvent peak; chemical shifts (in ppm from TMS) are known for all of the common solvents.

SAMPLE SPINNING

In order to average the magnetic fields produced by the spectrometer in the sample, the sample is spun inside the instrument at about 20 Hz, or 20 revolutions per second, while acquiring the spectrum. This produces a vortex in the tube which can interfere with acquisition, giving erratic, nonreproducible spectra. The effect of the vortex is minimized if the sample tube is filled to a minimum depth of about 2.5–3.0 cm, which keeps the vortex above the area where the radiofrequency receiver coil aligns with the sample tube (**Figure 7.26a**).

SAMPLE PREPARATION

After checking that a sample is soluble in the solvent of choice, a typical ^{1}H NMR sample is prepared by placing 5–20 mg of sample into a 5-mm-diameter tube. *CAUTION: NMR sample tubes are expensive and delicate!* The NMR solvent is then added to bring the level up to about 2.5–3.0 cm depth in the bottom of the tube (about 0.5–0.7 mL). If a sample contains some insoluble materials, solids floating around in the NMR tube will cause a very erratic spectrum. If necessary, solids should be removed by filtering the sample through a small plug of cotton in a Pasteur pipette. If you use the desired NMR solvent during this filtration, the filtration can be done directly into the NMR tube (**Figure 7.26b**).

 NMR Sample Preparation

(a)

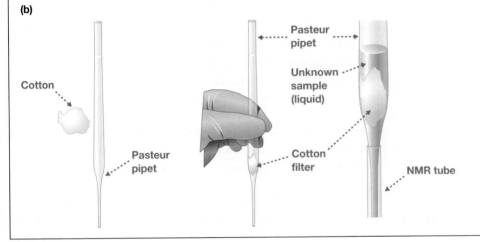

FIGURE 7.26

(a) The sample solution should be a minimum of 2.5–3.0 cm depth in the NMR tube. (b) Filtering a solution into an NMR tube.

(b)

7.1M Interpretation of ¹H NMR Data: Characterization of a Known Compound

1. Draw the structure of the compound that you are trying to characterize. Count the total number of protons, and identify those that are equivalent.

2. Determine roughly what you would expect to see (number of peaks, chemical shift, integration, and multiplicity) for the structure you've drawn. Consider also any special circumstances, such as proton exchange or overlapping peaks, that could affect the appearance of the spectrum.

3. Identify solvent and calibration peaks (e.g., solvent, H_2O, TMS) in your spectrum and omit them from the analysis.

4. Convert peak integration data to an integer hydrogen count.

5. Determine the chemical shift of each peak in your spectrum.

6. Identify the multiplicity (splitting) of each resonance in the spectrum. You may also want to determine coupling constants (*J* values) for each set of resonances. This information can be used to determine which groups of protons are adjacent to each other in the structure.

7. Use all of this information to account for the resonances you see in the NMR spectrum. It may be helpful to consult a correlation table of chemical shifts.

8. If the spectrum of this compound is available in the chemical literature, compare your data to the literature data.

7.1N Reporting NMR Data

A single NMR spectrum provides a great deal of information that must be clearly interpreted and reported in an organized manner. Include a drawing of the compound (number the atoms as necessary) in the Results and Discussion section of your lab report to facilitate identification and interpretation of specific resonances. In addition to the spectrum itself, a table is also useful to convey the information clearly to the reader. Prepare a table, and refer back to the table as needed, to clarify the discussion in the body of the report.

REPORTING NMR DATA IN A TABLE

When displaying ^{1}H NMR data in a report or journal article, each peak is presented along with its assignment to a particular set of hydrogens in the structure, as well as the chemical shift (in ppm), the integration (the number of protons), and the multiplicity (s, d, t, q, m, dd, etc.) as shown in **Table 7.2**. When sufficient data are available to determine the coupling constants, these may be provided along with the multiplicity.

All hydrogens that can be observed should be accounted for in the table. Occasionally, hydrogen bonding protons on heteroatoms (usually O, N) may be very broad or may not appear in the spectrum at all. This can be noted in the body of the report.

If there are other peaks (aside from the sample, TMS, and NMR solvent peaks), comment about the meaning of this in the body of the report. For example, if a sample was obtained by evaporation of ethyl acetate, then ethyl acetate peaks in the spectrum would indicate that the sample was not completely free of the solvent. While it may not be possible to assign every set of protons to a specific resonance (they may be grouped together, overlapping, or difficult to distinguish), every proton should be accounted for in the integration. The integration must be reported in whole numbers, and it should agree with the number present in the compound being evaluated. If these values do not agree, you must explain why.

TABLE 7.2

Peak Assignments and ^{1}H NMR Data for *n*-Butyl Vinyl Ether (See Figure 7.20)

CHEMICAL SHIFT	INTEGRATION	MULTIPLICITY	ASSIGNMENT
0.9 ppm	3	t (J_{fg} = 7 Hz)	H_g
1.4 ppm	2	m (apparent sextet, J = 7 Hz)	H_f
1.6 ppm	2	m (apparent quintet, J = 7 Hz)	H_e
3.7 ppm	2	t (J_{de} = 7 Hz)	H_d
3.9 ppm	1	dd (J_{ac} = 6.8 Hz, J_{bc} = 1.8 Hz)	H_c
4.2 ppm	1	dd (J_{ab} = 14.3 Hz, J_{bc} = 1.8 Hz)	H_b
6.4 ppm	1	dd (J_{ab} = 14.3 Hz, J_{ac} = 6.8 Hz)	H_a

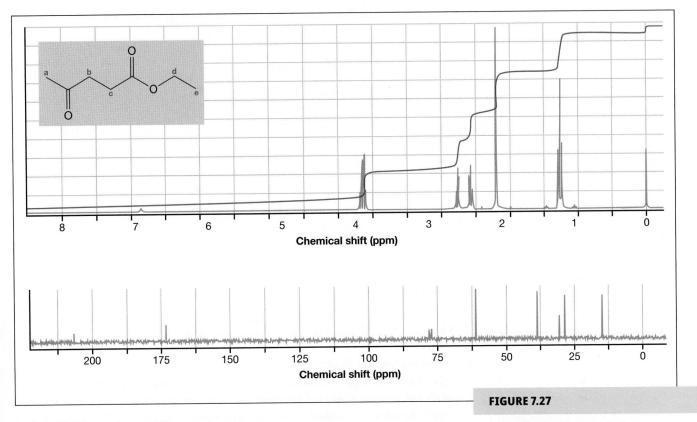

FIGURE 7.27

^{1}H and ^{13}C NMR spectra of ethyl levulinate. The ^{13}C NMR spectrum (bottom) has a solvent peak (CDCl$_3$) appearing as three peaks centered at 77 ppm.

© Sigma-Aldrich Co. LLC. Reproduced with permission from Merck KGaA, Darmstadt, Germany and/or its affiliates.

Example

The ^{1}H and ^{13}C NMR spectra (300 MHz, CDCl$_3$) of ethyl levulinate are presented in **Figure 7.27**. **Table 7.3** and **Table 7.4** show how these NMR data should be tabulated in your report.

TABLE 7.3

^{1}H NMR Data for Ethyl Levulinate

CHEMICAL SHIFT	INTEGRATION	MULTIPLICITY	ASSIGNMENT
1.25 ppm	3H	t	—CH$_3$ (e)
2.19 ppm	3H	s	—(C=O)CH$_3$ (a)
2.55 ppm	2H	t	—CH$_2$(C=O) (b or c)
2.74 ppm	2H	t	—CH$_2$(C=O) (b or c)
4.15 ppm	2H	q	—OCH$_2$ (d)

TABLE 7.4

^{13}C NMR Data for Ethyl Levulinate

CHEMICAL SHIFT	ASSIGNMENT
14.2 ppm	—CH$_3$ (e)
28.1 ppm	—CH$_2$(C=O) (c)
29.8 ppm	—(C=O)CH$_3$ (a)
38.0 ppm	—CH$_2$(C=O) (b)
60.5 ppm	—OCH$_2$ (d)
172.7 ppm	C=O, ester
206.6 ppm	C=O, ketone

Example 3

An unknown aldehyde has the molecular formula C_4H_8O and exhibits the ^1H NMR spectrum in **Figure 7.28**. What is its structure?

FIGURE 7.28

^1H NMR spectrum of an aldehyde with the formula C_4H_8O (300 MHz, $CDCl_3$).

© Sigma-Aldrich Co. LLC. Reproduced with permission from Merck KGaA, Darmstadt, Germany and/or its affiliates.

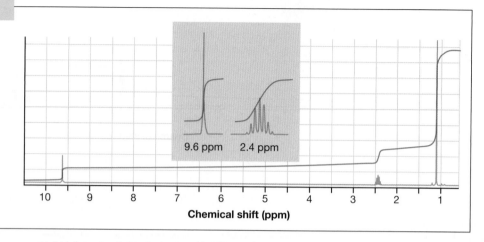

9.6 ppm   2.4 ppm

Chemical shift (ppm)

Problem-Solving Strategy

The overall strategy is outlined in **Figure 7.29**. First, draw all isomers of aldehydes with the formula C_4H_8O. Each structure will serve as a hypothesis, and you will apply the scientific method to test each hypothesis, using the NMR evidence to systematically disprove all but one. To do this, predict the number of signals, chemical shift, integration, and multiplicity for each structure. Then, compare these predictions with the observed spectrum. If there is any inconsistency between prediction

FIGURE 7.29

Problem-solving strategy for identifying an unknown. Here the unknown is an aldehyde of formula C_4H_8O having the spectrum shown in Figure 7.28. In this case only the number of signals was needed to identify the structure by process of elimination.

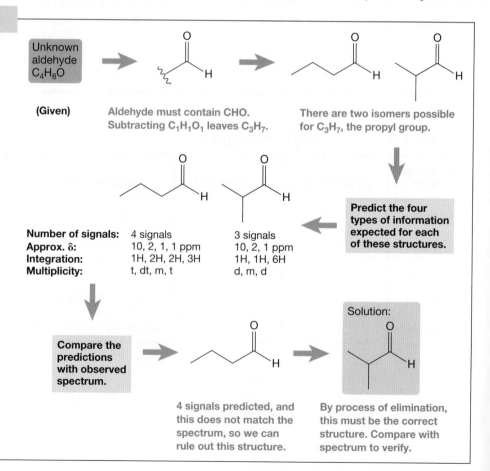

(Given)

Aldehyde must contain CHO. Subtracting $C_1H_1O_1$ leaves C_3H_7.

There are two isomers possible for C_3H_7, the propyl group.

Predict the four types of information expected for each of these structures.

Number of signals:	4 signals	3 signals
Approx. δ:	10, 2, 1, 1 ppm	10, 2, 1 ppm
Integration:	1H, 2H, 2H, 3H	1H, 1H, 6H
Multiplicity:	t, dt, m, t	d, m, d

Compare the predictions with observed spectrum.

4 signals predicted, and this does not match the spectrum, so we can rule out this structure.

Solution:

By process of elimination, this must be the correct structure. Compare with spectrum to verify.

and observation, that structure can be ruled out (disproven). When you are left with only one structure that cannot be ruled out, you have solved the problem. Disproving a hypothesis requires only one inconsistency between prediction and observation. Thus, you may not need to use all four types of information that the NMR spectrum provides.

Example 4

An unknown alcohol has the molecular formula $C_8H_{10}O$ and exhibits the ^1H NMR spectrum in **Figure 7.30**. What is its structure?

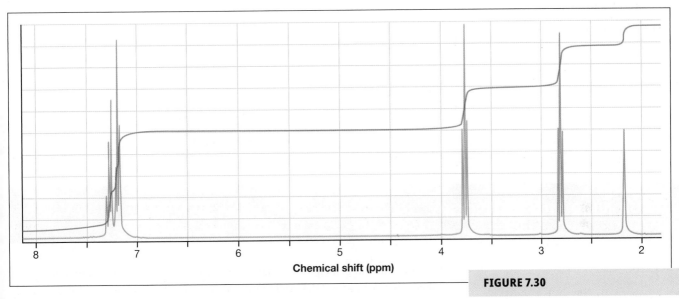

Chemical shift (ppm)

FIGURE 7.30

^1H NMR spectrum of a compound of an alcohol with the formula $C_8H_{10}O$ (300 MHz, CDCl$_3$).

© Sigma-Aldrich Co. LLC. Reproduced with permission from Merck KGaA, Darmstadt, Germany and/or its affiliates.

Example 5

An unknown ester has the molecular formula $C_5H_8O_2$ and exhibits the ^1H NMR spectrum in **Figure 7.31**. What is its structure?

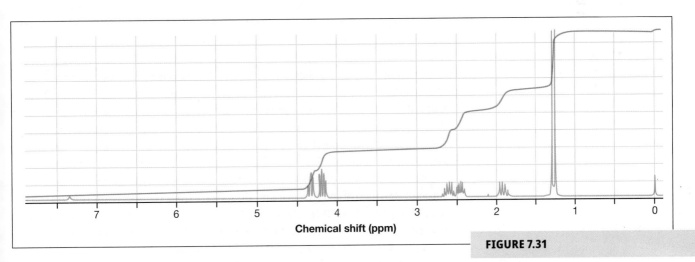

Chemical shift (ppm)

FIGURE 7.31

^1H NMR spectrum of an ester with the formula $C_5H_8O_2$ (300 MHz, CDCl$_3$).

© Sigma-Aldrich Co. LLC. Reproduced with permission from Merck KGaA, Darmstadt, Germany and/or its affiliates.

Example 6

An unknown compound, insoluble in water and soluble in 10% aqueous HCl, has the molecular formula $C_6H_8N_2$ and exhibits the ^1H NMR spectrum in **Figure 7.32**. What is its structure?

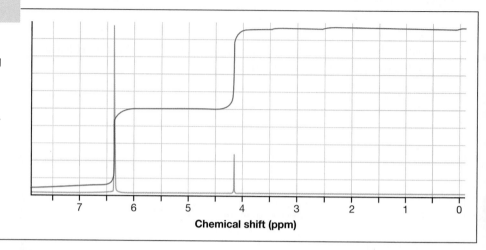

Chemical shift (ppm)

7.2 ^{13}C NMR SPECTROSCOPY

The fundamental principles you have learned so far about nuclear spin, energy differences between spin states, and the absorption of radiofrequency energy also apply to ^{13}C NMR spectroscopy. However, there are some key differences.

First, absorption of energy by ^1H and ^{13}C nuclei in a magnetic field occurs in different areas of the radiofrequency range, so ^1H and ^{13}C peaks do not appear on the same spectrum. Compared to ^1H NMR, the peaks in ^{13}C NMR appear over a much wider δ range, generally from –10 ppm to 220 ppm.

Second, there is a big difference in the signal strength between ^1H and ^{13}C NMR spectroscopy. ^1H is the most abundant isotope of hydrogen, whereas ^{13}C is a minor isotope of carbon. Around 99% of carbon atoms are ^{12}C, which has nuclear spin $I = 0$ and is invisible to NMR spectroscopy. An isotope with one more neutron, ^{13}C, has nuclear spin $I = 1/2$, and therefore gives an NMR signal, but it is only present as 1.1% of the carbons in a sample. To get enough signal strength for a useful ^{13}C NMR spectrum, the NMR instrument acquires many copies of the ^{13}C radiofrequency emission data by scanning repeatedly over a period of time. These data are added together by the instrument's computer. In an individual scan, the signal is accompanied by a lot of random noise (**Figure 7.33a**). When many scans are added together, the signals grow in intensity through constructive interference, while the random noise is averaged out through destructive interference and disappears from the spectrum (**Figure 7.33b**). So, although a ^1H NMR spectrum can be acquired in a couple of minutes, the time needed to acquire a ^{13}C NMR spectrum is typically in the range of 30–60 minutes. Much more time may be required if the sample concentration is low.

Third, ^1H NMR differs from ^{13}C NMR in coupling and multiplicity. Among the 1.1% of carbons that are ^{13}C, there is only a 1.1% chance of having another ^{13}C nucleus as its neighbor. Thus, there is no significant coupling between carbons. When a ^{13}C nucleus is attached to another carbon, it is almost always the more abundant ^{12}C isotope that has no nuclear spin. On the other hand, nearby ^1H nuclei may be coupled with ^{13}C, and the coupling constants are large enough that the signal splittings can cause the peaks to overlap throughout the spectrum. This makes it very difficult to

FIGURE 7.33

The ^{13}C NMR spectrum of 1-chloropropane. Peaks at 0 ppm and 77 ppm are from the solvent and the calibration standard.

Karty, J. *Organic Chemistry: Principles and Mechanisms*, 3rd ed.; W. W. Norton: New York, 2022; p 854.

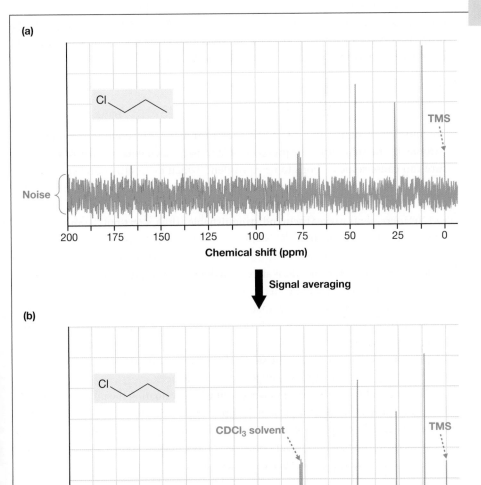

(a)

Noise

TMS

Chemical shift (ppm)

Signal averaging

(b)

CDCl$_3$ solvent

TMS

Chemical shift (ppm)

get useful information from the spectrum. To avoid this, ^{13}C NMR spectra are usually acquired with the ^{1}H and ^{13}C nuclei decoupled. This is accomplished by applying radiofrequency irradiation in the frequency range of the ^{1}H nuclei while simultaneously acquiring data in the ^{13}C range. All of the peaks in the resulting ^{13}C NMR spectrum appear as sharp singlets. Another benefit of decoupling is the improved signal-to-noise ratio when the peaks are condensed into singlets.

A fourth difference has to do with the intensity of ^{13}C signals within a spectrum of a given compound. Within a molecule, some nuclei relax from the excited β state to the lower energy α state more rapidly than others. If a spectrum was based on only one scan, this wouldn't be a problem. However, if another scan takes place before one type of nucleus has relaxed to its original α and β population distribution, then that nucleus gives less of a response in the instrument in the next scan, and its signal intensity will be weaker. In a spectrum that requires many scans, the differences are magnified in the ^{13}C NMR spectrum, and signal intensity from one carbon to the next can be vastly different. The signal intensities are related to how many hydrogens are attached at the carbons. Those carbons with no hydrogens attached, such as the 3° carbon in *tert*-butyl alcohol or the carbonyl carbon in a ketone, tend to have greatly diminished signal intensity relative to a CH$_2$ or CH$_3$. This can be helpful in assigning peaks in the ^{13}C NMR spectrum. Quantifying these intensity differences is generally not useful, so integration is not commonly employed in ^{13}C NMR.

Because of the four differences outlined here, multiplicity and integration are not generally included in the interpretation of ^{13}C NMR spectra. Fortunately, the other two types of information—the number of signals and chemical shift—can offer useful insight for organic structure determination.

7.2A Number of Signals

The number of signals in a ^{13}C NMR spectrum may correspond to the number of carbons in the structure. The presence of symmetry elements in parts of the structure can cause some carbons to be equivalent, which lowers the number of signals, just as it does in ^{1}H NMR. Some examples are given in **Figure 7.34**. For instance, the three methyls of a *tert*-butyl group are equivalent because rotation of a C—C bond in 120° increments makes them indistinguishable. Thus, a *tert*-butyl group will contribute only two signals to a ^{13}C NMR spectrum rather than four. A benzene ring can have a plane of symmetry cutting across it, reducing the number of signals. In a monosubstituted benzene derivative, ethylbenzene (Figure 7.34) for example, the two carbons labeled c are equivalent, as are the two carbons labeled b.

As with ^{1}H NMR, if you are unsure whether two carbons are equivalent, draw the structure twice to replace each of the two carbons with another atom label, and analyze whether these two new structures are rendered the same by rotating, flipping, and so on. If they are the same, then the two carbons will be equivalent in the ^{13}C NMR spectrum and will appear as a single peak.

Solvent peaks must be taken into account when evaluating the number of signals. For example, CDCl$_3$ gives a peak at 77 ppm (Figure 7.33b) that appears as a 1:1:1 triplet because of coupling of the carbon with deuterium. This relative intensity of the peaks within the triplet occurs because D has a nuclear spin of 1, with three spin states, −1, 0, +1, unlike the H nucleus, which has only two spin states, −1/2 and +1/2.

FIGURE 7.34

Examples illustrating the number of nonequivalent carbons in a variety of structures. Nonequivalent carbons are labeled in red letters. Blue letters indicate carbons that are equivalent to those with the same letter labeled in red.

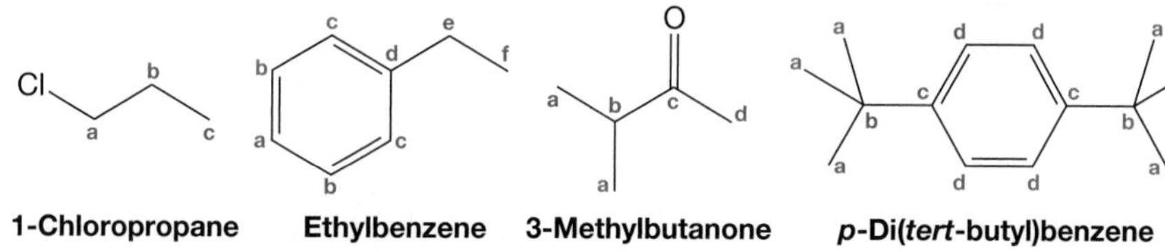

1-Chloropropane   Ethylbenzene   3-Methylbutanone   *p*-Di(*tert*-butyl)benzene

7.2B Chemical Shift

Chemical shift in ^{13}C NMR correlates to shielding and deshielding, as in ^{1}H NMR, and rough estimates of chemical shifts of various functional groups are summarized in **Figure 7.35**. The effects of electronegativity on deshielding can be observed in the comparisons of chemical shifts of alcohols (60–80 ppm) to amines (30–60 ppm). Oxygen is more electronegative than nitrogen, and oxygen shifts the carbon absorbance farther downfield. These deshielding effects are additive, as in ^{1}H NMR, so two oxygens combine to shift acetal carbons even farther downfield (80–100 ppm). Hybridization and magnetic anisotropy combine with electronegativity to impact the chemical shifts of π systems such as alkenes, aromatic rings, and carbonyl compounds.

Certain areas of the ^{13}C NMR spectrum are exceptionally useful in diagnosing functional groups. For example, the carbonyl carbon of aldehydes and ketones (190–220 ppm) is readily differentiated from that of carboxylic acids and their

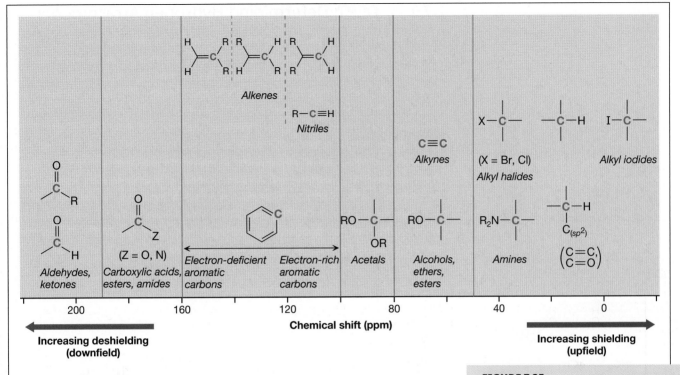

FIGURE 7.35

Typical ^{13}C NMR chemical shift ranges for some common functional groups.

derivatives (160–180 ppm, includes esters, amides, acyl halides, and anhydrides), complementing the information available from ^{1}H NMR, which does not directly display a peak for the C=O.

Drawing resonance structures can be very helpful in assigning signals to carbons within conjugated π systems such as unsaturated carbonyl compounds. For 2-cyclohexen-1-one (**Figure 7.36**), carbons at positions 2 and 3, though both alkene carbons, have a dramatic 21 ppm difference in their chemical shifts. The partial charges in the resonance hybrid reveal why this is so. There is greater electron density at C-2, and more shielding, because of its partial negative charge; C-2 appears at 129.8 ppm. On the other hand, C-3 has a partial positive charge, and this electron-deficient character causes it to be deshielded in comparison with C-2; C-3 appears at 150.9 ppm. Similar effects can be seen in a variety of substituted π systems with significant contributions of charged resonance structures, including benzene derivatives.

FIGURE 7.36

Effects of charge distribution on the chemical shift in the π system of 2-cyclohexen-1-one.

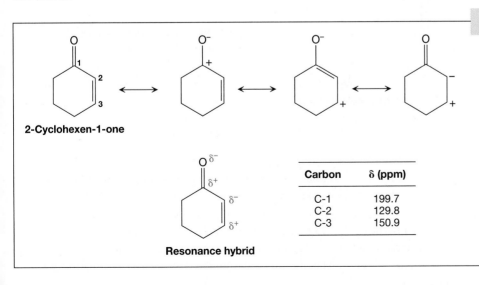

2-Cyclohexen-1-one

Resonance hybrid

Carbon	δ (ppm)
C-1	199.7
C-2	129.8
C-3	150.9

DEPT: Determining Hydrogen Attachments

All of the peaks in a typical ^{13}C NMR spectrum are singlets because the ^{13}C nuclei are decoupled from the ^{1}H nuclei. It can be difficult, then, to assign some peaks to specific carbons in the structure. A helpful tool for this job is *distortionless enhancement by polarization transfer* (DEPT) NMR spectroscopy. The most commonly used format of DEPT NMR is known as DEPT-135. This is a variant of the ^{13}C NMR experiment, and carbons show up as singlets as usual, but are phased differently so that peaks appear above or below the baseline depending on how many hydrogens are attached. Carbons with an odd number of hydrogens attached (CH or CH$_3$) appear above the baseline, and those with an even number of hydrogens (CH$_2$) appear below the baseline. Carbons with no hydrogens attached do not appear in the spectrum. All of the chemical shifts in the DEPT spectrum match those of the ^{13}C NMR spectrum. So, by comparing the standard ^{13}C NMR spectrum with the DEPT-135 spectrum, you can identify the types of substitution that are associated with each peak.

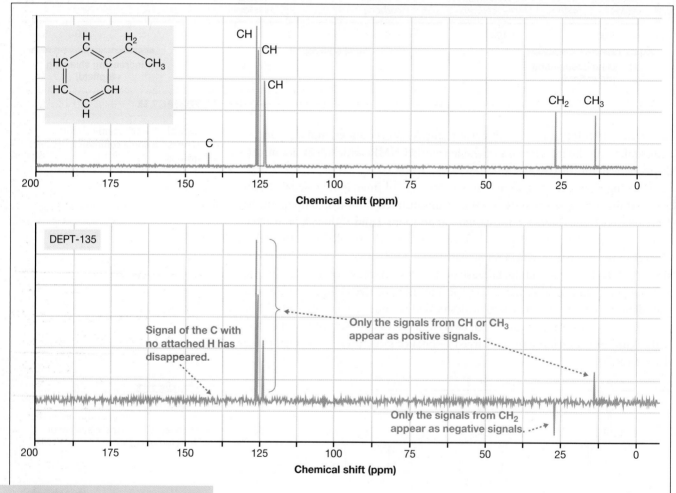

FIGURE 7.37

^{13}C NMR and DEPT-135 NMR spectra of ethylbenzene.

Karty, J. *Organic Chemistry: Principles and Mechanisms*, 2nd ed.; W. W. Norton: New York, 2018; p 810. Karty, J. *Organic Chemistry: Principles and Mechanisms*, 3rd ed.; W. W. Norton: New York, 2022; p 858.

For example, the ^{13}C NMR spectrum of ethylbenzene (**Figure 7.37**) has four signals in the 120–140 ppm range and two in the 10–30 ppm range. Based on the chemical shifts, these can be assigned to the aromatic carbons and the ethyl group, respectively. But how can we determine which of the ethyl group signals belongs to the CH$_2$? In the DEPT-135 spectrum, there is only one peak below the baseline, so that peak corresponds to the CH$_2$ group. Moreover, the ring carbon at the point of attachment of the ethyl group has disappeared from the DEPT-135 spectrum, because it has no hydrogens attached.

8

LEARNING OBJECTIVES

- Describe how mass spectrometry can be used to determine a molecular formula.

- Predict fragmentation processes associated with structural features of organic compounds.

- Distinguish among possible molecular formulas, using high-resolution mass spectrometry.

- Use combustion analysis data to determine molecular formulas.

- Evaluate purity of an organic compound, using combustion analysis.

- Identify degrees of unsaturation from a molecular formula and apply it to structure determination.

- Use molecular mass and formula for structure determination in combination with spectroscopy.

Determination of Molecular Mass and Formula

MASS SPECTROMETRY INSTRUMENTS

Mass spectrometers are traditionally cost- and space-intensive tools for identifying organic compounds, but newer instruments continue to become more widely accessible.

RICHARD NOWITZ/Science Source.

I n the identification of an organic compound's structure, a critical piece of information is its molecular formula, or the ratio of the elements present in the compound. Two example formulas are C_6H_{14} for hexane and $C_{21}H_{22}N_2O_2$ for strychnine (**Figure 8.1**), a natural alkaloid known for its pesticide properties.

FIGURE 8.1

Selected structures and molecular formulas.

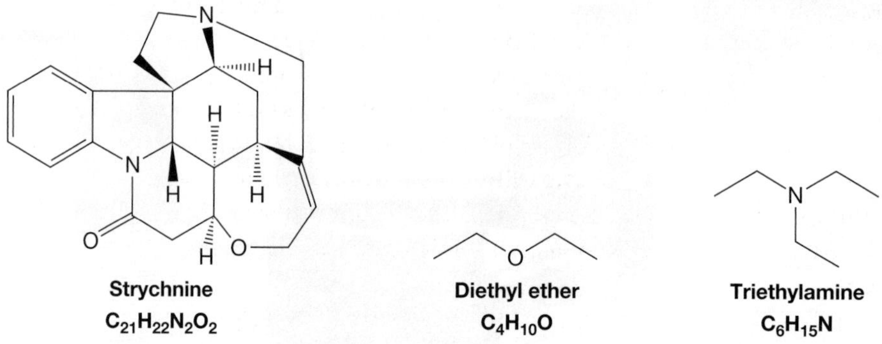

Strychnine
$C_{21}H_{22}N_2O_2$

Diethyl ether
$C_4H_{10}O$

Triethylamine
$C_6H_{15}N$

What can we do with this molecular formula? Given some sample history, such as the source of the compound and the conditions of its reactions or processing, there may be a short list of possible compound structures, and determining the molecular formula can often distinguish between those possibilities. It should be noted that isomers will have the same molecular formula, and cannot be distinguished in this manner. For example, ethanol and dimethyl ether each have the formula C_2H_6O, but have different connectivity of the atoms, and the enantiomers (R)-2-butanol and (S)-2-butanol are different configurations of the same formula and connectivity. Neither of these pairs can be distinguished by molecular formula. Still, there are many situations where the molecular formula will provide a key to identifying the compound in question.

Without any sample history, some additional data from spectroscopy (for example, IR and NMR) would generally need to accompany the formula in order to provide enough information for a confident structural assignment. In some cases, the formula is the deciding factor. Consider two colorless liquids that exhibit IR and NMR spectra that are quite similar: diethyl ether and triethylamine (Figure 8.1). Outside of the fingerprint region, neither has obvious **diagnostic peaks** in the IR spectrum. In their ^{1}H NMR spectra, both would exhibit the readily recognized pattern of an ethyl group attached to an electronegative atom—an upfield triplet and a downfield quartet. But the molecular formulas of these two compounds are very different; one contains nitrogen and the other oxygen. Determining the molecular formula experimentally would easily distinguish between these two compounds.

In this chapter we will examine two methods to determine a molecular formula experimentally. First, we will see how ions may be generated from organic molecules in the gas phase, how the masses of these ions are measured via mass spectrometry, and how the data are correlated to molecular formula. Second, we will learn about a technique known as combustion analysis, in which the organic compound is burned under very controlled conditions that allow the measurement of the percent by weight of carbon, hydrogen, and nitrogen in the sample. This too can be correlated to molecular formula. With the molecular formula in hand, the chemist is armed with the information needed to either confirm a structure or propose a short list of possible structures for further evaluation.

diagnostic peaks >>
In spectroscopy, peaks that are characteristic of a compound and can be used for its identification or quantification. In comparing reactants and products, the diagnostic peaks are usually those that are closely associated with the functional group where the reaction occurred.

8.1A ## Introduction: Separation of Ions by Mass

Organic compounds can be converted into positively or negatively charged ions in the gas phase. When these charged particles are generated in a vacuum chamber, their travel through the chamber can be accelerated and directed by electrostatic and magnetic fields, causing the ions to pass through the chamber toward a detector. The **relative abundance** of different ions that reach the detector, sorted by mass, can be recorded in a spectrum. The ion current or quantity of ions detected is plotted on the y axis versus the mass-to-charge ratio (m/z) of the ions on the x axis. Although an ion may have a net charge of two or more, either positive or negative, in typical organic chemistry applications we will consider particles that have a charge $z = +1$. In this case the mass-to-charge ratio m/z simplifies to just mass. Although we will refer to mass of ions in this chapter, it is good to be aware that *the mass spectrometer is actually measuring m/z.*

At the core of the mass spectrometry instrument is an analyzer device that separates ions as a function of m/z, so that ions of differing mass are detected separately and their relative abundance can be recorded in a **mass spectrum (Figure 8.2b)**. There are several different types of analyzers, including magnetic sector, quadrupole, ion trap, and time-of-flight (TOF). As this chapter is focused on interpretation of mass spectra for structure determination, we will not include details on the physics

<< **relative abundance**
In mass spectrometry, the ratio of the intensity of a smaller peak to the intensity of the largest peak, usually expressed as a percentage.

<< **mass spectrum**
A plot of masses of molecular ions (or more precisely, their mass-to-charge ratio m/z) versus their relative abundance. Such a plot is useful in identification and quantification of organic compounds.

FIGURE 8.2

(a) A mass spectrometer, circa 1989, and (b) an example of a mass spectrum.

Gado Images/Alamy Stock Photo. Karty, J. *Organic Chemistry: Principles and Mechanisms*, 3rd ed.; W. W. Norton: New York, 2022; p 742.

(a)

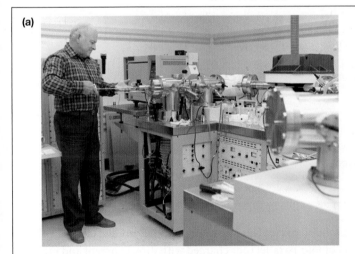

(b)

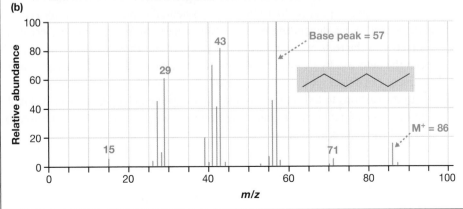

behind all of these analyzers. However, some explanation of magnetic sector and quadrupole analyzers will aid in understanding the principles of mass spectrometry.

The *magnetic sector analyzer* (**Figure 8.3a**) uses electrostatic fields to focus and accelerate ions into a specific flight path, or ion beam. A magnetic field then deflects the ion beam, generating a semicircular flight path that passes through a curved tube leading to a detector that measures the ion current. This deflection is related to the m/z and velocity of the particles as well as the strength of the applied magnetic field. If those variables yield an ion beam that matches the curvature of the tube, the ions reach the detector, where an ion current is registered. By systematically varying electrostatic or magnetic fields, the instrument can scan the ion current across a range of m/z. The instrument then processes the signal into a plot of ion current versus m/z to yield the mass spectrum.

FIGURE 8.3

(a) Schematic of electron impact mass spectrometer with magnetic sector analyzer, and (b) a quadrupole analyzer, in which the parallel charged rods cause ions to travel in helical paths. In both cases, the paths vary so that ions of specific m/z are detected separately.

Karty, J. *Organic Chemistry: Principles and Mechanisms*, 3rd ed.; W. W. Norton: New York, 2022; p 740. Taouatas, N.; Drugan, M. M.; Heck, A. J. R.; Mohammed, S. Straightforward Ladder Sequencing of Peptides Using a Lys-N Metalloendopeptidase. *Nat Methods* © **2008**, *5*, 405–407. https://doi.org /10.1038/nmeth.1204, reprinted by permission from Springer Nature.

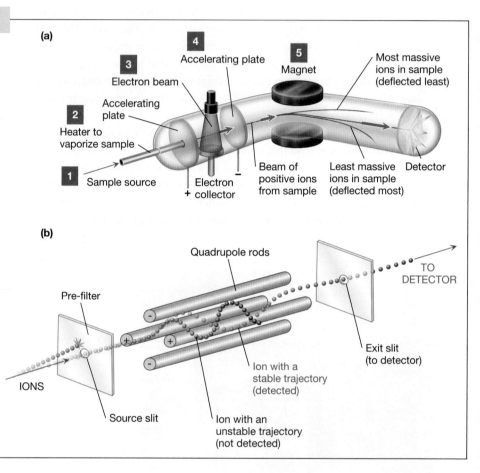

A *quadrupole analyzer* passes the ion beam through an electric field that is created between a set of four parallel rods charged with direct current and radiofrequency voltages (**Figure 8.3b**). In this environment, the ion beam takes on a helical pattern that varies depending on this electric field. Most ions will be lost upon contact with the rods or escape from the analyzer, but those of a specific m/z will reach the detector. By varying the electric field in the quadrupole analyzer, the instrument can detect ions across a range of m/z and obtain a mass spectrum.

In the sections to follow, we will first examine how the organic chemist can obtain this type of data from a sample, and then we will discuss how to interpret the data and correlate it to structure.

Ionization Techniques

There are many different ways to generate ions from a sample of organic material. In early mass spectrometers, ions were generated by electron ionization (EI), also known as electron impact, by subjecting the sample to an electron beam. In collisions with the organic molecules, the high-energy electrons can knock an electron out of the bonding orbitals within the organic molecule, forming an ion known as a molecular ion (M^+):

$$M + e^- \rightarrow M^+ + 2\,e^-$$

The high energies involved usually cause the ions to fragment into many smaller ions and radicals; the ones that retain a charge continue on through the instrument. The masses that are detected are those of the fragments derived from the molecular ion. For an unknown compound, the mental exercise of putting these fragments back together can enable the chemist to derive structural information.

To better preserve the intact molecular ion, a softer ionization technique known as chemical ionization (CI) can be advantageous. Here, electron ionization takes place first on a reagent gas that then causes ionization of the sample. With methane as the reagent gas, CH_4^+ is formed, and this reacts with another methane molecule, transferring a hydrogen atom to give CH_5^+. This in turn transfers a proton to an analyte molecule, leading to an ion of formula $[M+H]^+$:

$$M + CH_5^+ \rightarrow [M+H]^+ + CH_4$$

Although this ion contains an extra hydrogen and is one mass unit more than M^+, it is still called the molecular ion because no fragmentation has occurred. Numerous reagent gases, in addition to methane, can be used for CI, including isobutane, ammonia, water, and others.

Electrospray ionization (ESI) is a method common to many newer instruments that does an outstanding job of retaining the molecular ion. Here, a solution of the sample passes through a metal capillary, causing charged **aerosol** particles to form. As these travel into the vacuum chamber, the solvent is removed, leaving behind charged particles of the sample that fly through the analyzer. As these are generated from solutions containing aqueous sodium ions, the molecular ions generally contain an extra proton ($[M+H]^+$) or sodium ion ($[M+Na]^+$). A mass spectrum using ESI often produces only molecular ions, so it is very useful for more complex samples that would give hopelessly complicated spectra if fragmentation occurred. As such, it is very useful for the analysis of mixtures, including biological analytes containing large proteins and other sensitive biopolymers. This type of instrument is often directly interfaced with high-performance liquid chromatography (HPLC); as the components of a mixture emerge from the chromatographic separation, they are detected by the mass spectrometer.

Finally, matrix-assisted laser desorption ionization (MALDI) is a method that permits analysis of very large ions up to m/z 300,000. The sample is placed within a matrix of crystalline material, often a carboxylic acid, that can absorb UV light. A UV pulse from a laser then causes rapid heating of the matrix and sample, forming gas-phase ions in a controlled fashion with little fragmentation. Since the materials used to form the matrix obscure the region from m/z 100 to m/z 500, this technique is more suited to polymers including biopolymers such as DNA, proteins, etc., and less often used for smaller molecules typically encountered in the organic chemistry lab.

<< **aerosol**
Finely divided liquid droplets, as in a spray or mist.

Practical Aspects of Sampling and Instrumentation

One of the great advantages of mass spectrometry is that vanishingly small samples can be analyzed effectively. A spectrum can be obtained on less than 0.1 mg of sample, a mass too small to accurately measure with a typical analytical balance. This is fortunate, because it is a destructive technique—sample recovery is generally not possible.

Sample preparation depends on the type of mass spectrometer you will use. In some larger laboratories or universities, mass spectrometers are operated by dedicated staff, and you may simply place a few milligrams of a sample in a labeled vial and submit it for analysis. If you'll be using the instrument yourself, ask your instructor what type of sample preparation will be needed. Some mass spectrometers allow for solid or liquid samples to be placed on a probe that is inserted directly into the mass spectrometer. Many modern instruments connect mass spectrometry with separations tools such as gas chromatography (GC-MS) or high-performance liquid chromatography (HPLC-MS), so that the eluent from the separation is directly passed into the mass spectrometer, affording a molecular formula of each separated component of a mixture. For these instruments, simply prepare the sample as you normally would for the separation. For GC, this generally involves dissolving the sample in a volatile solvent such as pentane or diethyl ether. For HPLC, check the type of column and its recommended mobile phase solvent, and prepare a solution using that solvent. Concentrations in the range of 1–10 mg/mL are normal, and only a few microliters of this solution will typically be used.

8.1D ## The Mass Spectrum

What information do most organic chemists seek in the mass spectrum? In most cases, the molecular ion is generally the primary source of information in the spectrum because it allows a determination of the molecular weight of the compound—information that is not available from IR or NMR spectra. Several other types of information can be accessed from routine mass spectra. These include extra peaks of higher m/z that indicate the presence of certain elements having heavier isotope(s) in natural abundance, such as chlorine, bromine, or sulfur. Mass spectra are also used to detect and measure heavier isotopes used to synthetically label compounds for studies of reaction mechanisms or biosynthetic pathways, such as ^{2}H (deuterium, D) or ^{13}C. Third, the fragmentation of organic ions during mass spectrometry can allow for identification of connectivity in small organic molecules, and thus distinguish between isomers. Fragmentation can also reveal the sequence of amino acids or nucleic acids in biopolymers such as proteins and DNA.

THE MOLECULAR ION

In an example of an electron impact mass spectrum (**Figure 8.4**), peak intensity on the y axis is normalized to the highest peak, which is assigned a relative abundance of 100%, and is called the *base peak*. Other peaks in the spectrum have intensities expressed as a percentage in relation to the base peak. Depending on how much fragmentation is occurring under the given ionization conditions, the base peak may or may not be the molecular ion. For a pure substance, usually the molecular ion is found in a cluster of peaks located at the highest mass in a mass spectrum, and corresponds to the species present after ionization of a molecule M but before any fragmentation occurs. Depending on the ionization method, this can take several forms, most commonly M^+, $[M+H]^+$, or $[M+Na]^+$ (see the section above on ionization techniques).

FIGURE 8.4

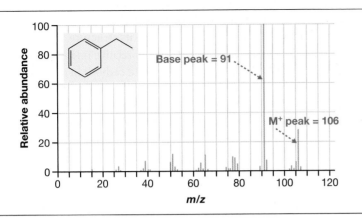

An electron impact ionization (EI) mass spectrum of ethylbenzene. The base peak at *m/z* 91 corresponds to $C_7H_7^+$, and the M^+ peak is observed at *m/z* 106 (relative abundance 28% of the base peak).

Karty, J. *Organic Chemistry: Principles and Mechanisms*, 3rd ed.; W. W. Norton: New York, 2022; p 744.

ISOTOPE PEAKS

The molecular ion peak is accompanied by several smaller peaks of masses one or two units higher than the main molecular ion peak; these are isotope peaks. Different isotopes result from extra neutron(s) in the nucleus of an atom. Some elements have significant amounts of more than one isotope. In typical samples of carbon found in nature, 98.9% of the carbon atom is ^{12}C, with atomic mass of 12 g/mol, while 1.1% of the carbon atoms is ^{13}C, one mass unit higher because it carries an extra neutron in its nucleus (**Table 8.1**). Thus we say the natural abundance of the ^{13}C isotope is 1.1%.

TABLE 8.1

Isotope Exact Masses and Their Relative Abundance

ELEMENT	ATOMIC MASS	ISOTOPE	ISOTOPE EXACT MASS	RELATIVE ABUNDANCE
Hydrogen (deuterium)	1.00797	^1H	1.00783	100
		^2H (D)	2.0141	0.015
Carbon	12.01115	^{12}C	12.00000	100
		^{13}C	13.0034	1.11
Nitrogen	14.007	^{14}N	14.0031	100
		^{15}N	15.0001	0.37
Oxygen	15.999	^{16}O	15.9949	100
		^{18}O	17.9992	0.20
Sulfur	32.064	^{32}S	31.9721	100
		^{33}S	32.9715	0.79
		^{34}S	33.9679	4.43
Chlorine	35.453	^{35}Cl	34.9689	100
		^{37}Cl	36.9659	31.98
Bromine	79.909	^{79}Br	78.9183	100
		^{81}Br	80.9163	97.3
Iodine	126.904	^{127}I	126.9045	100

These two forms of carbon give separate peaks in the mass spectrum, one mass unit apart. Thus the molecular ion peak is accompanied by an $[M+1]^+$ peak that corresponds to a molecular ion bearing one ^{13}C within its structure. If a mass spectrum of methane (CH_4) has a base peak at m/z 16 (100% relative intensity), it will also exhibit a peak at m/z 17 (1.1% intensity). If there are 10 carbons in a molecule, then the chances of the molecule containing one ^{13}C are 10 × 1.1, or 11%, and its M^+ peak will be accompanied by an $[M+1]^+$ peak that is 11% of the intensity of M^+. Thus, the relative abundance of the $[M+1]^+$ ions is related to the number of carbons in the structure times the natural abundance of ^{13}C. For a pure sample, the precision of this carbon count estimate can be within one or two carbons in a typical small organic compound. This is sufficient to distinguish whether one or more heavier atoms such as phosphorus or iodine may be accompanying the carbons within the structure.

For compounds containing S, Cl, and Br, there is a significant isotope peak present at $m/z = M^+ + 2$ because the second-most abundant isotope of these elements contains two additional neutrons per atom (Table 8.1). Because these isotopes ^{34}S, ^{37}Cl, and ^{81}Br are much more abundant than ^{13}C, the $M^+ + 2$ peak is much more likely to come from one of these elements than from a molecule containing two atoms of ^{13}C. Note that the likelihood of having two ^{13}C atoms within the same molecule is the natural abundance squared, times the number of carbons; for a 5-carbon compound this likelihood is $0.0111^2 × 5 = 0.0006$, or 0.06%, and that is negligible. Therefore the $M^+ + 2$ peak is a useful diagnostic for the presence of S, Cl, or Br (**Figure 8.5**). On the other hand, the natural abundance of ^{81}Br is almost the same as its most abundant isotope, ^{79}Br. Therefore, for a compound having more than one Br, there would be significant peaks at both $M^+ + 2$ and $M^+ + 4$.

FIGURE 8.5

An electron impact (EI) mass spectrum of a compound containing an $M^+ + 2$ isotope peak at m/z 92. We know that this compound contains Cl because the $M^+ + 2$ ions have a relative abundance of 10% of the base peak, while M^+ ions at m/z 90 are 30% of the base peak. The ratio $10/30 = 0.33$ corresponds closely to the 32% abundance of ^{37}Cl relative to ^{35}Cl.

Karty, J. *Organic Chemistry: Principles and Mechanisms*, 3rd ed.; W. W. Norton: New York, 2022; p 766.

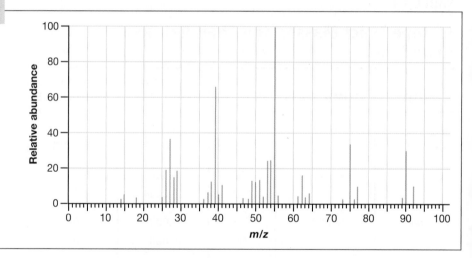

8.1E   Determination of Molecular Formula

The first step to determining a molecular formula from an unknown compound is to find the molecular ion. For a pure sample of typical organic small molecules, the molecular ion is often found in a cluster of peaks at the high end of the observed m/z range. The largest peak in that cluster will generally be used to determine the molecular formula. There are some cautionary notes to mention, however. If higher energy ionization processes (EI or CI) are used, the molecular ion may appear small or even undetectable, depending on how readily the molecular ion fragments to smaller ions. If the sample is impure, higher molecular weight impurities can also confuse the matter. The ideal scenario is when the molecular ion is easily found as the base peak, which often occurs when electrospray ionization (ESI) is used, as its mild ionization conditions induce minimal fragmentation.

If you are using CI or ESI ionization, you will need to subtract the mass of H from $[M+H]^+$, or Na from $[M+Na]^+$, in order to calculate the mass of M^+ before proceeding to determine the molecular formula. If the molecular ion is visible in an EI mass spectrum it will be M^+ and no adjustment is needed. For typical organic compounds containing C, H, N, O, S, and halogen, here is a procedure you can use, starting with M^+:

1. Evaluate the M^++2 peak for the presence of S, Cl, or Br.
    For S, the M^++2 peak will be 4.4% of the M^+ peak.
    For Cl, the M^++2 peak will be 32% of the M^+ peak.
    For Br, the M^++2 peak will be 98% of the M^+ peak (nearly equal intensities).

2. Determine if the mass of M^+ is even or odd.
    If M^+ is even, there is an even number of nitrogens (0, 2, 4, etc.).
    If M^+ is odd, there is an odd number of nitrogens (1, 3, etc.).

3. Estimate the carbon count (C_n) from the intensity of the M^++1 peak (relative to M^+).
    General formula: $n = $ (intensity of M^++1/intensity of M^+) $\times$ 100/1.1

4. Add hydrogens and oxygens to the formula.
    Start with C_n plus any N, Cl, Br, or S.
    Add $2n + 2$ hydrogens.
    Add the minimum number of oxygens needed for the formula mass to exceed the observed m/z.
    Subtract hydrogens as needed to match the exact mass.

Solved Problem

Identify a possible formula for a pure compound that gives the following mass spectral data:

m/z	RELATIVE INTENSITY
86 (M$^+$)	10.00%
87	0.56%
88	0.04%

Solution

1. Because the M^++2 peak is very small (only 0.4% relative to the M^+ peak), there is no Cl, Br, or S.

2. From M^+ we determine the molecular weight is 86 g/mol. An even or odd number for M^+ indicates the number of nitrogens is even or odd, respectively. In this case, m/z of M^+ is even (86), indicating an even number of nitrogens (0, 2, 4, . . .).

3. The M^++1 peak gives us an estimate of the carbon count:

    $$\text{Number of carbons} = (0.56/10) \times 100/1.1 = 5$$

4. The formula C_5 by itself corresponds to 60 g/mol, so more atoms must be added. Start with C_5 and try various formulas to find one which adds up to 86, adding N, H, and O as necessary.
    Try $C_5N_2 = 60 + (2 \times 14) = 88$ g/mol (2 N gives MW > 86; must be 0 N)
    Try $C_5H_{12} = 60 + (1 \times 12) = 72$ g/mol (saturated, but MW too low; try adding O)
    Try $C_5H_{12}O = 88$ g/mol (MW is close; can adjust by adding unsaturation)
    Try $C_5H_{10}O = 86$ g/mol → SOLVED!

In this solved problem, $C_5H_{10}O$ is one possible formula for this compound. Because the number of carbons is not always very precise, you may wish to use the same procedure to determine formulas with carbon counts that are adjacent to C_5 (such as $C_4H_{10}N_2$, $C_4H_6O_2$, or C_6H_{14}). These can serve as alternative formulas that could potentially be distinguished by further information such as IR or NMR spectroscopy. As mass numbers get larger, there may be many more formulas that fit the mass spectrometry data.

8.1F | Fragmentation and Structure

With ionization methods that impart high energies to the ions, especially EI and, to a lesser extent, CI, the ions that reach the detector include smaller fragments of the molecule in addition to the original molecular ion. Thus a mass spectrum generated in this way can be quite complex. Historically, this kind of spectrum has been analyzed carefully to determine formulas of various fragments, which can be used to reconstruct a possible structure. This is possible because the fragmentation can be predictable. It usually occurs at weaker bonds that can produce relatively stable cationic fragments. Therefore the masses of fragments can be correlated to the positions of various organic functional groups within the molecular structure, helping to illuminate its connectivity. A selection of common fragmentation processes is illustrated in **Figure 8.6**. However, for small molecules this kind of fragmentation analysis has been largely displaced by powerful NMR spectroscopy methods that give more direct information on structural connectivity.

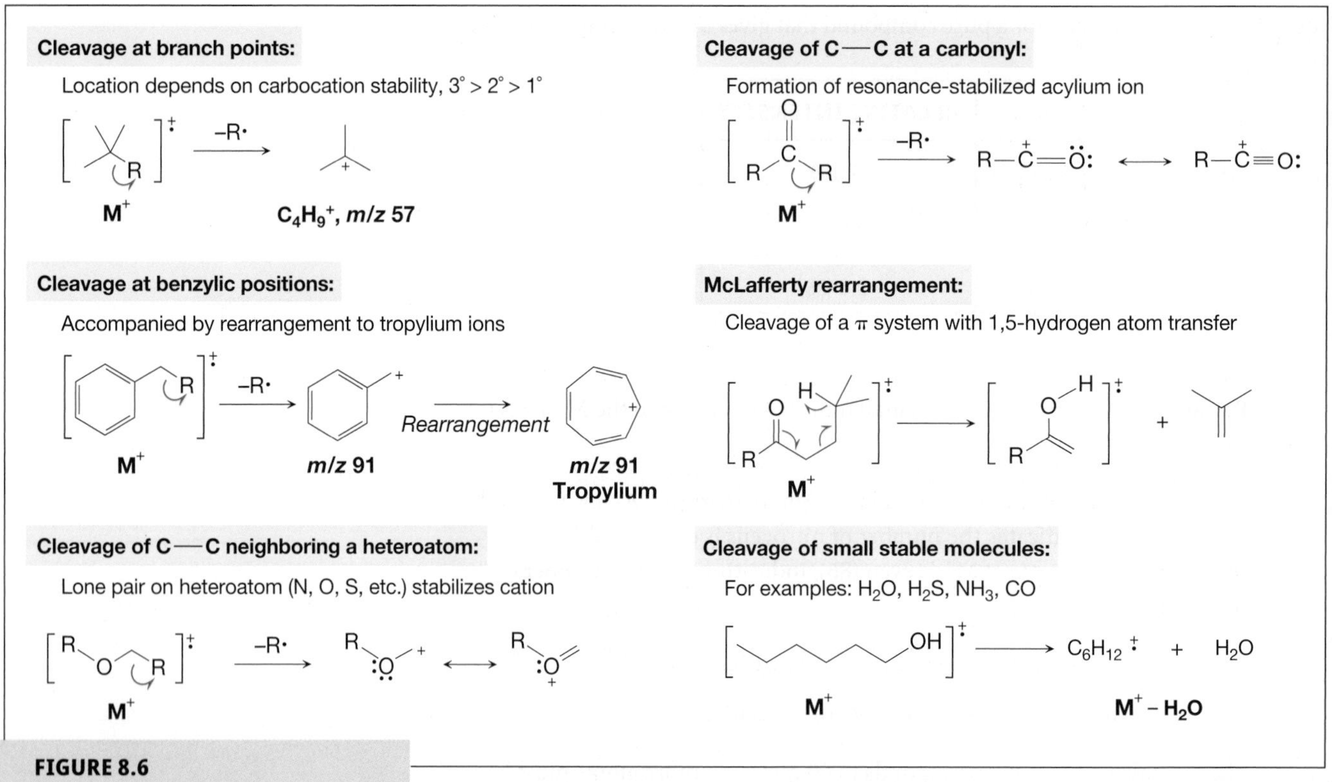

Cleavage at branch points:

Location depends on carbocation stability, 3° > 2° > 1°

Cleavage at benzylic positions:

Accompanied by rearrangement to tropylium ions

m/z 91   *m/z* 91 Tropylium

Cleavage of C—C neighboring a heteroatom:

Lone pair on heteroatom (N, O, S, etc.) stabilizes cation

Cleavage of C—C at a carbonyl:

Formation of resonance-stabilized acylium ion

McLafferty rearrangement:

Cleavage of a π system with 1,5-hydrogen atom transfer

Cleavage of small stable molecules:

For examples: H_2O, H_2S, NH_3, CO

FIGURE 8.6

Several common modes of fragmentation observed in electron-impact (EI) mass spectrometry.

8.1G | High-Resolution Mass Spectrometry

Some mass spectrometers are able to differentiate compounds that would appear to have the same molecular weight, such as ethene ($H_2C\!=\!CH_2$) and dinitrogen (N_2). For both of these we would predict a M^+ of *m/z* 28 amu (atomic mass units) in the mass spectrometry methods discussed so far, which are considered low-resolution mass spectrometry. However, some instruments are capable of resolutions of 0.001 amu, 0.0001 amu, or even higher, and these allow for *high-resolution mass spectrometry* (HRMS). With this level of resolution, we must use very high-precision atomic masses of elements from the periodic table to calculate the *m/z* of the molecular ion. To do this, we use the "exact masses" of the most abundant isotopes (see Table 8.1) to calculate the "exact mass" *m/z* of the molecular ion composed of the most abundant isotopes. Using the exact masses 1.00783, 12.00000, and 14.0031 for hydrogen, carbon, and nitrogen, respectively, we can see that a high-resolution mass spectrum with resolution of 0.0001 could easily distinguish between ethene (exact mass 28.0313) and N_2 (exact mass 28.0062).

The high-resolution mass spectrum is a powerful way to determine the molecular formula of an unknown compound.

Worked Example

In an HRMS analysis of an unknown sample, a molecular ion M^+ of *m/z* 100.0890 was found. Is this compound 1,2-diaminocyclopentane (100 g/mol), cyclohexanol (100 g/mol), or heptane (100 g/mol)?

Solution

First, use the most abundant isotope exact masses from Table 8.1 to calculate the exact mass of the molecular ion in each case (**Table 8.2**). Then, compare the calculated M^+ exact masses to the observed experimental data. The smallest difference between the calculated and observed masses is with cyclohexanol (Δ 0.0001 amu), so we can say that this is likely the compound.

A final caveat about HRMS: In most cases this analysis technique gives information about identity, but with no indication of purity. A sample of 95% or 5% purity may yield a molecular ion peak of the same exact mass, and it may be difficult to know if the resulting data are from a minor impurity or from the major component in the sample. However, like other mass spectrometry methods, HRMS can be combined with gas chromatography (GC) or other separation techniques, so that HRMS data are obtained from each peak eluted during the chromatography. Information about identity and purity may then be available from a single analysis.

TABLE 8.2

Calculated Exact Masses (Worked Example)

COMPOUND, FORMULA	ISOTOPE EXACT MASSES	M^+ EXACT MASS, CALCULATED
1,2-Diaminocyclopentane, $C_5H_{12}N_2$	^{12}C: 5 × 12.00000 = 60.00000 ^1H: 12 × 1.00783 = 12.09396 ^{14}N: 2 × 14.0031 = 28.0062	For $[C_5H_{12}N_2]^+$, *m/z* 100.1002
Cyclohexanol, $C_6H_{12}O$	^{12}C: 6 × 12.00000 = 72.00000 ^1H: 12 × 1.00783 = 12.09396 ^{16}O: 15.9949	For $[C_6H_{12}O]^+$, *m/z* 100.0889
Heptane, C_7H_{16}	^{12}C: 7 × 12.00000 = 84.00000 ^1H: 16 × 1.00783 = 16.12528	For $[C_7H_{16}]^+$, *m/z* 100.1253

8.2A Elemental Ratios from Combustion By-Products

Prior to any sort of spectroscopy, molecular formulas of organic compounds were determined by combustion and analysis of the combustion products, and this remains a valuable method today. Combustion of a precisely weighed sample of an organic compound at high temperatures with an oxygen source produces carbon dioxide (CO_2) and water (H_2O), and their amounts can be precisely measured. Similarly, if nitrogen gas (N_2) is produced by the combustion process, its amount can be measured. Together, these quantitative measurements of combustion gases allow the chemist to obtain a *CHN analysis* of a sample, which consists of a percentage by weight of the elements carbon, hydrogen, and nitrogen in the sample. Although there are variations on this method that can determine the percentages of sulfur, phosphorus, and other elements, CHN analysis is the preferred method for typical organic chemistry samples.

There are three main purposes of CHN analysis:

- Propose a complete or partial molecular formula of an unknown organic compound. Using the molecular weight, along with independent experimental evidence of other elements that may be present (such as halogen), CHN analysis can be translated into a molecular formula.
- Independently confirm a molecular formula suggested by other evidence. If a molecular formula has been proposed on the basis of other evidence, the calculated elemental ratios for that formula may be compared with the observed elemental ratios, and if they match, this confirms the formula.
- Confirm the purity of a known substance. For a known substance, if the calculated and observed elemental ratios are in good agreement, this confirms that the substance is of high purity, e.g., >95%.

8.2B Historical Importance of Combustion Analysis

Because combustion analysis is a very old technique that predates routine IR and NMR spectroscopy by at least a century, it is worth noting its importance in the development of organic and biological chemistry. In the early 1800s, the combustion elemental analysis method played a key role in rudimentary investigations of botany and physiology, leading to the classification of animal and plant substances into carbohydrates, lipids, and proteins based on their differing ratios of carbon, hydrogen, and nitrogen. By 1831, Justus von Liebig had refined combustion analysis to a level of precision sufficient to propose the beginnings of a chemical understanding of nutrition, respiration, and metabolism as presented in his 1842 book, *Animal Chemistry*.[1] Liebig popularized the general concept that physiological processes can be related to organic chemical reactions observed in the laboratory. Identifying the specific structures and reactions involved would take many decades of organic chemistry research efforts, but Liebig's proposals, born of combustion analysis, comprised a visionary leap that inspired such efforts.

---

[1] Liebig, J. *Animal Chemistry; or, Organic Chemistry in Its Applications to Physiology and Pathology*; Gregory, W., Ed.; Johnson Reprint Corp.: New York, 1964. (A facsim. of the Cambridge edition of 1842.)

In Liebig's version of the combustion analysis techniques, the combustion gases were trapped within tubes containing chemical adsorbents, and the change in mass of these tubes was measured to determine the amount of each gas.[2] This required combustion of significant amounts (0.5–1.0 g) of sample for accurate measurements. Advances made by Fritz Pregl in the early 1900s lowered the sample size to 2–4 mg. This was an important advance in the analysis of organic compounds, leading to a Nobel Prize awarded to Pregl in 1923. R. M. Wilstätter, another Nobel Laureate honored for his studies on natural pigment compounds, having isolated very small quantities of a pigment from 10,000 cattle ovaries, was able to use Pregl's improved combustion analysis technique to show that this compound was carotene, the same pigment found in carrots; thus demonstrating that pigments from the animal and plant kingdoms were chemically identical.[3]

With modern methods, about 0.5–1 mg sample sizes are introduced into analyzer instruments that automate the combustion process and exploit gas chromatography for precise measurement of the gaseous products. At least four significant figures are needed in the mass measurement of the sample in order to obtain meaningful data, so smaller samples require high-precision analytical balances (± 0.1 mg or better). The instrument is calibrated using known compounds of high purity in order to ensure the output elemental ratios are accurate.

8.2C | Proposing a Molecular Formula of an Unknown

A completely unknown compound cannot often be identified solely by elemental ratios, so additional information is generally required. Laboratory qualitative tests can determine the presence of certain elements such as sulfur or halogens. Sample history may suggest possible structures if the compound came from a known starting material and a routine type of organic reaction.

Worked Example

For an unknown compound, qualitative lab tests showed there was no halogen or sulfur present, and the following elemental mass ratios were determined by combustion analysis, in % by weight: C 58.53, H 4.09, N 11.38. Propose a molecular formula for this compound.

Solution

First, subtract all the C, H, N percentages from 100% in order to find the percentages of other elements present. We assume the remainder is oxygen since we are provided independent information that there is no halogen or sulfur present (**Table 8.3**). Divide the element percentages by their respective atomic masses to obtain the mole ratios of these elements. Next, convert these mole ratios to integers by dividing each of these by the smallest one to obtain relative amounts.

The results in the right-hand column all conform very closely to integers, so we can propose the molecular formula as $C_6H_5NO_2$. However, it should be noted the combustion analysis gives an empirical formula, and the same results would be obtained from, for example, $C_{12}H_{10}N_2O_4$.

TABLE 8.3

Formula from Mass % (Worked Example)

ELEMENT, MASS %	MOLE RATIO	INTEGER FORM
C 58.53%	58.53/12 = 4.88	4.88/0.813 = 6.00
H 4.09%	4.09/1 = 4.09	4.09/0.813 = 5.03
N 11.83%	11.38/14 = 0.813	0.813/0.813 = 1
O (100 − C, H, N) = 25.55%	25.55/16 = 1.60	1.60/0.813 = 1.97

---

[2]Holmes, F. L. Elementary Analysis and the Origins of Physiological Chemistry. *Isis* **1963**, *54* (1), 50–81.
[3]Pregl, F. Quantitative Micro-Analysis of Organic Substances. Nobel Lecture, December 11, 1923.
https://www.nobelprize.org/prizes/chemistry/1923/pregl/lecture (accessed May 2022).

If the problem-solving method above leads to numbers in the right-hand column that appear to include a half of an atom, this is nonsensical, as it is impossible to have a half of an atom. Multiply all elements by two to make them all integers in the formula. Similarly, if there are results such as 1.33 or 2.67, with 1/3 or 2/3 of an atom, then multiply all elements by three to make them integers.

8.2D | Confirmation of Molecular Formula

Often the chemist comes to a combustion analysis with a pure compound and some knowledge of its formula, and perhaps also the structure. Some of these situations may involve two or more alternative formulas/structures that seem consistent with whatever other information is at hand, such as IR or NMR spectroscopy. In such situations, the role of combustion analysis is to confirm one formula by comparison of predicted and observed elemental ratios. Typically, such comparisons are considered a match if the observed percentage of an element is within 0.4 of the calculated percentage.

8.2E | Assessment of Purity of a Known Compound

The presence of some impurities in a sample, such as solvents or leftover reactants, may be detected readily by NMR spectroscopy, and their amounts measured by integration. The presence of impurities (but not their identity) can also be detected if combustion analysis does not match the calculated ratios. If the difference between calculated and observed elemental percentage is more than 0.4, then the compound fails the assessment of purity.

In some cases impurities are undetectable by NMR. Combustion analysis takes on extra importance when such impurities could be present. For example, solvents without protons (e.g., CCl_4), various inorganic salts used in workups (e.g., NaCl), and adsorbents used in chromatography (e.g., silica gel) are all lacking signals in a ^1H NMR spectrum. These all happen to have low carbon percentages also, so they contribute to the mass of the sample, but do not produce CO_2 on combustion, making the carbon percentage appear lower than what is calculated for the pure compound. So a combustion analysis of a sample containing any of these impurities will not give a satisfactory CHN analysis; that is, the observed percentage of one or more elements will likely be more than 0.4% off the calculated percentage.

8.3 STRUCTURAL CLUES FROM MOLECULAR FORMULA

degrees of unsaturation >>
A measurement of the amount of hydrogen that is absent from a structure due to the presence of pi bonds or ring connections, in comparison with the saturated hydrocarbon formula (C_nH_{2n+2}). Each ring and/or pi bond introduced to a structure corresponds to a decrease of two hydrogens from the saturated hydrocarbon formula.

Whether the molecular formula is obtained from the mass spectrum, combustion analysis, or other means, there are clues within the formula that can be used to narrow down the range of possible structures. Some of these are related to the functional group. For example, if there is no nitrogen, you can rule out amines, amides, nitro compounds, and any other nitrogen-containing functional group. If there is only one oxygen, you can rule out carboxylic acids and esters. These simple considerations can get you on the right track to proposing a reasonable structure.

To refine the structural possibilities further, you can use a systematic assessment of the hydrogen content of the compound, which reflects whether it is saturated or unsaturated and to what degree (this is the organic chemistry version of oxidation states). The result of such an assessment is the **degrees of unsaturation** (also known as *double bond equivalents* or *index of hydrogen deficiency*) of the compound, and it gives information about how many π bonds or rings are present in the structure. To allow for one π bond or ring, a saturated hydrocarbon (C_nH_{2n+2}) requires the removal of two hydrogens from its structure in order for carbon to not exceed its valence of 4. A formula of C_nH_{2n} is missing two hydrogens in comparison with the saturated formula, and therefore has one degree of unsaturation.

For example, consider an unknown compound with the formula C_6H_{12}. The saturated hydrocarbon of the same carbon count has the formula C_nH_{2n+2} or in this case C_6H_{14}. Our unknown has two hydrogens fewer than the saturated compound, so we say it has one degree of unsaturation, and each possible structure we propose must have exactly one π bond or one ring in order to match the formula C_6H_{12}.

This type of analysis can be systematically applied to various formulas beyond hydrocarbons by subtracting the number of H's in the formula from the number of H's in the corresponding saturated hydrocarbon of the same carbon count, and then dividing the number of remaining H's by two. When heteroatoms (N, O, halogens, etc.) are present, some adjustments may be needed to account for the standard valences of the various heteroatoms:

- For compounds containing O, there is no adjustment to the H count.
- For every halogen X, add an H (then ignore the X).
- For every N, subtract an H (then ignore the N).

Worked Example

(a) Determine the degrees of unsaturation for the formula C_4H_4NOBr, and (b) propose a structure that is consistent with this formula.

1. The oxygen can be ignored.

2. For 1 N, subtract 1 H = C_4H_3.

3. For 1 X, add 1 H = C_4H_4.

4. Saturated hydrocarbon of the same carbon count = C_nH_{2n+2} = C_4H_{10}.

5. $H_{10} - H_4 = H_6$, and dividing by two we get 3 degrees of unsaturation.

With 3 degrees of unsaturation, any proposed structure must contain 3 π bonds, 3 rings, or any combination of rings and π bonds that totals 3. Note that triple bonds count as 2 π bonds. Two possible structures are shown in **Figure 8.7a**.

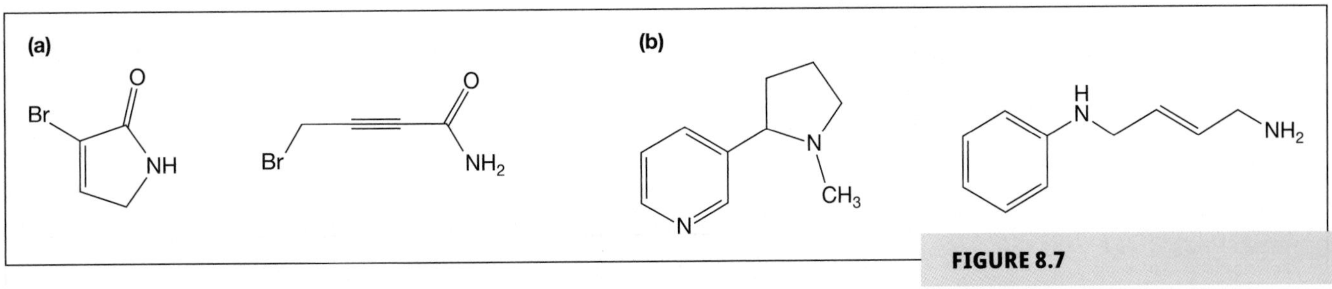

FIGURE 8.7

Possible structures corresponding to formulas (a) C_4H_4NOBr and (b) $C_{10}H_{14}N_2$.

Worked Example

Propose a structure having the formula $C_{10}H_{14}N_2$.

First, determine the degrees of unsaturation.

1. For 2 N, subtract 2 H = $C_{10}H_{12}$.

2. Saturated hydrocarbon = C_nH_{2n+2} = $C_{10}H_{22}$.

3. $H_{22} - H_{12} = H_{10}$, and dividing by two we get 5 degrees of unsaturation.

Second, consider the constraints on the proposed structure. With 5 degrees of unsaturation, any proposed structure must contain 5 π bonds, 5 rings, or any combination of rings and π bonds that totals 5. For the purposes of this analysis, benzene and other 6-membered aromatic rings are considered to have 3 π bonds, so a benzene ring would account for 4 degrees of unsaturation (one for the ring and the rest for the 3 π bonds). Two possible structures are shown in **Figure 8.7b**.

8.4 COMBINED SPECTROSCOPY PROBLEMS

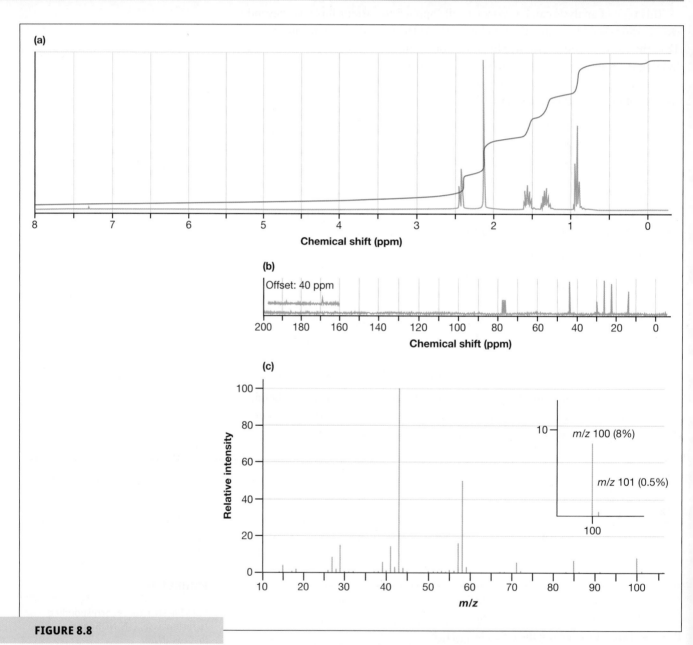

FIGURE 8.8

(a) ^{1}H NMR and (b) ^{13}C NMR data in CDCl$_3$ and (c) mass spectrum.

1. Identify the compound from the following ^{1}H NMR, ^{13}C NMR, mass spectrometry, and combustion analysis data. The ^{13}C and ^{1}H NMR spectra and mass spectrum are shown in **Figure 8.8**. Note that the chemical shift of the ^{13}C NMR peak shown in the offset at left is 40 ppm higher, at approximately 208 ppm.

   Combustion analysis, %: C, 71.91; H, 12.10

2. Identify the compound from the following ^{1}H NMR, ^{13}C NMR, IR, and HRMS data. The ^{13}C and ^{1}H NMR spectra are shown in **Figure 8.9**.

   IR spectrum, cm^{-1}: 2927, 2880, 1687

   High-resolution mass spectrum: m/z 99.0682 (M$^+$)

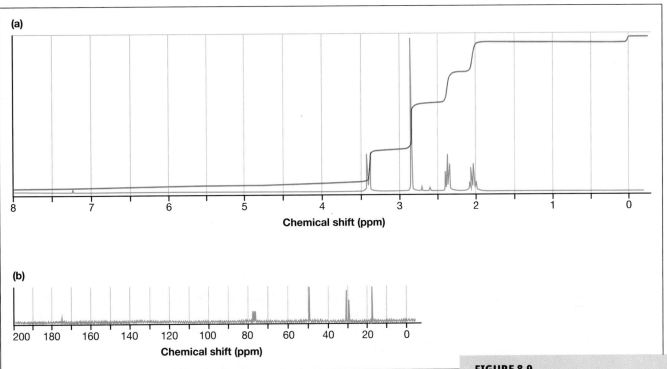

(a)

(b)

Chemical shift (ppm)

FIGURE 8.9

(a) ^{1}H NMR and (b) ^{13}C NMR data in CDCl$_3$.

© Sigma-Aldrich Co. LLC. Reproduced with permission from Merck KGaA, Darmstadt, Germany and/or its affiliates.

3. Identify the compound from the following ^{1}H NMR, ^{13}C NMR, IR, and combustion analysis data. The IR data are shown in **Figure 8.10**, and the ^{13}C and ^{1}H NMR spectra are in **Figure 8.11**.

   Combustion analysis, %: C, 66.64; H, 6.73

FIGURE 8.10

Infrared spectrum data.

Spectral Database for Organic Compounds, SDBSWeb. National Institute of Advanced Industrial Science and Technology. https://sdbs.db.aist.go.jp/ (accessed August 2022). Reprinted by permission.

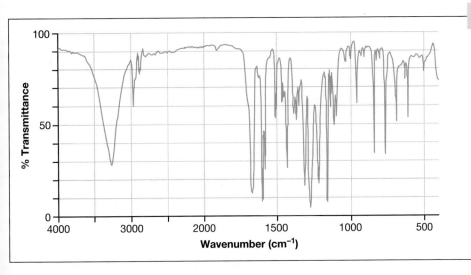

% Transmittance

Wavenumber (cm^{-1})

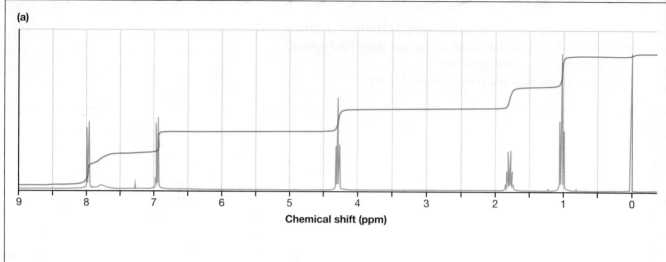

FIGURE 8.11

(a) ^{1}H NMR and (b) ^{13}C NMR data in CDCl$_3$.

© Sigma-Aldrich Co. LLC. Reproduced with permission from Merck KGaA, Darmstadt, Germany and/or its affiliates.

4. Identify the compound from the following IR, ^{1}H NMR, ^{13}C NMR, and mass spectrometry data. The IR spectrum showed three broad peaks above 3000 cm^{-1}. The ^{13}C and ^{1}H NMR spectra are shown in **Figure 8.12**. *Hint:* An isomer assignment is revealed by four ^{1}H peaks in the aromatic region of the ^{1}H NMR, including a singlet.

Mass spectrum: *m/z* 137.08 (100%), 138.09 (8.7%)

Combustion analysis, %: C, 70.02; H, 8.06; N, 10.20

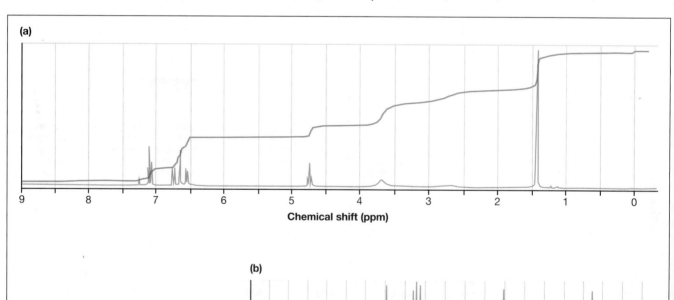

FIGURE 8.12

(a) ^{1}H NMR and (b) ^{13}C NMR data in CDCl$_3$.

Pouchert, C. J.; Behnke, J. *The Aldrich Library of ^{13}C and ^{1}H FT NMR Spectra*; Aldrich Chemical Company: Milwaukee, 1993.

PART B
EXPERIMENTS

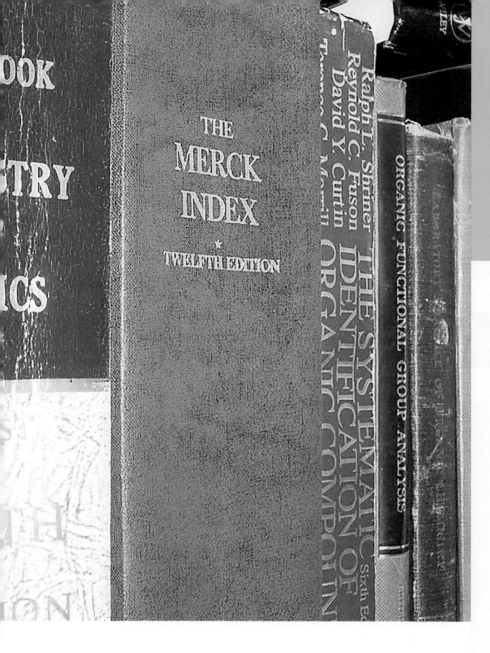

9

Literature of Organic Chemistry

FINDING WHAT'S KNOWN

Information about organic compounds and their properties can be found in vast repositories of print volumes, most of which are now accessible online.

Courtesy of Gregory K. Friestad.

Regardless of the scientific field you may be studying, discoveries made in the lab need to be reported to scientists around the world. Publishing the results of laboratory studies allows scientists elsewhere to have access to the details so they can stay informed on research progress in their fields, verify the findings, and use them as a springboard for future work on new ideas. In the organic chemistry lab, we don't want to reinvent the wheel, so we need to know what's been done before. To do this, we need to know how to search in the vast historical records of all organic chemistry research, and it can be like finding a needle in a haystack. Here you'll learn about some of the search tools, and get a bit of experience using them.

9.1 PRIMARY SOURCES

There are many *types* of chemistry literature, but they are generally categorized as *primary* or *secondary sources*. We look first at the primary sources, which are written by the people who actually performed the research.

Journals contain new work. There are thousands of journals in all fields. About 50–100 of them pertain to organic chemistry. Journals contain *articles* (also called *full papers*) and *communications*. Communications (also called *letters*) are brief reports intended for rapid publication, and generally contain preliminary data that can be republished in a more complete form later. Data in articles or full papers cannot be republished.

In organic chemistry, native English speakers are lucky because the most important journals are published or are translated into English worldwide. Nevertheless, it can be a great advantage to read German, because Germany was the chemistry leader prior to World War II, and many of the most important organic chemists of that era published their research in German. Articles in Russian, French, and Japanese are common as well. At most universities, it is not hard to find people with some scientific vocabulary who can read these languages, and often they are willing to translate an article in exchange for a batch of cookies or a cold beverage.

A few of the most important journals that publish organic chemistry research are as follows:

Angewandte Chemie International Edition
Bulletin of the Chemical Society of Japan
Chemical Communications
European Journal of Organic Chemistry (formerly *Justus Liebigs Annalen der Chemie*)
Journal of the American Chemical Society
Journal of Medicinal Chemistry
Journal of Organic Chemistry
Journal of Organometallic Chemistry
Nature
Organic and Biomolecular Chemistry (formerly *Journal of the Chemical Society, Perkin Transactions 1*)
Organic Letters

Science

Tetrahedron Letters

In this list, the standardized journal abbreviations are indicated in bold, with a period following each abbreviated term in the citation (e.g., *J. Am. Chem. Soc.*). Some of the journals cover a range of scientific fields (e.g., *Nature* and *Science*), some are specific to chemistry (e.g., *Chem. Commun.* or *J. Am. Chem. Soc.*), and others are even more specific to the discipline of organic chemistry (e.g., *Eur. J. Org. Chem.* or *J. Org. Chem.*).

Chemists can apply for patents for new compounds or a new method for making a known compound (amazingly, 20–30% of all patents are chemical patents). There will be considerable information in the publicly available documentation that accompanies the patent. Industry scientists may patent much of their work without ever publishing it, so some important discoveries will only be found by searching in the patent literature. Unfortunately, patents are not very easy to read, because they are written in legal terminology that sometimes seems like a different language. Also, even though patents are supposed to provide a full disclosure, they may conceal important information to maintain a competitive business advantage.

9.2 SECONDARY SOURCES

9.2A Early Print Databases

Extensive effort has gone into sorting key elements of organic chemistry primary literature sources into searchable form. One of the first important tools for this was the *Beilsteins Handbuch der Organischen Chemie*, which began in the 1880s. As it was regularly updated, it became an indispensable set of thick printed volumes filled with physical constants for organic compounds, as well as reliable procedures for their preparation. It was, however, written in German and sorted in a somewhat esoteric fashion.

In the United States, *Chemical Abstracts* emerged as an alternative way to sort through the primary literature. *Chemical Abstracts* provides short summaries, called abstracts, of many primary sources of information and is thus extremely useful for "hunting down" information. It was periodically published by the Chemical Abstracts Service (CAS), a branch of the American Chemical Society, and became the most comprehensive database available for all areas of chemistry.

To keep track of everything published about each unique organic compound, the Chemical Abstracts Service (CAS) assigns a *CAS registry number* to every compound when it is first reported in the literature. This CAS number is much like a social security number. Thus, while a compound may have different names in different languages, it has only one CAS registry number. There are over 70 million chemical substances with registry numbers, and about 15,000 are added each day. Other search tools and databases use the CAS registry number as an indexing term; this allows chemists to cross-reference compounds across different platforms and ensure precise searches for information.

9.2B Modern Electronic Databases

The data found in the printed form of *Beilstein* and *Chemical Abstracts* have been digitized, and have been available in various electronic formats for about 30–40 years, with user interfaces gradually transforming along with updated database software. In the late 1990s, some very user-friendly online interfaces emerged to take advantage of the explosion of internet capabilities.

SCIFINDER (https://scifinder.cas.org)

SciFinder is the current online interface to the Chemical Abstracts database, which includes abstracts of papers from thousands of journals dating from the 1800s to the present. It is updated daily as new papers are released. Research universities generally have subscription access to this database available via their sciences library websites. Other institutions may have access as well, and you should check with your instructor for SciFinder access details that are specific to your location. Access may be restricted to those who have library privileges.

Once in SciFinder, you will see a sidebar at left with choices of how to search for abstracts: *All, Substances, Reactions, References,* and *Suppliers.* Each of these expands to allow a range of different search types (see Figure 9.3). The following are a few examples of the variety of ways to search for abstracts on SciFinder:

1. Author name: Within the *References* search tool, you can look up an author, such as *Robert B. Woodward* (**Figure 9.1**).

FIGURE 9.1

Courtesy of scifinder.org.

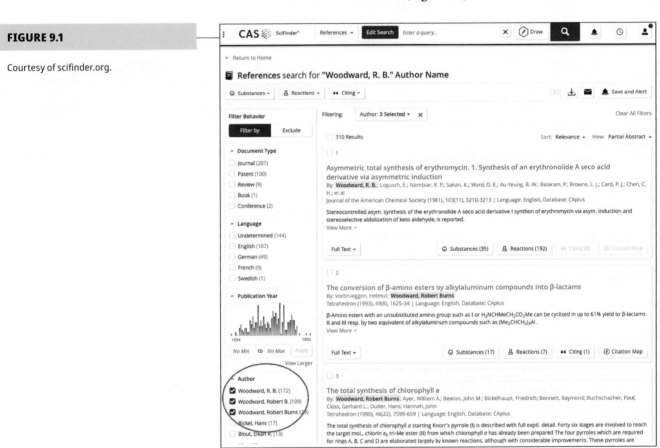

2. Research topic: You can look up keywords, such as *catalysis.* Multiple keywords can be entered to focus the search on a narrower topic, such as *gas chromatography of terpenes in citrus peels.*

    After clicking "Search," the various combinations of these terms will appear, along with statistics on how many references there are for each combination of terms. For example, you can select lists that include references with all of the terms closely associated, or lists with some of the terms found anywhere in the reference. After you select a list, the abstract for each reference in that list will appear, as shown in **Figure 9.2**. Sometimes the list of references is very large, and useful tools to narrow the list, such as *Filter by* and *Exclude* are found in a sidebar at the left. For example, clicking on "Document Type: Review" could shorten the list from 2804 to 105 references. These tools can

FIGURE 9.2

Courtesy of scifinder.org.

also narrow the search results to a selected author, specific years of publication, a certain university address of the authors, or other criteria.

Each abstract in SciFinder provides extensive information about what can be found in the reference, including the following:

- the title of the paper
- authors' names and addresses
- name of the journal, and the year, volume, and page number
- language in which the paper was written
- a short summary describing the key points of the paper

Clicking on the title gives detailed information to explore more deeply. This is how you can find the SciFinder abstract number (AN) that you use to cite the abstract.

3. Chemical substance: You can look up a chemical by its name (e.g., "cyclohexane"), CAS registry number (110-82-7), or formula (C_6H_{12}). Alternatively, SciFinder offers plug-in drawing applications (**Figure 9.3**) that will allow the structure to be recognized and identified for searching. Once a structure is drawn, it can be used to search for chemical substances or for reactions in which the structure is a component (e.g., as reactant or product).

FIGURE 9.3

Courtesy of scifinder.org.

4. Journal: You can view abstracts for the articles published in a particular journal and year.

The search types noted here offer an introductory starting point. The best way to learn more is to explore the options as you search.

The data found in *Beilstein* may now be accessed using Reaxys software. Various search formats are possible, but the most powerful is the structure searching. While this search tool may not be as comprehensive as searching with SciFinder, a comprehensive search is often more than you need. Chemical and physical properties and synthetic procedures for several million compounds may be found with Reaxys, and the data are provided in a convenient format with clean and easy-to-read structures.

WEB OF SCIENCE (https://clarivate.com/webofsciencegroup /solutions/web-of-science)

Web of Science includes data from a resource called *Science Citation Index (SCI)*. Searching by keyword, author, author's address, and the like are very easy with Web of Science, but unlike SciFinder or Reaxys, Web of Science does not allow searches based on the structures or physical properties of compounds. It provides abstracts and bibliographic data for all the important publications in the physical sciences. The data in Web of Science or Science Citation Index are linked by the references cited, so that once you find an early paper about a key discovery, you can then find all of the later articles that cite that earlier paper. Assuming the more recent authors properly cited the earlier paper, this can be a very valuable way to follow a research area from its origins right up to the cutting edge. SciFinder also allows citation searching.

SciFinder, Reaxys, and Web of Science all have active links to the papers they reference, so you can click and download the original article, as long as your library has a subscription to that resource.

9.2C Compendia of Tables and Information

Available both online and in printed format, some useful examples of these resources include the following:

1. *Combined Chemical Dictionary (CCD)* (https://ccd.chemnetbase.com). With CCD, you can search on a chemical name or draw a structure and do a search on that. CCD is also a quick way to obtain the *CAS registry number* for a chemical compound, a universal identifier which may make searches on other databases like SDBS (below) easier.

2. *AIST Spectral Database for Organic Compounds (SDBS)* (http://sdbs.db.aist .go.jp). SDBS is a web-based source of NMR, IR, and mass spectra data. It is useful for making comparisons with the spectra that you will collect in this course.

3. *Handbook of Chemistry and Physics* (http://hbcp.chemnetbase.com). This volume, often just called the *CRC*, is full of information, including the table on the "Physical Constants of Organic Compounds." Most libraries have copies of this available for consultation.

4. *The Merck Index* (http://www.rsc.org/merck-index). The Merck Index is a good source of information about chemicals of medicinal importance. Physical constants of the compounds, and references to papers about their synthesis, are provided, along with information about their clinical or laboratory uses.

5. *Handbook of Fine Chemicals* (http://www.sigmaaldrich.com). The Aldrich handbook is really a chemical sales catalog, but it contains physical constants on all the chemicals they sell. Most organic labs have a couple of these volumes available. If not, it's available online.

9.2D Reviews

These articles provide intensive surveys of rather narrow fields of research. A review provides an overview of a subject, so it can be a good starting point in an area of research, especially when you lack time to read all of the primary literature. Be aware, however, that you are getting the author's (or authors') view of a subject, and they may not be telling you everything you really want to know. Journals that include reviews are

*Accounts of **Chem**ical **Res**earch*

Angewandte Chemie, International Edition in English

Chemical Reviews

Chemical Society Reviews

9.2E Popular Press

A few other important periodicals that scientists read are

Discover Magazine

Scientific American

Chemical and Engineering News (weekly chemistry news from the American Chemical Society)

9.2F Serial Publications

These resources are like reviews but are published irregularly and are hardbound. Examples include

Advances in Organic Chemistry

Advances in Photochemistry

Advances in Protein Chemistry

Organic Syntheses

Progress in Macrocyclic Chemistry

9.2G Textbooks and Reference Books

These texts range from sophomore level to graduate level. Just a few examples of some commonly used advanced books are listed here. Some of the reference books have been converted into online database format.

Carey, F. A.; Sundberg, R. J. *Advanced Organic Chemistry*, 5th ed.; Parts A and B; Springer: New York, 2007.

Smith, M. B. *March's Advanced Organic Chemistry: Reactions, Mechanisms, and Structure*, 7th ed.; Wiley: New York, 2013.

Electronic Encyclopedia of Reagents for Organic Synthesis [Online]; Wiley. https://onlinelibrary.wiley.com/doi/book/10.1002/047084289X (accessed April 2022).

Larock, R. C. *Comprehensive Organic Transformations*, 3rd ed.; Wiley: New York, 2018.

Lowry, T. H.; Richardson, K. S. *Mechanism and Theory in Organic Chemistry*, 3rd ed.; Harper and Row: New York, 1987.

Nicolaou, K. C.; Sorensen, E. J. *Classics in Total Synthesis*; VCH: New York, 1996.

Nicolaou, K. C.; Snyder, S. A. *Classics in Total Synthesis II*; VCH: New York, 2003.

Reich, H. J. *Total Syntheses* [Online]. https://organicchemistrydata.org/hansreich /resources/syntheses/ (accessed April 2022).

9.3 FORMATS FOR CITING THE CHEMICAL LITERATURE

Material in this section can be found in the *ACS Guide to Scholarly Communication* (https://pubs.acs.org/doi/book/10.1021/acsguide). An abbreviated companion resource, *ACS Style Quick Guide* (https://pubs.acs.org/doi/full/10.1021/acsguide .40303), provides easy access to some of the most useful content.

When citing a single work, generally follow one of these two examples:

- It is known that the Diels–Alder reaction provides a six-membered ring.[3]
- It is known that the Diels–Alder reaction provides a six-membered ring (3).

When citing multiple works, generally follow one of these two examples:

- Recent investigations indicate a link between temperature and reaction rate.[2,4–9]
- Recent investigations (2, 4–9) indicate a link between temperature and reaction rate.

Different subdisciplines may have different standards for the citation indicator. Most chemistry journals use superscript numerals, but the parenthetical style is often used in biochemistry. Your instructor will likely express a style requirement, and you should follow that carefully. Similarly, the style of the way bibliographic information is provided in the citation itself can vary; some journals require inclusive pagination (both start and end pages of an article or chapter), and some require the article title. The examples presented in this chapter are a standard style associated with the *Journal of the American Chemical Society*.

9.3A | Citing Books

In book citations, authors appear first, followed by the book title in italics. A single author or group of authors may have written the whole book, or the book may have been compiled by an editor, with a number of separate chapters written by different authors. The author names appear with family name (last name), followed by first and middle initials. Depending on the situation, you may need to cite a particular chapter, in which case the chapter author(s), chapter title, chapter page numbers, and the editor are all listed in the reference.

Citation of a whole book: Warren, S. G. *Organic Synthesis, the Disconnection Approach*; Wiley: New York, 1982.

Citation of this book: Friestad, G. K. *Techniques and Experiments in Organic Chemistry: Biological Perspectives and Sustainability*; W. W. Norton: New York, 2023.

Citation of a chapter within a book: Friestad, G. K. Asymmetric Radical Addition to Chiral Hydrazones. In *Topics in Current Chemistry: Radicals in Synthesis III*; Gansauer, A.; Heinrich, M., Eds.; Springer-Verlag: Berlin, 2012; Vol. 320, pp 61–92.

9.3B   Citing Journal Articles

Journal article authors are cited in the same order given in the article. Titles of journals also appear in italics. For journals, the year appears in bold and the volume number in italics, followed by page numbers, as shown in the following example:

> MacGillivray, L. R.; Reid, J. L.; Ripmeester, J. A. *J. Am. Chem. Soc.* **2000**, *122*, 7817–7818. (This means year 2000, volume 122, pages 7817 to 7818.)

9.3C   Citing Theses or Dissertations

> Grubbs, R. H. 1. Cyclobutadiene Derivatives II. Studies of Cyclooctatetraene Iron Tricarbonyl Complexes. Ph.D. Dissertation, Columbia University, New York, 1968.

9.3D   Citing Personal Communications

> Flintstone, F. Prinstone University, Bedrock, CT. Personal communication, 1998.

9.3E   Citing Material from the Internet

Use the title provided on the website, and if it is unclear whether the title refers to an internet location, add the words "Home Page" to clarify. Include the date accessed.

> *Library of Congress Home Page.* https://www.loc.gov (accessed May 2022).

> Frontier, A. J. *Not Voodoo X.4: Demystifying Synthetic Organic Chemistry since 2004.* http://www.chem.rochester.edu/notvoodoo (accessed April 2022).

9.3F   Citing Computer Programs

Computer programs are typically cited using the book format:

> Deppmeier, B.; Driessen, A.; Hehre, T.; Hehre, W.; Klunzinger, P.; Ohlinger, S.; Schnitker, J. *Spartan'18* (18 2, v 1.4.6); Wavefunction, Inc.: Irvine, CA, 2019.

9.4 LITERATURE ASSIGNMENT

This assignment consists of three parts:

1. Table of chemicals and their properties

2. SciFinder search and one-paragraph summary of a primary research publication

3. Summary of a review/popular article (one page)

9.4A   Creating a Table of Chemicals and Their Properties

Your instructor will assign a list of 3–5 compounds, similar to the number of materials you would be using in a typical laboratory experiment. Use the handbooks and databases listed previously in this chapter to prepare a table of properties of the compounds, including hazard information. Similar information should be gathered before each experiment.

For each chemical compound, provide:

1. Name

2. Formula

3. Formula weight

4. Structure (draw the structure)

5. Melting point or boiling point (note the pressure if it is not 760 mmHg)

6. Physical state at room temperature (gas, solid, liquid, solution in solvent)

7. Any hazardous indications (e.g., toxic, flammable, carcinogen, etc.)

8. Source, including page number

An additional column can be added for comments to clarify any of the above points.

Warning: Do *not* get *all* of your information from one website or book. Use at least *two* sources—namely, one web source and one non-web source. If you discover a second web source that is useful, then you may use it in place of a non-web source. You may also use a source that is not listed, as long as you cite it properly. Some chemical compounds appear to be listed multiple times in some resources, such as in the Aldrich catalog. These might not be the same—they could be compounds of the same name but containing isotopic substitutions (e.g., deuterium in place of hydrogen, as in CD_3OD or deuterated methanol, where D represents ^2H), or there may be different levels of purity offered by a commercial supplier. Use information about the material of highest purity whenever possible. Finally, most print resources have a formula index or CAS registry number index in the front or back that may be useful when the compound is named in some unexpected way.

9.4B | Performing a SciFinder Search

Use SciFinder to look up one of the following chemistry authors based on the letter that *your* last name begins with.

YOUR LAST NAME	CHEMISTRY AUTHOR
A–C	Phil S. Baran
D–F	Abigail G. Doyle
G–I	Neil K. Garg
J–K	Jeffrey S. Johnson
L–N	James P. Morken
O–P	Nicola L. B. Pohl
Q–R	Sarah E. Reisman
S–T	Richmond Sarpong
U–Z	M. Christina White

Use the following general plan for locating an article (expect this to take a while):

1. Register for SciFinder through your library website. After registering, you can use the website anywhere. Once you link to SciFinder you will be taken to a screen with a variety of search options. SciFinder defaults to the keyword search, so you must change the search to "author" for this part.

2. Search for your author. You may search by the author's first and last name, last name only, last name and first initials, etc. For this assignment, perform your search based on the name given above (first, middle initial, last). SciFinder will provide a list of abstracts that match your search. You have been assigned an author based upon your last name. One problem is that you need to consider alternative names that could be used. For instance, some of Sarah E. Reisman's papers could be listed under S. E. Reisman or Sarah Reisman, or there could be another person of the same or similar name.

3. Find a journal article that deals with some topic in organic chemistry.

4. Most citations will include links to PDF (or HTML) versions of the full text article. Click on the "Full-Text" icon given with each entry; then, you will be directed to a website where you can view/download the paper. If the article that you would like is unavailable on SciFinder, your best bet will be to ask one of the librarians to help you locate the journal on the library website.

5. Print both the abstract and the first page of the journal article. *Be prepared to cite the article using the journal article citation format. Include a copy of the abstract and first page of the paper with your assignment.*

Once you find a journal article, provide a one-paragraph summary of the work. Do *not* simply rewrite the abstract. Use your own words, and use good English with complete sentences. You need not analyze the paper in detail, because much of the paper may be unfamiliar to you. Along with the one paragraph, include one or more chemical structures or a reaction sequence. *The whole of this (text plus pictures) should* not *be more than one page long or you will lose points.* You must cite the journal article. An example is given at the end of this chapter.

Preparing a Summary of a Review Article or Popular Journal Article

Look through the following review and popular journals, and find a review article that sounds interesting, even if you do not understand it thoroughly. *It does not have to be hard-core chemistry, just chemistry-related. The article you choose should be at least two pages long.* If you feel at all uncertain about your choice, feel free to ask questions. Simply start scanning the following journals and magazines until you find something interesting.

*Accounts of **Chem**ical **Research***

*Journal of **Chem**ical **Educ**ation*

*Chemistry and **Ind**ustry (**London**)*

Discover Magazine

*Chem**istry in **Brit**ain*

*Scient**ific **Am**erican*

*Chem**ical and **Eng**ineering **News***

Print the first page of the article and summarize the article. Your summary cannot be longer than one page, and it does not need to include pictures unless you wish to do so. *Attach the* first page *to your assignment. Make sure you cite the article properly.*

9.4D | Assignment Examples

If you are unsure about any information in your table, include a question mark or a comment. Your table can be turned 90° on the page if you wish, and it can be more than one page long if necessary. Note that units are provided in the column headings.

CHEMICAL TABLE

The following table contains all of the information required for each compound.

NAME	FORMULA	FORMULA WEIGHT, g/mol	STRUCTURE	mp OR bp, °C	STATE	HAZARDS	SOURCE	COMMENTS
Nitric Acid	HNO_3	63.01		–	Liquid, fuming	Corrosive, oxidizer	1	No mp or bp given
Paclitaxel	$C_{47}H_{51}NO_{14}$	853.92		mp 213–216	Solid	Antineo-plastic	2,3	Other names: Taxol
3-Ethylpyridine	C_7H_9N	106.16		bp 166	Liquid	Corrosive, toxic	4	"Technical grade" (contains an impurity)

Sources: (1) *Handbook of Fine Chemicals*; Sigma-Aldrich: Milwaukee, WI, 1996–1997; pp 1073–1074. (2) *Combined Chemical Dictionary* [Online]; Taylor & Francis Group. https://ccd .chemnetbase.com (accessed April 2022). (3) *MilliporeSigma Home Page*. https://www.sigmaaldrich.com (accessed April 2022). (4) *Handbook of Fine Chemicals*; Sigma-Aldrich: Milwaukee, WI, 1996–1997; p 708.
© Sigma-Aldrich Co. LLC. Reproduced with permission from Merck KGaA, Darmstadt, Germany and/or its affiliates.

REVIEW/POPULAR ARTICLE SUMMARY

After your SciFinder search, you will need to prepare a summary of a primary research publication that you find. *Please follow the general format presented in the following example.* Number your pictures, schemes, tables, or graphs so that you can refer to them easily in the text. It is not necessary to go into exhaustive detail. Try to pick out at least one feature that you understand and highlight it.

A synthesis of strychnine by a longest linear sequence of six steps. Strychnine is an alkaloid that has been synthesized by a few different chemists, including Nobel Laureate R. B. Woodward. In this paper, D. Martin and C. Vanderwal describe a synthesis of strychnine that is much shorter than any of the prior ones. There are a couple of key steps in this synthesis highlighted in Scheme 1. First, the authors developed a way to open the ring of a pyridine cation (compound **1**) to generate a diene (compound **2**) for a Diels–Alder reaction. The Diels–Alder reaction forms a cyclohexene that is part of a polycyclic ring system in compound **3**. The alkene's position has moved from its initially expected location so that it is in conjugation with the aldehyde. The other key step is a conjugate addition of a carbanion to the beta position of the alpha,beta-unsaturated aldehyde **4**, forming another ring. This reaction took advantage of a Brook rearrangement that transfers a silicon group to an alkoxide ion, in order to make the carbanion. After another step, the synthesis of strychnine was completed. The overall synthesis required a total of nine steps—three steps for one branch of the synthetic pathway, and six steps in the main sequence. Most of the other syntheses of strychnine (including Woodward's) have required 20 to 30 steps, so this was a big improvement.

Scheme 1: **Figure 9.4**.

Sources:

1. Martin, D. B. C.; Vanderwal, C. D. A Synthesis of Strychnine by a Longest Linear Sequence of Six Steps. *Chem. Sci.* **2011**, *2*, 649–651.

2. Web of Science. https://clarivate.com/webofsciencegroup/solutions /web-of-science/ (accessed April 2022).

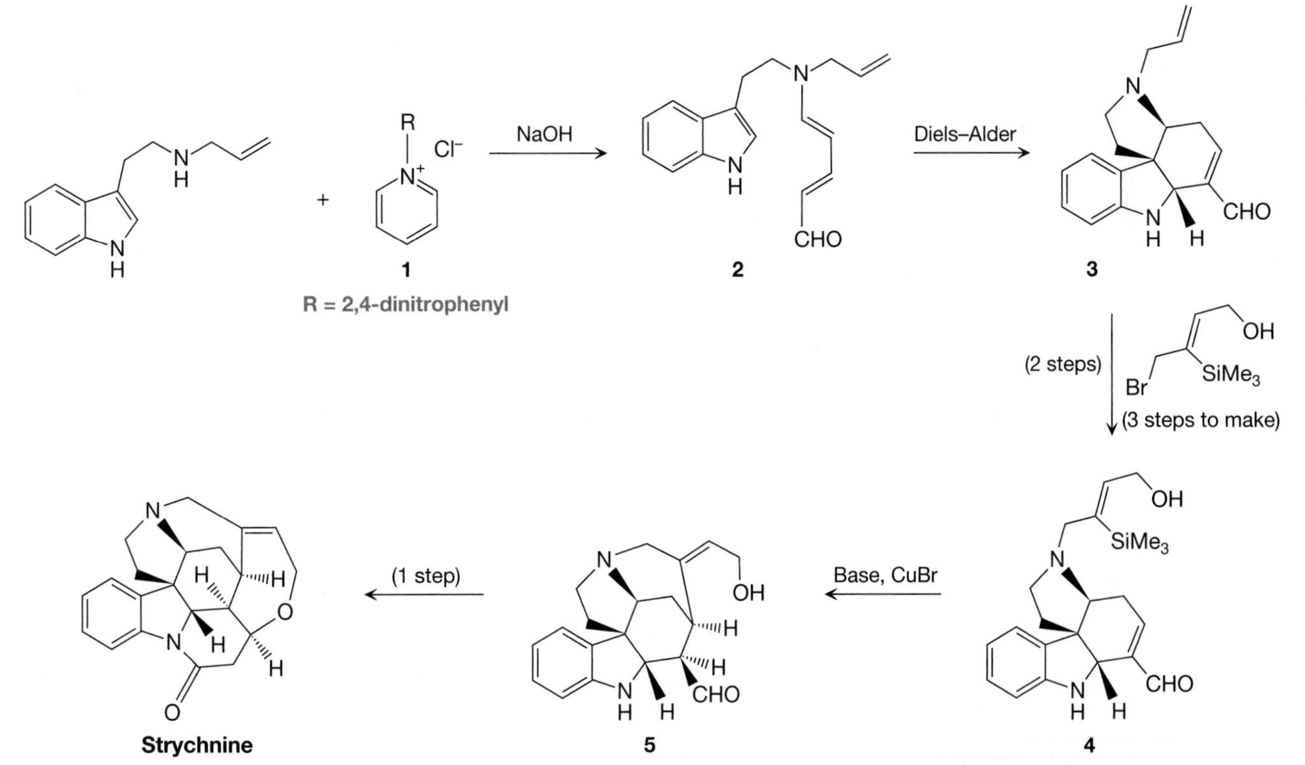

FIGURE 9.4

10

LEARNING OBJECTIVES

- Use liquid CO_2 as an environmentally benign solvent to isolate terpenes from a natural source.

- Recognize phase behavior of CO_2 at different temperatures and pressures.

- Identify and quantify organic compounds, using gas chromatography and spectroscopy.

- Use organic–aqueous extraction to separate organic compounds according to their acidity.

- Purify an organic solid by recrystallization.

- Implement melting point measurement to obtain evidence of identity and purity of an organic solid.

Extractions of Organic Compounds

ESSENTIAL CONTENT

Organic compounds are separated, or extracted, from more complex mixtures like these citrus fruits, using various types of organic solvents.

Shutterstock/Maria Uspenskaya.

What's in that stuff? This question goes back to the very origins of organic chemistry as a discipline, and it's one we still ask when we investigate the organic chemistry of naturally occurring flavors and fragrances or the medicinal properties of plants used by Indigenous peoples. The first step to answering it is to isolate the different components so that they may be studied in more detail. To isolate the individual components requires separation techniques, and a variety of these are available, including methods of extraction.

In part 1 of the extraction experiment, we use liquid carbon dioxide to extract essential oils (terpenes, in this case) from citrus peels. In part 2, an aqueous–organic extraction sequence is used to separate a three-component mixture based on the acidities of the components. The techniques learned in part 2 are used throughout the semester to separate acidic or basic compounds.

Both of these extraction techniques exploit the fact that different compounds have different intermolecular interactions. This allows selected components of a mixture to be transferred into a separate liquid phase.

10.1 PART 1: LIQUID CARBON DIOXIDE EXTRACTION OF CITRUS PEEL

10.1A Pre-Lab Reading Assignment

Review the following before attempting part 1 of this experiment:

- Experiment: Chapter 10 (this experiment)
- Technique: Gas chromatography (Chapter 5, section 5.5B)
- Technique: Infrared (IR) spectroscopy (Chapter 6, section 6.2)

10.1B Background

Essential oils are organic compounds extracted from natural sources, and often are prized for their medicinal properties, or as flavorings or fragrances. Many of the purified components from these oils are classified as terpenes (**Figure 10.1**), a general class of compounds that are derived from units of isoprene (C_5H_8, 2-methyl-1,3-butadiene). These compounds have carbon counts that are generally multiples of five: a monoterpene has 10 carbons, a sesquiterpene has 15 carbons, a diterpene has 20 carbons, and so on. They may be cyclic or acyclic and they may be saturated or unsaturated. They also are frequently oxidized at various locations, which can give them alkene, alcohol, or ketone functional groups. They also show signs of skeletal restructuring via carbocation rearrangements like hydride shifts and alkanide migration. These processes are biosynthetic, happening inside organisms with enzymes as the catalysts, and they lead to a tremendous diversity of structures. Terpenes offer a wealth of untapped potential in new drug discovery and development.

Oil of caraway is an essential oil from caraway seed, a dried fruit mainly grown in Finland. Approximately 1–3% of the seed weight contains two main compounds, called carvone and limonene (Figure 10.1). Both compounds are terpenes, with carbon skeletons composed of isoprene units joined together, and both are chiral. The

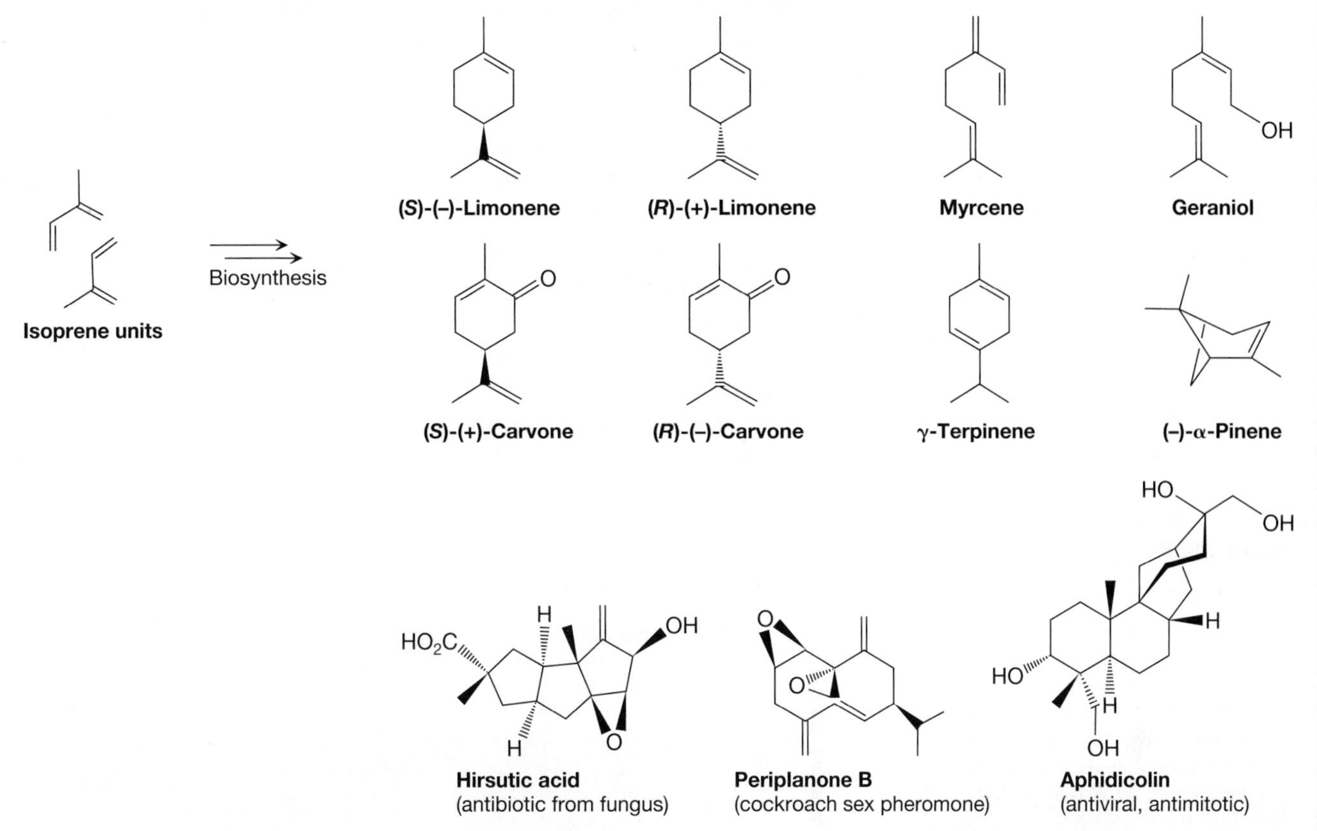

Isoprene units

Biosynthesis

(S)-(–)-Limonene   **(R)-(+)-Limonene**   **Myrcene**   **Geraniol**

(S)-(+)-Carvone   **(R)-(–)-Carvone**   **γ-Terpinene**   **(–)-α-Pinene**

Hirsutic acid
(antibiotic from fungus)

Periplanone B
(cockroach sex pheromone)

Aphidicolin
(antiviral, antimitotic)

FIGURE 10.1

Structures of some terpenes found in nature. Combinations of two, three, four, or more isoprene units, plus oxidations and rearrangements at various locations, lead to a diversity of structures.

two enantiomers of carvone have drastically different smells—one smells like caraway, while the other smells like spearmint.

Many terpenes are naturally occurring materials with interesting biological activities. Limonene and geraniol are known to have cancer chemopreventive properties, while others such as aphidicolin are antimitotic (stop cell division) and have been examined for their potential in cancer chemotherapy development.[1] Others have interesting roles as secretions with attractant or defensive purposes. Periplanone B, for example, is the sex pheromone of the American cockroach.

The traditional ways to extract essential oils involve steam distillation or extraction with an organic solvent. In steam distillation, the volatile oils distill along with the water, and separate as they condense back to the liquid phase. While this is effective, it is an energy-intensive process to heat water to make steam, and the high temperatures can decompose some sensitive organic compounds. Extraction with organic solvents can be handled at lower temperatures, but there are additional measures needed to prevent release of the volatile organic compounds into the atmosphere, and traces of the organic solvents may remain in the product.

A more environmentally benign approach involves the use of liquid carbon dioxide (CO_2) or supercritical carbon dioxide ($scCO_2$). A supercritical fluid is a form of matter that is neither liquid nor gas, but has properties of both. In the liquid or supercritical form, CO_2 is a valuable nonpolar organic solvent that is excellent for extracting essential oils. Importantly, it is nonflammable (unlike many organic solvents), readily available, relatively nontoxic, and easy to remove from the product. Therefore, its use in the food products industry has become well-established,

---

[1]Wang, G.; Tang, W.; Bidigare, R. Terpenoids as Therapeutic Drugs and Pharmaceutical Agents. In *Natural Products: Drug Discovery and Therapeutic Medicine*; Zhang, L., Demain, A. L., Eds.; Humana Press: Totowa, NJ, 2005; pp 197–228.

especially for the decaffeination of coffee. Even though CO_2 is a greenhouse gas, supercritical CO_2 used industrially is recaptured and reused, so that very little is released to the atmosphere.

The advantages of CO_2 make it an excellent alternative from the perspective of *green chemistry*. Green chemistry is a way of designing and managing chemical processes and products so as to minimize environmental impact. "The 12 Principles of Green Chemistry" provide a useful framework for these considerations (see Chapter 1).[2]

Normally CO_2 is either a solid ("dry ice") or a gas, but the liquid phase can exist at the combinations of temperature and pressure shown in the phase diagram in **Figure 10.2**. At pressures below the triple point (about 5 atm), solid CO_2 sublimes directly to a gas without any liquid phase. As the pressure is increased, the liquid phase is increasingly available at a wider range of temperatures and pressures, and beyond the critical point (74 atm and 31°C), the supercritical fluid phase may be found. In this experiment, we work with CO_2 in the liquid phase, very close to the triple point.

FIGURE 10.2

The phase diagram for carbon dioxide, showing the solid, liquid, and vapor phases at various temperatures and pressures. The supercritical fluid phase is the region above and right of the critical point.

Gilbert, T.; Kirss, R.; Bretz, S.; Foster, N. *Chemistry*, 6th ed.; W. W. Norton: New York, 2020; p 524.

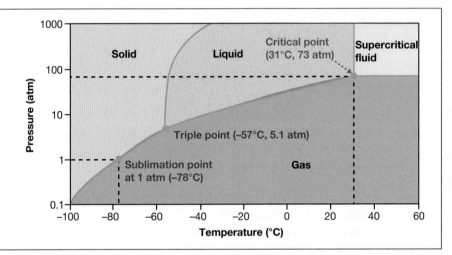

Limonene can be obtained via solid–liquid extraction with pentane or dichloromethane, or via steam distillation. In this experiment, however, we use liquid CO_2 to extract limonene from citrus peels in a more environmentally benign fashion. We then use gas chromatography to measure the terpene content of the essential oils that are obtained.

10.1C Experimental Procedure

You will work in pairs. One person will carry out the procedure[3] with orange peels, whereas the other will use lemon peels. You will exchange your data with your lab partner, so that each of you can prepare your own separate report which compares the results from orange and lemon. Be sure to insert a footnote to properly credit your partner whenever you refer to your partner's data in your report.

1. Grate the colored portion of the citrus peel, using the fine grating surface of a standard kitchen grater, and then collect the gratings in a pre-weighed weighing boat until you have at least 2.5 g of gratings.

---

[2]Anastas, P.; Warner, J. *Green Chemistry: Theory and Practice*; Oxford University Press: New York, 1998.
[3]This procedure is adapted from McKenzie, L. C.; Thompson, J. E.; Sullivan, R.; Hutchison, J. E. Green Chemical Processing in the Teaching Laboratory: a Convenient Liquid CO_2 Extraction of Natural Products. *Green Chem.* **2004**, *6*, 355–358.

2. Label a 15-mL centrifuge tube and cap with your initials and lab section, and record the exact mass.

3. In order to keep citrus peels from reaching the bottom of the tube, a 10-cm length of copper wire is prepared as follows (**Figure 10.3a**): One end of the wire is coiled about three times around an object such as a pencil, so that the diameter of the coil matches the inside diameter of the centrifuge tube. The remaining wire is straightened to serve as a handle. A small piece of screen mesh is inserted between the coils in the coiled section of wire. Place this into the centrifuge tube (with the mesh at the bottom of the tube), replace the cap, and record the exact mass.

4. Add about 2.5 g citrus peel gratings, and tap the centrifuge tube gently on the benchtop to remove any large air pockets in the gratings. The gratings should not be tightly packed, and should stay on top of the wire coil and screen, leaving an empty space at the conical bottom of the tube (**Figure 10.3b**). Replace the cap and record the exact mass. Subtract the tare weight from step 3 to determine the mass of citrus peel gratings.

5. Fill a large *plastic* graduated cylinder to a depth of about 15–20 cm with warm tap water (about 40–50°C), and place it in the hood.

CAUTION: There are safety hazards associated with increased pressure inside the tube during the extraction. Read all the following steps thoroughly before proceeding!

6. Add crushed dry ice to fill the remaining space in the centrifuge tube, tapping the bottom of the tube gently on the benchtop to remove any large air pockets. Large pieces of dry ice should be avoided.

7. Screw the cap onto the centrifuge tube as tightly as a normal adult human would screw on a cap; it should reach a point where it stops turning, but do not attempt to use all of your strength to tighten beyond that point. If it is too loose, no liquid CO_2 will form. If it is too tight, the centrifuge tube will rupture.

8. Immediately after screwing on the cap, drop the centrifuge tube into the plastic graduated cylinder so that it is immersed in the warm tap water (**Figure 10.3c**), place it in a fume hood, and close the hood sash. Pressure will build up in the capped centrifuge tube at this point. The plastic graduated cylinder serves as a secondary container in the event a tube ruptures, directing any materials upward and away from people. *CAUTION: Do not put hands or face or anything else above the opening of the graduated cylinder while there is liquid CO_2 present.*

9. Observe the centrifuge tube from the side of the graduated cylinder. After about 15 seconds, liquid CO_2 should appear. (If no liquid appears after 1 minute, remove the tube from the cylinder and attempt to tighten the cap further. If repeated trials fail to produce liquid CO_2, get another cap and tube and try again.) The liquid CO_2 should percolate down through the citrus peel gratings, extracting terpenes and carrying them to the bottom of the tube. The gas phase will slowly escape through the threads of the cap over a period of 2–3 minutes, and the liquid CO_2 will slowly disappear, leaving about 0.1 mL of essential oil at the bottom of the centrifuge tube. (If the liquid does not reach the bottom of the tube, the citrus peel gratings may be too tightly packed. Loosen them with a spatula.)

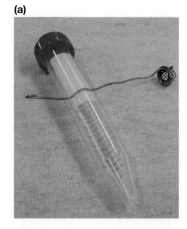

(a)

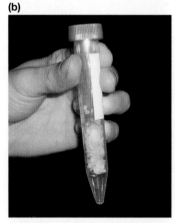

(b)

(c)

FIGURE 10.3

(a) Centrifuge tube for liquid CO_2 extraction, and filtration assembly made of coiled copper wire with a mesh screen inserted between the coils. (b) Tube loaded with citrus peels, before adding CO_2. (c) Extraction in progress.

Courtesy of Gregory K. Friestad.

10. *CAUTION: Wait until all liquid CO$_2$ is gone before proceeding.* After all liquid CO$_2$ is gone and gas is no longer escaping from the tube, remove the tube from the cylinder and open the cap slowly.

11. Repeat steps 6–10 again with the same batch of citrus peel gratings. If the citrus peel was frozen by the dry ice, it is not necessary to thaw it between extractions. If the water in the graduated cylinder cools to room temperature or below, replace it with warm tap water. *CAUTION: Do not add hot water while a capped centrifuge tube is still in the graduated cylinder. A sudden temperature increase can cause the tube to rupture.*

12. Carefully remove the solids from the centrifuge tube using a pair of tweezers to pull on the vertical segment of copper wire. Discard the citrus peel in the designated container and return the copper wire and screen assembly for cleaning and reuse.

13. Dry the outside of the centrifuge tube plus cap with a paper towel, then record its mass. Subtract the tare weight measured in step 2 to find the mass of citrus oil.

14. Obtain a gas chromatogram of the citrus oil. Print two copies—one for you and one for your lab partner. By comparison with standard chromatograms for limonene, γ-terpinene, α-pinene, and β-myrcene, determine which of these components are present and calculate their ratios. This can be done outside of lab.

15. Obtain an infrared spectrum of the product. Print two copies—one for you and one for your lab partner.

16. You will have a complete set of data for either orange oil or lemon oil. Your partner will have the other set. Each of you will need to discuss and compare both data sets in your lab report. Before leaving the lab, make sure you have shared these data sets with your partner.

10.2 PART 2: SEPARATION OF A THREE-COMPONENT MIXTURE BY LIQUID–LIQUID EXTRACTION

10.2A | Pre-Lab Reading Assignment

Review the following before attempting part 2 of this experiment:

- Experiment: Chapter 10 (this experiment)
- Technique: Filtration (Chapter 5, section 5.1)
- Technique: Extraction (Chapter 5, section 5.2)
- Technique: Recrystallization (Chapter 5, section 5.4)
- Technique: Melting point measurement (Chapter 5, section 5.6B)

10.2B Background

The liquid–liquid two-phase extraction technique used in this experiment involves partitioning compounds between immiscible (mutually insoluble) liquid phases. When placed together, the immiscible liquids form two layers, with the liquid of higher density on the bottom. These layers can be shaken together to allow solutes to transfer from one liquid solution to the other. The layers then separate again, so that they can be drained off into separate flasks. This is almost always done with an aqueous phase and an organic solvent (e.g., water–ethyl acetate), but other immiscible pairs may also be used. Examples include acetonitrile–hexane and methanol–perfluorohexane (C_6F_{14}).

In aqueous–organic extractions, inorganic salts are soluble in the aqueous phase, while organic substances are soluble in an organic phase, such as hexane, dichloromethane, or *tert*-butyl methyl ether. If you have a mixture of organic substances with a salt such as NaCl, you can add water and *tert*-butyl methyl ether, shake up the mixture, and wait for the layers to separate. The ionic compound, being very polar, will dissolve in the more polar aqueous phase, whereas the organic substances will dissolve in the ether. The ionic compound can then be removed from the organic substances by separating the layers. Organic reactions often involve various inorganic salts as reagents or by-products, so the reactions are often followed by an aqueous "workup," which involves this kind of partitioning to separate water-soluble components from the organic product mixture.

A useful variation on this procedure makes it possible to separate two organic compounds with different functional groups, if the functional groups have acidic or basic behavior. Imagine, for example, that you have a solution of cyclohexylamine (an amine, which is basic, like ammonia) and naphthalene (a nonbasic organic compound) in an organic solvent. The basic compound can react with aqueous hydrochloric acid to make a salt (**Figure 10.4**). This changes its solubility, because its salt form is an ionic compound and is no longer soluble in the nonpolar organic phase. In this scenario, the cyclohexylamine salt is in the polar aqueous phase, while the naphthalene remains behind in the organic phase because it has no basic functional group to be converted to a salt. After the layers are subsequently separated into two flasks, the two compounds may be processed separately. The cyclohexylamine can be converted back to its "free base" by neutralizing the hydrochloric acid.

This type of extraction was a key to the advancement of organic chemistry in the 1800s. During that time, many chemists began to unlock the mysteries of traditional medicinal plants—namely, chemists learned they could use extraction to separate, purify, and identify the components responsible for the medicinal properties of plants. Cocaine, for example, is a base, and it can be separated from the rest of the organic materials in coca leaves (its plant source) by extracting with aqueous acid. This converts the basic cocaine to a salt that dissolves in the aqueous phase, allowing it to be separated from the other plant materials.

In this experiment, you use the differences in acid strengths of carboxylic acids, phenols, and amides to separate *p*-toluic acid, *p-tert*-butylphenol, and acetanilide from one another. The *p*-toluic acid is converted to a salt by aqueous sodium bicarbonate ($NaHCO_3$) or sodium hydroxide (NaOH). The *p-tert*-butylphenol is converted to a salt by aqueous NaOH, but not by $NaHCO_3$. Acetanilide has a neutral functional group, which cannot be converted to a salt in aqueous solution. Using sequential extraction with aqueous $NaHCO_3$, then aqueous NaOH, you can then separate the three components from an organic phase solution.

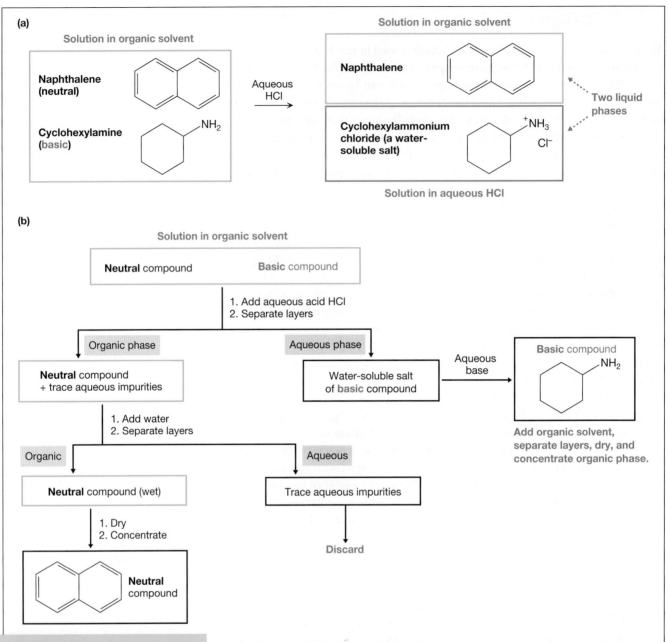

FIGURE 10.4

(a) Addition of aqueous acid to a solution of naphthalene and cyclohexylamine (two organic compounds) in organic solvent converts the basic cyclohexylamine to a salt. When the aqueous and organic phases separate, the salt is found in the aqueous phase, whereas naphthalene remains in the organic phase. (b) A flowchart shows the operations involved in separating and recovering neutral and basic compounds from a mixture.

 Aqueous–Organic Extractions and Drying Organic Solutions

10.2C Experimental Procedure

In this procedure,[4] you will generate several colorless liquids that might be difficult to distinguish later. They could be incorrectly identified, inadvertently discarded, or mixed by mistake. If any of these things happen, you may have to restart the experiment. To avoid this, *label the flasks*, and *save all of the fractions* of any extraction until the final products are purified and identified.

1. Weigh out 2.0–2.7 g of the mixture of *p*-toluic acid, *p*-*tert*-butylphenol, and acetanilide provided by your lab instructor into a 125-mL Erlenmeyer flask. Assume this mixture is made up of a 1:1:1 mass ratio of *p*-toluic acid, *p*-*tert*-butylphenol, and acetanilide. Record the mass.

---

[4]This procedure is adapted from Harwood, L. M.; Moody, C. J. *Experimental Organic Chemistry: Principles and Practice*; Blackwell Scientific Publications: Oxford, 1989; pp 114–127.

2. Add 25 mL of *tert*-butyl methyl ether to the mixture, and swirl the mixture to ensure complete dissolution. If solid is still present, use a glass rod to break up any chunks. You may also use a hot plate to briefly heat the mixture, but do not boil it.

EXTRACTION WITH AQUEOUS NaHCO₃: SEPARATION OF *p*-TOLUIC ACID

3. Slowly add 10 mL of aqueous $NaHCO_3$ solution to the mixture.

    CAUTION: $NaHCO_3$ reacts with carboxylic acids to produce carbon dioxide (CO_2) gas, which will bubble and foam out of the mixture.

    Notice whether the new aqueous layer forms above or below the organic (*tert*-butyl methyl ether) solution. *Be sure the stopcock is tightly secured and in the closed position* before *pouring solutions into the separatory funnel.* Check this by adding a few milliliters of distilled water (the separatory funnel does not need to be dry before use). Pour the mixture into a 125-mL separatory funnel (**Figure 10.5**) supported by an iron ring clamped to a vertical support bar or ring stand in the fume hood. Place the stopper in the top of the separatory funnel, and invert the funnel while holding the stopper in place. Point the stopcock into the back of the hood and open it to vent the CO_2 gas that builds up. Close the stopcock and gently mix the two layers by rocking and shaking the separatory funnel back and forth. Repeat this mixing and venting several times until no more gas forms. Place the funnel in the iron ring, remove the stopper, and allow the layers to separate.

4. Open the stopcock to allow the aqueous layer to drain into a clean, labeled 125-mL Erlenmeyer flask. *Note:* The bottom layer will *not* drain smoothly from the separatory funnel if the stopper is left in the top. When the interface between the layers just reaches the top of the stopcock, close the stopcock to retain the organic layer in the funnel.

5. Repeat steps 3 and 4 two more times.

EXTRACTION WITH AQUEOUS NaOH: SEPARATION OF *p-tert*-BUTYLPHENOL

6. Using the organic layer remaining in the separatory funnel, extract with three 10-mL portions of aqueous NaOH, following the same procedure described in steps 3 and 4. The NaOH extracts should be drained off into a clean, labeled 125-mL Erlenmeyer flask, separate from the $NaHCO_3$ extracts.

7. Add 5 mL of distilled water to the ether remaining in the separatory funnel and mix. Allow the layers to separate. Drain the water layer into the flask containing the three NaOH extracts.

RECOVERY OF ACETANILIDE

8. Drain the organic phase into a clean, labeled 50-mL Erlenmeyer flask. This organic phase should contain acetanilide, the remaining nonacidic compound.

You should now have three fractions in Erlenmeyer flasks—one containing aqueous $NaHCO_3$ extracts (sodium salt of *p*-toluic acid), one containing aqueous NaOH extracts (sodium salt of *p-tert*-butylphenol), and one containing the organic phase (acetanilide). The next steps of the procedure describe how to isolate the purified solid components from these solutions.

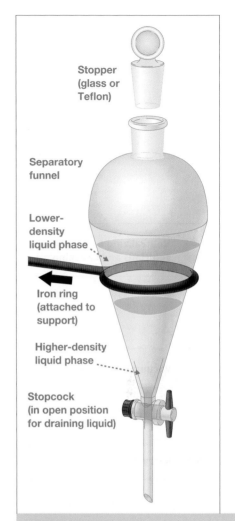

Stopper
(glass or
Teflon)

Separatory
funnel

Lower-
density
liquid phase

Iron ring
(attached to
support)

Higher-density
liquid phase

Stopcock
(in open position
for draining liquid)

FIGURE 10.5

Separatory funnel. The relative densities of liquids will predict the locations of organic and aqueous phases.

Steps 9, 10, and 11 may be done in any order. Step 11 (isolation of acetanilide using a rotary evaporator) typically has some waiting time, so some of you should do that step first.

Note: It is usually not possible to complete all of the steps in one lab period. A good stopping point is after steps 9b and 10 have been completed. Steps 9c and 12 may be reserved for a later lab period.

9. **Isolating *p*-toluic acid.** In this step, the sodium salt of *p*-toluic acid is converted back to the acid and is purified.

   a. To the NaHCO$_3$ extracts, carefully add aqueous 6 M HCl dropwise. Foaming will occur as the NaHCO$_3$ is neutralized, and a precipitate of *p*-toluic acid will form. Continue adding 6 M HCl dropwise with swirling until the solution tests acidic (pH < 3) with pH paper. (To test for acidity, dip a stirring rod into the solution and touch the stirring rod to a piece of pH test paper. Do *not* dip the pH paper into the solution.)

   b. Separate the solid from the solution using vacuum filtration with a Büchner funnel. Allow the vacuum to pull air through the solid for a few minutes while you work on step 10 or 11. When the solid is dry, transfer it to a tared (weighed while empty) 25-mL Erlenmeyer flask and measure its mass. Transfer about 1 mg of this crude *p*-toluic acid to a capillary tube for a melting point determination at a later time.

   c. Recrystallize from ethanol to purify the *p*-toluic acid. (This may be done in a separate lab period.) Turn on a hot plate to medium heat. Place about 20 mL of ethanol in a 50-mL Erlenmeyer flask, and warm it on the hot plate (**Figure 10.6a**). When the ethanol begins to boil, slowly add the hot ethanol to the flask containing the precipitated *p*-toluic acid. Add a couple of mL of hot ethanol at a time, swirling between each addition, and check to see if the solid dissolves. Keep both flasks warm by placing them on the hotplate between additions of ethanol. Continue adding in small portions until the solid is just dissolved. Too much solvent will markedly reduce the amount of *p*-toluic acid that you recover. Set the solution aside to cool, and allow it to stand undisturbed while crystallization occurs (**Figure 10.6b**). When the flask reaches room temperature, place it in an ice bath to complete the crystallization, then separate the crystals by vacuum filtration using a small Hirsch funnel (**Figure 10.6c**). Allow the crystals to dry while you work on steps 10 or 11, then transfer them to a tared vial and record the mass.

▶ **Neutralizing Acidic or Basic Solutions and Checking pH**

▶ **Vacuum Filtration**

▶ **Recrystallization**

FIGURE 10.6

(a) Crude solid *p*-toluic acid, with ethanol heating on a hotplate.
(b) *p*-Toluic acid recrystallizing after cooling in an ice-water bath.
(c) Clamped filter flask for vacuum filtration with a small Hirsch funnel.

Courtesy of Gregory K. Friestad.

(a)

(b)

(c)

10. **Isolating *p-tert*-butylphenol.** In this step, the sodium salt of *p-tert*-butylphenol is converted back to the neutral phenol and is purified.

   a. Heat the NaOH extracts to about 60°C on a hot plate in order to remove any remaining traces of *tert*-butyl methyl ether that might inhibit the crystallization of the *p-tert*-butylphenol. Cool the solution for a few minutes in an ice bath. Carefully add 3 M aqueous HCl dropwise to the cooled solution. Crystalline *p-tert*-butylphenol should begin to form. Continue dropwise addition of 3 M HCl until the solution tests acidic (pH < 3) with pH paper. (To test for acidity, dip a stirring rod into the solution and touch the stirring rod to a piece of pH test paper. Do *not* dip the pH paper into the solution.)

   b. Separate the crystalline *p-tert*-butylphenol by vacuum filtration. Allow the vacuum to pull air through the crystals for a few minutes, then transfer the crystals to a dry filter paper and allow them to dry while you work on another step. After they are dry, transfer them to a tared vial and record the mass.

   c. To obtain a higher purity, the *p-tert*-butylphenol could be recrystallized from petroleum ether. Whether or not you perform this step is up to your instructor.

11. **Isolating acetanilide.** In this step, the acetanilide in the organic phase is dried and isolated.

   a. To the 50-mL Erlenmeyer flask containing the organic phase, add approximately 0.75 g of anhydrous Na_2SO_4. This is a drying agent that should remove any traces of water from the organic solution. Allow the mixture to stand for 5 min with occasional swirling. As the anhydrous Na_2SO_4 absorbs water, it becomes clumpy. If it is all clumped, there may be insufficient drying agent. Add a little more so that some of the drying agent particles remain unclumped and swirling around freely. Prepare a fluted filter paper for gravity filtration (see Figure 5.1 for guidance on the preparation of the fluted filter paper). Place the fluted filter paper in a conical stem funnel and insert the stem into a tared 100-mL round-bottomed flask (**Figure 10.7a**). Pour the organic solution through the fluted filter paper and into the flask. Using a pipet, add 2 mL *tert*-butyl methyl ether to the remaining drying agent, swirl, and pour this through the fluted filter paper as a rinse to ensure complete transfer.

   b. Remove the solvent using a rotary evaporator ("rotovap," **Figure 10.7b**), which rapidly distills volatile liquids under vacuum. The acetanilide may crystallize while the flask is attached to the rotovap, but it may also remain as an oily residue. Remove the flask from the rotovap and cool it in an ice bath to induce crystallization. If necessary, scratch the bottom of the flask with a glass rod, or add a seed crystal. Transfer the acetanilide crystals to a tared vial and allow them to dry while you work on another step. Record the mass of acetanilide.

12. Measure melting points of your crude *p*-toluic acid, recrystallized *p*-toluic acid, *p-tert*-butylphenol, and acetanilide. Record the melting range, from first visible liquid to complete melting.

Your instructor may also request that you obtain additional data, for example from IR and/or NMR spectroscopy, to strengthen your evidence of identity and purity for one or more of the isolated compounds.

 Gravity Filtration

Rotary Evaporation

 Melting Point Measurement

(a)

(b)

FIGURE 10.7

(a) Gravity filtration with a fluted filter paper for removal of drying agent from an organic solution.
(b) Rotary evaporator for rapid removal of organic solvent from solutions.

Courtesy of Gregory K. Friestad.

Your lab report should include separate results and discussion sections for both part 1 and part 2.

For part 1, discuss the amount of oil extracted from each citrus peel sample (orange and lemon), comparing the different citrus peels in terms of the percentage of terpenes by mass. Use GC to identify the components of the orange extract and lemon extract, referring to retention times observed by GC analysis of the standards. Using integration data from the GC analysis, compare the relative amounts of each terpene component in orange and lemon oil. Interpret the IR spectra of the orange and lemon oils, and use the IR data to identify the major component. Label the peaks in the IR and GC data.

For part 2, how well did the separation work in terms of quantity, based on percent recovery of each of the three components? How well did the separation work in terms of purity, based on melting points? Discuss the results of your recrystallization based on percent recovery and melting point comparisons. Did the recrystallization produce a higher purity?

Submit your data sheet along with your report.

10.3A Data to Include in the Laboratory Report

A. PART 1: LIQUID CARBON DIOXIDE EXTRACTION OF CITRUS PEEL

Mass of orange peel placed in extraction tube (g):

Mass of orange oil extracted (g):

Essential oil content of orange peel (%):

Mass of lemon peel placed in extraction tube (g):

Mass of lemon oil extracted (g):

Essential oil content of lemon peel (%):

Gas chromatography data for terpene components in orange oil and lemon oil:

OIL	LIMONENE	β-MYRCENE	α-PINENE	γ-TERPINENE
ORANGE				
Retention time:				
Integration (area):				
Area percent:				
LEMON				
Retention time:				
Integration (area):				
Area percent:				

For the preceding table, retention time and integration are given in the GC output. Identify the peak corresponding to each compound, and place those data in the appropriate column here. Then calculate the area percent of limonene (L), myrcene (M), pinene (P), and terpinene (T) in the following way, as illustrated for limonene:

$$\text{Area percent } L = \frac{L}{(L + M + P + T)}$$

where L, M, P, and T are the integrations (areas) of the peaks from the GC output.

B. PART 2: SEPARATION OF A THREE-COMPONENT MIXTURE BY LIQUID–LIQUID EXTRACTION

Starting mass of 1:1:1 mixture, amount used (g):

Starting mass of each component (g):

Note: Divide starting mass of mixture by 3; this is the theoretical maximum recovery of each component, used as the denominator in calculating percent recovery of each component.

Mass of isolated crude *p*-toluic acid (g):

Percent recovery of crude *p*-toluic acid:

Melting point range of crude *p*-toluic acid (°C):

Mass of *p*-toluic acid after recrystallization (g):

Percent recovery of *p*-toluic acid recrystallization:

Note: Use mass of isolated crude *p*-toluic acid as the denominator here.

Melting point range of recrystallized *p*-toluic acid (°C):

Mass of isolated crude *p-tert*-butylphenol (g):

Percent recovery of crude *p-tert*-butylphenol:

Melting point range of *p-tert*-butylphenol (°C):

Mass of isolated acetanilide (g):

Percent recovery of acetanilide:

Melting point range of acetanilide (°C):

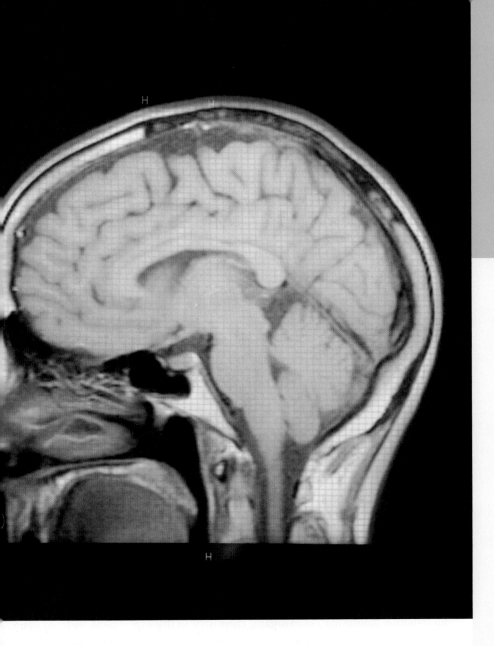

LEARNING OBJECTIVES

- Recognize how theory aids in a *practical* understanding of how NMR spectroscopy works.

- Implement basic instructions for acquisition of an NMR spectrum.

- Demonstrate how various features of NMR spectra are associated with organic chemical structures.

- Identify an unknown compound, using NMR spectroscopy.

NMR Spectroscopic Analysis

MAGNETIC RESONANCE IMAGING

Magnetic resonance is used in organic structure determination as well as tissue imaging for health care, as in this example of MRI data.

Donna Beeler/Shutterstock.

INTRODUCTION

Nuclear magnetic resonance (NMR) is a physical phenomenon in which atomic nuclei absorb radiofrequency energy. It is of great importance in noninvasive medical imaging because it is the foundation of magnetic resonance imaging (MRI). The MRI instrument (**Figure 11.1**) can detect varying concentrations of analytes at locations in three-dimensional space, and use those data to construct a picture of different tissues that may not show up with x-ray imaging.

FIGURE 11.1

An instrument for magnetic resonance imaging (MRI).

Diana Deak/Depositphotos.

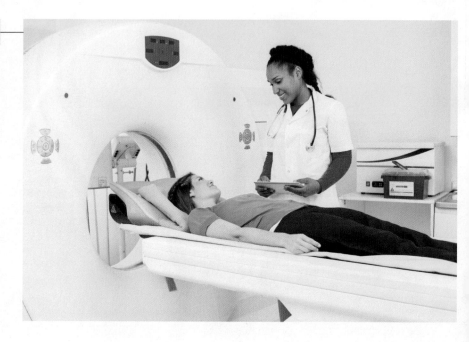

The same magnetic resonance phenomenon used as the basis of MRI can be used to unravel the structural details of organic compounds, providing a powerful tool for structure determination in organic chemistry. In this experiment, we build upon the theory you learned in your lecture course by incorporating practical aspects of interpreting NMR spectroscopic data. This provides a foundation for your use of the NMR technique throughout the semester.

11.1 PART 1: NMR SAMPLE PREPARATION AND ACQUISITION OF A SPECTRUM

11.1A Pre-Lab Reading Assignment

Review the following before attempting part 1 of this experiment:

- Experiment: Chapter 11 (this experiment)
- Technique: Nuclear magnetic resonance (NMR) spectroscopy (Chapter 7, section 7.1)

11.1B | Background

Once you prepare an organic compound in a laboratory experiment, then you need to substantiate the identity of the purified product. Organic chemists can gather information about the product by a variety of instrumental analysis methods, and NMR spectroscopy is probably the most valuable of these. In many of the other experiments in this book, you will use data from NMR spectroscopy to validate what you have made, provide evidence of its purity, or determine ratios of components within a mixture of products. To provide a foundation for this, you will start by preparing a sample of an unknown compound and determining its structure.

▶ **NMR Sample Preparation**

SAMPLE PREPARATION

A typical ^{1}H NMR sample is prepared by placing 5–20 mg of sample into a 5-mm-diameter 18-cm-long glass tube (**Figure 11.2a**). A sufficient amount of a solvent that is NMR-silent (see below) is then added to bring the level up to about 2.5–3.0 cm depth in the bottom of the tube (about 0.5–0.7 mL). The tube is capped and inverted several times to ensure the solution is homogeneous. If there are solid materials floating around in the NMR sample, this can diminish the quality of the spectrum that is obtained. Any solids should be removed by filtering the sample through a small plug of cotton in a Pasteur pipet (**Figure 11.2b**). The NMR data will not provide any information about the solid that was removed.

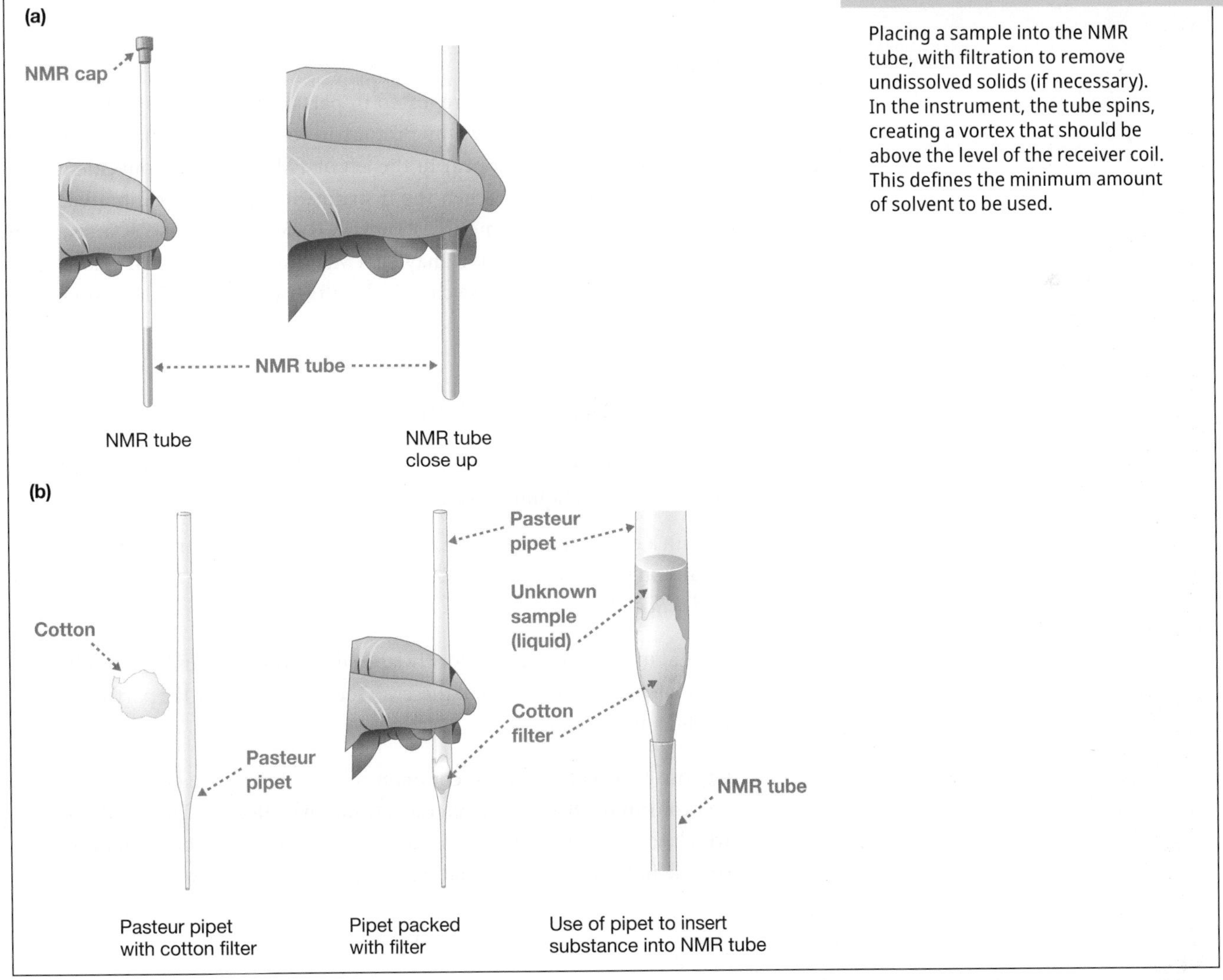

FIGURE 11.2

Placing a sample into the NMR tube, with filtration to remove undissolved solids (if necessary). In the instrument, the tube spins, creating a vortex that should be above the level of the receiver coil. This defines the minimum amount of solvent to be used.

(a)

NMR cap

NMR tube

NMR tube

NMR tube close up

(b)

Cotton

Pasteur pipet

Pasteur pipet

Unknown sample (liquid)

Cotton filter

NMR tube

Pasteur pipet with cotton filter

Pipet packed with filter

Use of pipet to insert substance into NMR tube

SOLVENT CHOICE

Solvents are chosen so that they do not interfere with the analyte peaks in the spectrum. Usually this means that all the hydrogens in the solvent are replaced with deuterium (^2H, or D), an isotope of hydrogen that contains an extra neutron. The nucleus of a deuterium atom gives no peaks in the range of the ^1H NMR spectrum. Deuterochloroform ($CDCl_3$) is one of the most widely used solvents in NMR because it dissolves a wide range of compounds. Very small amounts of $CHCl_3$ in the $CDCl_3$ will always give a small peak at 7.26 ppm. A wide variety of other deuterated solvents are commercially available, including deuterated dimethylsulfoxide (DMSO-d_6, CD_3SOCD_3), deuteroacetone (CD_3COCD_3), and deuterobenzene (C_6D_6). In each case, a small peak is visible from the solvent because not all solvent molecules have had all the hydrogens replaced with deuterium. Generally there will also be a small peak from traces of water that may be in the NMR solvent source bottle, or that may have been introduced by condensation while transferring the sample.

SAMPLE SPINNING

In order to average the magnetic fields produced by the spectrometer in the sample, the sample is spun inside the instrument at 20–40 rpm while the spectrum is acquired. This produces a vortex in the tube that can interfere with acquisition, giving erratic, nonreproducible spectra. The effect of the vortex is minimized if the sample tube is filled to a minimum depth of about 2.5–3.0 cm, keeping the vortex above the area where the radiofrequency receiver coil aligns with the sample tube (Figure 11.2).

CALIBRATION

Chemical shifts of hydrogens are reported in parts per million (ppm) relative to the position of a sharp peak given by protons in tetramethylsilane (TMS). Most peaks in the NMR spectrum appear to the left of TMS, so it is a convenient standard which is set to 0.00 ppm when the instrument is properly calibrated. Stock solutions of NMR solvents sometimes contain 0.05–3% TMS, so if a spectrum shows a small peak at 0.00 ppm, it may be TMS. The spectrum can also be calibrated to the solvent peak, because chemical shifts (in ppm from TMS) are known for all the common solvents.

11.1C | Experimental Procedure (Part 1)

Work with a partner on this procedure. One of you will prepare a solid sample, and the other will prepare a liquid sample.

1. Obtain a clean, dry NMR tube and cap. Your lab instructor will provide details on where to find one. *CAUTION: The NMR sample tubes are expensive and delicate!*

2. Several unknowns, each uniquely labeled, are provided in the lab. With your lab partner, pick two of these, so that you and your lab partner have a solid and a liquid.

3. Choose a solvent. The most common solvent choices are deuterated chloroform ($CDCl_3$) and deuterated dimethylsulfoxide (DMSO-d_6), in which the hydrogens are replaced by deuterium (D). If the compound is soluble in $CDCl_3$, this is usually preferred for two reasons: First, it can be

easily evaporated in order to recover the sample, whereas DMSO has a very high boiling point. Secondly, $CDCl_3$ generally contains only traces of water, whereas DMSO readily absorbs atmospheric moisture, leading to a bothersome water peak that may interfere with peaks from the analyte. Other solvents such as D_2O, acetone-d_6, benzene-d_6, or acetonitrile-d_3 are also sometimes used. First, test the solubility of your compound in $CDCl_3$. Place a tiny amount (just the tip of a spatula, no more than 1–2 mg) of a solid unknown, or one drop of a liquid unknown, into a vial. Add three drops of $CDCl_3$, and carefully observe whether the compound dissolves.

    a. If the compound dissolves in $CDCl_3$, then use $CDCl_3$ as your NMR solvent.

    b. If the compound does *not* dissolve in $CDCl_3$, then use DMSO-d_6 as your NMR solvent.

4. Add enough of the unknown to the NMR tube so that its depth at the bottom of the tube is about the same as the diameter of the tube. This amount will be approximately 5–20 mg, depending on the density of the sample, but you don't need to know the exact mass.

5. Add the NMR solvent to a depth of 3 cm (this corresponds to about 0.6–0.8 mL of solvent). Cap the tube and invert it several times until the sample is homogeneous.

6. Acquire the ^1H NMR spectrum. Your lab instructor will provide further information about sample submission and/or instrument operation. If you are to run the instrument yourself, detailed instructions on instrument operation will be provided separately; be sure to follow these instructions carefully.

7. Interpret the NMR data and identify the structure of the unknown. For the ^1H NMR spectrum, prepare a table with chemical shifts, integrations, multiplicities, and peak assignments. Be sure to include a structure of the compound with your table, and make sure that you have clearly indicated which peak belongs to each proton or set of equivalent protons in the structure. If the ^{13}C NMR spectrum is provided, prepare a similar table of chemical shift and peak assignments. Attach the printouts of the NMR spectra to your report.

11.1D   Tips on Identifying an Unknown Compound

Some additional information about an unknown compound[1] is usually available outside of NMR, such as the type of reaction used to produce the compound, its molecular formula from mass spectrometry or combustion analysis, or IR data that indicate what functional groups are present. This information will narrow the possible structures to just a few isomers. Then, the NMR spectrum can provide very powerful data to distinguish between the possibilities. The most efficient way to proceed is to predict the *number of signals* that each possible structure would exhibit, then rule out all those that are inconsistent with the NMR spectrum. Then, for those structures that are left, predict the *chemical shifts* and rule out any structures that are inconsistent with the spectrum. If necessary, do the same with *integration*, and finally *multiplicity* (signal splitting). If all but one structure has been ruled out, it may not be necessary to go through all of these steps, but they can strengthen your confidence that the structure is correctly identified.

---

[1]A list of some representative compounds and their formulas is available to instructors.

Provide answers to the following questions:

1. How many signals would each of the following compounds give in its ^{1}H NMR spectrum? Explain why. Draw structures and show all the different types of protons for each compound.

    a. Ethane

    b. Propane

    c. Butane

    d. *tert*-Butyl methyl ether

    e. 2,3-Dimethyl-2-butene

    f. (*Z*)-2-Butene

    g. (*E*)-2-Butene

    h. *sec*-Butyl alcohol (Hint: Consider diastereotopic hydrogens!)

    i. 4-Nitrotoluene

    j. *cis*-1,2-Dimethylcyclopropane

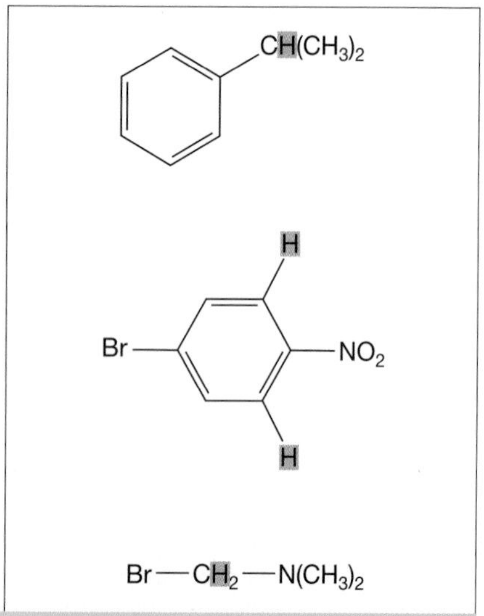

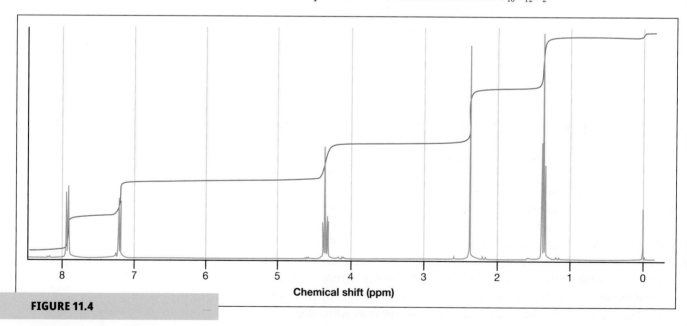

FIGURE 11.3

2. What splitting pattern in the ^{1}H NMR spectrum would you expect for the hydrogen atom(s) highlighted in the compounds shown in **Figure 11.3**? Explain your answer briefly for each compound.

    Your choices are **s** (singlet), **d** (doublet), **t** (triplet), **q** (quartet), or **m** (multiplet).

3. Answer questions a–d about the ^{1}H NMR spectrum in **Figure 11.4** of a compound with the molecular formula $C_{10}H_{12}O_2$.

Chemical shift (ppm)

FIGURE 11.4

The ^{1}H NMR spectrum of a compound with formula $C_{10}H_{12}O_2$. The peak at 0 ppm is tetramethylsilane (TMS).

© Sigma-Aldrich Co. LLC. Reproduced with permission from Merck KGaA, Darmstadt, Germany and/or its affiliates.

a. What is the multiplicity of the highest field signal from this sample?

b. What structural feature is suggested by the singlet at 2.4 ppm? Choose from the following options: CH_3—O—CO; —CH_2—; —O—H; —O—CH_3; $C_{(sp3)}$—CH_3; $C_{(sp2)}$—CH_3.

c. Using the integration trace and the formula of the sample, assign a whole number ratio to the protons as follows: 7.9 ppm _____; 7.2 ppm _____; 4.3 ppm _____; 2.4 ppm _____; 1.3 ppm _____.

d. Multiplet line separations (i.e., the coupling constants, J, in Hz) of the peaks are as follows: 7.9 ppm (7.2 Hz), 7.2 ppm (7.2 Hz), 4.3 ppm (6.8 Hz), 1.3 ppm (6.8 Hz). Which signal is coupled to the quartet at 4.3 ppm?

4. Given the formula and the ^{1}H NMR absorption peaks for several compounds, propose a structure that is consistent with each set of data. In some cases, characteristic IR absorptions and ^{13}C NMR data are given as well. Explain your answer by assigning the protons to the designated peaks and interpreting the IR and ^{13}C NMR data.

a. $C_4H_{10}O$; ^{1}H NMR spectrum: singlet at δ 1.28 (9H), singlet at δ 1.35 (1H).

b. C_4H_8O; ^{1}H NMR spectrum: triplet at δ 1.05 (3H), singlet at δ 2.13 (3H), quartet at δ 2.47 (2H). IR spectrum: strong peak near 1720 cm^{-1}.

c. $C_4H_7BrO_2$; ^{1}H NMR spectrum: triplet at δ 1.08 (3H), multiplet at δ 2.07 (2H), triplet at δ 4.23 (1H), singlet at δ 10.97 (1H). IR spectrum: broad peak in 2500–3000 cm^{-1} region and a strong peak at 1715 cm^{-1}.

d. A $C_8H_4N_2$ compound shows a sharp infrared absorption at 2230 cm^{-1}. Its ^{1}H NMR spectrum has a singlet at δ 7.6 ppm. The ^{13}C NMR spectrum shows signals at δ 132, 119, and 117 ppm.

5. Sketch the ^{1}H NMR spectrum of the structure in **Figure 11.5** using the scale provided. Do not worry about getting the chemical shift values perfectly correct. Instead, show the correct *relative* positions of the peaks and their multiplicities.

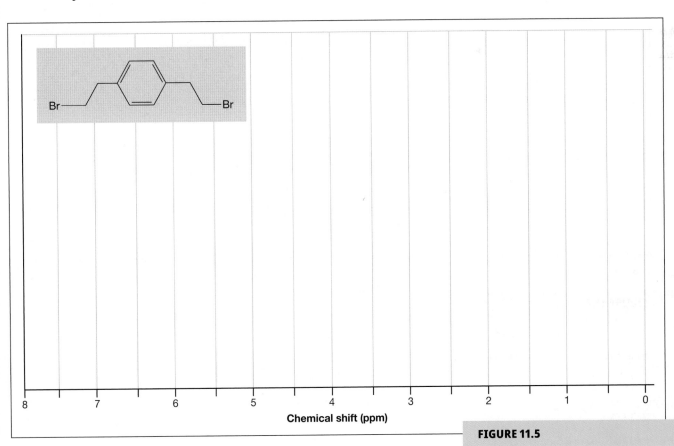

Chemical shift (ppm)

FIGURE 11.5

6. Propose a structure consistent with each of the ^{1}H NMR spectra provided. Explain how each peak corresponds to the structure you have proposed.

a. C_3H_7Br (see **Figure 11.6**)

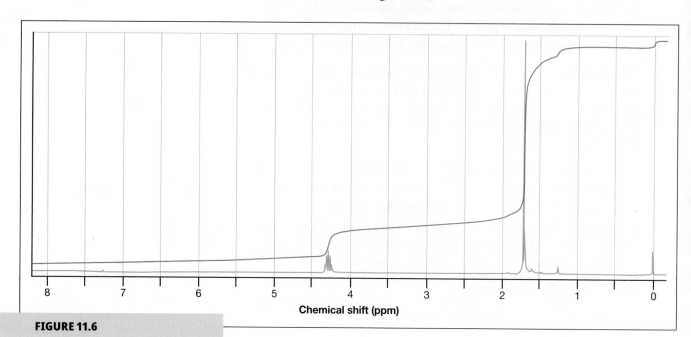

FIGURE 11.6

b. C_4H_7NO (see **Figure 11.7**); ^{13}C NMR peaks at 179.6, 42.4, 30.3, and 20.7 ppm

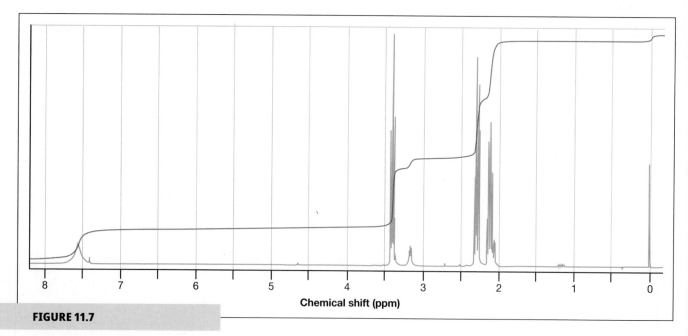

FIGURE 11.7

c. $C_7H_{14}O$ (see **Figure 11.8**); IR peak at 1717 cm^{-1}

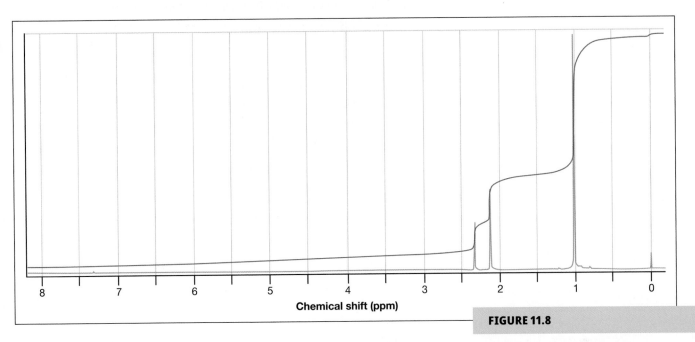

FIGURE 11.8

d. $C_9H_{11}NO_2$ (see **Figure 11.9**); broad ^1H NMR peak offscale at 12 ppm; ^{13}C NMR peaks at 168, 150, 132, 129, 117, 116, 113, and 40 ppm

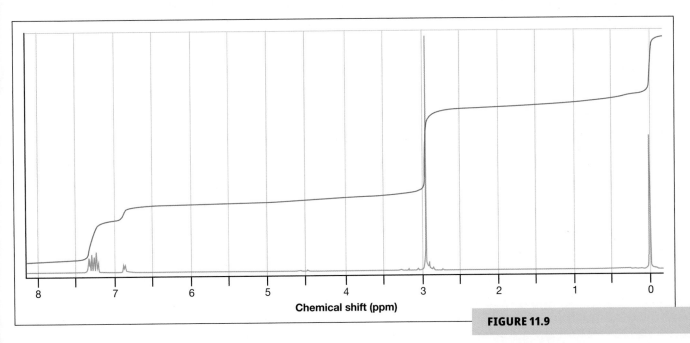

FIGURE 11.9

7. NMR spectra for an unknown compound will be provided to you (this may include ^{1}H, ^{13}C, and DEPT [Distortionless Enhancement by Polarization Transfer] spectra, depending on what is assigned by your instructor). Interpret the NMR spectra and identify the structure of the unknown.

For the ^{1}H NMR spectrum, prepare a table with chemical shifts, integrations, multiplicities (including coupling constants, if possible), and peak assignments. Be sure to include a structure of the compound with your table, and make sure that you have clearly indicated which peak belongs to each proton or set of equivalent protons in the structure.

If the ^{13}C NMR spectrum is provided, prepare a similar table of chemical shift and peak assignments.

Attach the NMR spectra to your report.

Synthesis of Acetaminophen and Analysis of Some Common Analgesics

ANCIENT MEDICINE

Willow trees are a traditional source of nonsteroidal antiinflammatory drugs (NSAIDs).

RSOrton Photograpy/Shutterstock.

analgesic >>
A drug that relieves pain without loss of consciousness.

There are a variety of simple organic compounds commonly available as over-the-counter pain relievers, or **analgesics**, and many of these operate by inhibiting the cyclooxygenase (COX) enzymes that are key to the biosynthesis of prostaglandins (**Figure 12.1**).

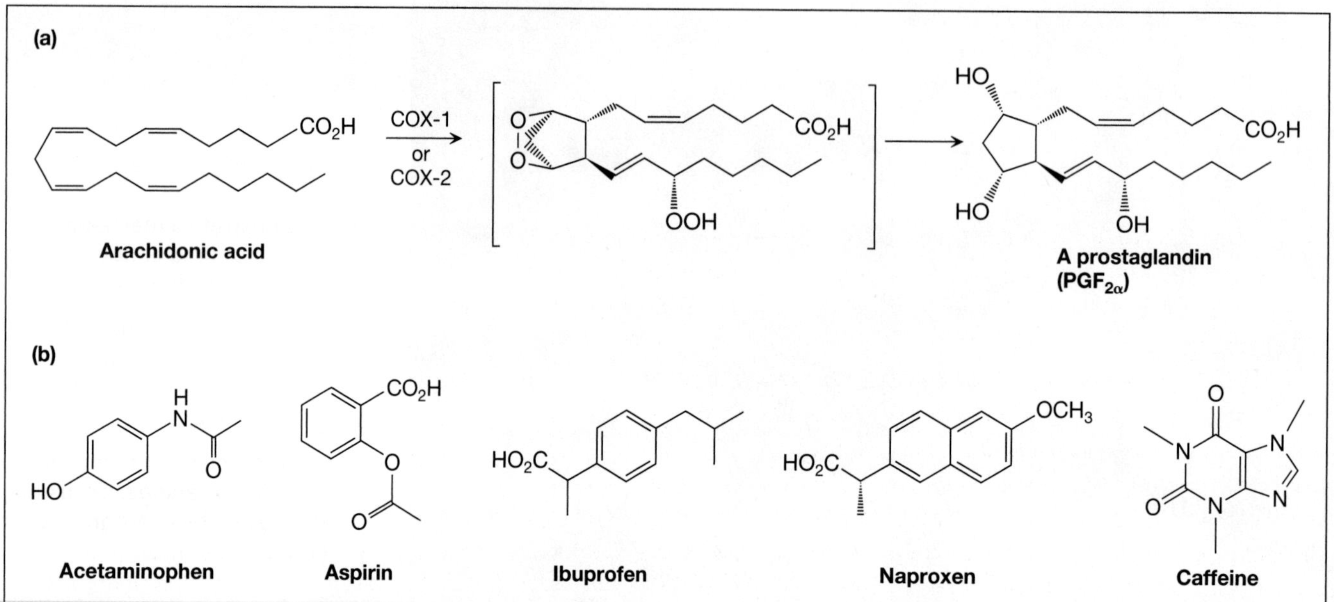

(a)

Arachidonic acid

COX-1 or COX-2

A prostaglandin (PGF$_{2\alpha}$)

(b)

Acetaminophen   Aspirin   Ibuprofen   Naproxen   Caffeine

(a) Prostaglandin biosynthesis from arachidonic acid through a sequence involving cyclooxygenase enzymes (COX-1 or COX-2). (b) Structures of compounds commonly found in commercial over-the-counter analgesics. Aspirin, ibuprofen, and naproxen are NSAIDs.

Prostaglandins are a family of closely related cyclopentane compounds bearing two lipophilic tails and varying degrees of oxidation to alcohol, ketone, and carboxylic acid functional groups. They were first discovered in seminal fluid and other male genital extracts, and are now recognized to be very broadly involved in chemical signaling events that control a number of biological processes. Among these are the stimulation of smooth muscle, lowering of blood pressure, induction of labor, inhibition of gastric secretion, and increase in ocular pressure.[1] They also mediate the inflammatory response, so inhibiting their synthesis eliminates some of the pain associated with inflammation. Compounds that act in this fashion, such as aspirin, ibuprofen, and naproxen, are referred to as nonsteroidal antiinflammatory drugs (NSAIDs). Acetaminophen also relieves pain by inhibiting COX enzymes and prostaglandin synthesis, but its mechanism of action is different, so it is not considered to be an NSAID.[2] In this experiment we synthesize one of these important agents while addressing green chemistry principles, specifically by using aqueous reaction conditions to minimize solvent wastes.

[1]*The Merck Index: An Encyclopedia of Chemicals, Drugs, and Biologicals*, 12th ed.; Merck and Company: Rahway, NJ, 1996.
[2]Aminoshariae, A.; Khan, A. Acetaminophen: Old Drug, New Issues. *J. Endod.* **2015**, *41*, 588–593.

12.1A Pre-Lab Reading Assignment

Review the following before attempting part 1 of this experiment:

- Experiment: Chapter 12 (this experiment)
- Technique: Extraction (Chapter 5, section 5.2)
- Technique: Recrystallization (Chapter 5, section 5.4)
- Technique: Infrared (IR) spectroscopy (Chapter 6, section 6.2)

12.1B Background

The structure of acetaminophen (**Figure 12.2a**) contains an amide functional group, which is a derivative of carboxylic acids. Amides have two main resonance structures and a resonance hybrid (**Figure 12.2b** and **Figure 12.2c**) that help to describe their reactivity.

Amides are generally prepared by a nucleophilic acyl substitution reaction. This class of reactions interconverts carboxylic acids and their derivatives by a two-step mechanism (**Figure 12.3**) involving *addition* to the carbon of the C=O bond to form a tetrahedral intermediate, then *elimination* of a leaving group to regenerate the C=O bond. Depending on the type of conditions, and on the specific reactants, there may also be proton transfers in the mechanism, but these are minor variations on the general theme of addition–elimination. A specific example of this general addition–elimination mechanism of nucleophilic acyl substitution is the reaction of an amine with acetic anhydride (Figure 12.3).

(a) Acetaminophen

(b) Amide bond resonance

(c) Amide resonance hybrid

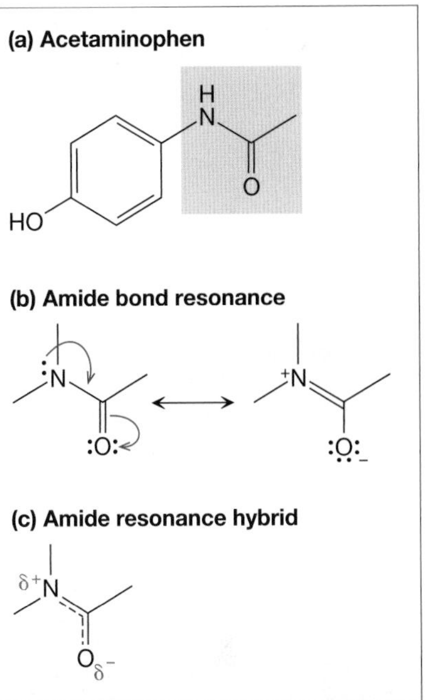

FIGURE 12.2

(a) Acetaminophen contains an amide functional group, highlighted in blue. (b) Two resonance structures of a typical amide. (c) Resonance hybrid composed of a combination of the resonance structures.

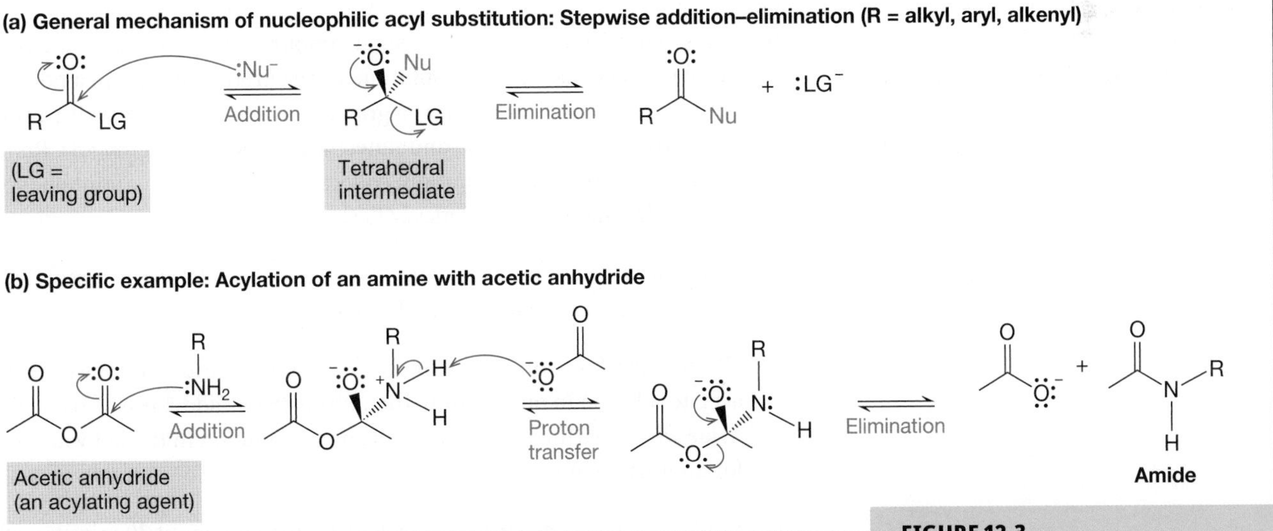

(a) General mechanism of nucleophilic acyl substitution: Stepwise addition–elimination (R = alkyl, aryl, alkenyl)

(b) Specific example: Acylation of an amine with acetic anhydride

FIGURE 12.3

Overview of nucleophilic acyl substitution reactions (a), with a specific example (b) illustrating amide bond construction. Acetate ion is employed to transfer a proton from the tetrahedral intermediate.

Upon nucleophilic addition to either of the C=O bonds of acetic anhydride, a tetrahedral intermediate forms, and after a proton transfer, acetate ion can then be ejected as a leaving group. The net result is the transfer of an acyl group, RC=O, (in this case acetyl, CH_3C=O) to the amine, thus forming an amide bond. Similar chemistry is involved in protein biosynthesis in nature, where amino acids are joined together via amide bonds to form polymeric protein structures.

In part 1 of this experiment,[3] *p*-acetamidophenol (acetaminophen) is prepared from *p*-aminophenol by reaction with acetic anhydride (**Figure 12.4**). In a preliminary step, the *p*-aminophenol (a basic compound) is purified by dissolving it in aqueous acid and filtering out insoluble impurities. In the reaction mixture, the salt form of the *p*-aminophenol is converted back to its original neutral form. The *p*-aminophenol contains both a phenol and an aniline, and both of these groups have the potential to react as nucleophiles in a nucleophilic acyl substitution reaction to form either an ester or an amide. The different nucleophilicities of the hydroxyl and amino groups can be distinguished under these conditions. In fact, the amide is selectively formed, and the product (acetaminophen) can be purified by recrystallization. The identity and purity of the synthetic acetaminophen is then analyzed by melting point, thin-layer chromatography (TLC), and infrared (IR) spectroscopy.

FIGURE 12.4

Preparation of acetaminophen via acylation under aqueous conditions.

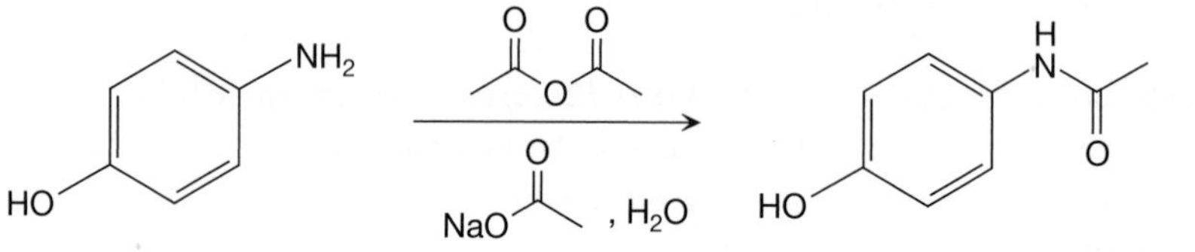

12.1C   Experimental Procedure (Part 1)

1. Set up a water bath (about 2 cm in depth) on a hot plate and begin heating.

2. Place 2.1 g of *p*-aminophenol (*CAUTION: Sensitizer; avoid contact*) and 35 mL distilled (or deionized) water in a 125-mL Erlenmeyer flask. Add concentrated hydrochloric acid (about 1–2 mL) dropwise with frequent swirling until the *p*-aminophenol dissolves completely. Avoid adding excess acid.

   CAUTION: Concentrated hydrochloric acid is corrosive and can damage skin and eyes. Avoid contact.

Gravity Filtration

3. Using the tip of a spatula, add a small amount (about the size of a pea) of decolorizing charcoal to the solution, and swirl the mixture in a hot water bath for a few minutes. Prepare a gravity filtration apparatus with a funnel, a 125-mL Erlenmeyer flask, and fluted filter paper (see Figure 5.1). Remove the charcoal by gravity filtration, collecting the filtrate (*p*-aminophenol hydrochloride) in the Erlenmeyer flask.

4. While filtering out the charcoal, prepare a buffer solution by dissolving 2.5 g sodium acetate in 7.5 mL of water.

5. To the warm solution of *p*-aminophenol hydrochloride, add the sodium acetate solution in one portion, then immediately add 2 mL of acetic anhydride while swirling. Continue swirling the solution in the hot water bath for another 10 minutes.

6. Cool the reaction mixture in an ice-water bath for about 20 minutes. If crystallization does *not* occur, then scratch the walls of the flask with a glass rod.

[3]Pavia, D. L.; Lampman, G. M.; Kriz, G. S. *Introduction to Organic Laboratory Techniques*, 3rd ed.; Saunders Publishing: Philadelphia, PA, 1988; pp 29–46. Harwood, L. M.; Moody, C. J. *Experimental Organic Chemistry: Principles and Practice*; Blackwell Scientific Publications: Oxford, 1989; pp 127–132.

7. When crystallization is complete, collect the crude product by vacuum filtration using a Hirsch or a Büchner funnel. Using a small amount of ice-cold water (1 mL or less), wash the remaining solids out of the flask and into the funnel. Set aside 10 mg of this material to continue drying on a watch glass for later use in part 2 of this experiment.

 **Vacuum Filtration**

8. Recrystallize the crude product from water, using a 25-mL Erlenmeyer flask. Add the minimum amount of hot water needed to dissolve the crude product, adding in small portions while heating on the hot plate. Remove the flask from the heat and allow crystallization to occur while the solution slowly cools to room temperature. If crystals still have *not* formed, then scratch the walls of the flask with a glass rod or add a trace amount of seed crystals. When crystallization is complete at room temperature, cool the mixture further in an ice bath. Collect the recrystallized acetaminophen by vacuum filtration using your Hirsch or Büchner funnel, using a small amount of ice-cold water (1 mL or less) to wash the remaining crystals out of the flask and into the funnel. Allow the crystals to dry thoroughly by continuing to apply a vacuum and gently scraping them with a spatula. Transfer the recrystallized acetaminophen to a tared (preweighed) watch glass and allow to dry until a constant mass is observed.

Recrystallization

9. Record the mass of recrystallized acetaminophen.

10. Measure the melting point and infrared spectrum of your recrystallized acetaminophen.

 Melting Point Measurement

12.2 PART 2: TLC ANALYSIS OF ANALGESIC DRUGS

12.2A Pre-Lab Reading Assignment

Review the following before attempting part 2 of this experiment:

- Experiment: Chapter 12 (this experiment)
- Technique: Thin-layer chromatography (TLC) (Chapter 5, section 5.5D)

12.2B Background

Over-the-counter drugs, nutritional supplements, and the like often contain more than one component. Various brands or manufacturers might use different sets of ingredients in their formulations. Commercial over-the-counter pain relievers may contain acetaminophen, ibuprofen, aspirin, or other analgesic compounds, and some contain ingredients with other biological effects, such as caffeine. Known samples of these possible components can be used as reference standards for thin-layer chromatography (TLC) experiments to identify what's in a mixture. Here in part 2, the acetaminophen you prepared in part 1 is compared with commercial acetaminophen by TLC to assess the purity and identity of the prepared material. A series of reference standards and commercial analgesics are also subjected to TLC analysis, and the results are used to determine the active component(s) of unknown samples.

12.2C Experimental Procedure (Part 2)—TLC Analysis of Active Ingredients

Thin-Layer Chromatography

1. In *separate* test tubes, dissolve approximately 10 mg (exact weights are unimportant) of your crude acetaminophen and your recrystallized acetaminophen in 1 mL dichloromethane. Similar solutions of *p*-aminophenol, aspirin, and caffeine will be available in the lab for use as standards.

   CAUTION: Use dichloromethane in the fume hood. Avoid inhalation or skin contact.

2. Prepare a TLC plate, marking a faint pencil line about 0.5 cm parallel to one end of the plate, and on this line applying spots of samples of the crude acetaminophen, recrystallized acetaminophen, and *p*-aminophenol with a narrow glass capillary tube. Capillary tubes can be commercially supplied or made in the lab according to your instructor's directions. Elute the plate using ethyl acetate (EtOAc) as the mobile phase, marking the plate with pencil to indicate location of the solvent front. Visualize the results first using UV light, then using an iodine chamber (the order is important). Circle the spots with a pencil, using a solid line for the UV visualization and a dashed line for the iodine. Identify the spots observed, and calculate their R_f values. Trace the plate in your notebook and draw the spots, taking care to locate the spots accurately.

3. Prepare two TLC plates, each spotted with samples of your recrystallized acetaminophen, aspirin, caffeine, and a reference sample containing all three of these compounds. Elute the first plate with ethyl acetate and the second plate with CH_2Cl_2. Visualize the plates as before, being sure to record your observations. Note the effect of eluent polarity on R_f, as well as any variation in the shapes of the spots (i.e., tailing, elongation, etc.).

12.2D Experimental Procedure (Part 2)—TLC Analysis of Commercial Analgesics and Identification of Unknowns

4. Set up a water bath (about 2 cm in depth) on a hot plate and begin heating.

5. In two separate small test tubes, prepare solutions of each of two commercial powdered analgesics (e.g., Excedrin and Tylenol). Some labs may supply these in powdered form. If not, use a mortar and pestle to grind each analgesic into a fine powder, making sure it is clean and dry between each sample. Use approximately 10 mg (exact weights are unimportant) in about 2 mL of CH_3OH. Use the warm water bath to gently heat the tubes for about 5 minutes. Powdered tablets usually contain inert ingredients such as starch binders and inorganic buffering agents, so not all of your tablet will dissolve. Allow the sample to cool and the insoluble material to settle.

6. Prepare a TLC plate with five spots: the three-component reference solution (acetaminophen, aspirin, and caffeine), solutions of the two commercial analgesics, and solutions of two unknowns A and B. *Note:* Wider TLC plates may be needed to accommodate all five spots, or alternatively, more than one plate can be used. Elute with ethyl acetate. Visualize the TLC plate and note your observations as in steps 2 and 3, and record all R_f values. Determine and report the composition of the active ingredients of the analgesic drug you chose, as well as of the unknowns. Identify the unknowns.

12.3 THE LABORATORY REPORT

12.3A | Part 1: Preparation of Acetaminophen

From all the information that you have at your disposal (TLC, IR, melting point, and yield), discuss whether the acetaminophen synthesis was successful. The discussion should include evidence of both the identity and the purity of the product. Do not forget to include literature values where appropriate. Attach the IR spectrum to your report. Comment on what green chemistry principles are addressed in this procedure.

12.3B | Part 2: TLC Analysis of Analgesic Drugs

Report the composition of active ingredients in the analgesic you chose and the identity of the unknowns. Explain how the data you gathered from TLC allowed you to reach those conclusions. What was the optimal eluent for the TLC plates, and why?

12.3C | Data to Include in the Laboratory Report

A. PART 1: PREPARATION OF ACETAMINOPHEN

Amount of p-aminophenol used (g):

Amount of recrystallized acetaminophen (g):

Yield (%):

Melting point range of recrystallized acetaminophen (°C):

B. PART 2: TLC ANALYSIS OF ANALGESIC DRUGS

SAMPLE	R_f (EtOAc)	R_f (CH$_2$Cl$_2$)	TLC PLATE DRAWINGS
p-Aminophenol	_____		
Recrystallized acetaminophen	_____	_____	
Aspirin	_____	_____	
Caffeine	_____	_____	
Reference sample	_____	_____	
Tylenol	_____		
Excedrin	_____		
Unknown A	_____		
Unknown B	_____		

13

LEARNING OBJECTIVES

- Implement "closing the loop" and recognize its green chemistry implications.

- Execute a catalytic hydrogenation with a hydrogen transfer reagent.

- Examine the effects of conjugation on the infrared absorbance of a ketone.

Catalytic Transfer Hydrogenation: Closing the Loop

CARBON-FREE ENERGY

Technologies for producing hydrogen from water or food waste offer the potential for increased use of hydrogen as a carbon-free form of energy. In Chapter 13, we see that hydrogen is also a useful reagent for reductions of unsaturated organic compounds.

Audio und werbung/Shutterstock.

H ydrogenation of alkenes is a fundamental reaction type that converts unsaturated organic compounds to saturated ones by adding the equivalent of a molecule of H_2 across the carbons of the alkene π bond (**Figure 13.1a**). This process changes the structure's chemical properties. It also can change its physical properties, as in the conversion of vegetable oil to hydrogenated vegetable oil. Vegetable oils contain fatty acids with hydrocarbon chains that usually contain one or more alkenes (**Figure 13.1b**), and hydrogenation of the alkenes raises the viscosity and melting point of the material. The product becomes a solid at room temperature, has a longer shelf-life, and can be used as shortening, replacing animal fats, such as butter or lard, in various cooking and baking applications. Trans fats, which have C=C bonds in *trans* configuration, are side products of incomplete or partial hydrogenation of vegetable oils, and are implicated in adverse health impacts such as elevated risk of heart disease or stroke.

Hydrogenation reactions often involve pressurized hydrogen gas, which is a safety hazard for two reasons: containment of high-pressure gas in a reaction vessel, and extreme flammability. In this chapter, you will see this how this chemistry can be done by catalytic transfer hydrogenation at atmospheric pressure using formate ion (HCO_2^-) as a safer hydrogen source, which addresses some principles of green chemistry.

FIGURE 13.1

(a) A hydrogenation reaction of an alkene produces an alkane.
(b) Hydrogenation of vegetable oil changes its properties, raising its melting point so that it behaves more like an animal fat (solid at room temperature). This is a representative example of a glycerol ester found in fats and oils, which are complex mixtures of glycerol esters having slight variations in the fatty acid components.

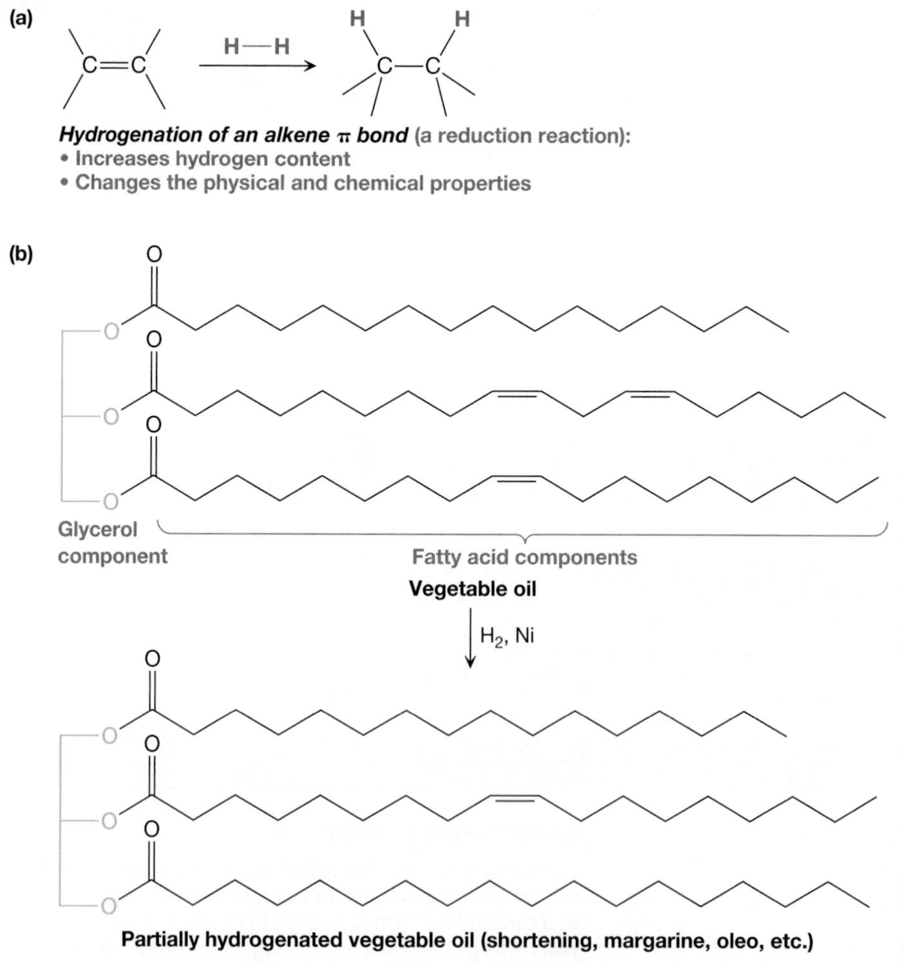

Pre-Lab Reading Assignment

- Experiment: Chapter 13 (this chapter)
- Technique: Infrared (IR) spectroscopy (Chapter 6, section 6.2)
- Technique: Extraction (Chapter 5, section 5.2)
- Technique: Recrystallization (Chapter 5, section 5.1)

13.1 BACKGROUND

13.1A Catalytic Hydrogenation

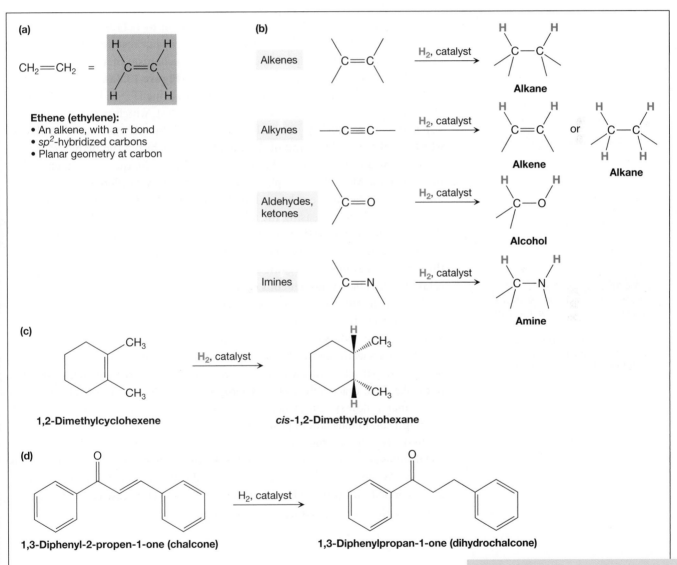

(a)

$CH_2{=}CH_2$ =

Ethene (ethylene):
- An alkene, with a π bond
- sp^2-hybridized carbons
- Planar geometry at carbon

(b)

Alkenes $\xrightarrow{H_2,\ catalyst}$ Alkane

Alkynes $\xrightarrow{H_2,\ catalyst}$ Alkene   or   Alkane

Aldehydes, ketones $\xrightarrow{H_2,\ catalyst}$ Alcohol

Imines $\xrightarrow{H_2,\ catalyst}$ Amine

(c)

1,2-Dimethylcyclohexene $\xrightarrow{H_2,\ catalyst}$ *cis*-1,2-Dimethylcyclohexane

(d)

1,3-Diphenyl-2-propen-1-one (chalcone) $\xrightarrow{H_2,\ catalyst}$ 1,3-Diphenylpropan-1-one (dihydrochalcone)

FIGURE 13.2

(a) Planar structure associated with the carbon–carbon double bond of an alkene. (b) Hydrogenation of various π-bonded functional groups. (c) Catalytic hydrogenation generally results in syn-addition of two hydrogens. (d) Hydrogenation of chalcone.

Many functional groups can be hydrogenated (**Figure 13.2b**), and all have in common the removal of one or more π bonds along with the addition of two hydrogen atoms, generating a saturated compound. Hydrogenation reactions of π bonds are also referred to as reductions, and these reductions result in an increase of the hydrogen content compared with the starting compound. Alkenes become alkanes, ketones and aldehydes become alcohols, imines become amines, and so on. Alkenes are typically reduced very easily, while hydrogenation of ketones may require heat and/or pressure. Reductions involve an addition reaction in which hydrogen gas adds across the unsaturated functional group; the two H atoms of the H_2 molecule become attached at opposite ends of the π bond. In order to do this, the H—H bond in H_2 must break, and this process has a high energy barrier. Thus a transition metal *catalyst* (commonly Ni, Pd, or Pt) is usually employed to lower the energy barrier, increasing overall reaction rates in a process called **catalytic hydrogenation**.

A catalyst lowers the energy barrier of a reaction, but is not consumed; it is recycled many times within the same reaction mixture until all of the reactants are consumed. Ideally, only a trace amount of a catalyst is required, and in some cases, it can be recovered for reuse. Therefore, ideal catalytic reactions generate little or no waste, and are highly desirable for environmentally benign chemical processing. Of course no case is perfectly ideal; in realistic situations, the catalyst can become inactive as it degrades by various side reactions, so small amounts of waste should be expected.

How does a catalyst like palladium, platinum, or nickel function in hydrogenation? Metal catalysts in the elemental form carry out reactions on the surface of the metal. First, it is helpful to consider the mechanism in a reaction happening at a single discrete metal catalyst molecule (**Figure 13.3a**). One critical part of the multi-step mechanism is the breaking of a strong H—H bond (432 kJ/mol, 104 kcal/mol) by the metal atom (M) of the catalyst, which occurs in *step i*. In *step ii*, the metal atom can share electrons donated by the π system of an alkene, which results in the binding of the metal above or below the plane defined by the sp^2-hybridized carbons of the alkene bond. In *step iii*, the alkene then inserts into one of the M—H bonds, placing the M and H at two different carbons at the former site of the alkene π bond. Finally, in *step iv*, the C—M bond is replaced with a C—H bond, releasing the alkane product from the metal atom, which is then ready to begin another cycle. Single metal atoms can have various ligands (L) attached that are not participating directly in the mechanism, but allow for the metal to be soluble in organic solvents. Thus the reactions with these types of catalysts are called **homogeneous catalysis**.

Now, at the surface of a metallic substance, there are many metal atoms, all in close proximity. These metallic catalysts are insoluble, so their reactions are classified as **heterogenous catalysis**. Although the mechanisms of heterogeneous catalysis are sometimes unclear because of the complications of studying reactions at the surface of insoluble particles, the advantages of easy separation and recycling of the catalyst make them practically useful. Heterogeneous catalysis is commonly employed in continuous flow systems, where a stream of reactants passes through the catalyst, yielding a continuous flow of products.

In the transfer hydrogenation of an alkene by formate ion, it is believed that the metal atoms at the surface of the catalyst react in a manner similar to homogeneous catalysts. When formate ion binds to the metal catalyst, loss of CO_2 occurs and a M—H bond forms (**Figure 13.3b**). The M—H bonds can release H_2, or if an alkene is present, they can transfer the hydrogen to the alkene.[1]

catalytic hydrogenation >>
Addition of hydrogen to an unsaturated organic compound, with the aid of a catalyst that lowers the activation energy for the reaction.

homogeneous catalysis >>
A reaction in which the catalyst is dissolved in the reaction medium.

heterogeneous catalysis >>
A reaction in which the catalyst is insoluble in the reaction medium.

---

[1]Representative examples of proposed mechanisms for transfer hydrogenation using formate: (a) Song, T.; Duan, Y.; Yang, Y. Chemoselective Transfer Hydrogenation of α,β-Unsaturated Carbonyls Catalyzed by Reusable Supported Pd Nanoparticles on Biomass-Derived Carbon. *Catal. Commun.* **2019**, *120*, 80–85. (b) Broggi, J.; Jurčík, V.; Songis, O.; Poater, A.; Cavallo, L.; Slawin, A. M. Z.; Cazin, C. S. J. The Isolation of [Pd{OC(O)H}(H)(NHC)(PR$_3$)] (NHC = N-Heterocyclic Carbene) and Its Role in Alkene and Alkyne Reductions Using Formic Acid. *J. Am. Chem. Soc.* **2013**, *135*, 4588–4591. (c) Brunel, J. Scope, Limitations and Mechanistic Aspects in the Selective Homogeneous Palladium-Catalyzed Reduction of Alkenes under Transfer Hydrogen Conditions. *Tetrahedron* **2007**, *63*, 3899–3906.

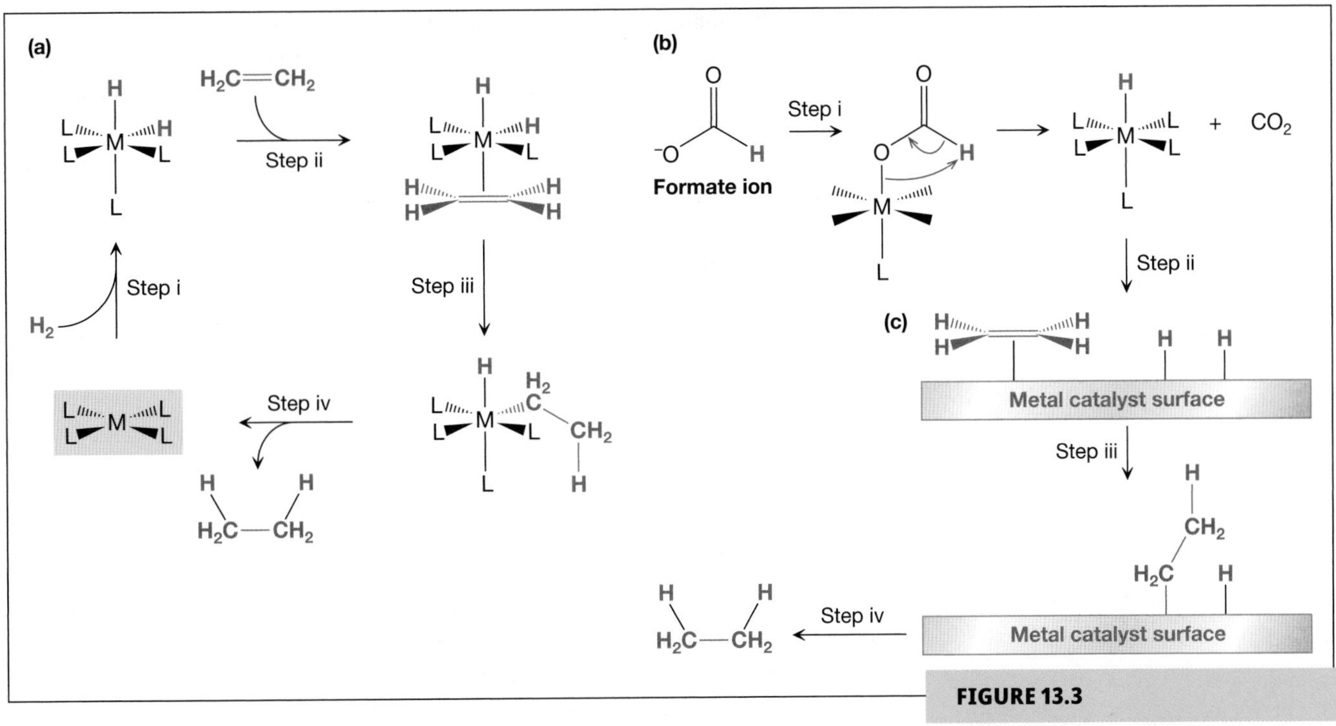

FIGURE 13.3

(a) Homogeneous catalysis: Intermediates involved in catalytic hydrogenation, viewed in a catalytic cycle with respect to a single metal atom. The reactants H_2 and $CH_2{=}CH_2$ enter the cycle, and the product $CH_3{-}CH_3$ leaves the cycle, but the metal is recycled many times. (b) Formate ion binds to a metal atom, depositing a hydrogen with loss of CO_2. (c) Heterogeneous catalysis: On the surface of a metal catalyst, hydrogen is transferred to the alkene in a fashion analogous to steps ii–iv depicted in (a).

13.1B Formate as a Source of Hydrogen

Formic acid, and its conjugate base HCO_2^- (formate ion), can be generated from CO_2 and H_2. In the catalytic hydrogenation of alkenes, the CO_2 contained in formate serves as a carrier of the hydrogen that is transferred to the alkene reactant via the catalyst. The CO_2 that is released in our experiment is a very small amount that is impractical to recover, but in a large-scale industrial application, chemists and chemical engineers would develop ways to capture the CO_2 for recycling.

This use of CO_2 as a carrier of hydrogen has important safety implications. The formate anion can be accompanied by various counterions such as sodium or ammonium cations, and in this salt form, formate is a shelf-stable, solid, nonflammable material. The release of hydrogen from formate does not occur until the time and place of the chemist's choosing. This is much safer than the direct use of hydrogen gas, which presents significant transport, storage, and handling hazards. It is stored as a pressurized gas, and any pressurized system can be hazardous due to the sudden release of gas by accident or from damage to the pressurized system. In addition, hydrogen gas is extremely flammable in the presence of air.

13.1C Closing the Loop

Using an output of a production activity as an input for further production is called **closing the loop**, and can decrease the environmental impact of the waste stream. With increasing concerns about impacts of CO_2 emissions, alternative ways to handle this inevitable by-product of human activity are a topic of great interest. One example is the capture of CO_2 before it can be released; this provides a source of carbon that can be used for other productive purposes. A host of innovative chemical processes can convert CO_2 to useful feedstocks for the chemical industry.[2] In this lab, we will use formate ion as a source of hydrogen for the hydrogenation of

<< closing the loop
Using the output of a production activity as an input for further production.

[2]Otto, A.; Grube, T.; Schiebahn, S.; Stolten, D. Closing the Loop: Captured CO_2 as a Feedstock in the Chemical Industry. *Energy Environ. Sci.* **2015**, *8*, 3283–3297.

an alkene. Formate ion is the conjugate base of formic acid, which in turn can be prepared in bulk quantities from a reaction of CO_2 and H_2. Thus, the CO_2 trapped from, say, a natural gas–fueled power plant, could be converted into a reagent for an organic chemistry transformation; thus, closing the loop.

The alkene that is to be hydrogenated in this lab is one of a family of compounds known as chalcones. These are produced by a base-catalyzed condensation of alkyl aryl ketones with benzaldehydes, a reaction known as the aldol condensation. The details of the aldol reaction are not important to learn for the purposes of this transfer hydrogenation experiment. If a chalcone is made using an aldol reaction later in the semester, that product can be used as the alkene reactant for a future group of students to do their hydrogenation experiment, closing the loop from the perspective of the organic starting material.[3]

13.1D | IR Spectra of Ketones and Unsaturated Ketones

After carrying out the reaction shown in **Figure 13.4**, we will obtain evidence for the identity and purity of the product using infrared spectroscopy. The infrared (IR) absorption peaks associated with carbonyl (C=O) bonds are found in the range from 1650–1750 cm^{-1}. These peaks have noticeably different frequencies depending on what is attached to the carbonyl, and such variations are correlated with structural features (**Table 13.1**). So, carbonyl peaks are useful for assigning the structures of aldehydes, ketones, and carboxylic acid derivatives.

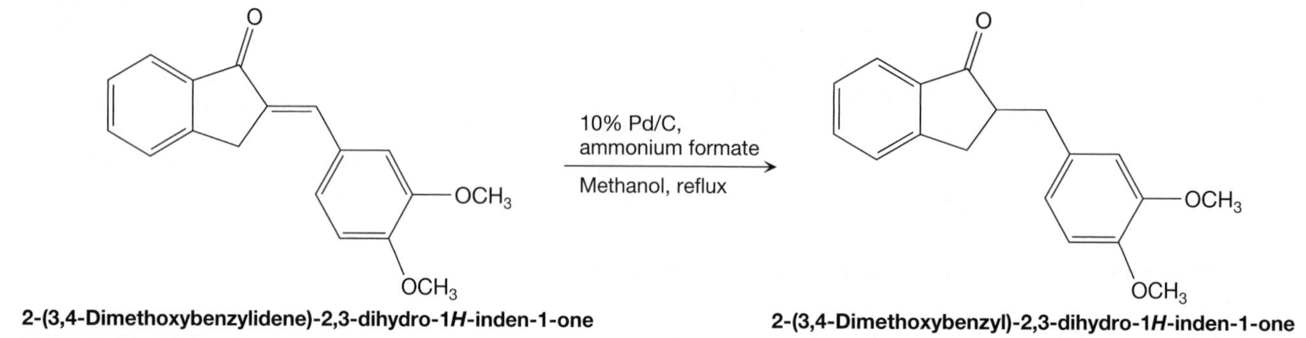

2-(3,4-Dimethoxybenzylidene)-2,3-dihydro-1*H*-inden-1-one → 10% Pd/C, ammonium formate, Methanol, reflux → 2-(3,4-Dimethoxybenzyl)-2,3-dihydro-1*H*-inden-1-one

FIGURE 13.4

Reduction of an alkene by catalytic transfer hydrogenation.

When there are unsaturated groups (alkenes or other sp^2-hybridized groups) directly attached to the carbonyl, this shifts the IR absorption to a lower frequency. This can be rationalized by the presence of some resonance structures that have a single bond between carbon and oxygen. The actual structure is a combination of all resonance structures, and when there are single bond contributors to the resonance hybrid, this is correlated to a weakening of the carbonyl bond. Weaker bonds absorb IR at lower frequency. Consequently, the frequency of the carbonyl peak in the IR spectrum of the product should be diagnostic for determining if the alkene bond has been successfully hydrogenated.

---

[3]This experiment could be adapted to various chalcones that are commercially available or prepared via simple aldol condensations. We have chosen this particular chalcone to illustrate the concept of closing the loop because it is a product from a later experiment in this book, Solventless Aldol. The task of making the chalcone is best suited for students who already have some understanding of enolate chemistry.

TABLE 13.1

Infrared Absorptions of C=O in Various Functional Groups

SATURATED		UNSATURATED	
STRUCTURE	C=O FREQUENCY	STRUCTURE	C=O FREQUENCY
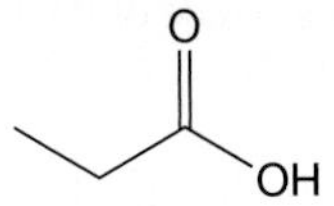 [propanoic acid structure]	1716 cm^{-1}	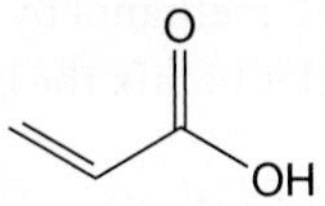 [acrylic acid structure]	1705 cm^{-1}
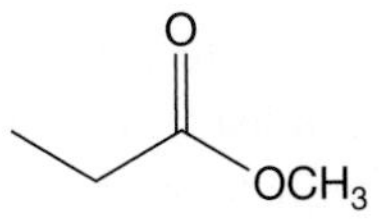 [methyl propanoate structure]	1741 cm^{-1}	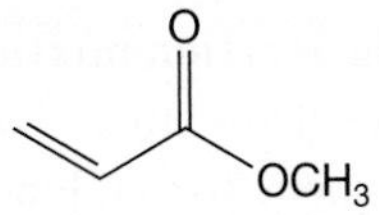 [methyl acrylate structure]	1732 cm^{-1}
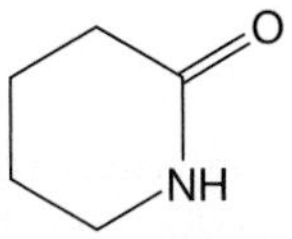 [piperidinone structure]	1673 cm^{-1}	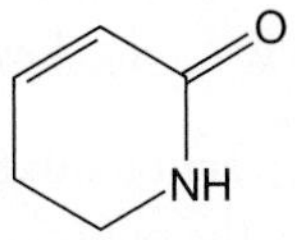 [dihydropyridinone structure]	1664 cm^{-1}
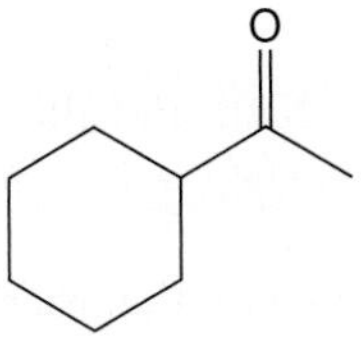 [cyclohexyl methyl ketone structure]	1709 cm^{-1}	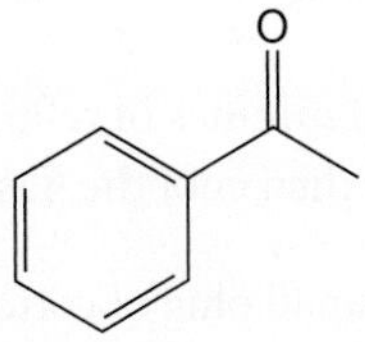 [acetophenone structure]	1686 cm^{-1}
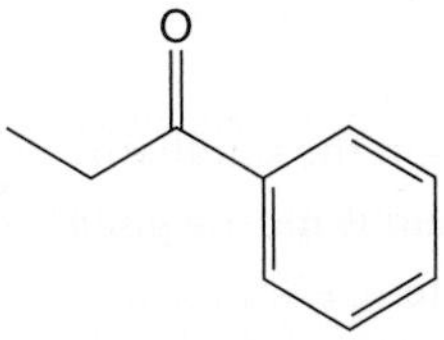 [propiophenone structure]	1688 cm^{-1}	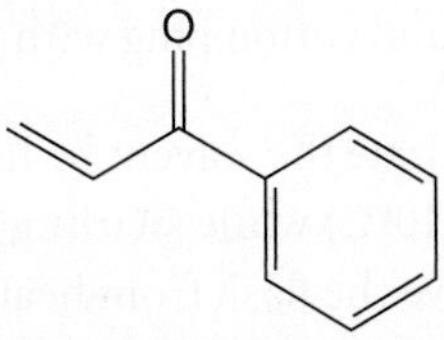 [phenyl vinyl ketone structure]	1678 cm^{-1}

13.2 EXPERIMENTAL PROCEDURE

13.2A Catalytic Transfer Hydrogenation[4]

DAY 1

1. Start a **sandbath** heating at a medium setting, so that the sand will reach a temperature of 90–110°C. Prepare a reflux condenser with water inlet and outlet hoses.

2. Into a 25-mL round-bottom flask equipped with a magnetic stir bar, place 0.28 g of chalcone and 0.47 g ammonium formate.

     CAUTION: Palladium on carbon must be combined with all of the solids before adding methanol. Adding Pd/C to methanol can cause methanol vapors to ignite.

<< sandbath
A common source of heat at stable temperature, used to heat the contents of laboratory glassware such as a distillation or reaction flask. Some laboratories use a heating mantle or a heated metal block for this purpose.

[4]This procedure is adapted from Hammond, C. N.; Schatz, P. F.; Mohrig, J. R.; Davidson, T. A. Synthesis and Hydrogenation of Disubstituted Chalcones: A Guided-Inquiry Organic Chemistry Project. *J. Chem. Educ.* **2009**, *86*, 234–239.

3. Using a waxed weighing paper, weigh 0.05 g of 10% palladium on carbon. Carefully roll the edge of the weighing paper into a funnel shape, and without getting the powder on the ground glass joint of the round-bottom flask, transfer the Pd/C into the flask. If any Pd/C spills, wipe it with a water-moistened lab tissue and deposit into the Pd waste.

4. Swirl the contents of the flask to mix the dry powders together.

5. Add 5 mL methanol by pouring it down along the inside surface of flask, and swirl it to mix the contents.

6. Clamp the flask securely and place a reflux condenser on top of the flask.

7. **Heat the reaction mixture at reflux**. Adjust sandbath heating to a temperature of 90–110°C. Lower the clamped flask into the sandbath and start the magnetic stirrer. Fine bubbles of hydrogen gas will begin to form within the reaction mixture. Start cooling the condenser by carefully turning on the water to a very slow trickle.

8. After the reaction mixture begins to boil, with solvent vapors visibly recondensing in the reflux condenser, note the time and continue the reflux for 20 minutes.

9. After 20 minutes of reflux, remove the flask from the sandbath, and allow to cool, then cool the flask in an ice-water bath.

10. Pack a small plug of cotton into a narrow-stem funnel, and filter the reaction mixture into a clean 50-mL Erlenmeyer flask, rinsing the round-bottom flask and cotton plug with 1 mL of methanol.

11. Evaporate the solvent by heating the Erlenmeyer flask on the sandbath (90–110°C) while swirling occasionally. When the liquid is mostly gone, remove the flask from heat and allow to cool to room temperature.

12. Add 5 mL water and 10 mL *tert*-butyl methyl ether (MTBE) and swirl the flask to mix the contents.

13. Pour into a 60-mL separatory funnel. Rinse the Erlenmeyer flask with 5 mL MTBE and add this to the separatory funnel. Stopper the separatory funnel and mix thoroughly by inverting it several times. Release pressure through the stopcock while pointing the stopcock up and into the hood.

14. Remove the stopper, and allow the layers to fully separate. Drain the aqueous phase into the previous Erlenmeyer flask, and the organic phase into a clean, dry 50-mL Erlenmeyer flask. Dry the organic phase over anhydrous Na_2SO_4.

15. Decant the organic phase into a dry, tared 50-mL round-bottom flask, rinsing the Erlenmeyer flask with 1 mL MTBE. Concentrate on the rotary evaporator for a few minutes, until the remaining material appears to be a constant volume.

Note: If a second class period is available, this is a suitable stopping point. Store the flask loosely stoppered so that the last traces of solvents will evaporate. If the entire procedure is to be completed in 1 day, use a vacuum adapter with stopper as shown in **Figure 13.5** to place the flask under vacuum for 20 minutes in order to remove the last traces of solvents.

Heating at Reflux

heating at reflux >>
Boiling a solvent with a condenser oriented so that the condensed solvent returns to the reaction vessel. When heating at reflux, the reaction temperature is held constant at the boiling point of the solvent. Heat is applied via heating mantle, sandbath, or heated metal block that is 20–30°C hotter than the boiling point of the solvent.

Aqueous–Organic Extractions

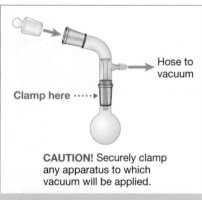

Hose to vacuum

Clamp here ·····▶

CAUTION! Securely clamp any apparatus to which vacuum will be applied.

FIGURE 13.5

Apparatus for removal of small amounts of residual solvent under vacuum. Always clamp the apparatus securely when vacuum is in use.

Rotary Evaporation

16. Record the mass of product.

17. Obtain the infrared spectrum of the product. Make sure to label the wave-numbers of all significant peaks in the C=O range (about 1650–1750 cm^{-1}).

18. *Optional:* Submit the sample for ^{1}H NMR spectrum acquisition. Check with your instructor about NMR sample preparation, and label the tube according to your instructor's procedures.

NMR Sample Preparation

19. *Optional:* Recrystallize the product from the minimum amount of hot 95% ethanol and water (3:1 ratio). It should take about 2–3 mL of hot solvent to dissolve the product, and crystals should form as the mixture slowly cools on the benchtop. Recover the crystals by vacuum filtration using a Hirsch funnel, washing them with about 1 mL of ice-cold 3:1 ethanol/water. Obtain the melting point.

Recrystallization

13.3 THE LABORATORY REPORT

In your report, comment on the infrared absorption(s) in the carbonyl region. How does its frequency (wavenumber) reflect whether the hydrogenation has been successful? Using your literature search knowledge, find IR data for chalcone and dihydrochalcone carbonyl peaks for comparison with your sample. If there are two peaks in the carbonyl region, what does this mean? If you have NMR and melting point data, comment on how these provide evidence of identity and purity of the product.

13.3A Post-Lab Questions

1. Absorbance of infrared by the C=O bond differs in propanoic acid (1716 cm^{-1}) and propenoic acid (1705 cm^{-1}). Which compound has a weaker C=O bond? Use resonance structures to explain your answer.

2. When a catalytic hydrogenation of vegetable oil is incomplete, so that some alkenes are hydrogenated and others are not, this produces "partially hydrogenated vegetable oil." This process can be accompanied by isomerization of *cis* alkenes to *trans* alkenes, forming so-called trans fats, which have been linked to cardiovascular disease and other health risks. Perform a literature search to answer the following questions.

   a. What are the reagents and catalysts commonly used in the food industry to carry out hydrogenation of vegetable oil?

   b. Using IR and NMR spectroscopy, what diagnostic peaks would enable you to detect whether a sample of vegetable oil was partially or fully hydrogenated?

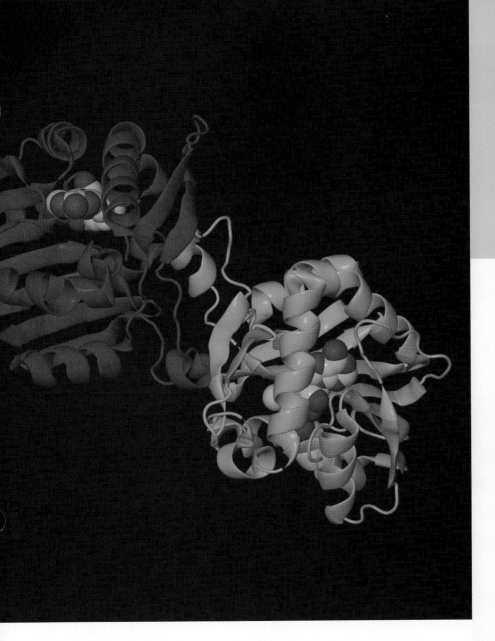

14

LEARNING OBJECTIVES

- Recognize practical aspects of nucleophilic substitutions and their mechanisms.

- Develop a mechanistic hypothesis, S_N1 or S_N2, and test it by experiment.

- Measure the ratio of two components in a mixture, using NMR spectroscopy.

Nucleophilic Substitution: A Mechanistic Inquiry

NATURE'S CATALYSTS

Enzymes known as methyltransferases, such as the one shown here, attach methyl groups to various biological molecules. The enzyme serves as a catalyst and lowers the activation energy to a nucleophilic substitution mechanism.

Courtesy of Dr. Angel Herráez, Universidad de Alcalá, Creative Commons cc-by 4.0.

rganic compounds bearing halogens at sp^3-hybridized carbon centers, also known as alkyl halides, undergo reactions that replace the halogen with another atom or group (**Figure 14.1**). These reactions are known as nucleophilic substitution reactions, and there are two common mechanistic pathways for them, termed S_N1 and S_N2. Each mechanism preferentially operates under specific circumstances. In this experiment, you will examine two different organic halides, develop a hypothesis about which mechanism will be operating, and use an experimental test to gain evidence to support or refute your hypothesis.

FIGURE 14.1

Generalized reaction of an alkyl halide (X = Cl, Br, I) via nucleophilic substitution.

$$\overset{|}{\underset{|}{C}} \!-\! X \quad + \quad :Nu^- \quad \xrightarrow{\ S_N1\ or\ S_N2\ } \quad \overset{|}{\underset{|}{C}} \!-\! Nu \quad + \quad X^-$$

Alkyl halide (1°, 2°, or 3°) · Nucleophile · Substitution product · Leaving group

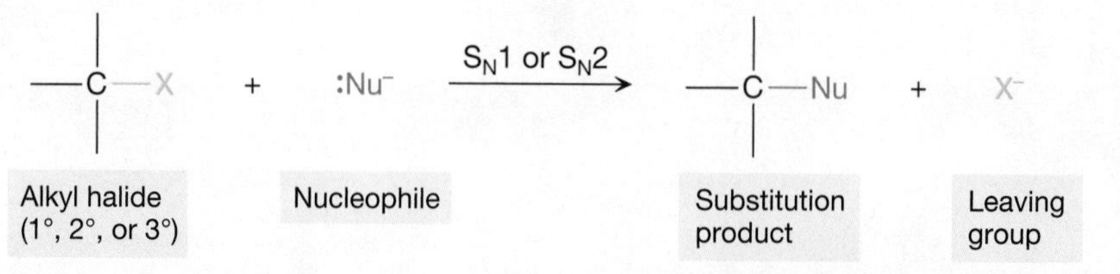

14.1 NUCLEOPHILIC SUBSTITUTION MECHANISMS

The mechanisms that operate in nucleophilic substitution reactions of alkyl halides are commonly either unimolecular (S_N1) or bimolecular (S_N2). The mechanism names derive from how the reaction rates vary with different concentrations of reactants. The unimolecular mechanism (S_N1) has a rate that depends on the concentration of only one reactant, while the bimolecular mechanism (S_N2) rate depends on concentrations of both reactants.

14.1A Pre-Lab Reading Assignment

- Experiment: Chapter 14 (this experiment)
- Technique: Nuclear magnetic resonance (NMR) spectroscopy (Chapter 7, section 7.1)

14.1B Unimolecular Nucleophilic Substitution (S_N1)

heterolysis >>
The cleaving of a σ bond, where both electrons from the cleaved bond reside with one of the product species.

In alkyl halides, the C—X bond can break, with the halogen atom leaving along with the electrons from the C—X bond (**Figure 14.2a**). This is called **heterolysis**, because both electrons of the C—X bond end up on one component, the leaving group X^-. When this happens, a full positive charge is left behind at the carbon in a reactive intermediate known as a *carbocation*. The carbocation has only six valence electrons at carbon, and is therefore sp^2 hybridized and planar, with the positive charge associated with an empty p orbital. Carbocations lack a complete octet and are highly electron-deficient, defining their reactivity; they react with electron-rich molecules.

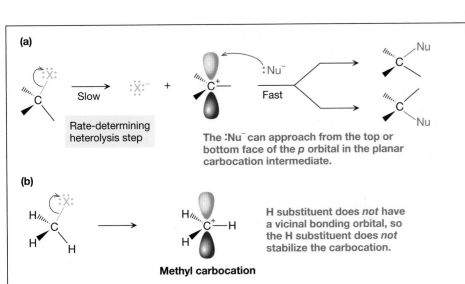

(a)

Rate-determining heterolysis step

Slow

Fast

The :Nu⁻ can approach from the top or bottom face of the *p* orbital in the planar carbocation intermediate.

(b)

Methyl carbocation

H substituent does *not* have a vicinal bonding orbital, so the H substituent does *not* stabilize the carbocation.

Ethyl carbocation

CH₃ substituent has three vicinal C—H bonding orbitals; their overlap with the carbocation *p* orbital stabilizes the carbocation (hyperconjugation).

FIGURE 14.2

Unimolecular nucleophilic substitution (S_N1) of alkyl halides. (a) Slow, rate-determining heterolysis to form a carbocation, followed by fast nucleophilic attack from top or bottom face. (b) Comparison of methyl and ethyl carbocations, showing the stabilizing overlap of a vicinal C—H bond with the empty *p* orbital of the carbocation, an effect known as hyperconjugation (shown here by the blue dotted line).

Carbocations are most likely to form when they are stabilized by the attached substituents. An alkyl substituent stabilizes the carbocation through an effect known as **hyperconjugation**, where the filled bonding orbitals of neighboring σ bonds overlap with the unfilled *p* orbital of the carbocation (**Figure 14.2b**). Vicinal σ bonds are sharing electron density with the carbocation in order to compensate for its electron deficiency. A hydrogen cannot function in this way, because there is no neighboring σ bond. The more alkyl groups connected directly to the carbocation, the greater the stabilizing effect, and this correlates with the carbocation's rate of formation. More stable carbocations form more easily, and at higher rates. Thus the relative reactivity of alkyl halides by the S_N1 mechanism is in the order 3° > 2° > 1° > methyl.

Once formed, carbocations are highly electrophilic, and will react with a variety of electron-rich nucleophiles, and one example of this is the hydroxide ion (OH⁻). This reaction is very rapid, partly because of the electrostatic attraction between the reactants, and also because the carbocation is planar, providing an unhindered pathway for a nucleophile to approach. Because this nucleophile–carbocation reaction is much faster than the formation of the carbocation, as soon as a carbocation forms it rapidly goes on to product. Thus the formation of the carbocation acts as a bottleneck to the rate of the overall reaction, and therefore the rate is determined only by the first step. Only one reactant is present in that rate-determining step, so we say that this mechanism is unimolecular.

For S_N1: Rate = $k[RX]$

k = rate constant, $M^{-1}s^{-1}$

[RX] = concentration of alkyl halide, M^{-1}

<< **hyperconjugation**
The stabilization of a carbocation by overlap of its empty *p* orbital with vicinal filled σ orbitals, such as those of neighboring C—H or C—C bonds.

14.1C Bimolecular Nucleophilic Substitution (S_N2)

In alkyl halides, the C—X bond (X = Cl, Br, I) is polarized by virtue of the higher electronegativity of the halogen atom, which causes the carbon to be partially positive. For X = I, the high polarizability of the large I atom facilitates its role as a leaving group. Electron-rich species will be attracted to this partial positive charge, and a nucleophile generally has either an unshared pair of valence electrons or a full negative charge, or both, making it electron rich. If the nucleophile approaches, a pair of electrons on the nucleophile enters the antibonding orbital of the C—X bond, approaching from the end opposite the halide (**Figure 14.3**). As this new pair of electrons begins to populate the antibonding orbital, the C—X bond weakens, causing X⁻ to be ejected from the other side, taking along the pair of electrons that was previously in the C—X bonding orbital. Thus the substitution occurs all in one step. When bond making and bond breaking occur in the same step we can say the reaction is *concerted*.

FIGURE 14.3

Bimolecular nucleophilic substitution (S_N2), in which both reactants, nucleophile and alkyl halide, are involved in the rate-determining step. Inversion of configuration occurs, which is only observable if the starting alkyl halide is a chiral non-**racemic** compound. For example, non-racemic 2-bromobutane will show inversion, but isopropyl bromide will not.

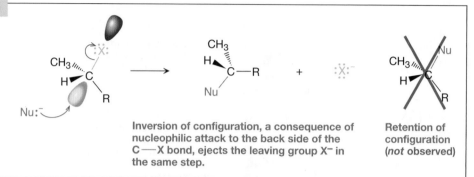

Inversion of configuration, a consequence of nucleophilic attack to the back side of the C—X bond, ejects the leaving group X⁻ in the same step.

Retention of configuration (*not* observed)

racemic >>
A mixture of two enantiomers that are present in equal amounts.

The mechanism is *bimolecular* because there are two reactants involved in the rate-determining step—in this case, the only step. This one-step reaction has a *transition state*, an energy maximum along the reaction coordinate, that includes both the nucleophile and the alkyl halide gathered together into a single congested structure. Because of this congestion, or **steric hindrance**, the rate of this process decreases when there are more alkyl substituents at the carbon of the C—X bond. Methyl bromide, CH_3—Br, with three small H substituents, is much more reactive than *tert*-butyl bromide $(CH_3)_3C$—Br. The latter has three methyl groups blocking the backside approach of the nucleophile, adding to the congestion in the transition state and raising its energy through **van der Waals repulsion**. The higher energy of the transition state translates into a slower reaction rate; fewer molecules have sufficient energy to get over the transition state barrier. Thus, the order of reactivity is methyl > 1° > 2° > 3° for the S_N2 mechanism.

steric hindrance >>
The energetically disfavored placement of two atoms or groups in very close proximity, within the distance range of van der Waals repulsion.

van der Waals repulsion >>
An increase in the energy of two species as their atoms become very close together.

For S_N2: Rate $= k[RX][Nu]$

k = rate constant, $M^{-1}s^{-1}$
[RX] = concentration of alkyl halide, M^{-1}
[Nu] = concentration of nucleophile

14.1D Our Reaction

We will generate the conjugate base of 2,6-dimethylphenol (pK_a 10.6) by deprotonation with a solution of NaOH in ethanol. Then, this nucleophile will be heated with a 1:1 mixture of 1-bromopropane and 2-bromopropane, generating a mixture of isomeric products (**Figure 14.4**).[1]

[1]This reaction and experimental conditions are adapted from Curran et al., who also developed a gas chromatography method for measurement of the product ratio. Curran, T. P.; Mostovoy, A. J.; Curran, M. E.; Berger, C. Introducing Aliphatic Substitution with a Discovery Experiment Using Competing Electrophiles. *J. Chem. Educ.* **2016**, *93*, 757–761.

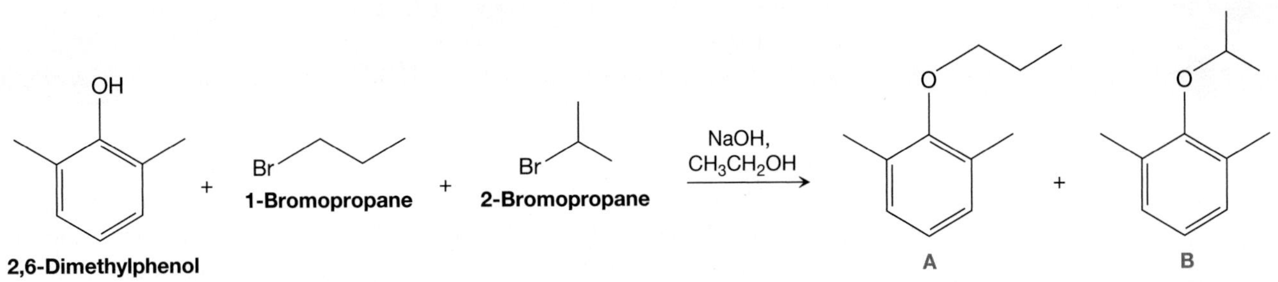

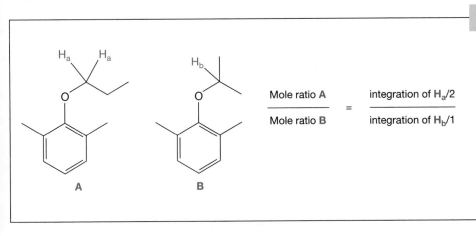

2,6-Dimethylphenol + **1-Bromopropane** + **2-Bromopropane** $\xrightarrow{\text{NaOH, CH}_3\text{CH}_2\text{OH}}$ **A** + **B**

FIGURE 14.4

With two different alkyl halides available for nucleophilic substitution, both products **A** and **B** will be produced in this reaction.

Ethanol is an environmentally benign solvent for this nucleophilic substitution reaction, and it can be produced in massive quantities from various agricultural feedstocks. This has several implications from the perspective of the 12 Principles of Green Chemistry. Not only is it a renewable resource, but compared to many other volatile organic compounds, ethanol is less hazardous to both laboratory occupants and the environment. However, its flammability and inhalation hazards are still of concern, so it should be handled in a fume hood.

14.1E Using NMR Spectroscopy to Measure Product Ratio

In a ^{1}H NMR spectrum of a two-component mixture, integration can be used to calculate the mole ratio of two components. In this experiment, a two-component mixture of *n*-propyl and isopropyl ethers will be generated from the 2,6-dimethylphenol, and their mole ratio will be used to assess the relative rate of the substitution reaction on the *n*-propyl bromide and isopropyl bromide.

For the two products formed in this reaction, there are peaks in the 3.5–4.5 ppm range that are diagnostic for the two ethers, as they correspond to the H's that are directly attached to the carbon that was involved in the substitution reaction. To determine the mole ratio, accurate integration data will be needed for these peaks (**Figure 14.5**).

FIGURE 14.5

$$\frac{\text{Mole ratio A}}{\text{Mole ratio B}} = \frac{\text{integration of H}_a/2}{\text{integration of H}_b/1}$$

First, in the range 3.5–4.5 ppm, assign the two peaks H$_a$ and H$_b$, one from each component **A** and **B** in the mixture. Then, convert the integrations to a "per hydrogen" basis by dividing each by the number of equivalent hydrogens giving rise to the peak. Then, the ratio of these equals the mole ratio of the compounds in the mixture, as shown in the equation here. For more details, see Dealing with Spectra of Mixture Samples, in Chapter 7.

In the two mechanisms S_N1 and S_N2, the rates of reaction depend on the structure of the alkyl halide (1° vs. 2°), but in different ways. In a competition experiment, where a mixture of 1° and 2° alkyl halides compete for reaction with the same nucleophile, the rates of reaction can be experimentally assessed by NMR integration of the ratio of products. If a product forms faster, there will be more of it in the product mixture. There are three possible outcomes that may be observed:

OBSERVATION	MECHANISM
1. The ratio A:B = 1 (the sample contains equal parts A and B)	
2. The ratio A:B > 1 (the sample contains more A than B)	
3. The ratio A:B < 1 (the sample contains less A than B)	

Review the mechanistic discussions in this chapter, and identify which of observations 1, 2, 3 above will serve as evidence for or against the S_N1 mechanism in the reaction of 2,6-dimethylphenol with n-propyl bromide and isopropyl bromide. Repeat this analysis for the S_N2 mechanism. In the spaces above, enter S_N1, S_N2, both, or neither.

14.3 EXPERIMENTAL PROCEDURE

14.3A Nucleophilic Substitution Competition Reaction

DAY 1

1. Start a heat source (sand bath, heating mantle, or metal block) at a medium setting so that it will reach a temperature of 90–110°C. Prepare a reflux condenser with water inlet and outlet hoses.

2. Into a 10-mL round-bottomed flask equipped with a magnetic stir bar, place 0.20 g 2,6-dimethylphenol, 0.5 mL ethanol, and 0.5 mL 20% aqueous NaOH solution.

3. Clamp the flask securely, place a condenser on top of the flask, and start the condenser cooling water at a very low flow rate.

4. Adjust sand bath heating to a temperature of 90–110°C. Lower the clamped flask into the sand bath, start the magnetic stirrer, and heat the mixture until it begins to reflux.

5. Add a mixture of 0.5 mL 1-bromopropane and 0.5 mL 2-bromopropane through the opening at the top of the condenser.

6. After 40 minutes at reflux, remove the flask from the heat and cool it in an ice-water bath.

Heating at Reflux

Aqueous–organic extraction: work up the reaction.

7. Transfer the reaction mixture to a 60-mL separatory funnel. Rinse the flask with three 5-mL portions of *tert*-butyl methyl ether (MTBE), adding each rinse to the separatory funnel. Extract the MTBE solution of the product mixture with three 5-mL portions of 1 M aqueous NaOH, then one 5-mL portion of water.

8. Drain the organic phase (ether solution) into a clean 50-mL Erlenmeyer flask and dry over anhydrous Na_2SO_4.

9. Decant the organic phase to a tared 25-mL round-bottomed flask, and concentrate using the rotary evaporator for a few minutes until the remaining oily liquid appears to be a constant volume (**Figure 14.6**). *Note:* If a second class period is available, this is a suitable stopping point. Store the flask loosely stoppered so that the last traces of solvents will evaporate. If the entire procedure is to be completed in 1 day, use a vacuum adapter with stopper to place the flask under vacuum for 20 minutes in order to remove the last traces of solvents.

DAY 2

10. Record the mass of the product.

11. Using a Pasteur pipet, add two drops of the product liquid to your NMR tube. Check with your instructor about whether to add a deuterated solvent. Label the tube in the manner suggested by your instructor, and submit for ^1H NMR spectrum acquisition.

 Aqueous–Organic Extraction and Drying Organic Solutions

 Link Rotary Evaporation

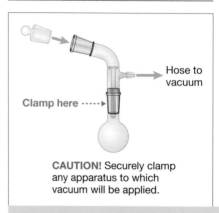

Hose to vacuum

Clamp here ----->

CAUTION! Securely clamp any apparatus to which vacuum will be applied.

FIGURE 14.6

Apparatus for removal of small amounts of residual solvent under vacuum. Always clamp the apparatus securely when vacuum is in use.

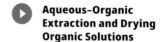 **Preparing NMR Samples**

14.4 THE LABORATORY REPORT

In your report, comment on the yield and compare the ^1H NMR spectrum to known spectra of 1-bromopropane and 2-bromopropane (these are available online) in order to conclude whether the reaction successfully achieved the nucleophilic substitution reaction. Note whether unreacted starting materials or solvents are present. Assign all peaks in the ^1H NMR spectrum that are associated with the nucleophilic substitution products. Using integration data, determine the ratio of 1-propyl to 2-propyl ethers, and use these data to determine whether the reaction proceeded by an S_N1 or S_N2 mechanism, or by both. Include answers to the following post-lab questions in your report.

14.4A Post-Lab Questions

1. Ethanol and 2,6-dimethylphenol both have hydroxyl groups that can be deprotonated by a base. Why does ethanol act as a solvent but not as a nucleophile under the conditions used in this experiment?

2. What is the purpose of washing the organic solution with 1 M aqueous NaOH during the workup procedure?

3. If 2-bromo-2-methylpropane (*tert*-butyl bromide) replaces the alkyl halides of this experiment, with all other conditions remaining the same, what would be the major product you would expect to observe?

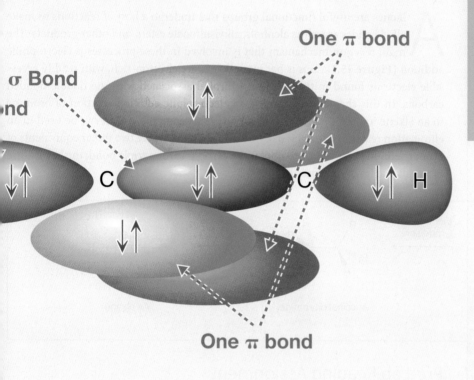

One π bond

σ Bond

One π bond

C C H

Alkene Bromination and Dehydrohalogenation

LEARNING OBJECTIVES

- Recognize key features of addition and elimination reactions.

- Execute an electrophilic addition reaction of an alkene using a safer bromine reagent that can be handled in the solid state.

- Implement a dehydrohalogenation reaction to synthesize an alkyne.

π BONDING

Alkynes contain *sp*-hybridized carbons in a triple bond, consisting of a σ bond and two π bonds that are oriented in different planes.

Karty, J. *Organic Chemistry: Principles and Mechanisms*, 3rd ed.; W. W. Norton: New York, 2022; p 126.

INTRODUCTION

A lkenes are useful functional groups that undergo a host of reactions to make alkyl halides as well as alcohols, alkylsulfonate esters, and other products. The main reaction mechanism that is involved in these processes is electrophilic addition (**Figure 15.1**). This is because alkenes are electron-rich, with readily accessible electrons found in the bonding π orbital associated with the double-bonded carbons. In this chapter, we examine an electrophilic addition reaction of bromine to an alkene to provide a vicinal dibromide. This compound then can be used in an elimination reaction known as dehydrohalogenation, where two molar equivalents of HBr are eliminated in order to obtain the carbon–carbon triple bond, the key feature of the alkyne functional group.

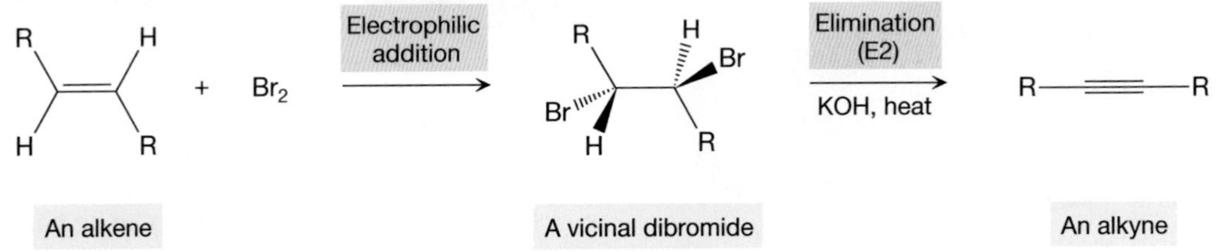

FIGURE 15.1

This experiment employs an electrophilic addition reaction of an alkene in part 1, using a less hazardous solid source of bromine, followed by double dehydrohalogenation in part 2 to yield an alkyne. Depending on the substituents on the starting alkene, a vicinal dibromide may have two stereogenic centers, and four stereoisomers could be possible.

Pre-Lab Reading Assignment

- Experiment: Chapter 15 (this chapter)
- Technique: Recrystallization (Chapter 5, section 5.4)
- Technique: Melting point determination (Chapter 5, section 5.6)
- Technique: Infrared (IR) spectroscopy (Chapter 6, section 6.2)

15.1 BACKGROUND

LUMO >>
Lowest unoccupied molecular orbital; the molecular orbital of lowest energy that is not populated by an electron.

HOMO >>
Highest occupied molecular orbital; the molecular orbital of highest energy that is populated by electron(s).

15.1A   Electrophilic Addition Reactions

The electrons that reside in a π bond come from a combination of two p orbitals from neighboring carbons (**Figure 15.2a**), each bringing one electron into the union. When these two orbitals combine, two new orbitals result, called π and π^* (**Figure 15.2b**). The two electrons go into the lower energy of these two orbitals (**Figure 15.2c**), and are shared by both carbons in a π bond. This filled π orbital is also referred to as the bonding orbital, while the π^* orbital is an antibonding orbital which is empty, as is every other orbital of higher energy than the π^*. As it is the lowest-energy empty orbital, this π^* is also known as the **LUMO**, or lowest unoccupied molecular orbital. The two electrons in the π bonding orbital are in the highest occupied molecular orbital, or **HOMO**. Because these electrons are the ones that are most accessible to oxidizing or electrophilic reagents, it is this orbital that is involved in electrophilic addition reactions of alkenes.

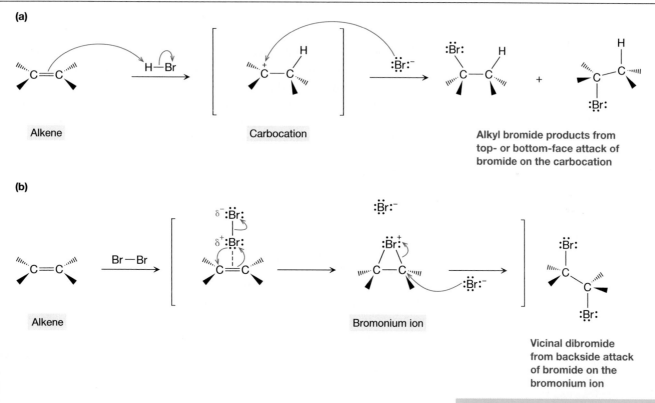

FIGURE 15.2

(a) The *p* orbitals of *sp*²-hybridized carbons are above and below the plane defined by the three bonds. **(b)** Additive, or constructive, combination of two *p* orbitals yields a new π bonding orbital. Subtractive combination yields π*, an antibonding orbital (not shown). **(c)** An orbital energy diagram shows that the combination of two *p* orbitals leads to two new orbitals, and the two electrons from each of the carbons are placed in the lower energy π bonding orbital.

Electrophilic means electron-loving. Electrophilic reagents are electron-deficient in some way, and will react with readily available electron sources like the HOMO of an alkene. One of the main ways they react is by using the two electrons from the alkene to make a new bond to the electrophilic reagent. Several examples of such

FIGURE 15.3

reagents include HBr, HCl, and H_3O^+, where in each case the electrophilic atom is a proton. The electrons from the alkene move to a new σ bond between the proton and one carbon of the alkene (**Figure 15.3a**). Now, the other carbon of the alkene is no longer sharing the pair of electrons, and it has three bonds and an empty *p* orbital. With only six electrons shared, short of the ideal octet, this structure therefore has a formal positive charge on carbon, and is called a *carbocation*. Carbocations can proceed through several kinds of reactions. They can rearrange into more stable carbocations, lose a proton to regenerate an alkene (the reverse of the reaction just described), or react with various types of nucleophilic species. After

(a) Electrophilic addition of a proton to an alkene generates a carbocation, which can react from the top or bottom face with various nucleophiles, in this case, bromide ion. **(b)** In electrophilic addition of Br_2, the electrophilic bromine atom possesses a lone pair, and thus a bromonium ion forms instead of a carbocation. The bromonium ion reacts with Br⁻ from the backside to generate a vicinal dibromide.

the proton from the electrophilic reagent had become bound to carbon, it released the other part of the reagent, such as Br^- (Figure 15.3a), which can act as a nucleophilic species to produce an alkyl bromide.

Another example of electrophilic addition occurs with elemental halogens such as Cl_2, Br_2, or I_2 (**Figure 15.3b**). Within Br_2 are two bromine atoms, each with seven valence electrons. They each share one with the other in order to achieve an octet for both. However, the Br_2 structure is an uneasy alliance between siblings; they find it even more favorable to each find their own extra electron, rather than share. Adding one electron to each bromine atom affords two bromide (Br^-) ions that have a complete octet on their own. Because of this tendency to acquire additional electrons from elsewhere, elemental halogens are strong oxidants.

Bromine can acquire electrons from the alkene HOMO. The reaction of bromine with an alkene involves an initial approach of the bromine to the alkene π bond, where the presence of electrons begins to polarize the Br—Br bond, so that one Br has a partial positive charge. This can then form a bond to the alkene, and release Br^- into the surrounding solvent. At the same time, the lone pair on the remaining bromine neutralizes the carbocation that is developing on the neighboring carbon. The resulting intermediate is a bromonium ion, a positively charged species with a strained three-membered ring. This is formed stereospecifically; substituents that are trans in the alkene will be trans in the bromonium ion. Like other positively charged species, the bromonium ion is an electrophile that reacts readily with nucleophiles. The Br^- released earlier can serve this role, attacking the bromonium ion from the backside of the C—Br bond, and this step is also stereospecific. The ring opening of the bromonium ion with a bromide ion places two bromines on the two carbons that were previously involved in the alkene π bond. The net result of the two stereospecific steps is an anti-addition product, where the two Br atoms have been attached to opposite faces of the alkene. This type of product is called a vicinal dibromide, where vicinal refers to two neighboring carbons. We will carry out this type of electrophilic addition to convert stilbene to a vicinal dibromide.

In our experiment, the two carbons of the bromonium ion are equivalent. However, if they have different substituents, it is the carbon bearing the carbocation-stabilizing groups that is generally attacked by the Br^- nucleophile.

15.1B | Dehydrohalogenation

Elimination reactions of alkyl bromides produce π bonds between vicinal carbons. When such reactions are initiated by a strong base, mechanistically, they generally involve the simultaneous deprotonation of one carbon while bromide ion is released from the neighboring carbon. When bond-forming and -breaking processes occur together in the same mechanistic step, they are said to be *concerted*. This concerted elimination of a proton and halide ion is called *dehydrohalogenation*. It is part of a class of reactions whose rates are dependent on concentrations of both the base and the alkyl halide, because both are involved in the rate-determining step (the only step, in this case) of the reaction. Such reactions are known more broadly as *bimolecular elimination*, or E2 reactions.

With a vicinal dibromide, two successive dehydrohalogenation reactions can occur upon treatment with strong base, removing a proton and bromide from neighboring carbons to afford a bromoalkene. The elimination occurs with the proton and bromide *antiperiplanar*; both of these atoms are in the same plane, but on opposite faces of the planar alkene that is forming. This then can further react to lose another proton and bromide to yield the carbon–carbon triple bond of the

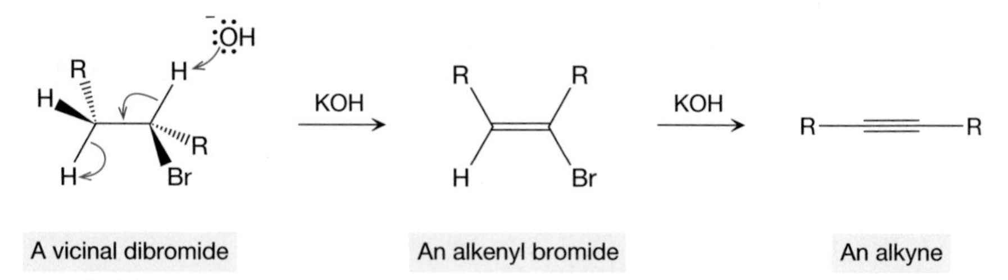

FIGURE 15.4

Double dehydrohalogenation of a vicinal dibromide proceeds via E2 elimination to give an alkenyl halide, followed by a second elimination to yield an alkyne. Two equivalents of KOH are used, which results in two equivalents of H_2O and two equivalents of KBr as by-products.

alkyne functional group (**Figure 15.4**). The mechanism of the second elimination is likely not E2, but a stepwise alternative.[1]

In part 1 of this experiment, we will use (*E*)-1,2-diphenylethene, commonly known as *trans*-stilbene, as the alkene reactant in an electrophilic addition reaction with the molecular elemental form of bromine, Br_2 (**Figure 15.5**). What is the structure of the product you expect to obtain? Note that this reaction consists of achiral reactants in an achiral environment. What stereoisomer(s) will be present?

In part 2, we will use the product you obtained in part 1 to produce an alkyne via two successive eliminations, or double dehydrohalogenation. This procedure affords an opportunity to observe a beautiful example of crystal growth.

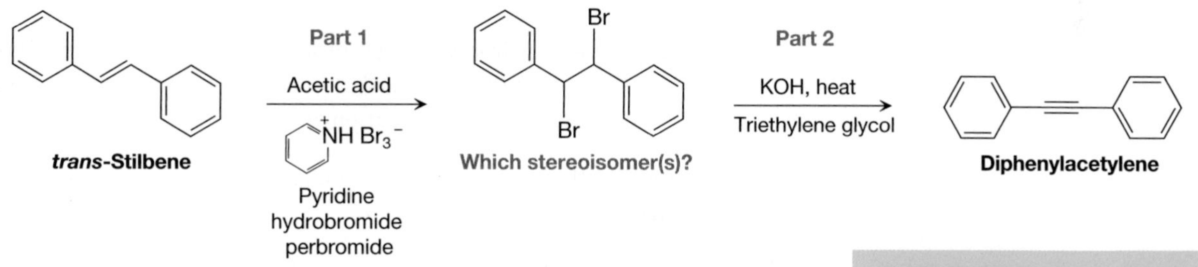

FIGURE 15.5

Parts 1 and 2 of this experiment: Electrophilic halogenation of an alkene followed by double dehydrohalogenation to obtain an alkyne.

15.1C Safer Reagents

In the context of synthesis, the principles of green chemistry dictate that safer reagents are advantageous, and should be chosen when possible. In this experiment, we will use a solid compound as a source of bromine, which avoids some hazards associated with this important reagent for synthesis. The elemental form of bromine, Br_2, is a liquid with an appreciable vapor pressure, while also being a powerful oxidant.

Oxidants are corrosive and need to be handled carefully and stored separately from organic compounds and reducing agents. For example, metallic objects or tools can often be oxidized; they could provide the electrons that an oxidant like bromine seeks. Oxidation (corrosion) can cause failure of metal components or structural weakness in a metal pipe; in larger scale production settings this could cause an accidental release of hazardous materials into the environment.

A volatile liquid oxidant like Br_2 presents special hazards because its liquid and vapor forms require more attention to keep them safely contained. One way to solve this problem is to replace Br_2 itself with a substitute, pyridine hydrobromide perbromide ($PyHBr_3$) (Figure 15.5). This is an easily contained solid material that is crystalline, nonvolatile, and odorless, and thus it can be conveniently handled in the lab. When an alkene is present, $PyHBr_3$ releases one molar equivalent of Br_2 so that electrophilic addition reactions can occur.

---

[1]A mechanism known as E1CB is likely operative in eliminations of vinyl halides. Kwok, W. K.; Lee, W. G.; Miller, S. I. The Kinetics, Isotope Rate Effect, and Mechanism of Dehydrobromination of *cis*-1,2-Dibromoethylene with Triethylamine in Dimethylformamide. *J. Am. Chem. Soc.* **1969**, *91*, 468–476.

| 15.2A | ## Part 1: Electrophilic Addition Reaction of an Alkene |

1. Turn on a sand bath or similar heat source to a medium setting, suitable to reach a temperature of 100–120°C.

2. Measure 0.27 g *trans*-stilbene into a 50-mL round-bottom flask, and clamp this reaction flask to a vertical support rod.

   (CAUTION: Acetic acid and pyridine hydrobromide perbromide are corrosive. Avoid contact. In case of contact, wash with water.)

3. Add 5 mL acetic acid to the reaction flask. Apply heat by lowering the clamped flask into the sand bath, and gently swirl occasionally until the *trans*-stilbene is dissolved.

4. Measure 0.55 g of pyridine hydrobromide perbromide into a weighing boat or paper. This can be plus or minus 0.05 g; record the actual amount used.

5. Add the pyridine hydrobromide perbromide to the reaction flask. If reagent sticks on the neck of the flask, wash it into the reaction mixture with a few drops of acetic acid from a Pasteur pipet.

6. Continue heating and swirling for about 5 minutes. The solid product may begin to separate from the mixture.

7. Cool the reaction mixture in a bath of cold tap water until it reaches the temperature of the cold tap water.

▶ **Vacuum Filtration**

8. Recover the dibromide product by vacuum filtration with a Hirsch funnel, using three 2-mL portions of methanol to wash the crystals out of the flask, adding each wash to the Hirsch funnel.

▶ **Melting Point Measurement**

9. Obtain the melting point of the product, and compare it with literature values for chiral and meso stereoisomers of stilbene dibromide.

| 15.2B | ## Part 2: Double Dehydrohalogenation to Form an Alkyne |

1. Set up a magnetic stir plate with a sand bath or similar heat source on top. Heat at a medium–high setting, suitable to reach a temperature of 160–170°C.

2. In a vial (about 5–10-mL capacity), place 0.32 g of stilbene dibromide (prepared in part 1) and a magnetic stir bar.

   (CAUTION: Potassium hydroxide is corrosive; avoid contact. In case of contact, wash with water.)

---

[2]This procedure was adapted from that of Williamson. The diphenylacetylene preparation was developed by Louis Fieser. Williamson, K. L. *Macroscale and Microscale Organic Experiments*, 3rd ed.; Houghton Mifflin: Boston, 1999.

3. Add 0.16 g potassium hydroxide. This is available in pellet form. There is no need to break the pellets; the number of pellets closest to 0.16 g should be used. Record the actual amount used.

4. Add 1 mL triethylene glycol. Place a clamp on the vial, start the magnetic stirrer, and lower the bottom of the vial into the sand bath.

5. Continue heating at 160–170°C with stirring for 10 minutes, during which some KBr may separate from the mixture as a colorless solid.

6. Remove from the heat, and allow to cool to room temperature.

7. Add 5 mL water and mix thoroughly. The KBr should dissolve, while the product diphenylacetylene precipitates.

8. Recover the product by vacuum filtration with a Hirsch funnel, using three 2-mL portions of water to wash the crystals out of the flask, adding each wash to the Hirsch funnel.

9. In the same vial, recrystallize by adding the minimum amount of hot 95% ethanol needed to dissolve the diphenylacetylene, adding dropwise. Allow the solution to cool while standing undisturbed on the benchtop.

10. Cool further in an ice bath to ensure complete crystallization, and then recover the product by vacuum filtration in a Hirsch funnel. With a 1-mL portion of ice-cold ethanol, wash the crystals from the vial into the funnel.

▶ **Recrystallization**

11. Measure the melting point of the diphenylacetylene.

12. Acquire the infrared spectrum of the diphenylacetylene.

15.3 THE LABORATORY REPORT

In your report, comment on the infrared absorption(s) in the region where alkyne carbon–carbon triple bonds are typically observed. Compare the melting point and the IR spectrum with literature data. Which technique provides stronger evidence of the identity and purity of the product?

15.3A Post-Lab Questions

1. What product would form in part 1 if acetate ion ($CH_3CO_2^-$) served as the nucleophile? Provide a drawing that clearly illustrates the configuration at each stereogenic center.

2. Would you expect the double dehydrobromination reaction to occur more readily (or at lower temperature) with racemic chiral dibromide **1** or meso, achiral dibromide **2** (**Figure 15.6**)? Draw the structures of the alkenyl bromides expected as intermediates in each case.

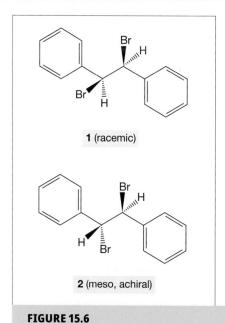

1 (racemic)

2 (meso, achiral)

FIGURE 15.6

LEARNING OBJECTIVES

- Identify the main features of acid-catalyzed dehydration and the E1 mechanism.

- Perform a synthesis involving two sequential reactions and purifications.

- Explore techniques of simple and fractional distillation.

- Recognize how the concept of catalysis connects with the principles of green chemistry.

Multistep Synthesis: Cyclohexanol to Adipic Acid

INNOVATOR IN STRENGTH

The material known as Kevlar was invented by Dr. Stephanie Kwolek in the 1960s. Kevlar is an extraordinarily strong polymer used in bulletproof vests, and is composed of repeating amide bonds. In Chapter 16 we will make adipic acid, a precursor to another polymer with amide bonds.

Pictorial Press Ltd/Alamy Stock Photo.

M aking an organic compound, whether it's done by people in the lab or biosynthetically in a living system, often involves multiple reaction steps. For example, the chemotherapy drug eribulin (**Figure 16.1**), also known as Halaven, requires over 60 synthetic steps to build all the individual pieces and put them together into the final structure.[1] It is a simplified analog of halichondrin b, an anticancer compound occurring naturally in a marine sponge. At the time of its approval for clinical use, eribulin was the most complex clinically approved drug to be produced by synthesis.[2]

FIGURE 16.1

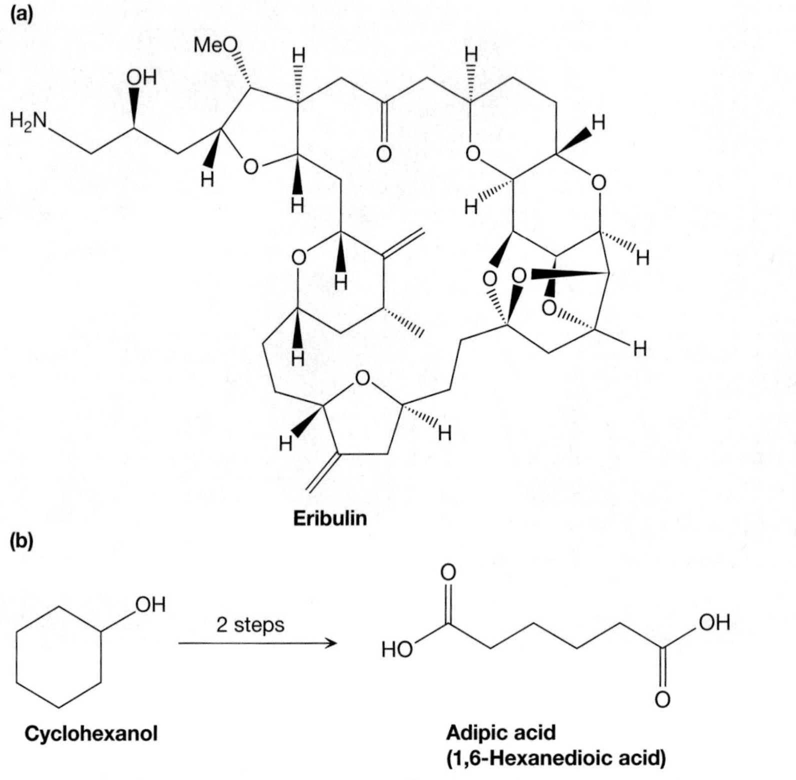

(a) Eribulin (trade name Halaven) is an anticancer drug that requires over 60 synthetic steps to make. (b) Adipic acid, the compound made in this experiment, will be made in two steps.

(a)

Eribulin

(b)

Cyclohexanol

**Adipic acid
(1,6-Hexanedioic acid)**

2 steps

Reactions of alcohol and alkene functional groups are particularly useful in organic chemistry, as they interconvert functional groups. Such reactions can be used to control the polarity and chemical behavior of organic compounds, and make it possible to link structures together in order to build up molecular architectures with specific functions.

In this experiment,[3] we learn how a synthesis sequence fits together by performing a two-step synthesis of adipic acid—that is, the product of the first step is used as the starting material for the second step.

---

[1]Yu, M. J.; Zheng, W.; Seletsky, B. M. From Micrograms to Grams: Scale-Up Synthesis of Eribulin Mesylate. *Nat. Prod. Rep.* **2013**, *30*, 1158–1164.

[2]Schreiber, S. L. Organic Synthesis toward Small-Molecule Probes and Drugs. *Proc. Natl. Acad. Sci. U.S.A.* **2011**, *108*, 6699–6702.

[3]This experiment is an adaptation from two experiments in *Green Organic Chemistry* by Doxsee and Hutchison, from which portions of this chapter are excerpted. Doxsee, K. M.; Hutchison, J. E. *Green Organic Chemistry: Strategies, Tools, and Laboratory Experiments*; Brooks/Cole: Belmont, CA, 2004.

16.1A Pre-Lab Reading Assignment

Review the following before attempting part 1 of this experiment:

- Experiment: Chapter 16 (this experiment)
- Technique: Distillation (Chapter 5, section 5.3)
- Technique: IR spectroscopy (Chapter 6, section 6.2)

16.1B Background

The first step of our synthesis of adipic acid is the dehydration of cyclohexanol (**Figure 16.2a**). This reaction proceeds through the E1 mechanism shown in **Figure 16.2b**. Acid-catalyzed dehydration is an equilibrium process—the reverse reaction is acid-catalyzed hydration of alkenes to make alcohols. By removing the alkene product as it forms, though, the reaction can be driven in the forward direction by Le Châtelier's principle.

The first step is a reversible proton transfer. Often sulfuric acid has been used for this protonation step, but we use phosphoric acid, which is less reactive and less corrosive than sulfuric acid, making it a less hazardous reaction to run in the lab. Concentrated phosphoric acid contains a significant amount of water (15% by weight), so the proton source could be either H_3PO_4 or H_3O^+, as shown. The protonated alcohol now bears an excellent leaving group—a neutral water molecule. After it leaves, along with the pair of electrons in the C—O bond, the remaining part of the structure retains a positive charge on carbon, making it a carbocation. At this stage, removal of a proton from the carbon next to the positively charged carbon completes the formation of the alkene product.

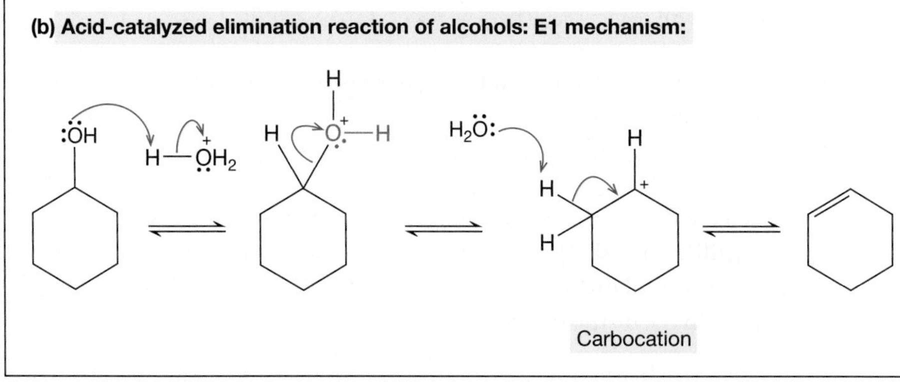

FIGURE 16.2

(a) Dehydration reaction to be carried out in this experiment.
(b) The E1 mechanism of acid-catalyzed dehydration of alcohols.

(a) Dehydration of cyclohexanol:

Concentrated H_3PO_4
(85%, aqueous)
Heat

$+$ H_2O

(b) Acid-catalyzed elimination reaction of alcohols: E1 mechanism:

Carbocation

FIGURE 16.3

Apparatus for fractional distillation. Water is circulated in the jacket of the condenser, but not the distilling column. Your instructor may have the distilling columns packed with glass beads or other porous inert materials to increase their efficiency. A Vigreux column or insert, with numerous glass protrusions, also performs this function. The drying tube may be attached via flexible tubing or a rubber stopper as shown. The apparatus may be set up with the receiving flask on the left (as shown here) or the right.

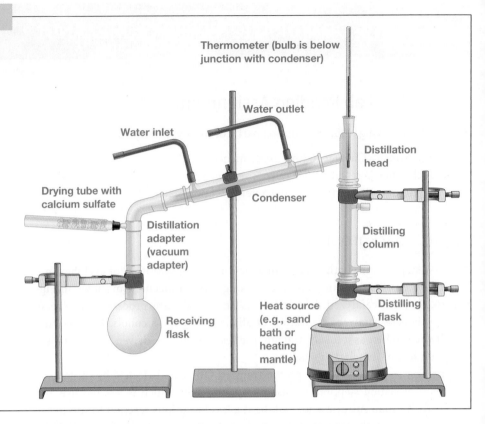

Thermometer (bulb is below junction with condenser)

Water outlet

Water inlet

Distillation head

Drying tube with calcium sulfate

Condenser

Distillation adapter (vacuum adapter)

Distilling column

Receiving flask

Heat source (e.g., sand bath or heating mantle)

Distilling flask

We will use a fractional distillation (**Figure 16.3**) to separate the cyclohexene as it forms. Fractional distillation employs a vertical distilling column to increase the efficiency of separating compounds having different boiling points (bp). This allows the cyclohexene (lower bp) to collect in the receiving flask while the cyclohexanol (higher bp) stays in the distilling flask until it has been dehydrated.

16.1C Experimental Procedure (Part 1)[3]

1. Set up the apparatus for fractional distillation (see section 5.3 and Figure 16.3). First, begin preheating the heat source at medium power, then assemble the glassware. Clamp the apparatus high enough to accommodate a laboratory jack or removable wooden block under the heat source.

2. To a 50-mL round-bottom flask containing a magnetic stir bar, add 8.0 mL cyclohexanol. *Note to instructors and staff:* Cyclohexanol may be partially solidified if the room temperature is below 25°C. Correct this by setting the bottle in a 30–35°C water bath prior to the start of the lab period.

3. Add 1.75 mL of concentrated H_3PO_4 (85% w/w, aqueous). Start the magnetic stirrer to mix the layers.

   CAUTION: Concentrated H_3PO_4 is corrosive; avoid contact.

4. Attach the flask to the fractional distillation apparatus and begin heating the distillation flask. If no distillate collects after 10–15 minutes, increase the heat a little until distillation begins. Keep distilling until only about 1 mL remains in the distillation flask.

5. Transfer the distillate to a separatory funnel and wash with approximately 5 mL water. Carefully separate the layers and transfer the organic layer to a 25-mL Erlenmeyer flask.

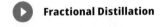

Fractional Distillation

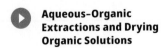

Aqueous–Organic Extractions and Drying Organic Solutions

6. Dry the organic phase over Na_2SO_4 by adding a pea-sized amount of anhydrous Na_2SO_4 and swirling the flask. If all the drying agent clumps together, it's not enough drying agent. Add another pea-sized amount and swirl again. When all the water is absorbed, a little of the drying agent should remain free flowing and the liquid should be clear. *Note:* This is a suitable stopping point. Store the crude cyclohexene and drying agent in a tightly sealed screw-cap vial until the next lab period.

7. Decant or pipet the organic liquid away from the drying agent and place it in a 25-mL round-bottom flask. Add a magnetic stirring bar.

8. Set up for a simple distillation (see Figure 5.10). The apparatus is the same as Figure 16.3 except without the distilling column. *Note:* The prior fractional distillation occurs with moisture present. Adding a simple distillation after use of a drying agent enables a more accurate boiling point to be measured, and produces higher purity cyclohexene that provides more reliable results in the oxidative cleavage (part 2).

 Simple Distillation

9. Distill slowly, monitoring the temperature carefully. When the temperature reaches 70°C, begin collecting the distillate. Most of the cyclohexene should be collected in the temperature range of 70–90°C, but be sure to record in your notebook the boiling range you actually observe.

10. Transfer the distilled cyclohexene to a tared screw-cap vial and record the mass.

11. Record the infrared spectrum of the cyclohexene. Your instructor may also assign gas chromatography to assess the purity of the cyclohexene.

12. You need the cyclohexene for the next synthetic step. It has a relatively low boiling point, and may evaporate before the next lab period unless you take care to seal the vial. Tightly screw the cap onto the vial, and wrap Parafilm around the gap between the cap and vial.

16.2 PART 2: OXIDATIVE CLEAVAGE OF CYCLOHEXENE

16.2A Pre-Lab Reading Assignment

Review the following before attempting part 2 of this experiment:

- Experiment: Chapter 16 (this experiment)
- Technique: Filtration (Chapter 5, section 5.1)
- Technique: Recrystallization (Chapter 5, section 5.4)
- Technique: Melting point determination (Chapter 5, section 5.6)

16.2B Background

Typical alkenes are electron-rich, so they react with electrophiles (such as a proton) or with oxidizing agents. Examples of these types of reactions, which include the hydration of alkenes, addition of Br_2 to alkenes, epoxidation of alkenes by *m*-chloroperbenzoic acid (MCPBA), and halohydrin formation, are shown in **Figure 16.4**.

Another example is 1,2-glycol formation, which adds a hydroxyl group to each of the two carbons previously involved in the carbon–carbon double bond. This is

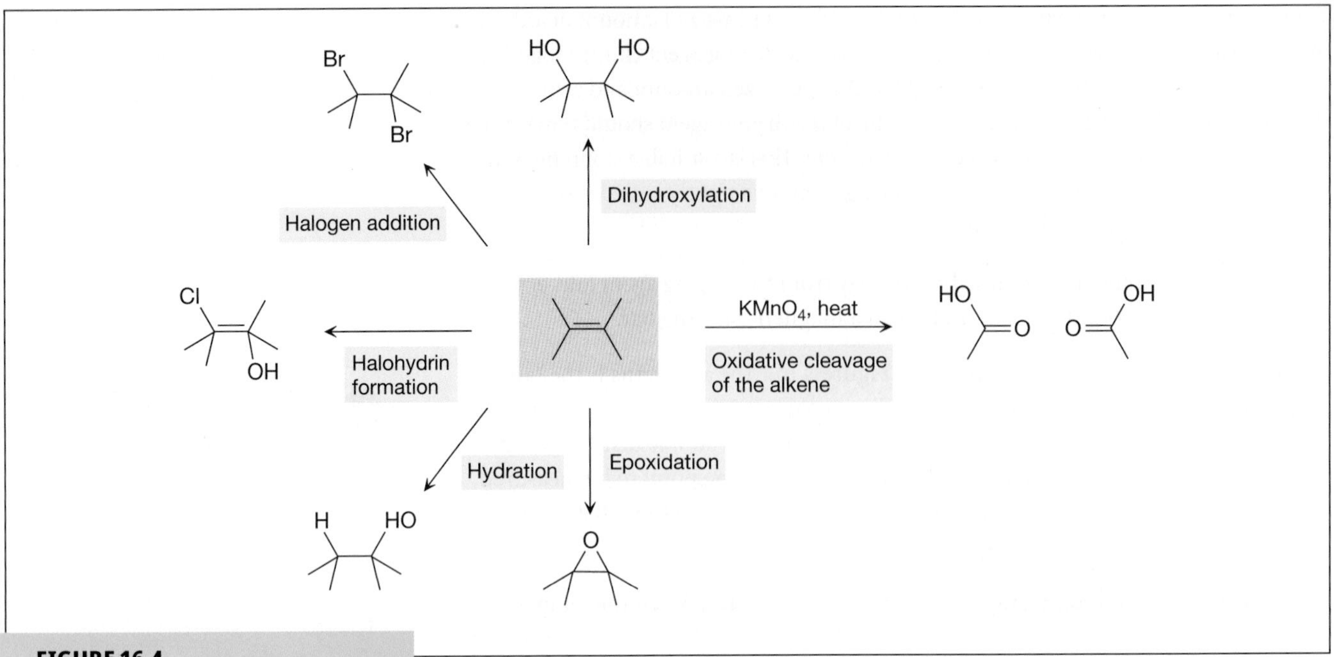

FIGURE 16.4

Representative functional group transformations of alkenes. The oxidative cleavage of a 1,2-disubstituted alkene by $KMnO_4$ leads to two carboxylic acids.

an oxidation reaction, and this is often accomplished by cold $KMnO_4$ under basic conditions. If the reaction with $KMnO_4$ is not kept cold, the C—C bond of the 1,2-glycol is cleaved between the two hydroxyl groups, leading to two aldehydes or ketones (Figure 16.4). Aldehydes are further oxidized to carboxylic acids, which in basic conditions are in the carboxylate salt form, then protonated on acidic workup.

In part 2 of this experiment, we carry out the second step of the synthesis—namely, the oxidative cleavage of cyclohexene to adipic acid (hexanedioic acid). The starting material is the cyclohexene produced in part 1.

The product, adipic acid, is used in the industrial synthesis of nylon-6,6, which is composed of repeating units of adipic acid and 1,6-diaminohexane (**Figure 16.5**).

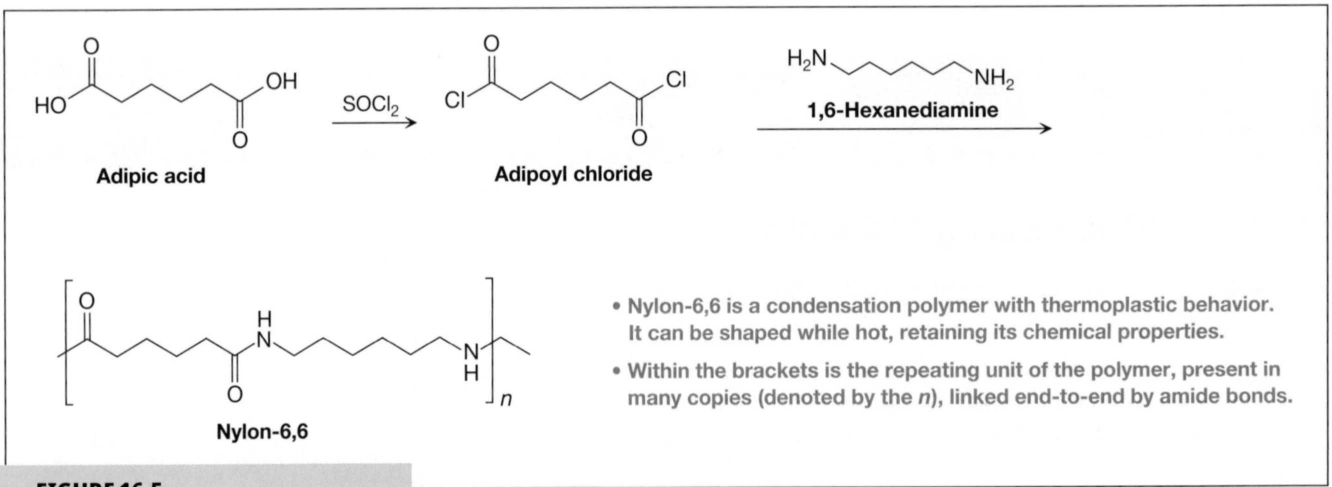

FIGURE 16.5

The preparation of nylon-6,6 involves the reaction of adipoyl chloride with 1,6-hexanediamine. Nylon was invented in 1935 and its remarkable strength and durability when formed into fibers led to its initial applications in toothbrush bristles, stockings, and parachute fabric.

- Nylon-6,6 is a condensation polymer with thermoplastic behavior. It can be shaped while hot, retaining its chemical properties.

- Within the brackets is the repeating unit of the polymer, present in many copies (denoted by the *n*), linked end-to-end by amide bonds.

The first commercial use of nylon-6,6 was for toothbrush bristles in the 1930s, followed by various replacements for natural silk, including in hosiery and parachute fabric. Today, injection-molded nylons (including nylon-6,6) are used when mechanical strength and rigidity are needed, such as in automotive parts. Nylons also continue to be widely used for long-lasting durable fibers found in luggage and carpet, and for stretchable fabrics found in swimwear and other athletic garments.

Using hot permanganate (MnO_4^-) for this reaction produces large amounts of metal oxide (MnO_2) waste material. Industrial production of adipic acid uses nitric acid in place of KMnO₄, which is highly corrosive and presents numerous safety hazards and environmental risks along with the potential for accidents due to violent reactions of nitric acid with some organic compounds. The use of nitric acid in the preparation of adipic acid results in the emission of nitrous oxide, a suspected greenhouse gas, and adipic acid production is believed to be the source of roughly 10% of all non-natural nitrogen oxide ("NO$_x$") emissions.[3]

In applying the principles of green chemistry to such oxidation reactions, Sato, Aoki, Tagaki, and Noyori found that the metal oxidant could be used as a catalyst, recycling it during the reaction, and thus avoiding the metal oxide waste. This entailed using a combination of sodium tungstate (Na₂WO₄) with hydrogen peroxide (H₂O₂).[4]

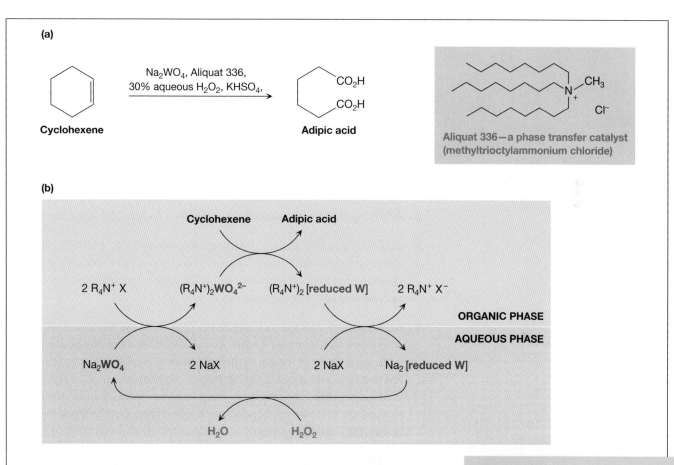

FIGURE 16.6

(a) The reaction carried out in this experiment. The Aliquat 336 phase transfer catalyst contains a tetraalkylammonium ion that can transfer a counterion into the organic phase as an ion pair. (b) Phase transfer catalysis in this reaction involves ion-exchange processes. When water-soluble tungstate ions (WO₄²⁻) exchange sodium counterions for organic-soluble tetraalkylammonium counterions (R₄N⁺), tungstate ions may transfer into the organic phase where the reaction takes place. The reduced form of tungsten can transfer back to the aqueous phase where it is recycled by oxidation with aqueous H₂O₂.

In this case, the tungstate is a catalyst and the metal oxide by-product is recycled back to tungstate by the hydrogen peroxide. Thus, no tungstate is consumed in the reaction, and the stoichiometric by-product from H₂O₂ is water. Furthermore, these reagents are soluble in water. We will use this reaction for oxidative cleavage of cyclohexene (**Figure 16.6**).

To make this process practical, however, we must address the fact that cyclohexene is insoluble in water, whereas sodium tungstate is insoluble in organic solvents. That is, the aqueous tungstate and alkene reactants are in two different phases. To bring both reactants into the same phase, where they can react with each other,

[4]Sato, K.; Aoki, M.; Takagi, J.; Noyori, R. Organic Solvent- and Halide-Free Oxidation of Alcohols with Aqueous Hydrogen Peroxide. *J. Am. Chem. Soc.* **1997**, *119*, 12386–12387. Review of the applications of this green oxidation system: Noyori, R.; Aoki, M.; Sato, K. Green Oxidation with Aqueous Hydrogen Peroxide. *Chem. Commun.* **2003**, 1977–1986.

we can apply a principle called **phase transfer catalysis** (Figure 16.6). The sodium ion of sodium tungstate can be exchanged with a tetraalkylammonium ion, such as methyltrioctylammonium (Aliquat 336), which is soluble in both phases. When the tetraalkylammonium ion transfers into the organic phase, it brings a negatively charged counterion along with it to balance the charge as part of an ion pair. Therefore, it transfers a tungstate ion from the aqueous phase into the organic phase, where the tungstate can react with the alkene. The tetraalkylammonium ion is not consumed—it takes the reduced tungsten species back into the aqueous phase, where it can be reoxidized by hydrogen peroxide for another catalytic cycle. Potassium bisulfate ($KHSO_4$) helps to control the acidity for optimal reactivity.

Because this reaction combines two types of catalysis, recycling both the tungstate ion and the Aliquat 336 during the reaction, both can be used in very small amounts, minimizing the waste generated.

16.2C Experimental Procedure (Part 2)[3]

1. Begin preheating a heat source (e.g., sand bath, heating mantle) to about 110–120°C. Place about 0.5 g Aliquat 336 into a 50-mL round-bottom flask. This is a sticky liquid that is best transferred with a pipet having the narrowest part of the tip removed to make a larger opening. It is not necessary to have exactly 0.5 g of this component.

2. Add 0.5 g of sodium tungstate dihydrate ($Na_2WO_4 \cdot 2H_2O$) and a stir bar.

3. Add 11.98 g of 30% aqueous hydrogen peroxide and 0.37 g $KHSO_4$. Start the stirrer, then add 2.0 g cyclohexene. *Note:* If you didn't obtain 2 g from the dehydration of cyclohexanol in part 1, supplement with commercial cyclohexene and note the amounts in your lab notebook.

4. Attach a water-cooled condenser to the flask. Heat the mixture at reflux with vigorous stirring for 1 hour. About halfway through the heating, add 2 mL water through the top of the condenser to rinse down any cyclohexene that may be trapped in the upper part of the condenser. The reaction is complete when it no longer separates into two layers after the stirrer is stopped. *Note:* If vapors of cyclohexene are escaping out the top of the condenser, the heating is excessive. To address this, turn down the heat a little.

5. Transfer the hot reaction mixture by pipet into a 25-mL Erlenmeyer flask, taking care to leave behind any oily material that may separate (this is the phase transfer catalyst). It is better to leave behind some aqueous solution than to contaminate the product with the phase transfer catalyst.

6. Cool the flask in an ice bath. A precipitate should form within 20 minutes.

7. Collect the solid by vacuum filtration. Set aside enough of the crude product for a melting point sample and allow it to dry further while you recrystallize the rest.

8. Purify the crude product by recrystallization from water, slowly adding the minimum amount of hot water needed to dissolve the solid. Adding too much water diminishes the yield. Collect the crystalline product by vacuum filtration and allow it to dry.

9. Measure the melting point of both the crude product and the purified product. If the materials are not completely dry by the end of the lab period, the melting points can be measured during a subsequent lab period.

▶ **Heating at Reflux**

▶ **Vacuum Filtration**

▶ **Recrystallization**

▶ **Melting Point Measurement**

In your results and discussion section, make sure you discuss the yield of each of the two steps of the synthesis, as well as the overall yield. Also, discuss your evidence for the identity and purity of the product of each step. Comment on the effectiveness of your recrystallization, both the mass recovered and a comparison of the purity of crude and recrystallized adipic acid.

Read an alternative published procedure for the oxidative cleavage of a similar alkene,[5] and compare it to your own. Considering the principles of green chemistry, provide a brief description of how catalysis and phase transfer have reduced the environmental impact of this type of reaction.

Complete the sheet on the next page by filling in the values you used or obtained in this synthesis. Attach this completed sheet to your report.

16.3A | Dehydration of Cyclohexanol

Amount of cyclohexanol used (g):

Amount of cyclohexene obtained (g):

16.3B | Oxidative Cleavage of Cyclohexene

Amount of cyclohexene used (g):

Mass of crude adipic acid (g):

Mass of recrystallized adipic acid (g):

Recrystallization percent recovery:

Melting range of crude adipic acid (°C):

Melting range of recrystallized adipic acid (°C):

---

[5]Scott, W. J.; Hammond, G. B.; Becicka, B. T.; Wiemer, D. F. Oxidation of (R)-(+)-pulegone to (R)-(+)-3-Methyladipic Acid. *J. Chem. Educ.* **1993**, *70*, 951.

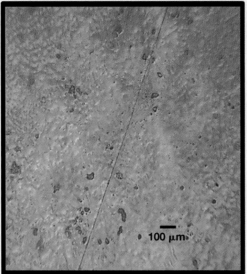

17

LEARNING OBJECTIVES

- Recognize how tandem reactions can contribute to greener organic synthesis.

- Use a metal hydride reagent to reduce a carbonyl compound.

- Implement a tandem Diels–Alder and acylation reaction under solvent-free conditions.

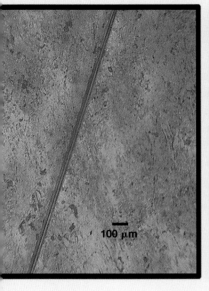

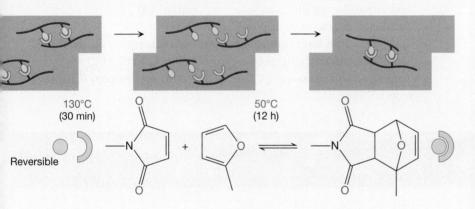

Tandem Reactions: Diels–Alder Cycloaddition and Intramolecular Acylation

SMART POLYMERS

(a) A self-healing polymer coating shows a scratch (diagonal dark line at left). At right, after self-healing by reversible Diels–Alder reactions, the scratch has mostly disappeared. The strength of films made from this material recovers after self-healing. (b) A schematic representation of the self-healing process.

Willocq, B.; Khelifa, F.; Brancart, J.; Van Assche, G.; Dubois, Ph.; Raquez, J.-M. One-Component Diels–Alder Based Polyurethanes: A Unique Way to Self-Heal. *RSC Adv.* **2017**, *7*, 48047–48053. DOI: 10.1039/C7RA09898G

Reactions producing high yields and high purities are particularly useful to practicing organic chemists, because these reactions are more efficient, generating more of the desired product while consuming fewer resources. When high yield and high purity are achieved, the reaction is usually more environmentally friendly, consistent with the green chemistry principles that pertain to efficiency, such as E-factor and atom economy. In this experiment, we use tandem reactions—that is, we combine more than one reaction into a single laboratory operation—to improve efficiency.

Pre-Lab Reading Assignment

Review the following before attempting this experiment:

- Experiment: Chapter 17 (this experiment)
- Technique: Recrystallization (Chapter 5, section 5.4)
- Technique: Extraction (Chapter 5, section 5.2)
- Technique: Thin-layer chromatography (TLC) (Chapter 5, section 5.5)
- Technique: Infrared (IR) spectroscopy (Chapter 6, section 6.2)
- Technique: Nuclear magnetic resonance (NMR) spectroscopy (Chapter 7, section 7.1)

17.1A Tandem Reactions

When a reaction sequence consists of multiple steps, often each product requires purification before proceeding to the next step. Purification procedures such as recrystallization, distillation, or chromatography add costs to the overall sequence; these costs are associated with labor, time, materials, and waste generation. Therefore, a great deal of innovation and effort has been devoted to tandem reactions, in which no purification is needed between steps. A *tandem reaction* has been defined by Tietze[1] as two or more bond-forming reactions in sequence, under the same conditions without any added reagents or catalysts, and where each reaction occurs as a consequence of the functionality generated by the previous step.

Tandem reactions are common in biology. For example, the biosynthesis of steroids proceeds via a series of carbocation cyclizations initiated by ring-opening of an epoxide in squalene oxide (**Figure 17.1a**); further carbocation rearrangements lead to the characteristic polycyclic ring system of steroids. Similar cationic polyene cyclization has been used by W. S. Johnson in a laboratory synthesis of the steroid hormone progesterone. A double Mannich reaction by the Nobel Laureate Robert Robinson led to a remarkably efficient synthesis of the alkaloid tropinone (**Figure 17.1b**).

Sometimes two different reactions can be combined into a tandem process. Tietze, for example, completed a synthesis of tetrahydrocannabinol using a tandem reaction consisting of an aldol condensation reaction and a Diels–Alder

---

[1]Tietze, L. Domino Reactions in Organic Synthesis. *Chem. Rev.* **1996**, *96*, 115–136.

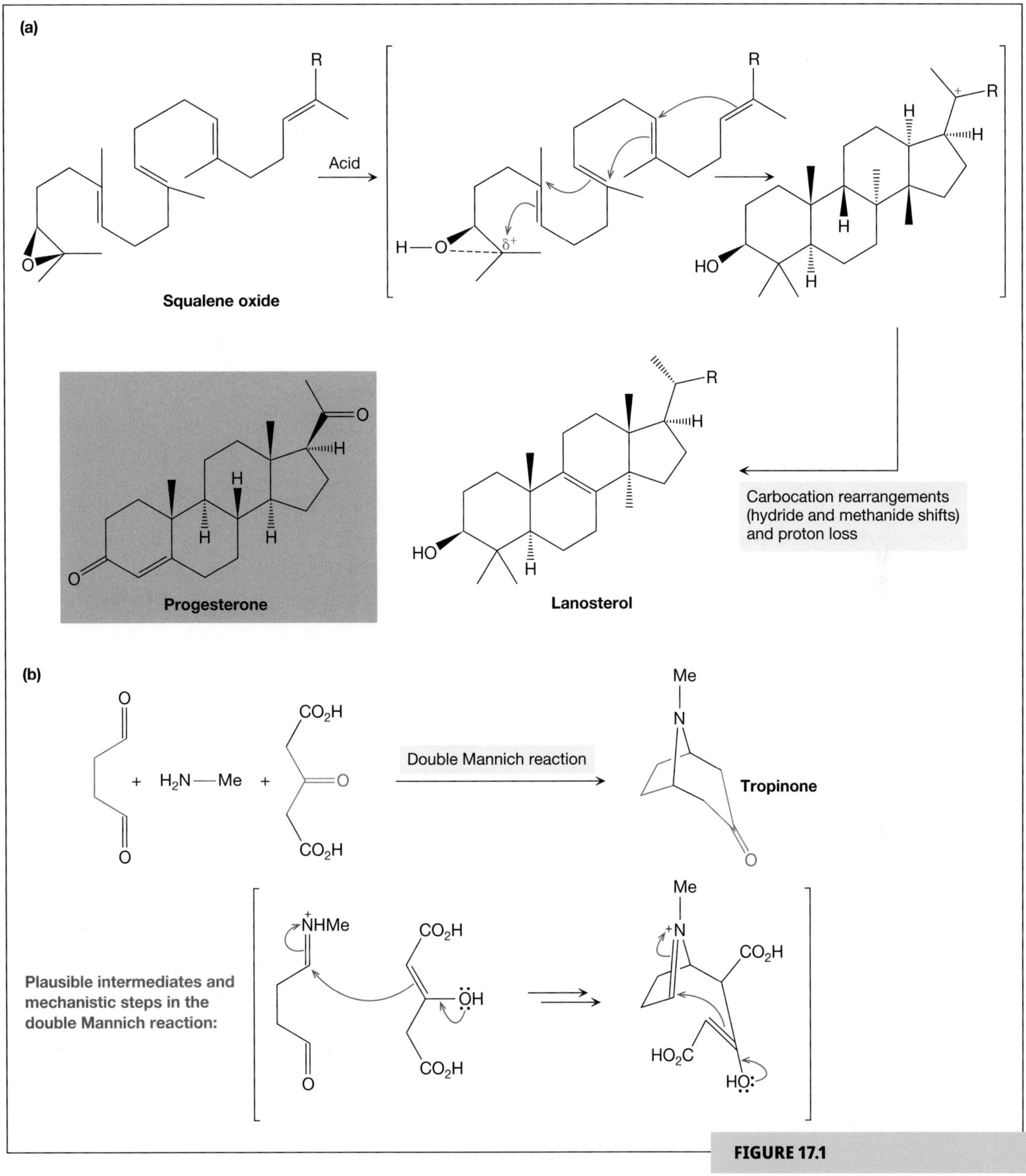

(a)

Squalene oxide

Progesterone

Lanosterol

Carbocation rearrangements (hydride and methanide shifts) and proton loss

(b)

Double Mannich reaction

Tropinone

Plausible intermediates and mechanistic steps in the double Mannich reaction:

FIGURE 17.1

Notable tandem reactions.
(a) Multiple carbocation cyclizations and carbocation rearrangements are involved in generating a polycyclic skeleton during steroid biosynthesis.
(b) Robinson's synthesis of tropinone.

cycloaddition (**Figure 17.2**).[2] Aldol addition product **A** undergoes dehydration to form the aldol condensation product, which can then undergo a Diels–Alder cyclo-addition to form **C**. An atom other than carbon (in this case, oxygen) is involved in the six-membered ring transition state for the cycloaddition; such cases are known as hetero-Diels–Alder reactions. Tetrahydrocannabinol is obtained after two addi-tional steps. The tandem reaction thus makes the overall sequence very efficient.

---

[2]This particular aldol condensation uses a 1,3-dicarbonyl compound as the nucleophile; this variation is called a Knoevenagel condensation. Diels–Alder and other cycloadditions are discussed in more detail in Chapter 18.

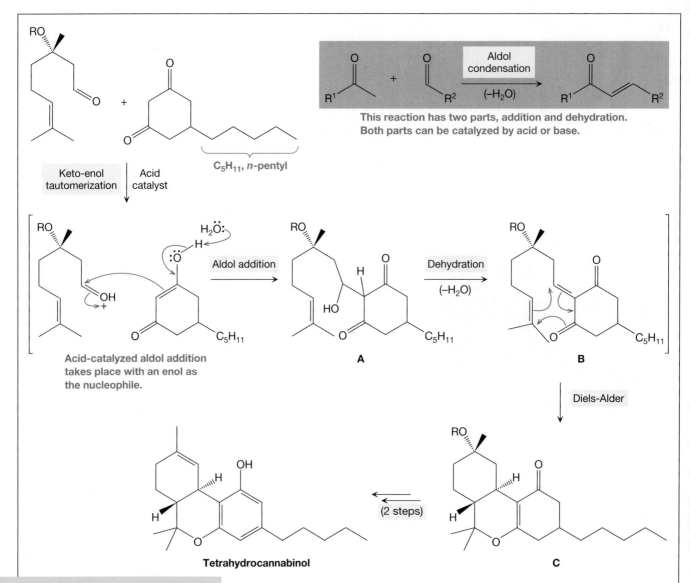

FIGURE 17.2

Tietze's three-step synthesis of tetrahydrocannabinol using a tandem reaction.

Despite the advantages of tandem reactions, they can be challenging to implement. Two or more reactive functionalities might need to be present, yet not react with each other until the time is right. All of the components have to be compatible with the steps involved, but sometimes the waste products of the first step may interfere with the next step. As a result, considerable experience and creativity may be needed to design a series of reactions that works as planned.

In this experiment, you will combine a Diels–Alder reaction with an acylation reaction (**Figure 17.3**). First you will prepare a diene that also has an alcohol functional group by reduction of the corresponding aldehyde. Then, in the tandem reaction, you will carry out a Diels–Alder reaction with maleic anhydride. After cycloaddition occurs, the alcohol and anhydride functional groups react with no additional reagents needed, forming an ester by nucleophilic acyl substitution. This tandem reaction occurs under solventless conditions without heating, minimizing energy consumption and waste solvent disposal.

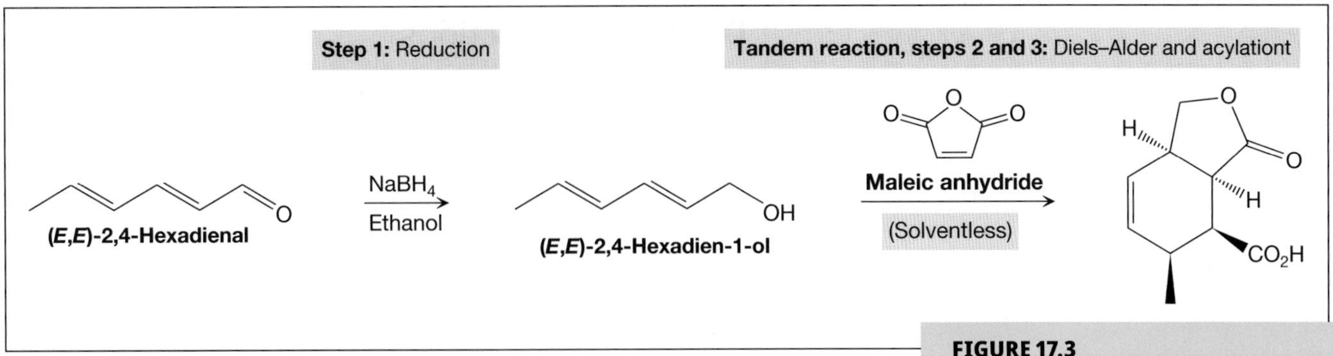

Step 1: Reduction

Tandem reaction, steps 2 and 3: Diels–Alder and acylationt

Maleic anhydride

(Solventless)

(*E,E*)-2,4-Hexadienal → NaBH₄ / Ethanol → (*E,E*)-2,4-Hexadien-1-ol

FIGURE 17.3

Hydride reduction of an aldehyde [(*E,E*)-2,4-hexadienal] to produce a primary alcohol [(*E,E*)-2,4-hexadien-1-ol] followed by a tandem Diels–Alder and acylation reaction.

17.1B Hydride Reduction of Aldehydes and Ketones

An irreversible nucleophilic addition to an aldehyde or ketone may be carried out with Grignard and organolithium reagents (carbon nucleophiles) and metal hydride reagents (hydrogen nucleophiles), among others. Several variants of the metal hydrides are used for nucleophilic addition reactions to convert aldehydes and ketones to alcohols, and most common among these are the hydrides of aluminum or boron (**Figure 17.4a**). These include lithium aluminum hydride (LiAlH₄), diisobutylaluminum hydride (DIBALH), borane (BH₃), and sodium borohydride (NaBH₄).

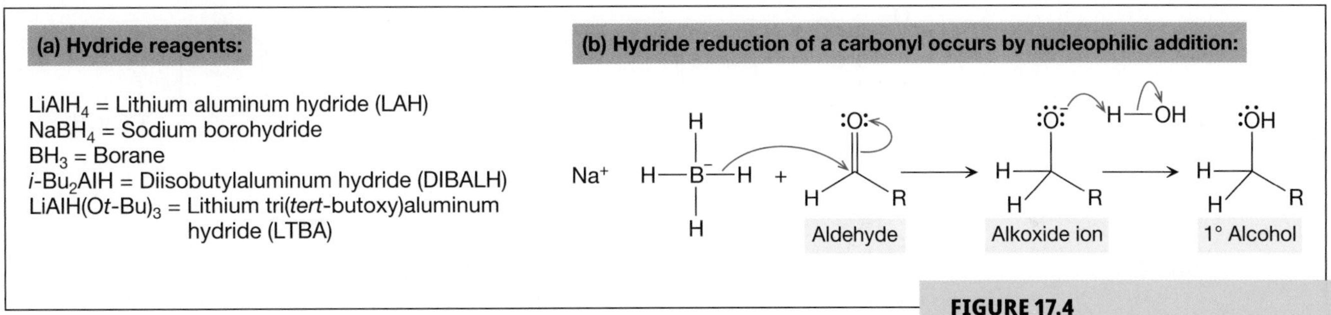

(a) Hydride reagents:

LiAlH₄ = Lithium aluminum hydride (LAH)
NaBH₄ = Sodium borohydride
BH₃ = Borane
i-Bu₂AlH = Diisobutylaluminum hydride (DIBALH)
LiAlH(O*t*-Bu)₃ = Lithium tri(*tert*-butoxy)aluminum hydride (LTBA)

(b) Hydride reduction of a carbonyl occurs by nucleophilic addition:

Aldehyde → Alkoxide ion → 1° Alcohol

FIGURE 17.4

(a) Various hydride reagents used as reductants in organic chemistry. (b) The mechanism of the borohydride reduction of an aldehyde to a primary alcohol begins with nucleophilic addition to the carbonyl group. The resulting alkoxide ion then accepts a proton from the water or ethanol solvent.

Like Grignard and organolithium reagents, metal hydride reagents are strongly nucleophilic and react with water or alcohols to produce hydrogen gas, so they can be quite hazardous. Fortunately, sodium borohydride is much safer than the others because it reacts very slowly with water, especially under neutral or basic conditions. In aqueous or ethanol solution, NaBH₄ behaves as a source of a hydride nucleophile, adding H⁻ to aldehydes and ketones (Figure 17.4). This generates an alkoxide ion that can accept a proton from water or ethanol to provide the product. The net result is an increase in the hydrogen content of the reactant by addition of two hydrogens (one as H⁻ and one as H⁺), while also decreasing the number of bonds to an electronegative atom (from two to one). Both of these are ways that organic chemists identify that a reduction has occurred. We will use this reaction to carry out the reduction of the aldehyde (*E,E*)-2,4-hexadienal to the corresponding primary alcohol (*E,E*)-2,4-hexadien-1-ol. This alcohol is then the diene starting material for our tandem reaction involving Diels–Alder cycloaddition and nucleophilic acyl substitution.

17.1C Diels–Alder Reactions

The Diels–Alder reaction is a powerful way to synthesize six-membered rings, which are commonly found in biologically active natural products and drug candidates in the discovery pipeline. This reaction (**Figure 17.5**) entails the concerted formation of two carbon–carbon bonds and up to four stereogenic centers in a single step through the overlap of a diene and an electron-deficient alkene (dienophile). In cases when new stereogenic centers form, mixtures of enantiomers and diastereomers may result.

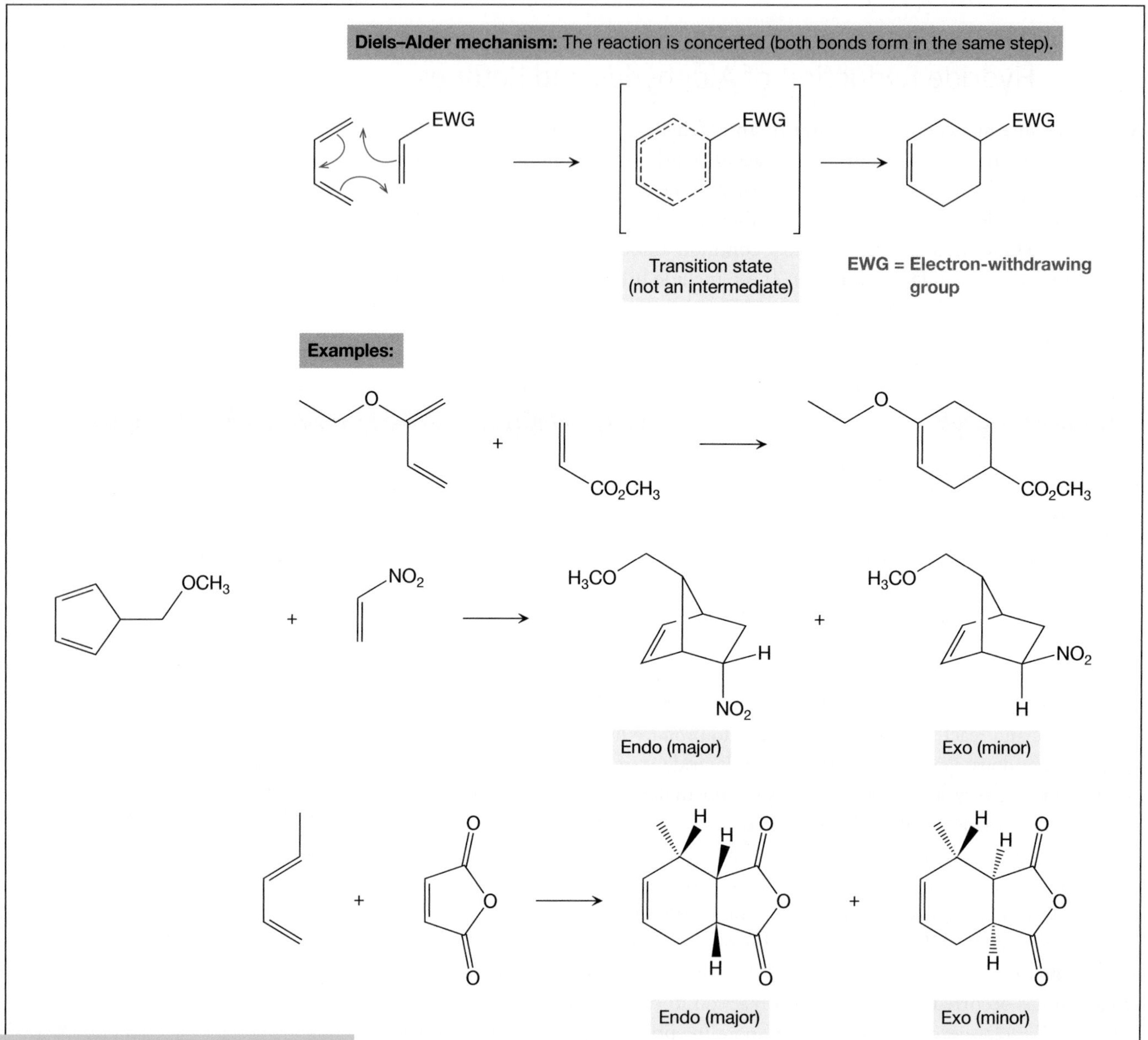

Diels–Alder mechanism: The reaction is concerted (both bonds form in the same step).

Transition state
(not an intermediate)

EWG = Electron-withdrawing group

Examples:

Endo (major)   Exo (minor)

Endo (major)   Exo (minor)

FIGURE 17.5

General Diels–Alder reaction and examples.

FIGURE 17.6

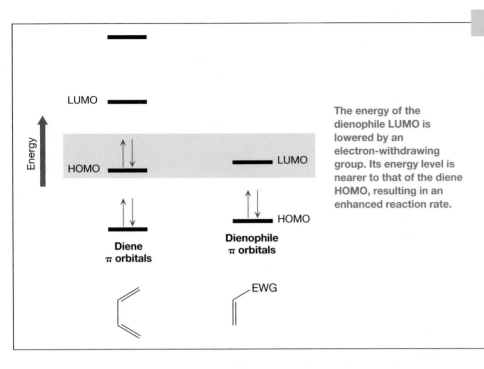

The energy of the dienophile LUMO is lowered by an electron-withdrawing group. Its energy level is nearer to that of the diene HOMO, resulting in an enhanced reaction rate.

The bond formation occurs through the productive overlap of the highest occupied molecular orbital (HOMO) of the diene π system and the lowest unoccupied molecular orbital (LUMO) of the dienophile π system (**Figure 17.6**).

Because the two bonds form simultaneously, the stereochemical information in the dienophile (*E* or *Z*) is preserved as a trans or cis orientation of substituents on the new six-membered ring (**Figure 17.7**). The following are some other notable observations about the reactivity and selectivity observed with various types of diene and dienophile structures:

1. The diene should easily access the *s*-cis conformation in order to react efficiently (Figure 17.7b).

2. The dienophile should be electron deficient, with the alkene generally bearing electron-withdrawing groups (EWGs), such as carbonyl, sulfonyl, or nitro. This lowers the LUMO energy, narrowing the energy gap between the LUMO and HOMO that overlap during bond formation, thereby increasing the reaction rate (Figure 17.6).

3. Reaction through an endo transition state is usually more rapid; in this pathway, the electron-withdrawing substituent on a dienophile can lower the transition state barrier through secondary orbital overlap (a stabilizing orbital interaction that does not lead to a bond; Figure 17.7c).

In our experiment, we will be able to see how all these factors play a role in the reactivity and stereoselectivity of a Diels–Alder reaction.

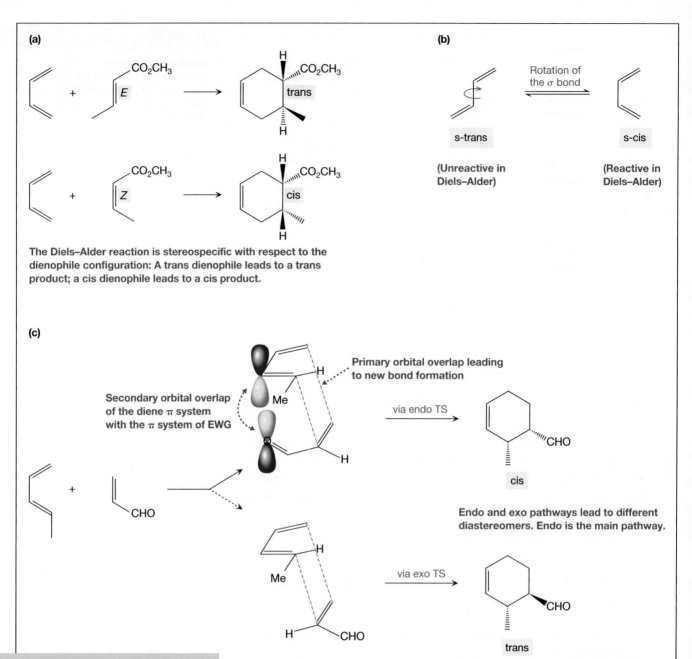

(a)

The Diels–Alder reaction is stereospecific with respect to the dienophile configuration: A trans dienophile leads to a trans product; a cis dienophile leads to a cis product.

(b)

Rotation of the σ bond

s-trans

(Unreactive in Diels–Alder)

s-cis

(Reactive in Diels–Alder)

(c)

Primary orbital overlap leading to new bond formation

Secondary orbital overlap of the diene π system with the π system of EWG

Me

via endo TS

cis

Endo and exo pathways lead to different diastereomers. Endo is the main pathway.

Me

via exo TS

trans

FIGURE 17.7

Some key features of the Diels–Alder reaction. (a) It is stereospecific with respect to the dienophile. (b) It requires an accessible s-cis conformation. (c) It is stereoselective, because the endo transition (TS) state is stabilized by a secondary orbital overlap.

17.1D Nucleophilic Acyl Substitutions

Recall from Figure 17.4 that nucleophilic addition to a carbonyl occurs in the hydride reduction of an aldehyde. The aldehyde has no leaving group, so the resulting anion is protonated by solvent and the reaction stops. However, when a leaving group is attached at the carbonyl carbon, the nucleophilic addition is often followed by the elimination of the leaving group. The net result of this two-step process is called *nucleophilic acyl substitution*. Often, the mechanism also involves one or more proton transfers. Typical mechanisms under basic and acidic conditions are shown in **Figure 17.8**.

In our reaction, this nucleophilic acyl substitution employs an alcohol as the nucleophile and an anhydride as the electrophile or acylating agent. Because the reaction occurs under neutral conditions, the mechanism is not clearly anionic or cationic, but may have some features of both. The alcohol adds to the carbonyl initially, as shown in **Figure 17.9**. Try to draw the rest of the mechanism on your own.

FIGURE 17.8

Nucleophilic acyl substitution: Stepwise addition–elimination mechanism

1 **Anionic mechanism** (usually observed with basic conditions, good Nu⁻, and/or excellent LGs)

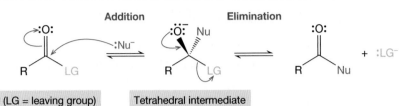

(LG = leaving group)   Tetrahedral intermediate

2 **Cationic mechanism** (usually observed with acidic conditions, weak NuH, and/or poor LGs)

FIGURE 17.8

Generic nucleophilic acyl substitutions by both an anionic mechanism (basic conditions) and a cationic mechanism (acidic conditions). In both cases the main steps are addition to the C=O bond to form a tetrahedral intermediate, followed by elimination of a leaving group. (Nu = nucleophile, LG = leaving group.)

Tetrahedral intermediates

Having some background on the reactions involved, we are now prepared to try the hydride reduction and tandem Diels–Alder/acylation reaction in the lab. For those who may be interested in seeing more examples of tandem reactions, several examples can be found in the Tietze review article.[1,3]

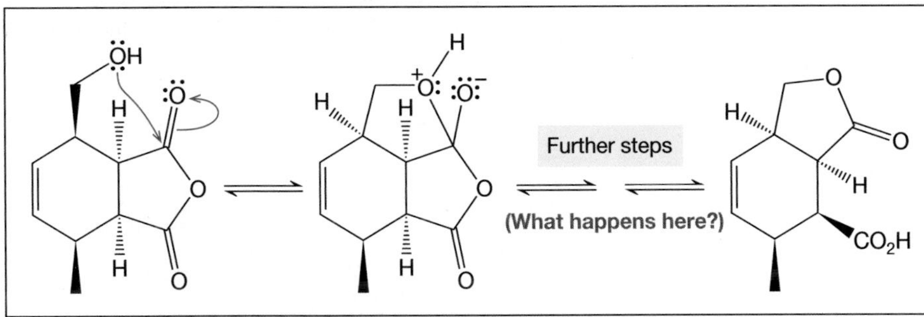

FIGURE 17.9

An initial step of the intramolecular nucleophilic acyl substitution involved in this experiment. Mechanisms of the remaining steps are left for you to propose.

Further steps

(What happens here?)

---

[3]Some examples in synthesis of biologically important compounds include a double Diels–Alder approach to taxanes (J. Winkler), morphine via a combination of Diels–Alder and retro-Diels–Alder reactions (E. Ciganek), double electrocyclic ring closure to access endiandric acids (K. C. Nicolaou), sequential radical processes en route to hirsutene (D. P. Curran), a tandem aza-Cope + Mannich sequence to make strychnine (L. E. Overman), a synthesis of vinblastine via condensation + cycloaddition (M. E. Kuehne), retro-Diels–Alder + ketene trapping + Diels–Alder for the synthesis of hirsutellone (E. Sorenson), and an especially spectacular series of tandem reactions in the synthesis of methyl homosecodaphniphyllate (C. Heathcock).

17.2A Part 1: Reduction of (*E,E*)-2,4-hexadienal[4]

1. Place 0.20 g of (*E,E*)-2,4-hexadienal into a 25-mL Erlenmeyer flask with a magnetic stir bar. Add 4 mL ethanol and stir gently until it is homogeneous.

2. Measure 2.4 mL of a solution of sodium borohydride (1.0 M in ethanol) in a vial. *Note to instructors/staff*: Solutions of sodium borohydride (1.0 M in ethanol) should be freshly prepared before the lab begins: Place 2.26 g NaBH$_4$ (60 mmol) in an Erlenmeyer flask or reagent bottle with 60 mL ethanol and stir or swirl until homogeneous. This is sufficient for about 20 students. For convenience, premeasured units of NaBH$_4$ and ethanol may be stored until needed, then mixed at the beginning of a lab section.

3. Over a period of 5 minutes, add the NaBH$_4$ solution in several small portions by Pasteur pipet to the (*E,E*)-2,4-hexadienal solution, with stirring or swirling. Allow the reaction to continue while proceeding to steps 4 and 5.

4. Prepare a thin-layer chromatography (TLC) plate and developing chamber containing 5:1 petroleum ether/ethyl acetate (**Figure 17.10**). In a vial, dissolve one drop of (*E,E*)-2,4-hexadienal in 0.5 mL ethanol to use as a TLC standard.

5. Use TLC to evaluate whether the aldehyde has been consumed. On the TLC plate, place a spot of (*E,E*)-2,4-hexadienal standard and a spot of the reaction mixture, side by side. Develop the plate using 5:1 petroleum ether/ethyl acetate.

▶ **Thin-Layer Chromatography**

FIGURE 17.10

Setup for thin-layer chromatography (TLC). This TLC plate is analyzing three separate compounds with a mixture of all three on the right. The width of the plate can vary according to the number of samples to be analyzed.

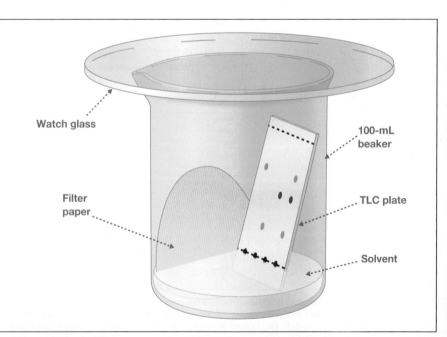

Watch glass

100-mL beaker

Filter paper

TLC plate

Solvent

---

[4]Pini, E.; Bertacche, V.; Molinari, F.; Romano; D.; Gandolfi, R. Direct Conversion of Polyconjugated Compounds into Their Corresponding Carboxylic Acids by *Acetobacter aceti*. *Tetrahedron* **2008**, *64*, 8638–8641.

6. If the reaction is complete, as judged by the absence of a spot matching the starting aldehyde, proceed to step 7. If it is incomplete, allow an additional 15–30 minutes of reaction time.

WORK UP THE REACTION

7. Place 10 mL water in a small separatory funnel. Using a Pasteur pipet, transfer the reaction mixture into the water and extract twice with 2 mL each time of *tert*-butyl methyl ether (MTBE). Return the combined organic fractions to the separatory funnel.

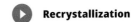

Aqueous–Organic Extractions and Drying Organic Solutions

8. Add 5 mL saturated aqueous sodium chloride solution (brine). Gently shake to mix the two phases and allow them to separate. Drain off the aqueous phase, combining it with the prior aqueous phase.

9. Drain the organic phase into a clean 25-mL Erlenmeyer flask and dry over about 0.5 g Na_2SO_4 for at least 10 minutes, swirling occasionally.

10. Label a clean vial with your full name, fume hood number, and the chemical name of the product [(*E,E*)-2,4-hexadien-1-ol]. Tare the vial (with cap) and record the tare weight in your notebook. Using a Pasteur pipet, transfer the solution of (*E,E*)-2,4-hexadien-1-ol to the tared vial.

11. Evaporate most of the solvent (MTBE) under a gentle flow of compressed air while warming the vial in your hand. *(CAUTION: Adjust the airflow before directing it into the vial; otherwise, a high flow rate of compressed air can cause the product to splash out.)* When most of the solvent is gone, the remaining oil will appear more viscous or syrupy. Store the vial in the hood with the cap loosened to allow remaining traces of MTBE to evaporate.

17.2B Part 2: Tandem Diels–Alder/Acylation[5,6]

1. Weigh the vial containing (*E,E*)-2,4-hexadien-1-ol, including the cap. Subtracting the tare amount, determine the mass of (*E,E*)-2,4-hexadien-1-ol.

2. Next, add powdered maleic anhydride to the vial. The amount you add must be calculated, so that the mole ratio of maleic anhydride with (*E,E*)-2,4-hexadien-1-ol is 1:1, plus or minus 1 mg.

3. Using a clean, dry metal spatula, vigorously stir the mixture for 10–15 minutes. During this time, the materials should first liquefy, then solidify again. At this time, the reaction should be complete.

Preparing NMR Samples

4. Submit a 10-mg sample of this solid (the crude product) for ^{1}H NMR spectroscopy. Use acetone-d_6 as the NMR solvent.

5. Recrystallize the product from toluene. *(CAUTION: Flasks containing boiling toluene should be handled with a clamp.)* Place your sample in a 25-mL Erlenmeyer flask on a hot plate set at medium-high heat, and slowly add toluene dropwise until the solid just dissolves. Stop adding toluene at this point and allow the solution to cool. If crystallization does not spontaneously occur,

Recrystallization

---

[5]McDaniel, K. F.; Weekly, R. M. The Diels-Alder Reaction of 2,4-Hexadien-1-ol with Maleic Anhydride: A Novel Preparation for the Undergraduate Organic Chemistry Laboratory Course. *J. Chem. Educ.* **1997**, *74*, 1465–1467.

[6]McKenzie, L. C.; Huffman, L. M.; Rogers, C. E.; Hutchison, J. E.; Goodwin, T. E.; Spessard, G. O. Greener Solutions for the Organic Chemistry Teaching Lab: Exploring the Advantages of Alternative Reaction Media. *J. Chem. Educ.* **2009**, *86*, 488.

scratch the inner walls with a glass rod. Complete the crystallization by cooling in an ice bath. *Note to instructors/staff:* If moisture is present during the reaction, or if the maleic anhydride has been exposed to moisture during long-term storage, maleic acid and related Diels–Alder by-products may complicate the recrystallization. Gravity filtration while hot, to remove insoluble materials (by-products and impurities from the reaction), may resolve this issue and provide for a higher purity product.

 Gravity Filtration

6. Isolate the crystalline product by vacuum filtration, rinsing the crystals with 1 mL of *cold* toluene.

7. Obtain the infrared spectrum of the product using direct sampling of the solid with attenuated total reflectance (ATR). Alternatively, use a $CHCl_3$ solution on a salt plate. The product is sparingly soluble; place a few crystals in a test tube and add a few drops of $CHCl_3$, then warm it for a few seconds on a sand bath. Transfer a drop of the supernatant liquid to the salt plate.

Melting Point Measurement

8. Determine the melting point of the recrystallized product (literature value is 159–161°C).[5]

17.3 THE LABORATORY REPORT

In writing your laboratory report, include the percent yield of each step and describe how your data provide evidence of identity and purity of the product. Also, address the following:

1. Estimate the amount of waste materials (including solvents, side products, and lost product) you produced during the purification of the product by recrystallization. Using this estimate, calculate the E-factor for the tandem reaction:

$$E\text{-factor} = \frac{\text{mass of waste (g)}}{\text{mass of product (g)}}$$

2. If purification was required between the Diels–Alder and acylation reactions, making it stepwise instead of a tandem reaction, how would this change the E-factor you just determined? Assume that the purification between steps is similar to the end purification.

3. The literature contains many examples of tandem reactions that rapidly increase molecular complexity.[3] Using keywords from reference 3, search for a primary literature report of one of these tandem reactions. Draw the reactants, reagents, and products of the key tandem reaction step.

LEARNING OBJECTIVES

- Recognize how solvent-free templated reactions in the solid state can minimize solvent use.

- Use a Wittig reaction to prepare the alkene for photocycloaddition.

- Execute a photocycloaddition reaction in the solid state.

- Determine percent conversion, using integration of the ^{1}H NMR spectrum.

Wittig Olefination and Solid State Photocycloaddition

UV AND DNA

Xeroderma pigmentosum is a rare genetic condition characterized by inability to repair DNA damaged by UV light. Because of their extreme susceptibility to skin cancer, people with this condition may need special measures to prevent exposure to UV light.

Anne Chadwick Williams/ZUMA Press/Newscom.

cycloaddition >>
A reaction in which two components add together to form a ring. If the two reactive components are tethered within the same compound, it is an intramolecular cycloaddition.

Diels–Alder reaction >>
A cycloaddition reaction of a 1,3-diene and an alkene, generating a cyclohexene product. This is also called a [4+2] cycloaddition because four π-bonded carbons in the diene and two π-bonded carbons in the alkene are added together to form a six-membered cyclohexene ring.

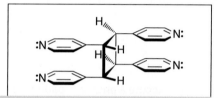

FIGURE 18.1

Structure of tetrakis(4-pyridyl)-cyclobutane (4,4′-tpcb) produced by [2+2] photocycloaddition of *trans*-1,2-bis(4-pyridyl)ethene in the solid state.

Cycloaddition reactions are among the most powerful reactions in organic synthesis because they prepare at least two bonds and up to four stereogenic centers in a single step. One of the most useful of these is the **Diels–Alder reaction**, which produces cyclohexene derivatives through a [4+2] cycloaddition between dienes and alkenes (see Chapter 17). There are also [2+2] cycloaddition reactions in which two alkenes react to make a four-membered cyclobutane ring, such as the one shown in **Figure 18.1**.

In this experiment you will carry out the [2+2] cycloaddition of *trans*-1,2-bis(4-pyridyl)ethene. The reaction is done in the solid state using resorcinol as a template to hold the alkenes in close proximity so that they can react efficiently and selectively. Ultraviolet light is used to induce the cycloaddition.[1] The starting material for this [2+2] photocycloaddition, *trans*-1,2-bis(4-pyridyl)ethylene, may be purchased commercially, or may be prepared using the Wittig olefination reaction as described in this chapter.

Pre-Lab Reading Assignment

Review the following before attempting this experiment:

- Experiment: Chapter 18 (this experiment)
- Technique: Filtration (Chapter 5, section 5.1)
- Technique: Extraction (Chapter 5, section 5.2)
- Technique: Thin-layer chromatography (TLC) (Chapter 5, section 5.5)
- Technique: Melting point measurement (Chapter 5, section 5.6)
- Technique: Ultraviolet-visible (UV–vis) spectroscopy (Chapter 6, section 6.1)
- Technique: Nuclear magnetic resonance (NMR) spectroscopy (Chapter 7, section 7.1)

18.1 BACKGROUND

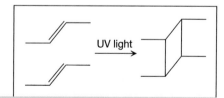

FIGURE 18.2

Photocycloaddition reaction involving two trans C=C bonds to give a cyclobutane ring.

dimerization >>
A reaction in which two molecules of the same compound add together to form the product.

18.1A   Selectivity in Cycloaddition Reactions

Unlike the Diels–Alder reaction, which occurs under thermal conditions, [2+2] cycloaddition reactions cannot generally occur without light, because one of the alkenes must be converted from its ground state electron configuration to an excited state. Absorption of a photon in the ultraviolet wavelength range (around 200–400 nm) provides appropriate energy to promote an alkene π electron from the highest occupied π molecular orbital (HOMO) to a higher-energy π molecular orbital. This is the same type of absorption process involved in UV–visible spectroscopy, and it produces the excited state that can react with another alkene to make a cyclobutane (**Figure 18.2**). This light-induced reaction is an example of a photochemical reaction. More specifically, it is a photocycloaddition because two separate reactants come together to make a ring.

This same [2+2] photocycloaddition can occur in deoxyribonucleic acid (DNA), the chemical storehouse of genetic information. Upon exposure to sunlight, neighboring thymines in the DNA can be **dimerized** through a [2+2] photocycloaddition

---

[1]This experiment was originally developed by Professor Len MacGillivray (University of Iowa).

(Figure 18.3). A combination of hydrogen bonds and other intermolecular forces holds the two thymines in a well-defined orientation and in close proximity so that they can undergo this reaction in response to UV light. The photocycloaddition reaction is destructive to DNA, and the resulting errors in genetic code cause skin cells to die, which is experienced as a sunburn. In some cases the disrupted genetic code can develop into skin cancer. Specialized DNA repair enzymes can excise the disrupted portion of the DNA code, helping to prevent cancer. Tomas Lindahl, Paul Modrich, and Aziz Sancar discovered the mechanisms of this DNA repair process, and were awarded the Nobel Prize in Chemistry in 2015.[2] Sunscreen works by absorbing UV light before it damages the DNA through [2+2] photocycloaddition.

FIGURE 18.3

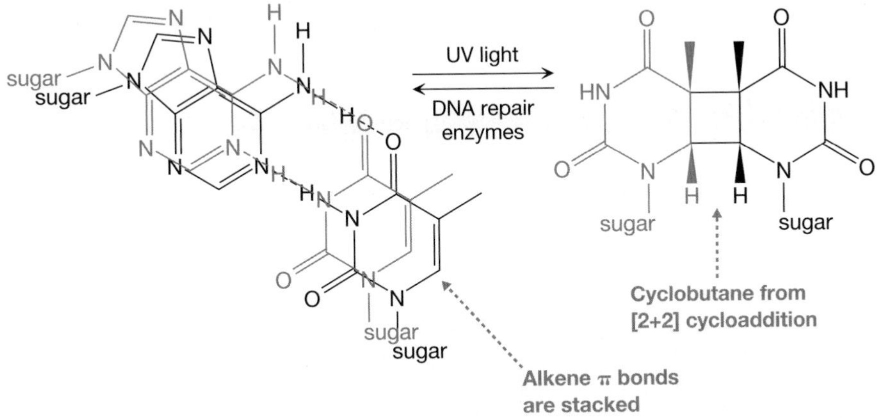

A [2+2] photocycloaddition of two neighboring thymines within a segment of DNA (left) can occur on exposure of DNA to ultraviolet radiation (e.g., from sunlight).[3]

Selectivity in cycloadditions can be quite complicated, producing mixtures of diastereomers and enantiomers, unless some special features are added to the reaction to control how the two reactants approach each other. In the Diels–Alder reaction, the number of possible stereoisomeric products is limited by the mechanism, which is stereospecific and often quite selective for endo products (**Figure 18.4a**). However, [2+2] photocycloadditions can often produce complex mixtures of products when they are done in the solution phase; the example shown in **Figure 18.4b** produced not only diastereomers, but also both head-to-tail and head-to-head orientations of the substituents (regioisomers), each of which is racemic, leading to a total of eight products.[4] The [2+2] cycloaddition mixtures are further complicated by the mechanism: stepwise construction of C—C bonds causes formation of additional isomers as a consequence of rotations prior to ring closure.

In contrast to the complex mixtures obtained in *solution*-phase [2+2] photocycloadditions, carrying out the reaction in the *solid* state can be *very* selective. You will carry out the [2+2] photocycloaddition in this experiment in an organic crystal. The crystal is constructed by **co-crystallization** of the reactant with an equal proportion of a chemical template. Within the crystal, the template positions the two reactant molecules in close proximity and well-defined orientations, using noncovalent forces such as hydrogen bonds. This concept is very similar to the biological process depicted in Figure 18.3, which positions two thymine units so that a single isomer of the thymine dimer is formed upon [2+2] photocycloaddition.

<< co-crystallization
Formation of crystalline solid from two different compounds, such that the regularly repeating structural unit making up the crystal contains two or more different molecules held together by attractive forces such as hydrogen bonds.

[2]NobelPrize.org. The Nobel Prize in Chemistry 2015. https://www.nobelprize.org/prizes/chemistry/2015/summary/ (accessed April 2022).

[3]Adapted from Molecule of the Day: Thymine Dimers and Skin Cancer. http://iverson.cm.utexas.edu/courses/old/310N/spring2008/MOTD%20Fl05/ThymineDimers.html (accessed April 2022).

[4]Sydnes, L. K.; Hansen, K. I.; Oldroyd, D. L.; Weedon, A. C.; Jorgensen, E. Photochemical [2 + 2] Cycloadditions. IV. Cycloaddition of 2-Cyclopentenone to Some (ω-1)-Alken-1-ols; Evidence for Regioselectivity due to Hydrogen Bonding. *Acta Chem. Scand.* **1993**, *47*, 916–924.

FIGURE 18.4

Comparison of selectivities in (a) [4+2] cycloaddition and (b) [2+2] photocycloaddition reactions. The excellent selectivities in Diels–Alder reactions are *not* observed in most solution-phase [2+2] photocycloadditions.

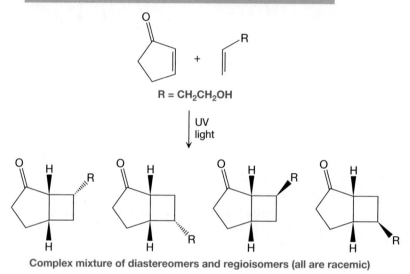

(a) Examples of selectivity in [4+2] cycloaddition (Diels–Alder reaction):

Heat

(Racemic)

• **Stereospecific:**
Groups trans in the dienophile are trans in the product

EWG = electron-withdrawing group

Heat

(Racemic)

• **Stereoselective:**
The endo approach is preferred with a cyclic diene, with EWG ending up on the side of the longer bridge.

(b) A typical outcome in solution-phase [2+2] photocycloaddition:

R = CH₂CH₂OH

R = CH_2CH_2OH

UV light

Complex mixture of diastereomers and regioisomers (all are racemic)

18.1B Co-Crystallization and Solid State [2+2] Photocycloaddition

From the perspective of green chemistry and environmental impact, the solid state is a desirable medium in which to conduct an organic reaction, because solvents, which are a major contributor to the waste output from organic reactions, are not used. Let's look in more detail at the tools for a solid state [2+2] photocycloaddition reaction.

Upon conversion from the liquid or gas phase to the solid phase, the space between molecules decreases dramatically, and the free movement of molecules known as diffusion no longer occurs. Molecules in the solid state are in a static environment, being "frozen" in position.

In the late 1960s, Gerhard Schmidt of the Weizmann Institute in Israel proposed two rules that describe how C=C bonds should be arranged in the solid state to undergo a [2+2] cycloaddition reaction:[5]

1. The C=C bonds must adopt a parallel orientation.

2. The C=C bonds must be within 4.2 Å of each other.

---

[5]Schmidt, G. M. J. Photodimerization in the Solid State. *Pure Appl. Chem.* **1971**, *27*, 647–678.

These geometric criteria place the C=C bonds close enough for the π orbitals of the two C=C bonds to overlap. Most organic molecules that possess C=C bonds do not crystallize with the C=C bonds in the special arrangement, outlined by Schmidt, that is needed in order for the molecules to react. One way to ensure that the molecules are organized in the solid state to enable [2+2] photodimerization is to use a chemical template. The template assembles two molecules into position for a chemical reaction, using noncovalent bonds (e.g., hydrogen bonds), yet is not part of the final product. A familiar example of a chemical template is DNA (Figure 18.3). The chemical template that we study in this experiment is 1,3-benzenediol (resorcinol), which can form hydrogen bonds to arrange two molecules of *trans*-1,2-bis(4-pyridyl)ethene (4,4′-bpe) for enhanced reactivity according to Schmidt's rules.

For a review of hydrogen bonding, consider the differences between gaseous H_2O (steam), liquid H_2O (water), and solid H_2O (ice). Hydrogen bonds hold together the H_2O molecules in ice and water (**Figure 18.5**). An O—H bond is strongly polarized due to the high electronegativity (EN) of oxygen (EN = 3.5) compared to hydrogen (EN = 2.1). A hydrogen atom of one H_2O molecule therefore bears a partial positive charge and can interact with a lone pair of an oxygen atom of a second H_2O molecule. This electrostatic attraction is the hydrogen bond. In effect, the hydrogen bond acts as "glue," allowing the H_2O molecules to "stick" together. In liquid water, the strength of a hydrogen bond is about 20 kJ/mol (about 5 kcal/mol), and other hydrogen bond strengths can range from about 5–40 kJ/mol.

The strength of hydrogen bonding depends on the polarization of the O—H bond; C—H bonds have very little polarization so they don't participate in hydrogen bonding. The intermolecular attractions are therefore much stronger in H_2O than in CH_4, and this is reflected in their boiling points: 100°C for H_2O versus −161°C for CH_4. In structural representations, the hydrogen bond can be denoted by a dashed line to differentiate it from covalent bonds: O—H---O (Figure 18.5). Examine the DNA in Figure 18.3 and locate the hydrogen bonds.

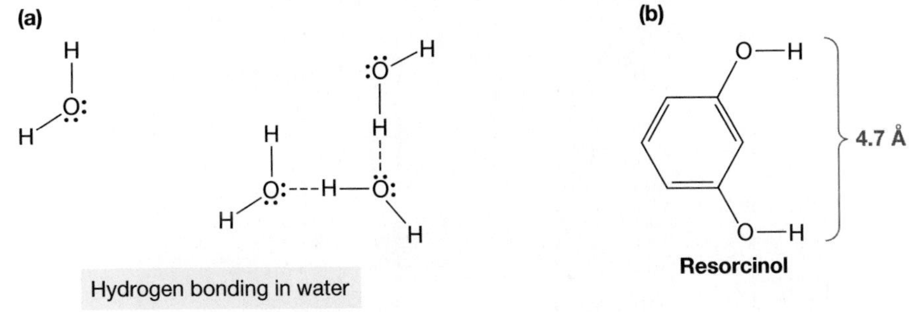

(a) Hydrogen bonding in water

(b) Resorcinol

4.7 Å

FIGURE 18.5

(a) Hydrogen bonds among three water molecules are shown as dashed lines. (b) In resorcinol, the distance between hydrogen bond donor groups is 4.7 Å. The lone pairs of oxygen or nitrogen atoms can form hydrogen bonds with the O—H groups of resorcinol.

Similar to H_2O, resorcinol (Figure 18.5) possesses two polarized O—H bonds, located at the 1- and 3-positions of the benzene ring. The separation distance between the two oxygen atoms of resorcinol is 4.7 Å. If two resorcinol molecules use their two O—H groups to form hydrogen bonds to the nitrogen atoms of two molecules of *trans*-1,2-bis(4-pyridyl)ethene (4,4′-bpe), as shown in **Figure 18.6**, then the two molecules of 4,4′-bpe will be held in close proximity, where the [2+2] photocycloaddition can readily occur according to Schmidt's rules.

To use resorcinol to assemble 4,4′-bpe in the solid state for a [2+2] photodimerization, we employ *co-crystallization*. The structures of these two compounds will be paired together through attractive noncovalent interactions at the molecular scale—in this case, hydrogen bonds. The co-crystallization of resorcinol and 4,4′-bpe can be conducted without the use of solvent by grinding the two components together in a mortar and pestle. This yields an assembly of two resorcinols bound

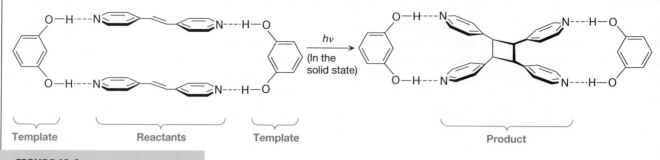

Template Reactants Template Product

FIGURE 18.6

Co-crystallization leads to the molecular assembly 2(resorcinol)·2(4,4′-bpe), which undergoes a [2+2]-photocycloaddition reaction in the solid state.

at opposite ends of two 4,4′-bpe molecules, wherein the C=C bonds of 4,4′-bpe are arranged in parallel and are separated by 3.7 Å—an ideal alignment for a [2+2] photocycloaddition. Irradiation of the co-crystal with ultraviolet light, using a mercury vapor lamp (**Figure 18.7**), produces the cyclobutane product tetrakis(4-pyridyl)cyclobutane (4,4′-tpcb; see Figure 18.1) in the solid state.[6]

FIGURE 18.7

A mercury vapor lamp used for the solid-state photocycloaddition reaction. The lamp is contained within a water-jacketed tube to keep it cool and mounted inside a cabinet to avoid exposure of workers to harmful UV light.

[6]MacGillivray, L. R.; Reid, J. L.; Ripmeester, J. A. Supramolecular Control of Reactivity in the Solid State Using Linear Molecular Templates. *J. Am. Chem. Soc.* **2000**, *122*, 7817–7818.

18.1C The Wittig Reaction

The starting material for the [2+2] photocycloaddition is *trans*-1,2-bis(4-pyridyl)-ethene (*trans*-4,4'-bpe). This compound can be prepared using the Wittig reaction (**Figure 18.8**), an invaluable method for converting an aldehyde or ketone into an **olefin**. This method involves the nucleophilic addition of a phosphorus-stabilized anion, known as a phosphorus **ylide** (pronounced ill'-id), to a carbonyl compound with subsequent elimination providing an alkene. Triphenylphosphine oxide is produced as a side product in the Wittig reaction.

<< olefin
A synonym for alkene.

<< ylide
A neutral compound that bears opposing charges on neighboring atoms within the structure.

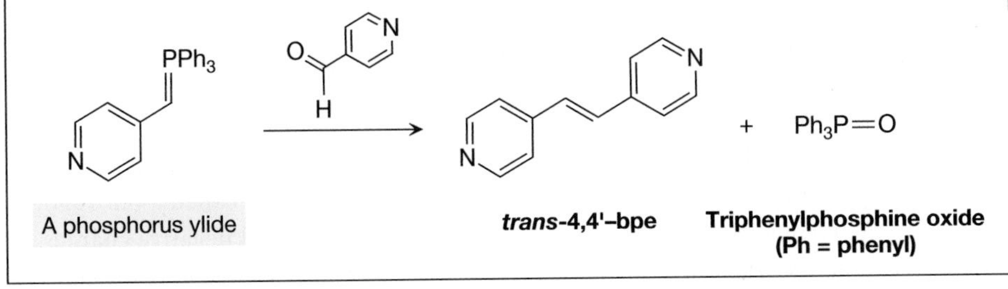

FIGURE 18.8

A phosphorus ylide   *trans*-4,4'-bpe   Triphenylphosphine oxide (Ph = phenyl)

A Wittig reaction of a phosphorus ylide and an aldehyde to furnish *trans*-1,2-bis(4-pyridyl)ethene. Phosphine oxides are waste by-products from these reactions.

The Wittig reaction is such a useful way to make alkenes that its principal developer, Georg Wittig, was awarded the 1979 Nobel Prize in Chemistry (with H. C. Brown). The reaction takes advantage of a positively charged phosphorus atom to allow the adjacent carbon to bear a negative charge, facilitating its reactions with aldehydes and ketones to furnish alkenes. A by-product that forms in this reaction, $Ph_3P=O$, accounts for a large quantity of waste material relative to the amount of alkene obtained. From the perspective of green chemistry, this is a big disadvantage of the reaction. Why, then, would chemists continue to use the Wittig reaction?

One reason for the importance of the Wittig reaction is its regioselectivity compared to other reactions that accomplish the same transformation (**Figure 18.9**). For example, a carbon needs to be added in order to prepare methylenecyclohexane from cyclohexanone, and one way to do this is by Grignard addition of CH_3MgBr to the ketone, followed by dehydration. This dehydration is somewhat regioselective, providing a 9:1 ratio of 1-methylcyclohexene and methylenecyclohexane. As a result, making methylenecyclohexane this way is undesirable because most of the product mixture is the wrong alkene, which must be separated and diverted into the waste stream. On the other hand, the Wittig reaction is selective for one alkene product.

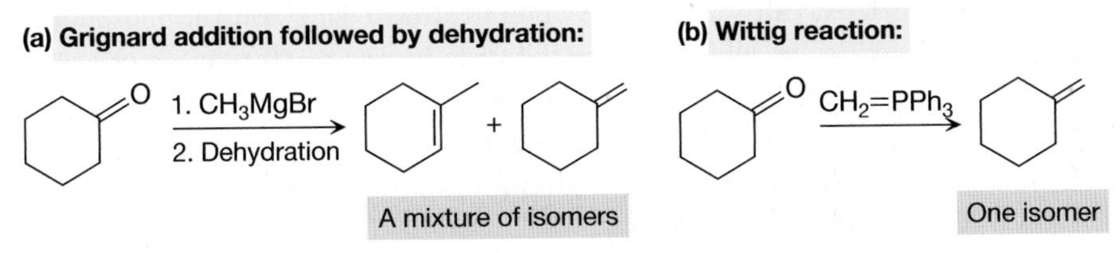

FIGURE 18.9

(a) Grignard addition followed by dehydration:

1. CH_3MgBr
2. Dehydration

A mixture of isomers

(b) Wittig reaction:

$CH_2=PPh_3$

One isomer

Comparison of two routes to alkenes. (a) Grignard addition and dehydration gives a mixture of isomers. The exocyclic alkene is the minor product. (b) The Wittig reaction furnishes the exocyclic alkene exclusively.

One of the starting materials used to perform the Wittig reaction is a phosphonium salt. Some simple phosphonium salts are commercially available, but often they are made from the corresponding alkyl halide via an S_N2 reaction (**Figure 18.10**). The phosphorus compound (triphenylphosphine, PPh_3, in this case) is neutral, but it has a lone pair of electrons, so it behaves as a nucleophile. In the course of the substitution reaction, the phosphorus becomes positively charged. Because there is no proton on phosphorus, the substitution product does not immediately lose a proton; thus, it retains the positive charge. The product is a phosphonium salt that may be isolated and purified by recrystallization.

FIGURE 18.10

The nucleophilic substitution reaction used to prepare a phosphonium salt.

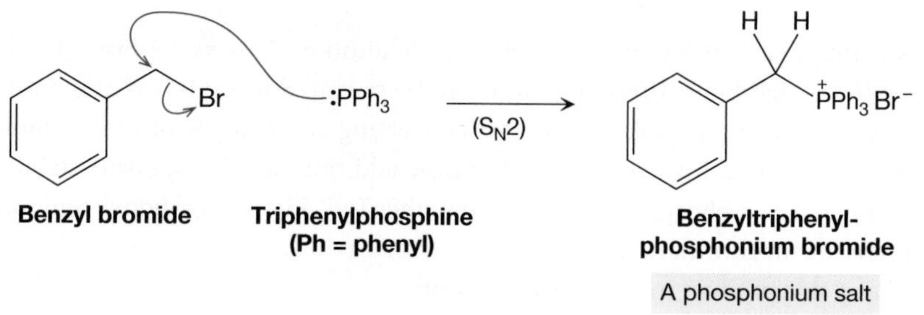

Benzyl bromide   Triphenylphosphine   Benzyltriphenyl-
                       (Ph = phenyl)   phosphonium bromide

A phosphonium salt

The mechanism of the Wittig reaction starts with treatment of the phosphonium ion with a strong base (**Figure 18.11**). Deprotonation at the position adjacent to the phosphorus leaves a lone pair and negative charge on carbon—that is, a carbanion. Because this carbanion is stabilized by the electron-withdrawing effect of the neighboring cationic phosphorus atom, this compound is called a phosphorus ylide. The term ylide refers to a structure that has positive and negative charges on adjacent atoms, where the positively charged atom stabilizes a neighboring carbanion. Here the cationic atom is phosphorus, but ylides can also involve nitrogen- or sulfur-centered cations. Most ylides are too unstable for routine purification and storage—instead, they are used immediately.

FIGURE 18.11

Deprotonation of a phosphonium salt to form a phosphorus ylide.

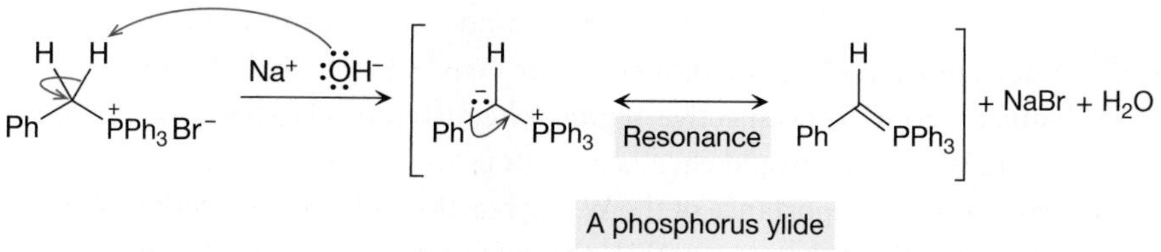

A phosphorus ylide

Once the phosphorus ylide is formed, the next step of the mechanism is a reaction with an aldehyde or ketone. The reaction occurs through a cycloaddition of C=O and C=P bonds, followed by elimination of the alkene (**Figure 18.12**). In contrast to the photocycloaddition of two alkenes, this cycloaddition can occur without light because it involves different orbital interactions. In some cases the cycloaddition is stepwise through a zwitterionic intermediate called a betaine.

The first part of this experiment is to prepare triphenyl(4-pyridylmethyl)phosphonium chloride via an S_N2 reaction. Upon treatment with base, the phosphonium salt is converted to the ylide. Because the ylide is very reactive, it is not isolated. Instead, it is generated in the presence of the 4-pyridine carboxyaldehyde (4-pyr-CHO) by treatment of the phosphonium salt with a strong base (NaOH) to generate the alkene. Although the reaction produces a mixture of cis and trans isomers of 4,4'-bpe, only *trans*-4,4'-bpe precipitates under the reaction conditions.

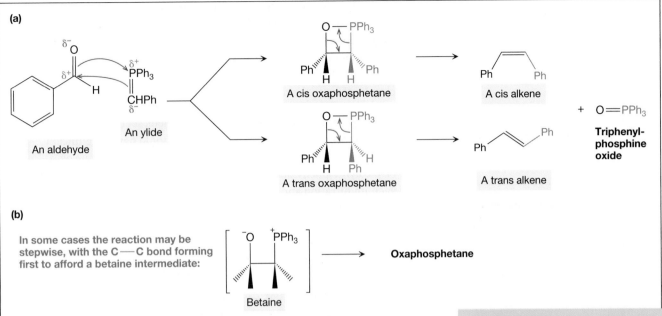

(a)

An aldehyde

An ylide

A cis oxaphosphetane

A cis alkene

A trans oxaphosphetane

A trans alkene

+ O=PPh₃

Triphenyl-phosphine oxide

(b)

In some cases the reaction may be stepwise, with the C—C bond forming first to afford a betaine intermediate:

Betaine

⟶ **Oxaphosphetane**

FIGURE 18.12

(a) Wittig reaction via a cycloaddition mechanism. Unlike the photocycloaddition of two alkenes, this cycloaddition does not require light. (b) There are some cases in which the cycloaddition may occur in a stepwise fashion via an intermediate called a betaine.

18.2 EXPERIMENTAL PROCEDURE

18.2A Day 1: Preparation of (4-Picolyl)-triphenylphosphonium Chloride

1. Preheat an oil bath at 80°C.

2. Weigh 2.25 g of 4-picolyl chloride [also known as 4-(chloromethyl)pyridine and 4-pyridylmethyl chloride] and 3.6 g of triphenylphosphine into a 50-mL Erlenmeyer flask. Add a stir bar and 15 mL of dimethylformamide (DMF) to the flask, plug the flask with glass wool, and stir the reaction mixture at 80°C for 45 minutes.

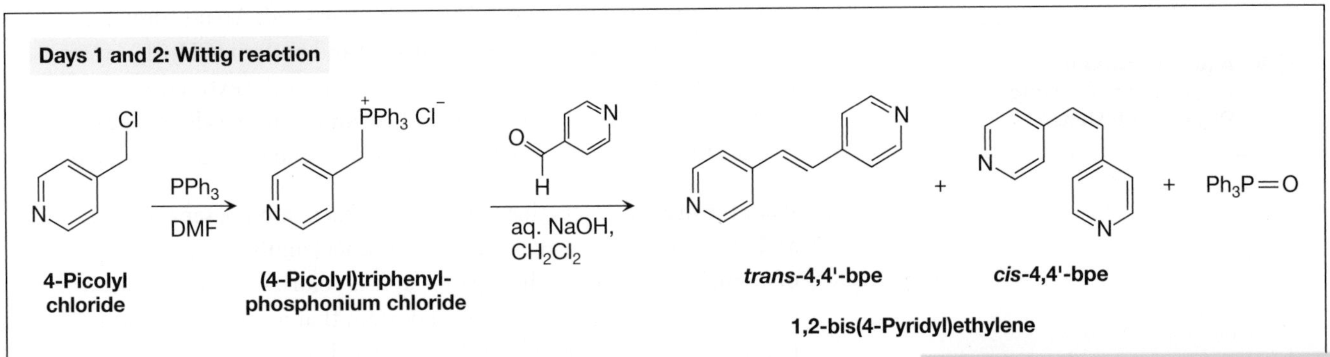

Days 1 and 2: Wittig reaction

4-Picolyl chloride

PPh₃
DMF

(4-Picolyl)triphenyl-phosphonium chloride

aq. NaOH,
CH₂Cl₂

trans-4,4'-bpe

cis-4,4'-bpe

+ Ph₃P=O

1,2-bis(4-Pyridyl)ethylene

FIGURE 18.13

3. While the reaction is proceeding, prepare an ice bath. After 45 minutes of stirring, cool down the reaction mixture to room temperature, then place the mixture in an ice bath and allow the product to sit undisturbed. A precipitate should begin to form within 15–20 minutes. If precipitation does not occur after 30 minutes, scratch the bottom of the flask and add small chunks of ice to the mixture to induce precipitation.

4. While waiting for the precipitation to complete, set up a vacuum filtration apparatus. After 30 minutes you should have a significant amount (i.e., a full flask) of a white-to-pale orange precipitate of (4-picolyl)triphenylphosphonium chloride. Recover the precipitate by vacuum filtration and allow it to dry.

▶ **Vacuum Filtration**

▶ **Melting Point Measurement**

5. Measure the mass and obtain a melting point for your phosphonium salt.

18.2B Day 2: The Wittig Reaction

6. Prepare a solution of 38% aqueous NaOH: Weigh 25 g of solid NaOH. (*CAUTION*: Do not allow NaOH to contact your skin or clothing.) Slowly add the solid to 40 mL of cold distilled water in a 100-mL beaker while stirring with a glass rod or spatula to avoid forming a glassy chunk of solid that is difficult to dissolve. When all of the NaOH is dissolved, cool the solution by placing the beaker in an ice bath.

7. Weigh 4.0 g of the (4-picolyl)triphenylphosphonium chloride that you prepared in section 18.2A in a 125-mL Erlenmeyer flask and clamp it securely over a stir plate. Suspend the solid in 10 mL of dichloromethane. Plug the neck with glass wool, then stir the suspension for 5 minutes.

8. Add 1.2 mL of 4-pyridinecarboxaldehyde (also known as isonicotinaldehyde) to the suspension while stirring. Slowly add the *cold* NaOH solution to the flask. Re-plug the flask with the glass wool, and stir the flask for 30 minutes.

9. While the reaction is proceeding, prepare for aqueous extraction using a 250-mL separatory funnel. You will need 35 mL of dichloromethane (CH_2Cl_2) and 100 mL of distilled water.

10. When the reaction is complete, transfer the solution to the separatory funnel, rinse the flask first with 100 mL of distilled water and then with 15 mL of CH_2Cl_2, adding both rinse solutions to the separatory funnel. To avoid an emulsion, do *not* shake the funnel—instead, gently invert it three to four times. If the layers do not separate, you have an emulsion. To help break the emulsion, gently swirl the contents with a glass rod. An oily brown third layer may form between the top (aqueous) and bottom (CH_2Cl_2) layers. Drain only the bottom (CH_2Cl_2) layer and repeat the extraction twice, using 10 mL of CH_2Cl_2 each time. Collect and combine all of your organic layers. Drain and dispose of the aqueous layer and any oily brown layer.

▶ **Aqueous–Organic Extraction and Drying Organic Solutions**

11. Place the organic layers that you collected back into the separatory funnel and add 30 mL of 10% HCl. Shake the funnel (emulsions are not as much of a problem here) until the organic layer is colorless. Your product is now present in the *aqueous* layer. It is important that none of the organic layer is present in the upper aqueous layer. Thus, drain *all* of the bottom organic layer and collect the aqueous layer in a clean 125-mL Erlenmeyer flask.

12. Neutralize the acidic aqueous solution that you collected with a 20% Na_2CO_3 solution (approximately 10–15 mL). Add the Na_2CO_3 solution *dropwise* to the acidic aqueous solution with gentle stirring, occasionally checking the pH using litmus paper. *Do not allow the pH to get higher than 7.* You will be close to the neutral point when you begin to see a precipitate forming and redissolving. Once the precipitate forms, do *not* add more Na_2CO_3 unless the pH is below 7.

▶ **Neutralizing Acidic or Basic Solutions and Checking pH**

13. Collect the precipitate by vacuum filtration, allow it to dry, and record the mass.

14. Place the solid into a 125-mL Erlenmeyer flask. Add a 20% NaOH solution, using 100 mL per gram of solid (e.g., for 0.3 g you would use 30 mL). Stir the mixture for approximately 5 minutes, then transfer the materials to a separatory funnel. Rinse the flask with 20 mL of chloroform ($CHCl_3$) and add the $CHCl_3$ rinse to the separatory funnel to perform an extraction. Save the organic layer. Repeat the extraction twice more using 20-mL portions of $CHCl_3$ each time. Combine all organic layers.

15. Dry with anhydrous Na_2SO_4. Separate the drying agent by carefully decanting the solution into a pre-weighed round-bottom flask. If the drying agent transfers along with the solution while you are decanting, carry out a gravity filtration to remove it.

▶ **Gravity Filtration**

16. Remove the organic solvent using a rotary evaporator until only a solid remains.

▶ **Rotary Evaporation**

17. Measure the mass of your *trans*-4,4'-bpe and obtain its melting point. Prepare a sample for ^1H NMR analysis.

18.2C Day 3: Co-Crystallization and the Photocycloaddition Reaction

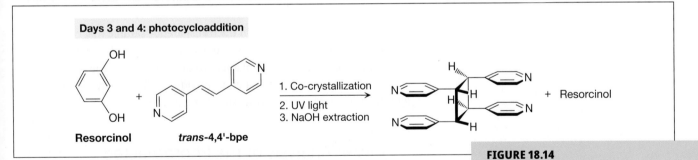

Days 3 and 4: photocycloaddition

Resorcinol   *trans*-4,4'-bpe

1. Co-crystallization
2. UV light
3. NaOH extraction

+ Resorcinol

FIGURE 18.14

18. Into a clean and dry mortar, place 120 mg of resorcinol and 180 mg of *trans*-1,2-bis(4-pyridyl)ethene (prepared by the Wittig reaction in section 18.2B, or commercial *trans*-4,4'-bpe). Using the pestle, grind the two solids together for a period of 25–30 minutes. Make sure the two solids are thoroughly mixed together. Collect the resulting fine powdery co-crystals of 2(resorcinol)·2(4,4'-bpe).

19. Obtain a melting point for your co-crystals of 2(resorcinol)·2(4,4′-bpe). If the melting point apparatuses are all in use, come back to this later.

20. Set aside an appropriate amount of your co-crystals of 2(resorcinol)· 2(4,4′-bpe) for TLC analysis (and melting point if you haven't done it yet). Place about 100 mg of your co-crystals on a pre-cut transparency film of about 25 cm^2 (provided by your lab instructor). Spread the sample evenly across the film, and as thinly as possible, then cover it with a second layer of transparency film. Close the edges by applying masking tape (**Figure 18.15**). Be sure to label your sample by writing your name and section number on the masking tape.

FIGURE 18.15

Courtesy of Mouna A. Maalouf.

21. Provide your lab instructor with the transparency sample that contains your co-crystals. Before your next lab period, your sample will be placed in a photoreactor for a period of approximately 15 hours. The photoreactor consists of a mercury lamp that emits broadband (i.e., many different wavelengths) UV energy of high intensity (Figure 18.7).

18.2D Day 4: Analysis of the Photoreacted Co-Crystals

22. Collect your photoreacted sample from your lab instructor.

23. Prepare four solutions for TLC analysis, using a 10:1 mixture of ethyl acetate and ethanol as the solvent: (a) co-crystals before photoreaction (day 3, step 20), (b) *trans*-4,4′-bpe (day 2, step 16), (c) resorcinol, and

(d) co-crystals after photoreaction (day 4, step 22). Analyze the four solutions by TLC, eluting the plate with a 10:1 mixture of ethyl acetate and ethanol. Locate the spots on the TLC plate using both UV light and iodine vapors.

● **Thin-Layer Chromatography**

24. Place about 10 mg of your photoreacted co-crystals into an NMR tube for ^{1}H NMR analysis using DMSO-d6 as the NMR solvent.

● **Preparing NMR Samples**

25. Start heating a water bath to 70–80°C. Transfer the remaining amount of your photoreacted sample into a tared 25-mL Erlenmeyer flask and record the mass. To the flask add a magnetic stir bar and 15 mL of 1 M NaOH solution. Heat the mixture with a hot water bath (70–80°C) for 20 minutes with vigorous magnetic stirring.

26. Cool the mixture to room temperature and transfer it to a separatory funnel. Extract the resulting yellow-green aqueous solution using three 20-mL portions of dichloromethane. Do *not* discard the aqueous phase (it contains the resorcinol template); retain it in the separatory funnel for later. Dry the organic phase over anhydrous Na_2SO_4. Decant the organic phase into a tared round-bottom flask. Remove the solvent using a rotary evaporator. Record the mass of the residual material in the flask, which is the cyclobutane photocycloaddition product 4,4′-tpcb.

27. Obtain a melting point for your 4,4′-tpcb (literature mp = 234–237°C).

28. To the aqueous phase that you retained in the separatory funnel, add 10% aqueous HCl dropwise until the solution tests acidic—namely, pH 2—with pH paper and becomes yellow or faintly red. Extract the resorcinol with three 20-mL portions of diethyl ether. Dry the organic phase over sodium sulfate, decant into a tared round-bottom flask, and remove the solvent using a rotary evaporator. Record the mass of the recovered resorcinol template.

29. Using TLC, compare your isolated photocycloaddition product and resorcinol template with standards provided in the lab, eluting the plate with a 10:1 mixture of ethyl acetate and ethanol. The TLC plate should have four lanes: (a) resorcinol standard, (b) your recovered template, (c) your photocycloaddition product, and (d) photocycloaddition product standard. Locate the spots on the TLC plate using both UV light and iodine vapors.

18.3 THE LABORATORY REPORT

Your results and discussion should include the following:

- Calculate the percent yield for the preparation of (4-picolyl)triphenylphosphonium chloride. What is the limiting reagent in the reaction?
- Calculate the percent yield of the Wittig reaction. What is the overall yield?
- Discuss both the overall yield and the purity of your product *trans*-4,4′-bpe based on ^{1}H NMR data and melting point.
- Why do we treat the crude product obtained from your Wittig reaction with 10% HCl? Clearly explain your answer.

From all the information that you have at your disposal (TLC, ^{1}H NMR, melting points), discuss whether or not the photocycloaddition reaction was successful, and explain how you made that determination. Do not forget to include literature values where appropriate. Report the percent conversion in the photocycloaddition and the percent recoveries of both the template and the photocycloaddition product. Attach the ^{1}H NMR spectrum to your report.

Calculation hints: A percent yield of the purified product cannot be calculated by its mass directly, because it cannot be separated from the leftover reactant in this procedure. Instead, we will use ^{1}H NMR to determine the percent conversion, or in other words, the extent to which the reaction has proceeded. The percent conversion of the photocycloaddition can be determined by integrating the peaks in the ^{1}H NMR spectrum of the photoreacted co-crystals. The integration ratios of reactant and product peaks equals the mole ratio of reactant and product. The theoretical recoveries of template and photoproduct are calculated from the mass of co-crystals used in the photoreaction. Percent recovery is simply mass recovered divided by theoretical maximum.

18.3A | Post-Lab Question

In this experiment, the resorcinol template used hydrogen bonds to organize the reactants in the solid state for the photoreaction. What is the typical strength (in kJ/mole) of a hydrogen bond? Compare this energy to that of a C—C bond. If a C—C bond linked the template and photoproduct, would it be possible to separate them by extraction? Why or why not? Attach these post-lab questions to your report.

18.3B | Data to Include in the Laboratory Report

A. PREPARATION OF (4-PICOLYL)TRIPHENYLPHOSPHONIUM CHLORIDE

Amount of 4-picolyl chloride used (g):

Amount of triphenylphosphine used (g):

Amount of (4-picolyl)triphenylphosphonium chloride obtained (g):

Melting range of (4-picolyl)triphenylphosphonium chloride (°C):

B. THE WITTIG REACTION

Amount of *trans*-1,2-bis(4-pyridyl)ethylene obtained (g):

Melting range of *trans*-1,2-bis(4-pyridyl)ethylene (°C):

C. CO-CRYSTALLIZATION AND THE PHOTOCYCLOADDITION REACTION

Melting range of 2(resorcinol)·2(4,4′-bpe) co-crystals (°C):

Amount of 2(resorcinol)·2(4,4′-bpe) co-crystals used in photoreaction (mg):

D. ANALYSIS OF THE PHOTOREACTED CO-CRYSTALS

Amount of isolated photocycloaddition product (mg):

Melting range of isolated product (°C):

Amount of resorcinol recovered (mg):

TLC R_f value of isolated product:

TLC R_f value of isolated resorcinol:

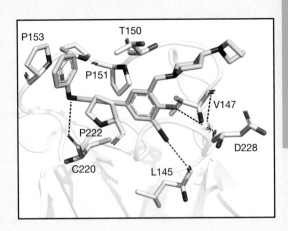

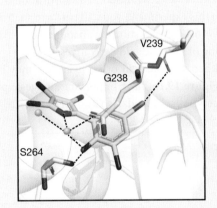

PK5196

(b)

Pentabromopseudilin

- Identify electrophilic aromatic substitution reactions and show their mechanisms.

- Implement an electrophilic aromatic halogenation reaction that adheres to green chemistry principles.

- Determine the location in which substitution occurred by spectroscopic analysis of the product.

Iodination of Salicylic Acid Derivatives

AROMATIC HALIDES IN PROTEIN–DRUG INTERACTIONS

(a) Structure of PK5196, an anticancer drug candidate that activates protein p53, resulting in tumor suppression. The iodine, shown in purple, contributes a stabilizing interaction (dotted line) that assists the binding of the drug to the protein. (b) Pentabromopseudilin uses a bromine to form a similar stabilizing interaction with an enzyme, resulting in inhibition of biosynthetic pathways that are specific to plants.

Wilcken, R.; Liu, X.; Zimmermann, M. O.; Rutherford, T. J.; Fersht, A. R.; Joerger, A. C.; Boeckler, F. M. Halogen-Enriched Fragment Libraries as Leads for Drug Rescue of Mutant p53. *J. Am. Chem. Soc.* **2012**, *134*, 6810–6818; PDB file 4AGQ. Kunfermann, A.; Witschel, M.; Illarionov, B.; Martin, R.; Rottmann, M.; Höffken, H. W.; Seet, M.; Eisenreich, W.; Knölker, H.-J.; Fischer, M.; Bacher, A.; Groll, M.; Diederich, F. Pseudilins: Halogenated, Allosteric Inhibitors of the Non-Mevalonate Pathway Enzyme IspD. *Angew. Chem. Int. Ed.* **2014**, *53*, 2235–2239; PDB file 4NAK.

alogenated aromatic rings are important components of innovative new functional materials, and are found in many biologically active compounds.[1] The halogen atom itself may play a key role in the biological activity. For example, in studies of a mutant tumor suppressor known as p53 that was no longer functional, it was found that an iodine substituent assisted in the binding of a drug that restored function (part a of the opening art).[2] The halogens of pseudilins (part b of the opening art) were found to be critical in the selective inhibition of a biochemical pathway in plants that is not used in animals,[3] a strategy that is important for safer herbicide design.

Attachment of substituents to aromatic rings usually involves reactions that replace one of the hydrogens at a carbon of the aromatic ring. Often these rings have π electrons available for bonding with electrophilic reagents, and this can initiate further steps that result in a substitution of one of the ring hydrogens (**Figure 19.1a**). In this experiment, you will perform such a reaction, known as *electrophilic aromatic substitution*, substituting an iodine for a hydrogen at a specific carbon of the benzene ring of salicylic acid derivatives. Two variations on the procedure are provided here; either salicylic acid or salicylamide may be used. Salicylic acid and its derivatives (**Figure 19.1b**) have important analgesic properties; one of the most well known is acetylsalicylic acid, also known as aspirin.

FIGURE 19.1

(a) Generalized reaction of benzene via electrophilic aromatic substitution. The mechanism involves two main steps—namely, addition of the electrophile followed by deprotonation. (b) Selected derivatives of salicylic acid.

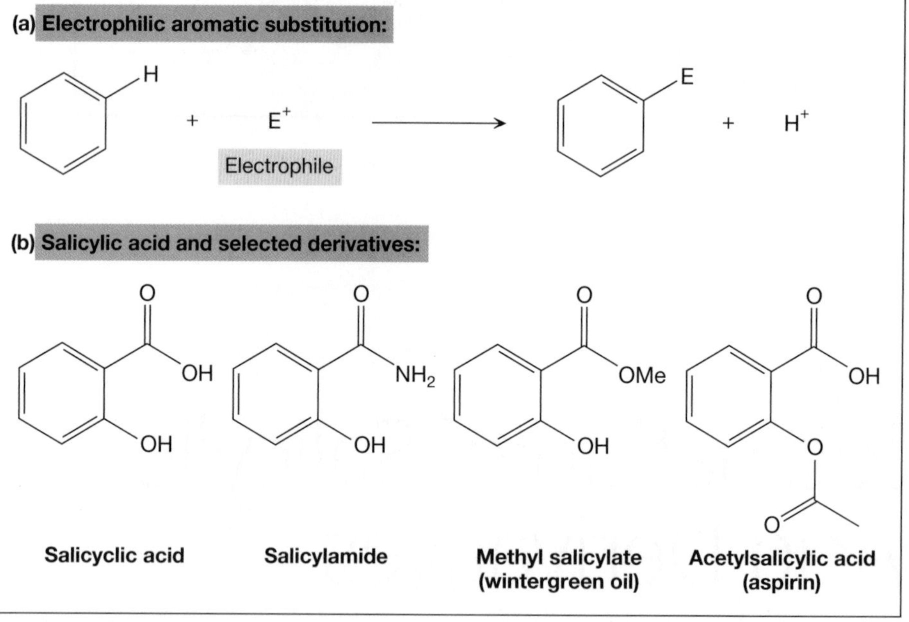

(a) Electrophilic aromatic substitution:

(b) Salicylic acid and selected derivatives:

Salicyclic acid Salicylamide Methyl salicylate (wintergreen oil) Acetylsalicylic acid (aspirin)

[1]Cavallo, G.; Metrangolo, P.; Milani, R.; Pilati, T.; Priimagi, A.; Resnati, G.; Terraneo, G. The Halogen Bond. *Chem. Rev.* **2016**, *116*, 2478–2601.

[2]Wilcken, R.; Liu, X.; Zimmermann, M. O.; Rutherford, T. J.; Fersht, A. R.; Joerger, A. C.; Boeckler, F. M. Halogen-Enriched Fragment Libraries as Leads for Drug Rescue of Mutant p53. *J. Am. Chem. Soc.* **2012**, *134*, 6810–6818.

[3]Kunfermann, A.; Witschel, M.; Illarionov, B.; Martin, R.; Rottmann, M.; Höffken, H. W.; Seet, M.; Eisenreich, W.; Knölker, H.-J.; Fischer, M.; Bacher, A.; Groll, M.; Diederich, F. Pseudilins: Halogenated, Allosteric Inhibitors of the Non-Mevalonate Pathway Enzyme IspD. *Angew. Chem. Int. Ed.* **2014**, *53*, 2235–2239.

Pre-Lab Reading Assignment

Review the following before attempting this experiment:

- Experiment: Chapter 19 (this experiment)
- Technique: Recrystallization (Chapter 5, section 5.4)
- Technique: Melting point measurement (Chapter 5, section 5.6)
- Technique: Infrared (IR) spectroscopy (Chapter 6, section 6.2)

19.1 BACKGROUND

19.1A Electrophilic Aromatic Substitution

The molecular formula of benzene is C_6H_6. A typical **saturated** hydrocarbon with six carbons has a formula of C_6H_{14} (more generally, this can be expressed as C_nH_{2n+2}). With eight fewer hydrogens, benzene is four molecules of H_2 away from being saturated—that is, it has four degrees of unsaturation. These four degrees of unsaturation can be represented by three carbon–carbon double bonds and one ring (**Figure 19.2**).

Alkenes, which are *unsaturated* hydrocarbons, generally undergo addition reactions with electrophiles, but benzene and related aromatic compounds don't—they generally undergo substitution instead (**Figure 19.3**). For example, when benzene reacts with bromine, usually aided by a catalyst such as $FeBr_3$, substitution yields C_6H_5Br plus HBr. The addition product, $C_6H_6Br_2$, is not formed.

<< saturated
A hydrocarbon compound or group that contains only sp^3-hybridized carbons, with no π bond functional groups such as alkenes or alkynes.

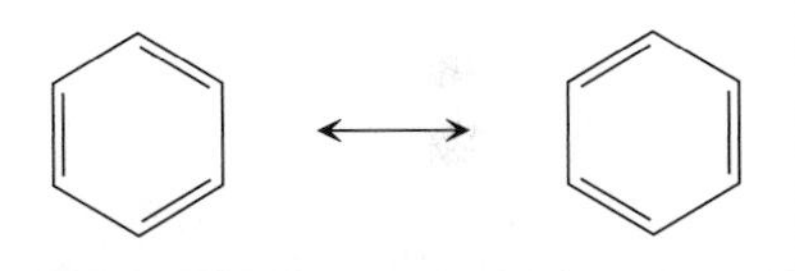

FIGURE 19.2

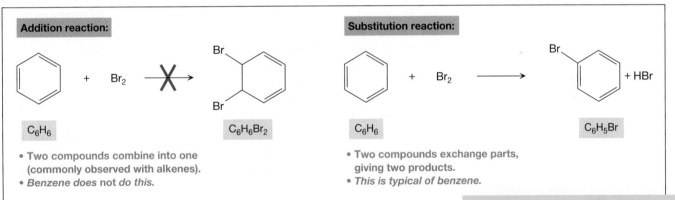

Addition reaction:

C_6H_6 + Br_2 ⤫ $C_6H_6Br_2$

- Two compounds combine into one (commonly observed with alkenes).
- *Benzene does not do this.*

Substitution reaction:

C_6H_6 + Br_2 ⟶ C_6H_5Br + HBr

- Two compounds exchange parts, giving two products.
- *This is typical of benzene.*

FIGURE 19.3

A comparison of addition and substitution. Alkenes undergo addition with electrophiles, whereas benzene undergoes substitution. Substitution reactions produce a by-product (in this example, HBr) that must be considered when planning for laboratory safety and waste disposal.

The reactivity of benzene has been known since the mid-1800s, but it took many years before chemists understood it. Although we can draw the structure of benzene with three carbon–carbon double bonds, they can't be the double bonds of typical alkenes, because benzene does not add Br_2 like an alkene. When the substitution occurs, the product retains all of its degrees of unsaturation. This is an important clue, and the theory of aromaticity allows us to explain why this occurs. The 6 π electrons in benzene are not behaving as they would for three alkenes; instead, they behave as a unit, circulating in the cyclic π system and causing a powerful stabilizing effect known as aromaticity. Because of aromaticity, benzene is roughly 150 kJ/mol (36 kcal/mol) more stable than what would be expected for a hypothetical "cyclohexatriene" compound with three alkenes in a ring. The substitution reaction allows the benzene to retain aromaticity and preserve that powerful stabilizing effect.

The currently accepted mechanism of the electrophilic aromatic substitution reaction is a two-step process involving addition followed by deprotonation (**Figure 19.4**).

FIGURE 19.4

The key addition and deprotonation steps common to the mechanisms of all electrophilic aromatic substitution reactions. There may be preliminary steps involved in generating the strong electrophile (E^+).

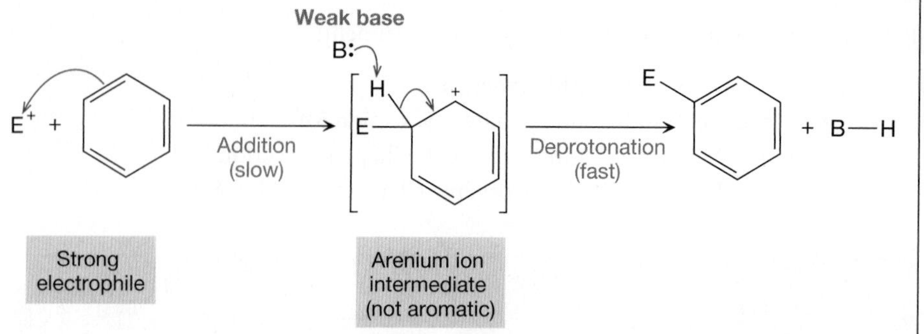

Weak base

Addition (slow)

Deprotonation (fast)

Strong electrophile

Arenium ion intermediate (not aromatic)

A variety of different electrophiles may be used for this purpose, and many of them involve harshly acidic conditions in order to generate the strong electrophile needed for the reaction. A strong electrophile is usually needed in order to break up the aromaticity to get to the arenium ion intermediate. In fact, most electrophilic halogenation reactions require Fe^{3+} salts as Lewis acid catalysts—they make the halogen more electrophilic.

On the other hand, when benzene rings have substituents that can stabilize positive charge through resonance (lone pairs of electrons at the atom directly attached to the benzene ring), the arenium is stabilized, facilitating the addition step. These groups are considered activating groups. In such cases, a Lewis acid catalyst may not be needed.

19.1B Green Aromatic Halogenation Conditions

Although they are commonly used to halogenate aromatic rings, the halogens Cl_2 and Br_2 present some significant inhalation toxicity hazards. In this experiment, we use a relatively benign source of electrophilic halogen—namely, household bleach, which is a 6% solution of sodium hypochlorite (NaOCl). By mixing this with sodium iodide, hypoiodous acid (HOI) is formed, and this can serve as a source of I^+ for electrophilic aromatic substitution, as long as the aromatic ring contains strongly activating substituents. Once HOI is formed, it can react with an aromatic ring via the general mechanism shown in **Figure 19.5**.

The only by-products in this reaction are NaCl and H_2O. By comparison with other electrophilic halogenation reactions (e.g., $FeCl_3$, Cl_2), the conditions are much less hazardous. Additionally, we will use water and ethanol as solvents in

FIGURE 19.5

(a) Reaction of bleach (6% aqueous NaOCl) with NaI. (b) Mechanism of electrophilic aromatic substitution with hypoiodous acid (HOI).

(a) Generating the electrophile:

$$NaOCl + H_2O \rightleftharpoons HOCl + NaOH$$

$$HOCl + NaI \rightleftharpoons HOI + NaCl$$

Electrophile

(b) Electrophilic aromatic substitution:

$$HO-I + \longrightarrow \longrightarrow + H_2O$$

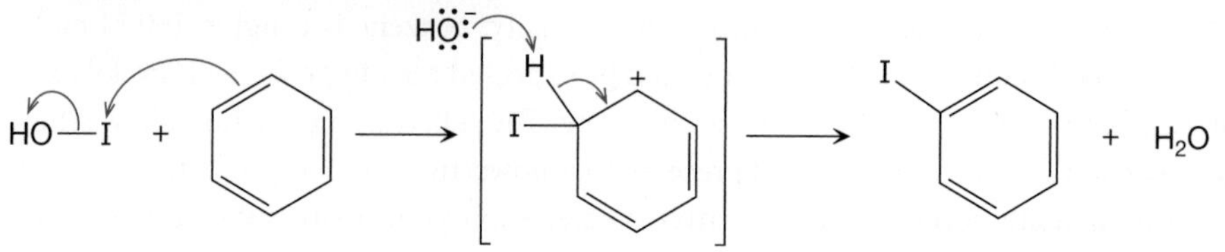

this experiment, eliminating volatile hydrocarbons or chlorinated solvents from the waste stream. Taken together, all these factors make the reaction much closer in accord with green chemistry principles.

In this mechanism, pre-existing substituents on the aromatic ring are not depicted. Your starting material, however, is salicylic acid or salicylamide (**Figure 19.6**), and these have a strong activating group (the OH group) that enables this reaction to occur under milder conditions.

Activating substituents—those that have lone pairs at the point of attachment that can stabilize the arenium ion intermediate—direct substituents to positions ortho and para to their location (**Figure 19.7**). Usually a mixture of ortho and para results. Substituents with a partial positive charge at the point of attachment to the ring destabilize the arenium ion, slowing the reaction and directing the electrophile to positions meta to the substituent of this type. When there are two pre-existing substituents (as in salicylamide), two additional features are important: the more activating group usually takes control of the directing effects, and having three adjacent substituents together is sterically disfavored.

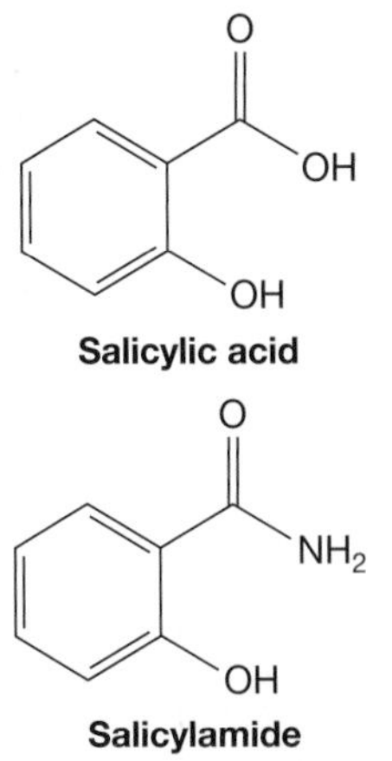

Salicylic acid

Salicylamide

FIGURE 19.6

ËDG
Ortho, o   Ortho, o

Para, p

EDG = electron-donating group

$\overset{\delta^+}{\text{EWG}}$

Meta, m   Meta, m

EWG = electron-withdrawing group

Y

X

Disfavored positions due to steric hindrance

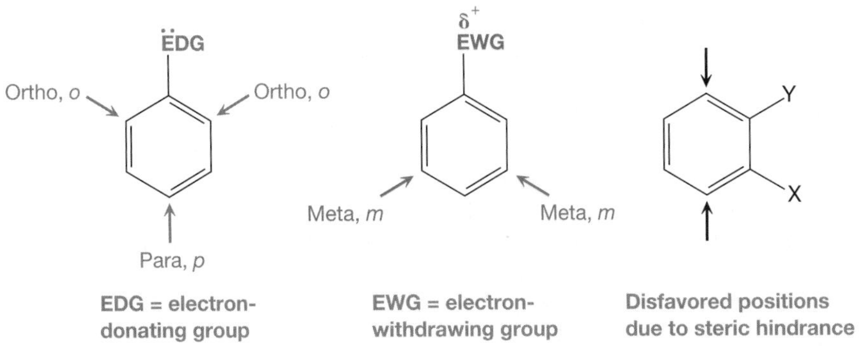

FIGURE 19.7

Summary of directing effects of aromatic ring substituents, indicating where the new substituent will be likely to attach in an electrophilic aromatic substitution reaction. Groups with an electron pair at the point of attachment generally direct substitution to the ortho and para positions. Groups with a partial (or full) positive charge at the point of attachment generally direct to the meta positions.

19.1C Determining Substitution Patterns using IR

Infrared spectroscopy is mainly used for functional group identification in organic compounds, especially with diagnostic peaks from 1600 to 4000 cm^{-1}. While this is an important area for information about organic functional groups, the fingerprint region of the IR spectrum (800–1600 cm^{-1}) can also provide specialized structural information. For example, the 700–900 cm^{-1} region contains useful information about the substitution pattern of aromatic rings, as listed in **Table 19.1**.

The electrophilic aromatic substitution you will perform in the lab is shown in **Figure 19.8**. The position of the iodine on the ring is not indicated in the product, so part of your job is to predict the structure of the product (your hypothesis) and to test that prediction in the lab.

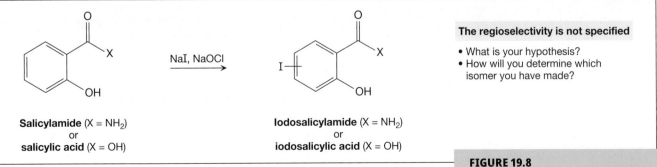

Salicylamide (X = NH$_2$)
or
salicylic acid (X = OH)

NaI, NaOCl

Iodosalicylamide (X = NH$_2$)
or
iodosalicylic acid (X = OH)

The regioselectivity is not specified

• What is your hypothesis?
• How will you determine which isomer you have made?

FIGURE 19.8

Iodination reaction of salicylic acid derivatives.

TABLE 19.1

Typical IR Peaks for Different Benzene Substitution Patterns

SUBSTITUTION PATTERN	EXAMPLE	TYPICAL IR ABSORBANCE (cm^{-1})
Monosubstituted		770–715
1,2-Disubstituted		770–730
1,3-Disubstituted		820–760
1,4-Disubstituted		870–800
1,2,3-Trisubstituted		790–750
1,2,4-Trisubstituted		850–800
1,3,5-Trisubstituted		910–830

19.2 EXPERIMENTAL PROCEDURE: IODINATION OF SALICYLIC ACID[4]

19.2A Day 1

1. Measure about 0.55 g of salicylic acid and record the exact mass. Place the salicylic acid into a 25-mL round-bottom flask equipped with a magnetic stir bar. Dissolve the salicylic acid in 10 mL of absolute ethanol, warming the flask with your hand to speed up the dissolution.

[4]This procedure is adapted from Eby, E.; Deal, S. T. A Green, Guided-Inquiry Based Electrophilic Aromatic Substitution for the Organic Chemistry Laboratory. *J. Chem. Educ.* **2008**, *85*, 1426.

2. Once the salicylic acid is completely dissolved, add 0.72 g of sodium iodide (NaI) to the reaction mixture, stirring until the solution is homogeneous.

3. Cool the reaction flask in an ice-water bath. After cooling for about 5 minutes, remove the reaction vessel from the ice bath and quickly add 5.4 mL of 6% (w/v) sodium hypochlorite solution (ultra strength household bleach). Swirl the flask vigorously to completely mix the contents. The solution will change color from the initial colorless reaction mixture to a dark red-brown, then will fade to a yellow color as the reaction proceeds. When the yellow color of the solution stops fading, the reaction is complete. (Typically, this takes less than 5 minutes.) Allow the reaction vessel to sit on the benchtop undisturbed for another 10 minutes.

4. Quench any remaining oxidizing agents (iodine or hypoiodite or hypochlorite) by adding 5 mL of 10% (w/v) sodium thiosulfate to the reaction mixture, with thorough mixing.

5. Acidify the reaction solution by slowly adding 10% HCl. Monitor the acidity of the solution using pH indicator paper. When a white solid begins to form, the pH of the solution is near the desired acidity. Continue adding 10% HCl, but carefully monitor the acidity, stopping the addition when the mixture is pH 2–3. Cool the mixture in an ice bath for 10 minutes.

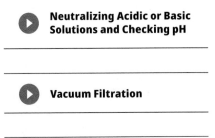

Neutralizing Acidic or Basic Solutions and Checking pH

6. Separate the solid product using vacuum filtration and a small Hirsch funnel.

Vacuum Filtration

7. Recrystallize the solid using the minimum amount of hot water and a little ethanol. Add 2 mL water to the solid in an Erlenmeyer flask. Bring this to a boil and add ethanol dropwise until the solid dissolves. Then cool slowly on the benchtop. After the flask reaches room temperature, place in an ice bath for 15 minutes to complete the recrystallization. With the aid of a spatula for transferring the product, collect the recrystallized product by vacuum filtration with a small Hirsch funnel. Rinse any remaining crystals from the flask into the filtration funnel using a small amount of cold water. Allow the vacuum to pull air through the crystals for a few minutes until they are fully dried. Store the iodosalicylic acid product in a vial until the next lab period.

Recrystallization

19.2B Day 2

8. Record the mass of the fully dried product.

9. Collect an IR spectrum of the product using attenuated total reflectance (ATR), if available. Alternatively, obtain a transmission IR spectrum using a $CHCl_3$ solution on a salt plate. The product may be sparingly soluble; place a few crystals in a test tube and add a few drops of $CHCl_3$, then warm it for a few seconds on the sand bath. Be sure to carefully label all peaks between 730 and 910 cm^{-1} on the spectrum.

10. Measure the melting point of the product. Under our typical conditions of melting point measurement, the iodosalicylic acid product decomposes by turning brown at around 130°C. Under different conditions, a literature report describes mp 197–198°C.[5]

Melting Point Measurement

[5]Cassebaum, H.; Herzberg, H. Über die Darstellung der 3- und 5-Jodsalicylsäure. *Z. Chem.* **1976**, *16*, 319–320.

19.3A Day 1

1. Measure about 0.55 g of salicylamide and record the exact mass. Place the salicylamide into a 25-mL round-bottom flask equipped with a magnetic stir bar. Dissolve the salicylamide in 10 mL of absolute ethanol, warming the flask with your hand to speed up the dissolution.

2. Once the salicylamide is completely dissolved, add 0.72 g of sodium iodide (NaI) to the reaction mixture, stirring until the solution is homogeneous.

3. Cool the reaction mixture in an ice-water bath. After cooling for about 5 minutes, remove the reaction vessel from the ice bath and quickly add 5.4 mL of 6% (w/v) sodium hypochlorite solution (ultra strength household bleach). Swirl the flask vigorously to completely mix the contents. The solution will change color from the initial colorless reaction mixture to a dark red-brown, then will fade to a yellow color as the reaction proceeds. When the yellow color of the solution stops fading, the reaction is complete. (Typically, this takes less than 5 minutes.) Allow the reaction vessel to sit on the benchtop undisturbed for another 10 minutes.

4. Quench any remaining oxidizing agents (iodine or hypoiodite or hypochlorite) by adding 5 mL of 10% (w/v) sodium thiosulfate to the reaction mixture, with thorough mixing.

5. Acidify the reaction solution by slowly adding 10% HCl. Monitor the acidity of the solution using pH indicator paper. When a white solid begins to form, the pH of the solution is near the desired acidity. Continue adding 10% HCl, but carefully monitor the acidity, stopping the addition when the mixture is pH 2–3.

6. Separate the solid product using vacuum filtration and a small Hirsch funnel.

7. Recrystallize the solid from 80:20 ethanol/water. Use the minimum amount of hot solvent to dissolve the product, then cool slowly. With the aid of a spatula for transferring the product, collect the recrystallized product by vacuum filtration with a small Hirsch funnel. Rinse any remaining crystals from the flask into the filtration funnel using a small amount of ice-cold 80:20 ethanol/water. Allow the vacuum to pull air through the crystals for a few minutes until they are fully dried. Store the product in a vial until the next lab period.

19.3B Day 2

8. Record the mass of the fully dried product.

9. Collect an IR spectrum.[6] Be sure to carefully label all peaks between 700 and 900 cm^{-1} on the spectrum.

10. Measure the melting point of the product (literature mp 228°C).

---

[6]An attenuated total reflectance (ATR) accessory is convenient for direct IR spectroscopic analysis of the solid. Similar results may also be obtained using a KBr pellet.[4]

In your report, use the data available (yield, IR, and mp) to conclude whether the reaction was successful, in terms of both identity and purity of the expected product. Using the IR data, determine which substitution pattern was obtained for the trisubstituted product.

In an appendix to your report, address the following: Use a database to search the chemical literature for a laboratory procedure for halogenation of benzoic acid ($PhCO_2H$) with Br_2 and $FeBr_3$. Referring to the principles of green chemistry, compare this literature procedure with your own experiment and comment on how well each addresses the principles of green chemistry.

20

LEARNING OBJECTIVES

- Execute a porphyrin synthesis via electrophilic aromatic substitution of pyrrole.

- Analyze reaction progress using thin-layer chromatography.

- Purify the product, tetraphenylporphyrin, using column chromatography.

- Use ultraviolet–visible spectroscopy to confirm the formation of tetraphenylporphyrin and determine its concentration in solution.

Synthesis of Tetraphenylporphyrin

LITTLE PAINT CREEK

Chlorophyll is a key component of the photosynthesis process in green plants and the conversion of carbon dioxide into the abundant biomass, as in this lush northeast Iowa forest. The modified porphyrin substructure of chlorophyll absorbs light to provide energy for this process.

Courtesy of Gregory K. Friestad.

A romatic heterocycles are cyclic compounds that meet all the criteria for aromaticity, but contain one or more noncarbon atoms in the ring. Some common examples are pyridine, pyrrole, furan, thiophene, and indole (**Figure 20.1**).

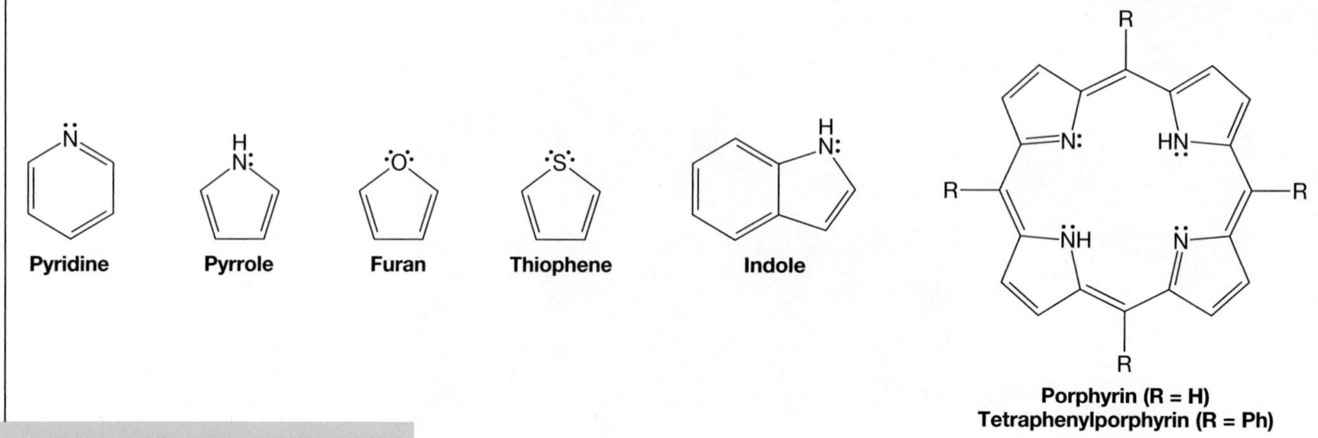

Pyridine   **Pyrrole**   **Furan**   **Thiophene**   **Indole**

Porphyrin (R = H)
Tetraphenylporphyrin (R = Ph)

FIGURE 20.1

Porphyrin and tetraphenylporphyrin are also members of this class. Although some of their bonds are drawn as localized double bonds, all of these are aromatic compounds, so their π electrons (sometimes including lone pairs) are actually delocalized in a fashion similar to benzene. This delocalization is associated with extra stability called resonance energy, which drives these unsaturated systems to maintain the aromaticity. Therefore, unlike alkenes, benzene and aromatic heterocycles undergo substitution reactions instead of addition reactions. In this experiment, we carry out electrophilic aromatic substitution reactions on pyrrole to synthesize tetraphenylporphyrin in the gas phase, and we purify the product by column chromatography.

Pre-Lab Reading Assignment

Review the following before attempting this experiment:

- Experiment: Chapter 20 (this experiment)
- Technique: Thin-layer chromatography (TLC) (Chapter 5, section 5.5)
- Technique: Column chromatography (Chapter 5, section 5.5)
- Technique: Ultraviolet-visible (UV–vis) spectroscopy (Chapter 6, section 6.1)

20.1 BACKGROUND

Porphyrin (Figure 20.1) is a complex heterocyclic compound composed of four pyrrole units, each of which is linked together by carbon–carbon bonds to form a macrocycle (a large ring). Because of the positions in which the carbon–carbon bonds are linked to each pyrrole, the four nitrogens point inward toward the interior of a cavity that can accommodate metal ions, such as the iron ion in heme B, the magnesium ion in chlorophyll a, or the cobalt ion in vitamin B$_{12}$ (**Figure 20.2**).

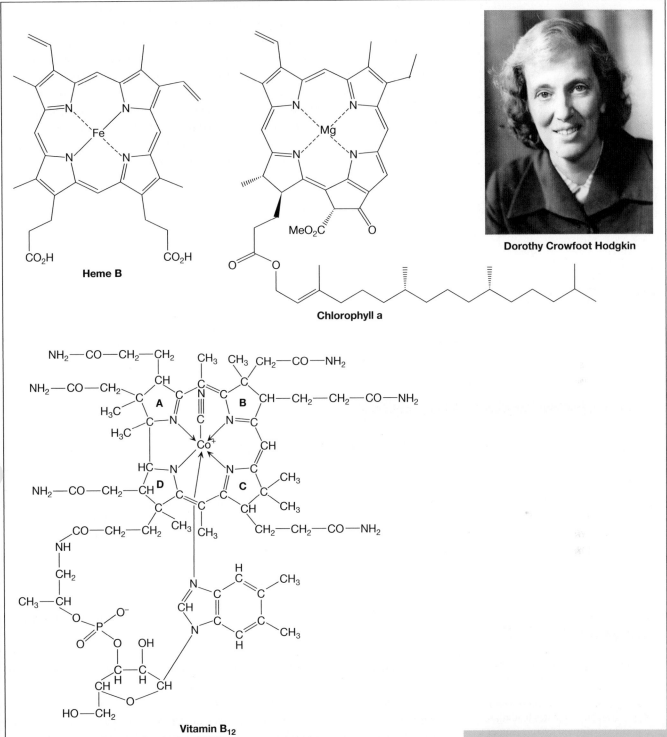

Dorothy Crowfoot Hodgkin

Heme B

Chlorophyll a

Vitamin B₁₂

FIGURE 20.2

Structures of several biologically important compounds that contain the porphyrin substructure. The historically important vitamin B_{12} structure shown here is drawn as originally shown in the landmark 1956 paper by Dorothy Crowfoot Hodgkin, a 1964 Nobel Laureate for crystallographic structure determination.[2]

Keystone Press/Alamy Stock Photo.

Heme is the mammalian oxygen transport chemical in blood, whereas derivatives of vitamin B_{12} are required in several biochemical processes, including the proper function of two essential enzymes in humans, called methionine synthase and L-methylmalonyl-CoA mutase. The porphyrin subunit forms the light-harvesting structure of chlorophyll, a natural compound from the plant world that has inspired many efforts to employ synthetic porphyrins in solar energy conversion.[1]

[1]Campbell, W. M.; Burrell, A. K.; Officer, D. L.; Jolley, K. W. Porphyrins as Light Harvesters in the Dye-Sensitised TiO₂ Solar Cell. *Coord. Chem. Rev.* **2004**, *248*, 1363–1379.
[2]Crowfoot Hodgkin, D.; Kamper, J.; Mackay, M.; Pickworth, J.; Trueblood, K.; White, J. G. Structure of Vitamin B₁₂. *Nature* **1956**, *178*, 64–66.

How would you synthesize a porphyrin-containing compound? Porphyrin consists of four pyrrole units, and pyrrole is an aromatic compound. The lone pair on its nitrogen atom participates in the π bonding, thus completing a set of six π electrons, as in benzene. So, if you need to link four pyrroles together through carbon–carbon bonds, then making the bonds through electrophilic aromatic substitution is feasible, just as it is for making carbon–carbon bonds on a benzene ring. The mechanism for the electrophilic aromatic substitution of pyrrole (**Figure 20.3a**) is the same as it is for benzene—namely, slow addition, which breaks aromaticity, then rapid deprotonation to return the system to aromaticity. With pyrrole, these substitutions generally favor the 2-substituted isomer (**Figure 20.3b**).[3]

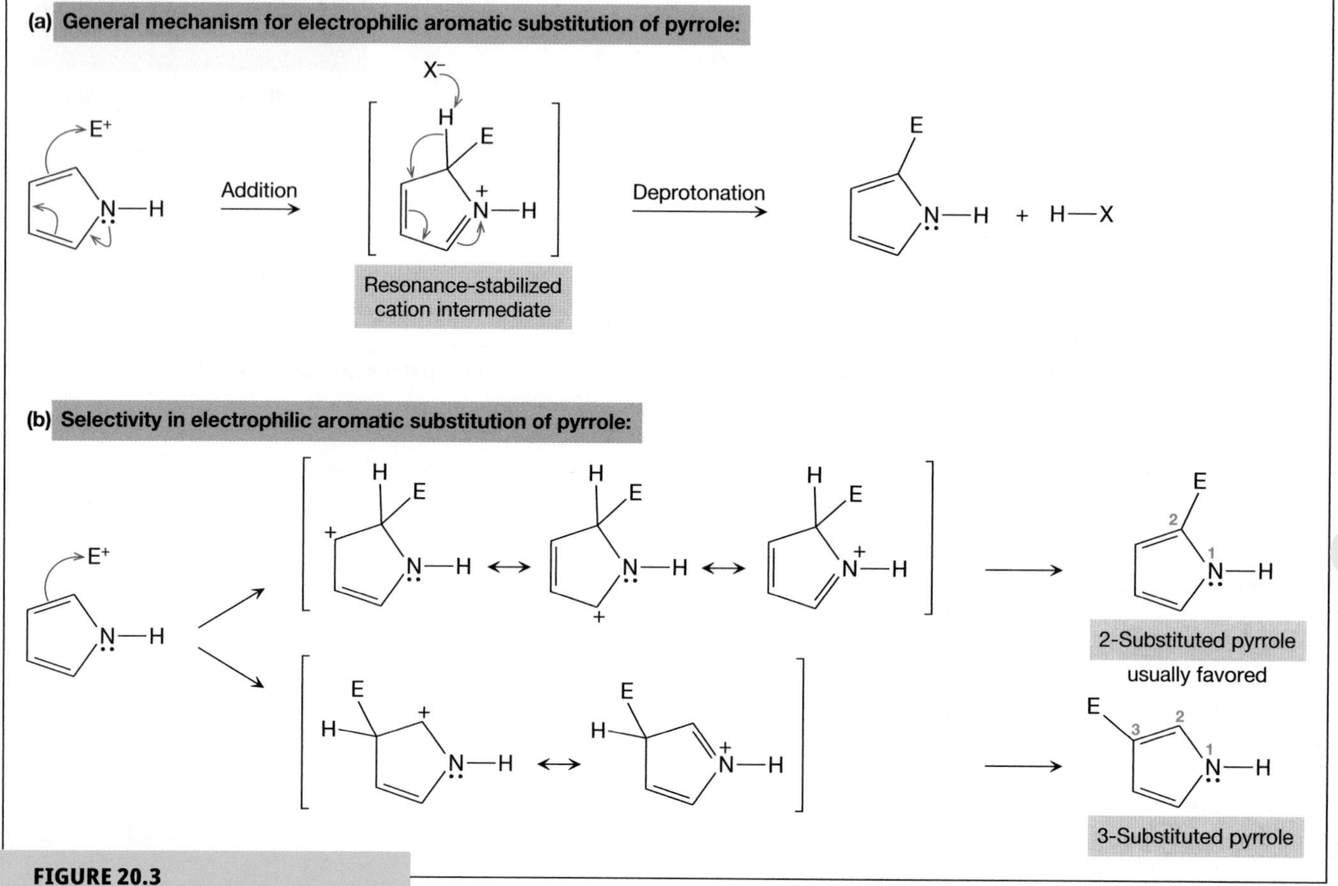

FIGURE 20.3

(a) Mechanism and (b) selectivity in electrophilic aromatic substitution of pyrrole. Note the difference in resonance stabilization for the two alternative intermediates leading to 2- and 3-substituted pyrrole isomers.

One method for the synthesis of porphyrins involves electrophilic aromatic substitution of pyrrole using various aldehydes, such as benzaldehyde, as the electrophile (**Figure 20.4**). We use this reaction to make tetraphenylporphyrin, a compound that has been used to make photoelectrodes for harvesting solar energy.[4] We start with benzaldehyde, a naturally occurring compound found in almonds, and pyrrole, which is one of the main components of a substance called bone oil, obtained from distilling charred bones.

[3]Tsuchimoto, T. Selective Synthesis of β-Alkylpyrroles. *Chem. Eur. J.* **2011**, *17*, 4064–4075.
[4]Durantini, E. N.; Otero, L. Solar Energy Conversion Using a Semiconductor Electrode Photosensitized by Tetraphenylporphyrin. *Chem. Educ.* **1999**, *4*, 144–146.

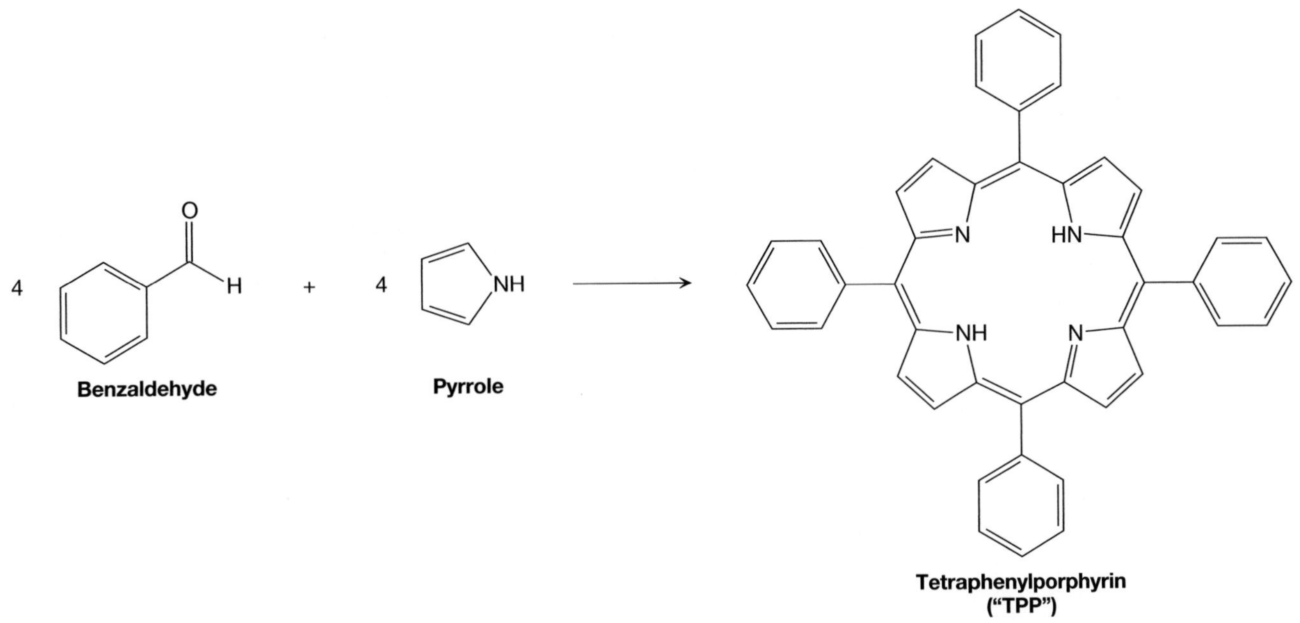

4 Benzaldehyde + 4 Pyrrole → Tetraphenylporphyrin ("TPP")

FIGURE 20.4

The reaction used in this lab to produce tetraphenylporphyrin from benzaldehyde and pyrrole.

Tetraphenylporphyrin has previously been prepared from benzaldehydes and pyrrole along with hazardous solvents such as pyridine or dichloromethane, and corrosive acids such as boron trifluoride or trifluoroacetic acid. In solution reactions, the acid catalysts are needed to protonate benzaldehyde (**Figure 20.5**), making a strong electrophile that could function in an electrophilic aromatic substitution. Further hazardous materials have often been used for the oxidation in the last stage of the synthesis.

In this experiment, we carry out the reaction in the gas phase at a temperature over 200°C—conditions that require no additional reagents and no solvent.[5] The oxygen naturally present in air is used as the oxidant. Thus, the environmental impact of the synthesis is considerably lower because the use of hazardous materials and solvents has been minimized, congruent with some of the 12 principles of green chemistry.

The mechanism for this reaction in the gas phase follows a path similar to that shown in Figure 20.5. Electrophilic aromatic substitution occurs eight times in sequence, using either protonated benzaldehyde or a carbocation generated by dehydration. The end result is that the four pyrroles are connected together with four intervening carbons derived from benzaldehyde. Although the source of acid needed to catalyze this reaction is unclear, there are likely small amounts of benzoic acid and water present that could serve this purpose. After the pyrroles and benzaldehyde units are connected, oxidation then completes the 18-electron aromatic ring system characteristic of porphyrins. The product can then be purified by column chromatography.

---

[5]Warner, M. G.; Succaw, G. L.; Hutchison, J. E. Solventless Syntheses of Mesotetraphenylporphyrin: New Experiments for a Greener Organic Chemistry Laboratory Curriculum. *Green Chem.* **2001**, *3*, 267–270.

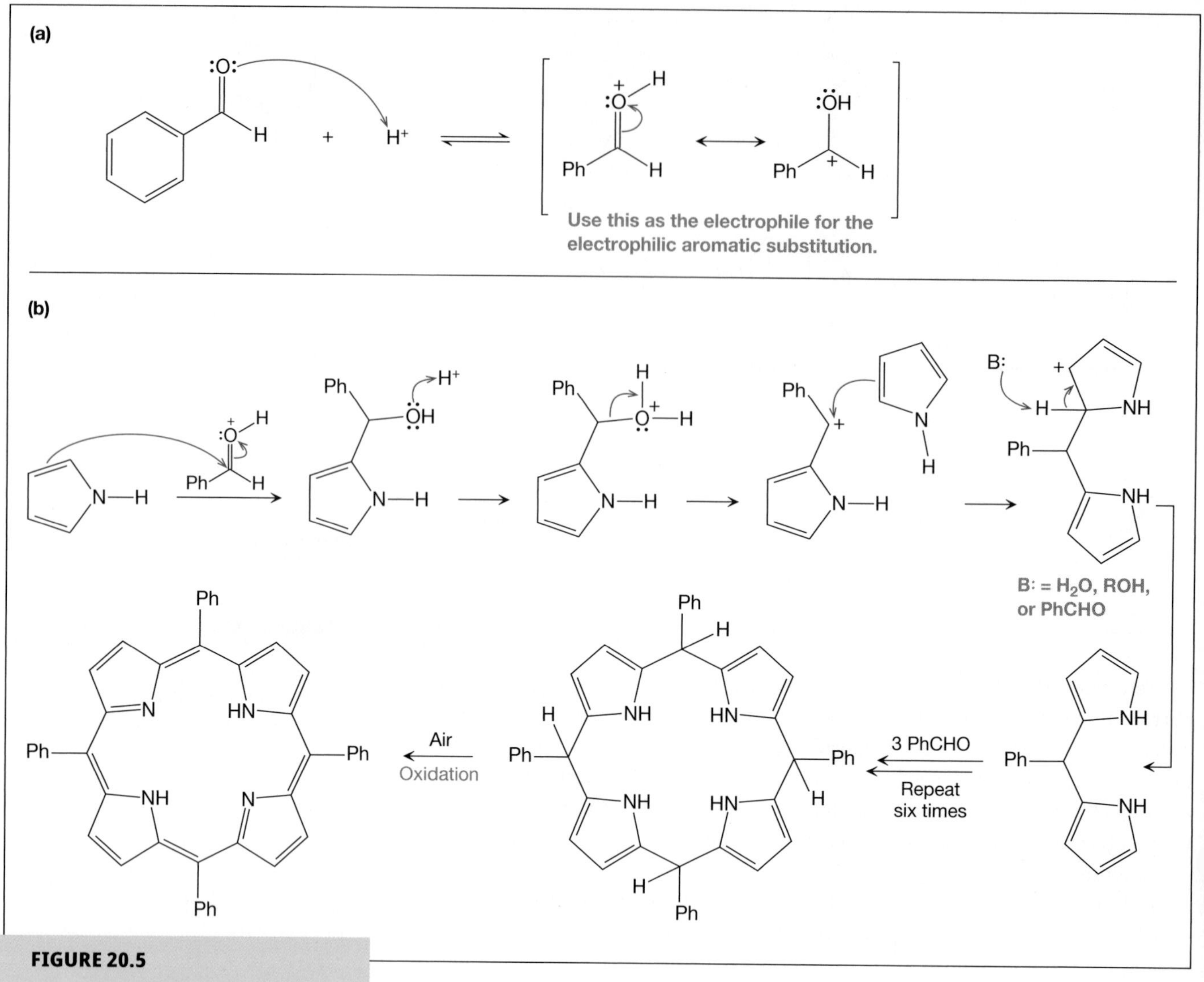

FIGURE 20.5

(a) Protonation of benzaldehyde makes a strong electrophile.
(b) The reaction to produce tetraphenylporphyrin involves repeated electrophilic aromatic substitutions, followed by oxidation by air.

20.2 EXPERIMENTAL PROCEDURE[6]

20.2A The Reaction

1. Place sand in a heating mantle to a depth of 2–3 cm, and turn on the heating mantle to medium-high heat. Immerse a septum-capped 5-mL conical vial into the sand bath to a depth of approximately 2 cm (**Figure 20.6a**), and monitor the temperature of the sand bath.

   Note: While a sand bath is described here, alternative heat sources may be used, such as a heated metal block with a well that can hold a vial.

[6]Doxsee, K. M.; Hutchison, J. E. *Green Organic Chemistry*; Brooks/Cole: Pacific Grove, CA, 2004.

2. When the heat source reaches 170°C, lower the heat to a setting that allows the temperature to continue rising slowly. Use a syringe to inject 10 microliters (μL) of benzaldehyde into the vial through the septum cap (**Figure 20.6b**).

3. As the temperature continues to rise to around 180°C, droplets of benzaldehyde form on the upper walls of the vial. Now, use a syringe to inject 7 microliters of pyrrole into the vial through the septum cap.

4. Continue heating, increasing the heat setting if needed, until the temperature has been above 250°C for about 15 minutes (**Figure 20.6c**).

(a)

(b)

(c)

FIGURE 20.6

(a) Preheating the reaction vial with a sand bath. (b) Adding the reagents via syringe. (c) Dark-colored material appears on the walls of the vial as the reaction proceeds.

Courtesy of Gregory K. Friestad.

5. Remove the vial from the heat and allow it to cool to room temperature.

6. When the vial has cooled to room temperature, add 1 mL of dichloromethane (CH_2Cl_2) to the vial, using the solvent to rinse the walls and cap liner.

7. Using thin-layer chromatography (TLC) on silica gel plates, analyze the product mixture. Spot benzaldehyde, pyrrole, and the product mixture on the same plate, then elute the plate with 7:1 petroleum ether/ethyl acetate (PE/EA). Tetraphenylporphyrin appears as a violet-colored spot at R_f 0.46, with impurities at lower R_f values.

 Thin-Layer Chromatography

20.2B Purification and Analysis

8. Purify the product mixture using column chromatography (Chapter 5, section 5.5). Throughout the column chromatography procedure, the level of solvent in the column should always be above the top of the silica gel—never let the column run dry.

a. *Pack the column.* Securely clamp a column of about 2 cm in diameter to the bars in the fume hood (or to a ring stand), close the stopcock, and plug the inside end of the stopcock tube with a loose plug of cotton (**Figure 20.7a**). Add about 1 cm of sand in a flat layer, followed by about 5 mL 7:1 PE/EA. Swirl together about 14 g of silica gel and 50 mL 7:1 PE/EA until a consistent slurry is obtained, and gently pour it in the column without disturbing the sand layer at the bottom. Open the stopcock and allow the solvent to drain into a clean Erlenmeyer flask as the silica gel settles. When the level of the silica gel stops settling, the height of the column should be about 15 cm. If not, add more of the slurry and allow it to settle again. Gently add about 1 cm of sand, allowing it to fall evenly

 Column Chromatography

through the solvent and settle in a flat layer on top of the silica gel. Allow the solvent to drain until its level falls just below the top of the sand. Close the stopcock. The prepared column is shown in **Figure 20.7b**. *Note:* The solvent that has drained out to this point is clean and can be reused. Be sure to change receiving flasks at this point.

(a)

(b)

FIGURE 20.7

Setup for column chromatography, (a) before adding the slurry of silica gel and solvent. (b) Prepared column with silica gel and solvent. The solvent will be drained to just below the level of the top layer of sand before adding the sample.

Courtesy of Gregory K. Friestad.

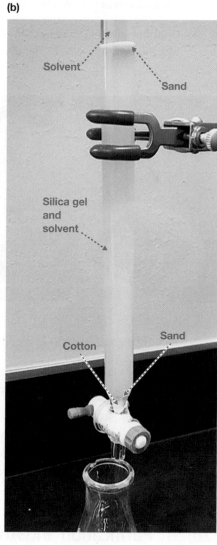

b. *Load the sample.* Using a pipet, add the product mixture solution from the vial to the column. Rinse the vial with a 0.5-mL portion of 7:1 PE/EA. Add this to the column, rinsing down any product mixture that may be on the inner walls of the column. Open the stopcock and allow the solvent to drain until its level falls just below the top of the sand. Close the stopcock and repeat the rinse with another 0.5-mL portion of 7:1 PE/EA. Open the stopcock and allow the solvent to drain until its level falls just below the top of the sand.

c. *Elute the column.* Using a pipet at first, carefully add about 5 mL 7:1 PE/EA without disturbing the sand. Then pour another 40 mL of 7:1 PE/EA into the column. Open the stopcock and allow the solvent to flow (elute the column). A purple band of tetraphenylporphyrin will migrate down the column (**Figure 20.8a**), and when it starts to emerge, begin collecting the solution (eluent) in a 100-mL graduated cylinder. After the purple band is collected, remove the graduated cylinder and record the volume of the purple tetraphenylporphyrin solution. Place a different container under the column and allow the rest of the solvent to drain out.

   d. *Analyze by TLC.* The purple fraction should be analyzed by TLC in the same way as before. Comparing against starting materials is unnecessary this time; you will just confirm whether the impurities have been removed. If so, there should be a single purple spot.

9. Obtain an ultraviolet–visible (UV–vis) spectrum of the tetraphenylporphyrin. Basic instructions are provided here, and your lab instructor may add further instructions that are specific to the instrument you have in your lab.

     Calibrate using exactly 3.0 mL 7:1 PE/EA as the blank solution in a glass cuvette. Use a plastic pipet to add one drop of your purple tetraphenylporphyrin solution to the cuvette. Locate the λ_{max}—the wavelength of maximum absorbance—in the UV spectrum; it should be around 410–420 nm (**Figure 20.8b**). While carefully monitoring the absorbance (*A*) at that wavelength, add more tetraphenylporphyrin solution dropwise, counting the drops carefully, until the maximum absorbance reads approximately 1.0. Record the exact *A* value at λ_{max}, and record the number of drops of tetraphenylporphyrin solution you added. Print the UV spectrum.

     The UV spectrum is used to calculate the concentration of product (instructions are given in section 20.3). It can also provide evidence to confirm the identity of the product (does the λ_{max} match the literature data?) and assess its purity (are there other large peaks present that have a different λ_{max}?).

(a)

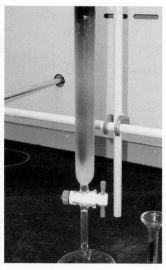

(b)

FIGURE 20.8

(a) Eluting the chromatography column. When the purple color begins to emerge from the stopcock at the bottom of the column, collect the purple eluent in a graduated cylinder. (b) Identifying the λ_{max} of tetraphenylporphyrin in the UV spectrum on-screen.

Courtesy of Gregory K. Friestad.

20.3 THE LABORATORY REPORT

20.3A | Report Content

In your report, comment on the percent yield of the reaction. Was the reaction successful? Did you get the expected product? How did you identify the product? Compare the TLC data before and after column chromatography, and comment on the effectiveness of the purification. Attach the UV–visible spectrum of your product.

20.3B Calculation Instructions

Calculate the moles of tetraphenylporphyrin in the UV–visible spectrometer cuvette using Beer's law, which relates UV–visible absorbance (A) to concentration (c):

$$A = \varepsilon \cdot b \cdot c$$

In this equation, b is the path length through the sample and ε is the extinction coefficient, also called molar absorptivity, which is characteristic for each individual compound at a given wavelength of absorbance. The following are some useful data for your calculation:

- The strongest tetraphenylporphyrin absorbance (λ_{max} = 410–420 nm) has the extinction coefficient $\varepsilon = 4.7 \times 10^5$ M^{-1} cm^{-1}.
- The path length of the cuvette is $b = 1.0$ cm.
- One drop of solution from a plastic 1-mL graduated pipet is 0.013 mL.[7]
- The volume of solution in the cuvette is 3.0 mL + (0.013 mL × number of drops).

Beer's law gives you the concentration in the cuvette (moles per liter), and you know the volume you put in the cuvette (liters), so you should be able to calculate the moles of tetraphenylporphyrin in the cuvette. From this, you can then calculate the concentration of the tetraphenylporphyrin solution you collected from the column chromatography (before it was diluted in the cuvette), the total amount of tetraphenylporphyrin produced in the reaction (in moles), and the percent yield.

When you do this calculation, remember to account for the dilution—there was blank solvent in the cuvette (3.0 mL) before you began adding drops of your tetraphenylporphyrin (TPP) solution.

20.3C Data to Include in the Laboratory Report

A. THE REACTION

1. Amount of benzaldehyde added (μL):

2. Amount of pyrrole added (μL):

B. PURIFICATION AND ANALYSIS

1. Volume of TPP solution collected from column chromatography (mL):

2. Number of drops of TPP solution added to cuvette:

3. Wavelength of λ_{max} (nm):

4. Absorbance at λ_{max}:

---

[7]There may be some variation in volume of drops depending on the supplier of the pipets, so the amount per drop should be verified with blank solvent by measuring the number of drops required to reach 1.0 mL in a 10-mL graduated cylinder.

LEARNING OBJECTIVES

- Recognize the role of organic chemistry in communications between organisms.

- Execute a two-step synthesis involving stereoisomers formed during C — C bond construction.

- Demonstrate appropriate handling of anhydrous Grignard reagents and diethyl ether.

- Determine adjustments to amounts of reagents at different reaction scales.

- Use catalysis in aqueous media to lessen the environmental impact of an oxidation reaction.

Two Insect Pheromones: Grignard Synthesis and Green Oxidation Methods

INSECT ATTRACTANTS

This damaged elm tree is infected with Dutch elm disease, caused by a fungus spread by elm bark beetles. These beetles aggregate in response to a pheromone, 4-methyl-3-heptanol.

Winston Fraser/Alamy Stock Photo.

U p to one-third of worldwide food production is destroyed by insects, and decades of insecticide use have not led to a sustainable decrease in the abundance of the pest populations.[1] Insects communicate using organic chemistry: They use chemicals called pheromones to send signals that control mating behavior, recruit other individuals to food sources, or alert to danger. In 1959, the first insect sex pheromone (bombykol, **Figure 21.1**) was identified,[2] and the concept of "integrated pest management" was proposed soon after: Could agriculture harness the chemical communication language of insects as part of an integrated approach to control pest populations?

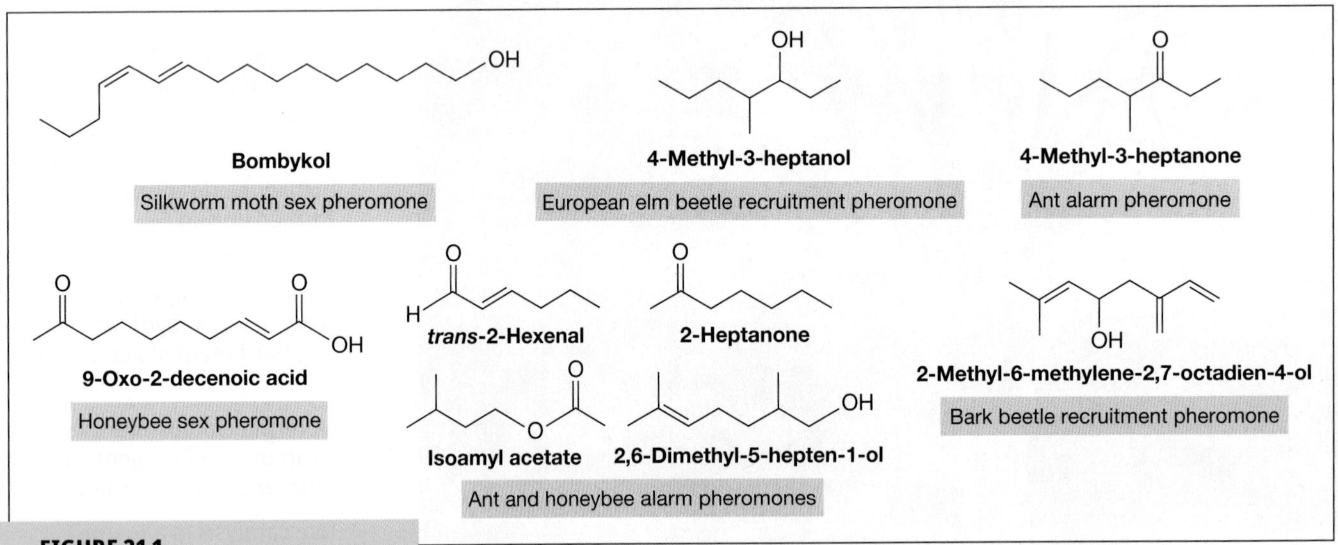

Bombykol
Silkworm moth sex pheromone

4-Methyl-3-heptanol
European elm beetle recruitment pheromone

4-Methyl-3-heptanone
Ant alarm pheromone

9-Oxo-2-decenoic acid
Honeybee sex pheromone

***trans*-2-Hexenal**

2-Heptanone

Isoamyl acetate   **2,6-Dimethyl-5-hepten-1-ol**
Ant and honeybee alarm pheromones

2-Methyl-6-methylene-2,7-octadien-4-ol
Bark beetle recruitment pheromone

FIGURE 21.1

Examples of insect pheromones and their functions.

Two insect pheromones, 4-methyl-3-heptanol and 4-methyl-3-heptanone, are prepared in this experiment, using Grignard synthesis and oxidation.

Pre-Lab Reading Assignment

Review the following before attempting this experiment:

- Experiment: Chapter 21 (this experiment)
- Experiment: Multistep synthesis: cyclohexanol to adipic acid (Chapter 16)
- Technique: Distillation (Chapter 5, section 5.3)
- Technique: Thin-layer chromatography (TLC) (Chapter 5, section 5.5)
- Technique: Infrared (IR) spectroscopy (Chapter 6, section 6.2)
- Technique: Nuclear magnetic resonance (NMR) spectroscopy (Chapter 7, section 7.1)

[1]Witzgall, P.; Kirsch, P.; Cork, A. Sex Pheromones and Their Impact on Pest Management. *J. Chem. Ecol.* **2010**, *36*, 80–100.
[2]Butenandt, A.; Beckmann, R.; Stamm, D.; Hecker, E. Über den Sexual-Lockstoff des Seidenspinners Bombyx mori. Reindarstellung und Konstitution. *Z. Naturforsch.* **1959**, *14b*, 283–284.

21.1A | Pheromones

Pheromones are naturally occurring compounds that are released externally by organisms to act as chemical signaling agents between different individuals. These are distinct from hormones, which are released internally to send chemical signals within the individual. Communication between insects commonly involves compounds that can be categorized as sex pheromones, recruitment pheromones, or alarm pheromones (Figure 21.1).[3]

There are two main types of pheromones, called releasers and primers. *Releaser pheromones* elicit an immediate behavioral response, such as ants recruiting other colony members to a food source, or moths attracting other individuals for mating. *Primer pheromones* elicit physiological changes, which may result in observed behaviors after some delay. For example, recently mated female mice fail to become pregnant when they are exposed to odors of a foreign male. Instead, a primer pheromone blocks implantation of the fertilized eggs, and many of the females then become receptive to mating again within a week. A pheromone used by the queen honeybee, 9-oxo-2-decenoic acid (Figure 21.1), suppresses the reproductive functions of all the other females in the hive, which exhibit atrophied ovaries (primer pheromone function).

Because of their remarkable effects on insect behavior, some pheromones have generated commercial interest as tools for environmentally friendly pest management, such as by disrupting insect mating or by luring insects to traps. But identifying and acquiring the compounds can be challenging because they're found in very small quantities in individual insects. To extract just a few milligrams of an insect pheromone for structure determination, hundreds of thousands or even millions of insects may be required. Once organic chemists have determined the structure, they can design an economically feasible laboratory synthesis that can provide sufficient quantities for pest management programs.

Two different insect pheromones can be prepared by a two-step synthesis involving a Grignard addition reaction followed by oxidation of the resulting alcohol to a ketone. The first reaction furnishes 4-methyl-3-heptanol, an aggregation or recruitment pheromone of the European elm beetle. Oxidation of this compound produces 4-methyl-3-heptanone, an alarm pheromone used by various ant species.

21.1B | The Grignard Reaction

The Grignard reaction is a valuable method for forming carbon–carbon bonds, so it occupies an important place in organic synthesis. In the Grignard synthesis, alkyl or aryl halides are first converted to organomagnesium reagents, commonly referred to as Grignard reagents, through reaction with magnesium in anhydrous ether (**Figure 21.2**).

Grignard reagents are examples of organometallic compounds (compounds with bonds between carbon and metallic elements), and they have a highly polarized

---

[3]Regnier, F. E.; Law, J. H. Insect Pheromones. *J. Lipid Res.* **1968**, *9*, 541–551.

FIGURE 21.2

Preparation of a Grignard reagent, with depiction of its polarized carbon–magnesium bond. Because of its partial negative charge, the carbon atom is both nucleophilic and basic.

carbon–magnesium bond, which causes the reagent to behave as a carbanion. Because carbanions are among the strongest bases encountered in organic chemistry, Grignard reagents react with weak acids, including water. They must not be allowed to come into contact with moisture, including the humidity present in air.

The more useful feature of Grignard reagents is their behavior as carbon nucleophiles. This allows them to form carbon–carbon bonds by reacting with carbon electrophiles such as ketones, aldehydes, and esters (**Figure 21.3**). Through the formation of carbon–carbon bonds, simple **precursors** can be used to generate a larger or more complex carbon framework.

precursors >>
A synonym for reactants.

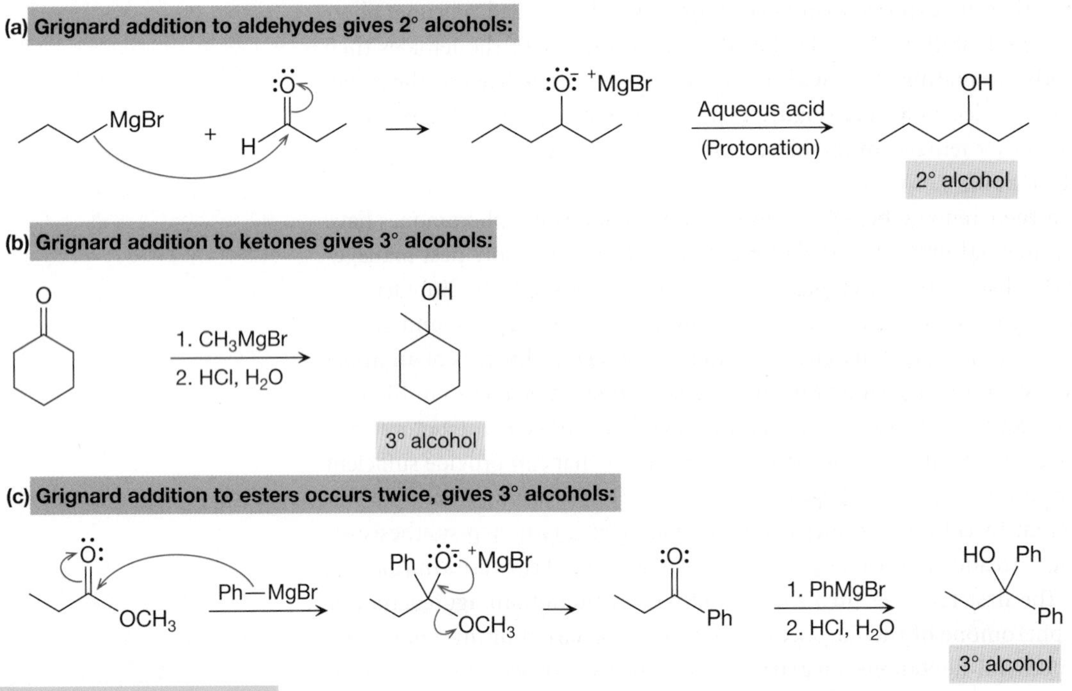

(a) Grignard addition to aldehydes gives 2° alcohols:

(b) Grignard addition to ketones gives 3° alcohols:

(c) Grignard addition to esters occurs twice, gives 3° alcohols:

FIGURE 21.3

(a) Nucleophilic addition of a Grignard reagent to an aldehyde. The product is an alkoxide ion, with its charge balanced by a magnesium bromide counterion. When the reaction is completed, dilute aqueous acid is added to the reaction mixture to protonate the alkoxide ion, resulting in a secondary alcohol. (b) Example of a Grignard reaction with a ketone, affording a tertiary alcohol. (c) Addition of a Grignard reagent to an ester occurs twice; after the first addition, an elimination of an alkoxide ion gives a ketone, which reacts as in part (b) to give a tertiary alcohol.

In this experiment, the simple unbranched carbon chains of an aldehyde (propanal) and an alkyl halide (2-bromopentane) are coupled to generate a secondary alcohol (4-methyl-3-heptanol, **Figure 21.4**). Two new stereogenic centers are formed, and thus a mixture of stereoisomers will be present.

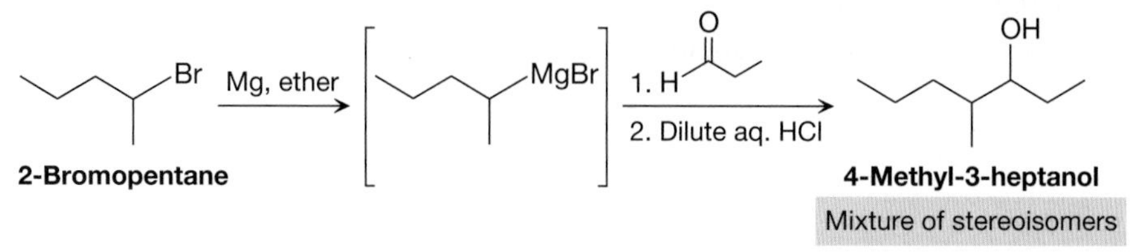

FIGURE 21.4

Grignard reaction affording the insect pheromone 4-methyl-3-heptanol.

21.1C Methods for Greener Oxidations

Oxidation reactions are frequently used to adjust functional groups within multistep syntheses. In recent years, much attention has been focused upon making these reactions more environmentally friendly. In oxidations of alcohols to carbonyl compounds, chromium(VI) reagents (such as $CrO_3 + H_2SO_4$ in the Jones oxidation) have long been used. These require stoichiometric amounts of the metal reagent, which generates toxic heavy metal waste in an amount similar to that of the desired product. Newer methods use smaller amounts of a heavy metal—usually as a catalyst in combination with stoichiometric amounts of safer oxidants. The stoichiometric oxidant can allow the heavy metal catalyst to be recycled many times during the course of the reaction, thereby minimizing the heavy metal waste generated and lowering the environmental impact of the oxidation reaction. One such combination is perruthenate ion (RuO_4^-) and a tertiary amine *N*-oxide, which usually employs dichloromethane as solvent.[4]

A choice of two oxidation methods is provided in this experiment. One is the use of sodium hypochlorite (NaOCl, household bleach) under aqueous conditions. The other is tungstate ion (WO_4^{2-}) and hydrogen peroxide. Because both of these reactions occur in aqueous media, the use of organic solvents can be minimized, diminishing the environmental impact. Tungstate ion serves as a catalyst which is not consumed; thus it can be used in a very small amount that is recycled during the reaction, decreasing the amount of tungsten waste that is generated. The water-soluble tungstate ion is used in conjunction with a phase-transfer catalyst, which allows this catalyst to pass between the organic and aqueous phases.

During the sodium hypochlorite oxidation (**Figure 21.5a**), hypochlorite (OCl^-) is converted to chloride (Cl^-), producing table salt (NaCl) as a relatively safe byproduct. Additionally, the aqueous reaction conditions reduce the amount of volatile organic compounds used as solvents, minimizing air pollution and flammability hazards. In the mechanism of this reaction, the oxidant NaOCl causes a good leaving group to form at the oxygen of the alcohol. Then removing a proton from the carbon that will become the carbonyl carbon causes the electrons in the O—Cl bond to reduce the chlorine atom to Cl^-. This mechanism is common to many alcohol oxidations.

A quite different mechanism is observed in tungstate oxidation of alcohols (**Figure 21.5b**). In this case the electrons that are transferred to the oxidant are taken from the C—H bond, while the electrons from the O—H form the C═O π

[4]For a review of perruthenate oxidation, see Ley, S. V.; Norman, J.; Griffith, W. P.; Marsden, S. P. Tetrapropylammonium Perruthenate, $Pr_4N^+RuO_4^-$, TPAP: A Catalytic Oxidant for Organic Synthesis. *Synthesis* **1994**, 639–666.

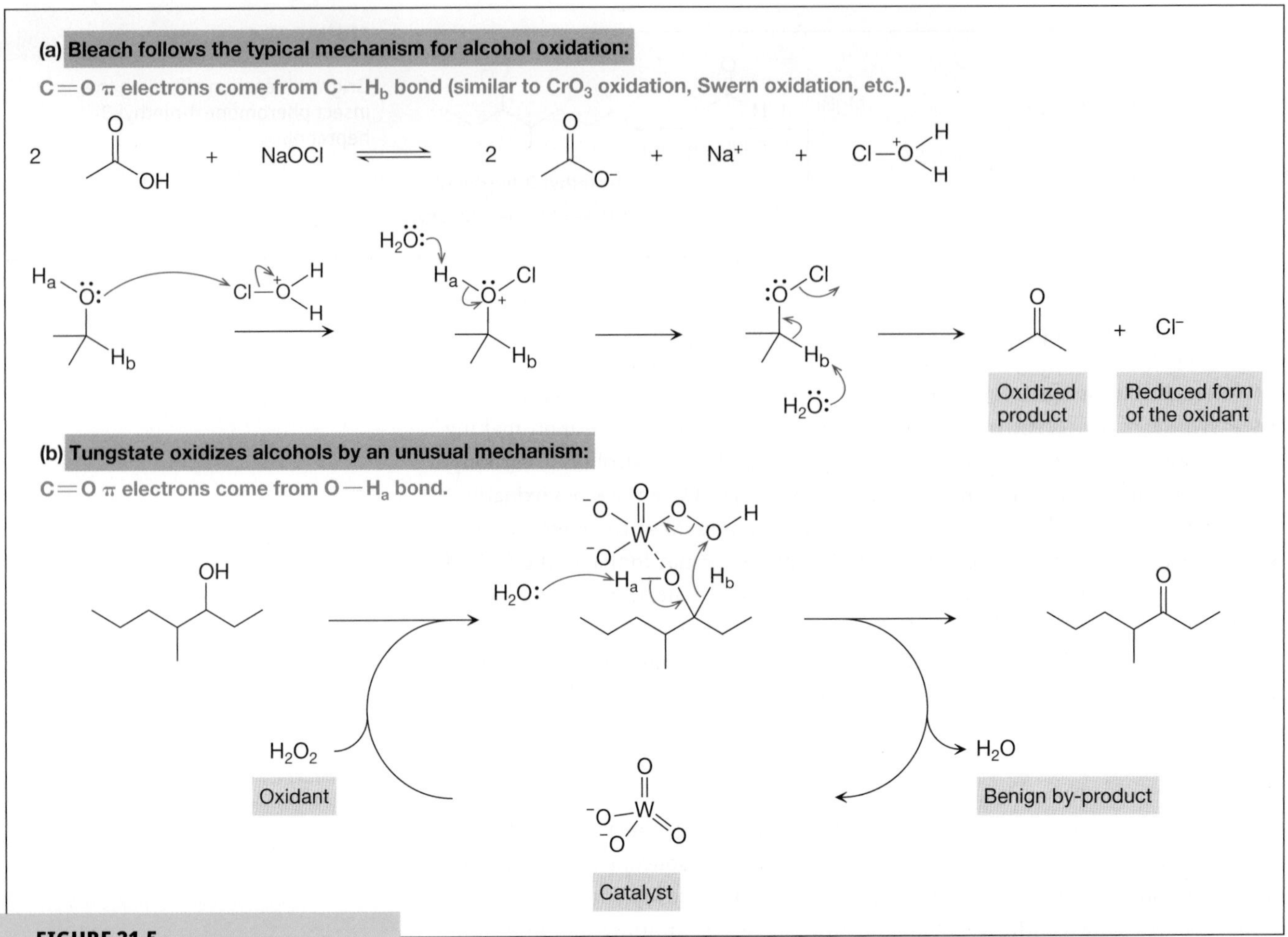

(a) Bleach follows the typical mechanism for alcohol oxidation:

C=O π electrons come from C—H_b bond (similar to CrO_3 oxidation, Swern oxidation, etc.).

(b) Tungstate oxidizes alcohols by an unusual mechanism:

C=O π electrons come from O—H_a bond.

FIGURE 21.5

(a) Alcohol oxidation mechanism for bleach. (b) Plausible catalytic cycle and mechanism of oxidation for tungstate ion. Most alcohol oxidations proceed via a base removing H_b, but for this reagent it has been proposed that H_a is removed by the base (water, in this example).[5] In this mechanism, the H_2O_2 is reduced by cleavage of the O—O bond, so the metal ion doesn't require an oxidation state change. This may allow for more efficient recycling of the metal ion, or higher "turnover number" in the terminology of catalysis.

bond. This is not the usual mechanism for the oxidation of alcohols, but it's worth developing new reaction types with unusual mechanisms when there are significant green benefits. Here the metal oxidant is used in very small amounts because it is recycled in the reaction, and the only by-product is water.

Procedures for both the tungstate-catalyzed oxidation and the hypochlorite oxidation are provided in this experiment to convert 4-methyl-3-heptanol (a secondary alcohol) to 4-methyl-3-heptanone (a ketone) (**Figure 21.6**). Your lab instructor will assign which oxidation procedure you should use.

FIGURE 21.6

Synthesis of two insect pheromones: One pheromone is a secondary alcohol resulting from a Grignard addition reaction (see Figure 21.4) and is converted to a second pheromone by oxidation of the secondary alcohol to a ketone.

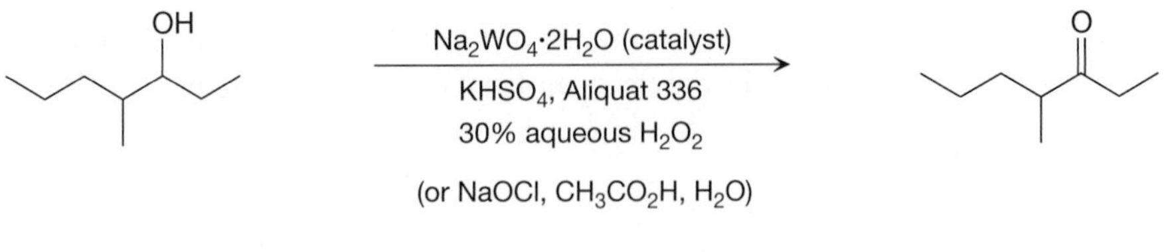

[5]The mechanism proposed here for the tungstate-catalyzed peroxide oxidation of alcohols is supported by the finding that a methyl ether is also oxidized, suggesting there is no requirement to form a tungstate ester analogous to chromate ester intermediates in the Cr(VI) oxidations. Venturello, C.; Ricci, M. Oxidative Cleavage of 1,2-Diols to Carboxylic Acids by Hydrogen Peroxide. *J. Org. Chem.* **1986**, *51*, 1599–1602.

21.2A ## Part 1: Grignard Synthesis of
4-Methyl-3-heptanol[6] (Day 1)

1. Load a drying tube with fresh anhydrous calcium sulfate (Drierite). Securely attach a strong clamp to the neck of a dry 100-mL round-bottom flask. *Note:* If there is moisture in the flask, the reaction will likely fail. If a drier flask is available, change to a dry flask and proceed to step 2. Instructors may decide whether to permit students to flame-dry flasks, but in most cases it is not necessary. *CAUTION:* Students should *NEVER flame dry while diethyl ether is in use* because diethyl ether is a serious fire hazard. If you are instructed to flame-dry the flask, make sure there is no diethyl ether in use, then clamp the flask securely to a ring stand or other support and apply a Bunsen burner flame to all parts of the outside of the flask, continuously moving the flame around it for 20–30 seconds. Applying vacuum during this part is helpful, but not necessary. Allow the glass to cool for a few minutes. Do not touch the hot glass—remember, it looks exactly like cold glass.

2. Add magnesium turnings (2.4 g) to the flask. Immediately attach a dry reflux condenser on top of the flask and place the drying tube on top of the condenser.

3. Prepare a cooling bath of ice and water, for use at a later stage.

4. Measure 2-bromopentane (6.2 mL) into a vial with screw-cap closure. The 2-bromopentane should remain tightly capped except when removing portions. *CAUTION:* Do *not* proceed to the next step until you have confirmed that there are no open flames in the lab. Diethyl ether is highly flammable and must not be opened while Bunsen burners or other ignition sources are in use.

5. After the apparatus is cooled to room temperature, add a magnetic stir bar and 30 mL anhydrous diethyl ether (freshly purchased, without inhibitors), being careful to limit the exposure of the diethyl ether to air during measurement and transfer (±5 mL is okay). Vigorously stir the mixture for 5 minutes, then stop the stirrer. Attach water hoses to the condenser, and cautiously turn on the condenser cooling water to a very low flow rate (just a slow trickle is sufficient).

6. With the magnetic stirrer turned off, add about 1 mL of the 2-bromopentane through the top of the condenser, then replace the drying tube. Wait patiently, without stirring or swirling the mixture. After 5–20 minutes, the reaction will begin, causing slight cloudiness and a few bubbles to appear.

---

[6](a) This procedure was adapted from that of Einterz, R. M.; Ponder, J. W.; Lenox, R. S. The Synthesis of 4-Methyl-3-heptanol and 4-Methyl-3-heptanone. *J. Chem. Educ.* **1977**, *54*, 382. (b) This synthesis of 4-methyl-3-heptanol may also result in a minor amount of 4,5-dimethyloctane (ca. 20%) that is difficult to separate by distillation. Hoffman, R. V.; Alexander, M. D.; Buntain, G.; Hardenstein, R.; Mattox, C.; McLaughlin, S.; McMinn, D.; Spray, S.; White, S. A Reinvestigation of the Synthesis of 4-Methyl-3-heptanol. *J. Chem. Educ.* **1983**, *60*, 78. (c) A similar synthesis of 2-methyl-4-heptanol is reported to be less prone to this side reaction: de Jong, E. A.; Feringa, B. L. The Synthesis of 2-Methyl-4-heptanone. *J. Chem. Educ.* **1991**, *68*, 71–72.

Some hints to try if your Grignard reaction does not seem to start: Look for small bubbles emerging from around the magnesium, and some cloudiness, which are indications that the reaction has begun. If the reaction does not start within 15 minutes, crush the magnesium turnings with a clean, dry glass rod and add another 1-mL portion of 2-bromopentane, then let the reaction stand, unstirred, for 5 minutes. If it still doesn't start, add one small crystal of iodine and wait another 5 minutes. As a last resort, ask the lab instructor to help you add about 1 mL of Grignard reagent from someone else's reaction, noting the source in your notebook so that you can cite that in your report.

7. When a gentle boiling is underway and the bottom of the flask feels slightly warm, start the magnetic stirrer. The reaction may start to boil more vigorously. A gently boiling reflux, with the ether solvent condensing low in the condenser, is ideal, and will be maintained by the exothermic reaction. If the reaction appears to be refluxing too vigorously, so that liquid is being forced up into the condenser, briefly cool the bottom of the reaction flask with the ice-water bath you prepared in step 3. Slowly add the remaining 2-bromopentane through the top of the condenser in several small portions over 20 minutes, keeping the drying tube in place between additions. Continue stirring for 15 minutes after the addition is complete.

8. Cool the reaction flask with the ice-water bath, taking care to avoid getting water on the joints of the apparatus. Set the magnetic stirrer on a high speed. Using a pipet, slowly add propanal (2.9 mL, 0.040 mol) dropwise through the top of the condenser. *CAUTION: The propanal must be added drop by drop—this reaction is very vigorous!* If the reaction mixture begins to boil, stop adding the propanal and wait for the reaction to subside before continuing. After all the propanal has been added, continue the stirring for 5 minutes.

9. While stirring and cooling in the ice-water bath, use a pipet to add water (5 mL) dropwise. *CAUTION: Vigorous reaction!* Then, slowly add sufficient 10% aqueous HCl (around 10–20 mL) to make the mixture acidic (pH 2–4, as judged by pH paper); the mixture becomes clear. Decant the solution into a separatory funnel, leaving behind any excess magnesium metal, which can clog the stopcock. Shake briefly, drain off the lower aqueous layer, wash the organic layer with 5% NaOH (10 mL), and dry the organic layer over anhydrous Na_2SO_4. Concentrate using a rotary evaporator (with the water bath at room temperature), to a volume of about 5 mL—estimate this by comparing with another flask containing 5 mL water. The resulting unpurified product may be stored until the next lab meeting or, if time permits, it may be immediately purified by simple distillation.

▶ **Aqueous–Organic Extractions and Drying Organic Solutions**

▶ **Rotary Evaporation**

▶ **Simple Distillation**

21.2B Part 1: Grignard Synthesis of 4-Methyl-3-heptanol (Day 2)

10. Start heating a sand bath (or other heat source, such as a heating mantle or heated metal block) to about 180°C. Assemble a simple distillation apparatus. Place the alcohol in a 25-mL round-bottom flask with magnetic stir bar, attach it to the distillation apparatus, and apply heat. Distillate will begin to collect when the still head is about 40°C. When the temperature in the still head reaches 130°C, change to a clean, tared collecting flask and collect a second fraction. Stop heating just before the distillation flask is

empty. (*CAUTION:* Never distill to dryness!) Record the actual temperature ranges in which the two fractions of distillate were collected. Measure the mass of the second fraction, which contains the desired alcohol product.

11. Characterize the product. Obtain the infrared spectrum, and place 3 drops of this sample into your NMR tube. Follow your instructor's advice on further steps to prepare the NMR sample, then submit for acquisition of the NMR spectrum. Store the remainder of the sample in a labeled vial, tightly capped.

▶ **Preparing NMR Samples**

21.2C   Part 2: Oxidation

Two alternative oxidation procedures are provided here. Your lab instructor will tell you which one to use.

Both procedures are described for reactions employing 2 g (15 mmol) of the 4-methyl-3-heptanol reactant. Your amount of this reactant may be different. Adjust the scale of all the reaction components (i.e., the amounts of all reagents and solvents) accordingly.

21.2D   Part 2: Oxidation with Sodium Hypochlorite[7] (Day 3)

Skip the four steps of section 21.2D if you have been assigned the tungstate-catalyzed oxidation in section 21.2E.

Adjusting scale: If a student was unable to obtain the 4-methyl-3-heptanol, a larger sample (3–5 g) from another student may be divided into two portions so that both students may proceed. The source of this material must be cited in the lab notebook and in the final report.

1. Warm a sand bath or other heat source to 70°C. Measure the mass of the capped vial of distilled 4-methyl-3-heptanol, then transfer the sample into a clean 100-mL round-bottom flask, leaving behind a drop or two of the sample; this will be used later for TLC and IR analysis. Measure the mass of the capped vial after transfer. By subtraction, calculate the mass of 4-methyl-3-heptanol used (the difference of the two masses).

2. To the reaction flask containing 4-methyl-3-heptanol, add glacial acetic acid (4.0 mL, or 0.27 mL per mmol of alcohol), 5% aqueous sodium hypochlorite (60 mL, or 4 mL per mmol of alcohol), and a magnetic stir bar. Stir the mixture while applying heat at 70°C.

3. After 30 minutes, allow the reaction mixture to cool to room temperature, then add 10 mL saturated aqueous sodium thiosulfate solution. Pour the mixture into a separatory funnel. Rinse the flask with 10 mL diethyl ether, paying particular attention to the precautions for its use given earlier, transferring this to the separatory funnel also. Shake briefly, then separate the organic phase and dry over anhydrous Na_2SO_4.

4. Remove the diethyl ether by rotary evaporation to afford the ketone product. Transfer the product to a tared vial and measure its mass. Label the vial and store until the next lab period. (You may also store the organic phase prior to rotary evaporation.)

---

[7]This procedure was adapted from Williamson, K. L. *Macroscale and Microscale Organic Experiments*, 3rd ed.; Houghton Mifflin, 1999; pp 301–314.

Part 2: Tungstate-Catalyzed Oxidation with Hydrogen Peroxide[8] (Day 3)

Skip the seven steps of section 21.2E if you have been assigned the oxidation with sodium hypochlorite in section 21.2D.

Adjusting scale: If a student was unable to obtain the 4-methyl-3-heptanol, a larger sample (3–5 g) from another student may be divided into two portions so that both students may proceed. The source of this material must be cited in the lab notebook and in the final report.

1. Heat a sand bath (or other heat source, such as a heating mantle or heated metal block) at a medium heat setting, to reach a temperature of about 90°C.

2. Place about 0.25 g Aliquat 336 into a clean 100-mL round-bottom flask. This is a sticky liquid that is best transferred with a pipet having the narrowest part of the tip removed to make a larger opening. It is not necessary to have exactly 0.25 g of this component.

3. Add 0.25 g of sodium tungstate dihydrate ($Na_2WO_4 \cdot 2H_2O$) and a stir bar.

4. Add 2.0 mL of 30% aqueous hydrogen peroxide and 0.19 g $KHSO_4$. Start the stirrer.

5. Measure the mass of the capped vial of distilled 4-methyl-3-heptanol, then transfer the sample into the reaction flask, leaving behind a couple of drops of the sample (to be used later for TLC and IR analysis). Measure the mass of the capped vial after transfer. By subtraction, calculate the mass of 4-methyl-3-heptanol used.

6. Attach a water-cooled condenser to the flask. Heat the mixture at reflux with vigorous stirring for 1.5–2 h. Allow the reaction mixture to cool to room temperature, then add 10 mL saturated aqueous sodium thiosulfate solution. Pour the mixture into a separatory funnel. Rinse the flask with 10 mL diethyl ether, then transfer this rinsing to the separatory funnel. Shake briefly, then separate the organic phase and dry over anhydrous Na_2SO_4.

7. Remove the diethyl ether by rotary evaporation to afford the ketone product. Transfer the product to a tared vial and measure its mass. Label the vial and store until the next lab period. (You may also store the organic phase prior to rotary evaporation.)

Part 3: Short-Path Distillation, Boiling Point, and Thin-Layer Chromatography (Day 4)

8. Heat a sand bath (or other heat source, such as a heating mantle or heated metal block) to a temperature of about 180°C.

9. Using a pipet, place 0.5–1.0 g of the crude ketone into the bottom of a Hickman still (**Figure 21.7**), taking care to prevent contaminating the distillate collection area; the remaining crude ketone should be retained

Heating at Reflux

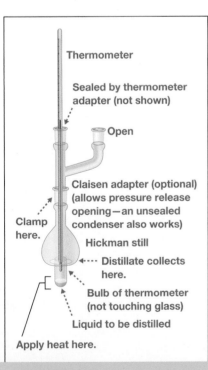

FIGURE 21.7

A Hickman still for small-scale distillation. Before heating, make sure there is an opening for pressure release.

Thermometer

Sealed by thermometer adapter (not shown)

Open

Claisen adapter (optional) (allows pressure release opening—an unsealed condenser also works)

Clamp here.

Hickman still

Distillate collects here.

Bulb of thermometer (not touching glass)

Liquid to be distilled

Apply heat here.

---

[8]This procedure was adapted from Hulce, M.; Marks, D. W. Organic-Solvent-Free Phase-Transfer Oxidation of Alcohols Using Hydrogen Peroxide. *J. Chem. Educ.* **2001**, *78*, 66.

as it will be needed at a later stage. Add a boiling stone. If used, attach a Claisen adapter, and place a thermometer in the straight neck so that the bulb is just below the distillate collection level. Leave the other neck open (*CAUTION:* Do not heat a sealed apparatus!). When the heat source reaches at least 180°C, apply it to the Hickman still. Distill until only a drop of liquid remains, then remove the heat source (*CAUTION: Never* distill to dryness!). Record the boiling point as a temperature range. After the apparatus cools, use a pipet to transfer the distillate to a tared vial and measure the mass. This sample will be used for TLC, IR, and NMR analysis; only a few drops of distillate is needed for this.

▶ **Simple Distillation**

10. Obtain the infrared spectra of both the undistilled and distilled ketone. Using about 3 drops of the distilled ketone to prepare the sample, acquire the ^{1}H NMR spectrum of the purified ketone.

11. Place two drops each of the undistilled and distilled ketones into separate vials and dilute with 10 drops diethyl ether. Add 10 drops of diethyl ether to the vial containing the reactant alcohol (4-methyl-3-heptanol) also. Perform TLC analysis (eluent 3:1 hexane:ethyl acetate, developed with iodine), comparing the reactant alcohol, undistilled ketone, and distilled ketone, all spotted on the same plate.

▶ **Thin-Layer Chromatography**

21.3 THE LABORATORY REPORT

In your results and discussion section, you should address the following questions.

1. Consider the stoichiometry of the Grignard reaction and the mole ratios used in this experiment.

   a. What is the mole ratio of 2-bromopentane to propanal?
   b. Write reactions to show what happens to the excess 2-bromopentane.
   c. At what stage is the excess 2-bromopentane (or its by-product) removed from the 4-methyl-3-heptanol?

2. This experiment starts with racemic 2-bromopentane. Consider the stereochemistry of the products.

   a. How many stereoisomers are present in the 4-methyl-3-heptanol you have synthesized? Draw each of them, clearly indicating their configurations.
   b. How many stereoisomers are present in the 4-methyl-3-heptanone that you made? Draw them.
   c. Which stereoisomers in parts a and b can be distinguished by their IR and ^{1}H NMR spectra?

3. A common impurity in the product of this experiment is 4,5-dimethyloctane.

   a. Propose how 4,5-dimethyloctane is formed in this reaction.
   b. Using the data you acquired for your ketone sample, how would you be able to tell if this impurity is present?

Data to Include in the Laboratory Report

A. PART 1: GRIGNARD SYNTHESIS OF 4-METHYL-3-HEPTANOL

1. Amount of 2-bromopentane used (g):

2. Amount of propanal used (g):

3. Yield of distilled 4-methyl-3-heptanol (g):

4. Percent yield of 4-methyl-3-heptanol:

B. PARTS 2 AND 3: OXIDATION AND SHORT-PATH DISTILLATION

5. Amount of 4-methyl-3-heptanol used (g):

6. Amount of crude 4-methyl-3-heptanone obtained (g):

7. Amount of crude 4-methyl-3-heptanone distilled (g):

8. Yield of distilled 4-methyl-3-heptanone (g):

9. Percent recovery in distillation [amount in (8) ÷ amount in (7)]:

10. Yield of 4-methyl-3-heptanone (g) [percentage in (9) × amount in (6)]:

11. Percent yield of 4-methyl-3-heptanone (for oxidation step):

12. Overall percent yield (for both steps):

22

LEARNING OBJECTIVES

- Examine the use of renewable resources for organic synthesis.

- Define the role of acetal functional groups in polysaccharide biopolymers.

- Implement an acid-catalyzed hydrolysis of corn cobs to obtain furfural.

- Use biomimetic catalysis to synthesize furoin from furfural, forming a C — C bond via the benzoin condensation reaction.

Renewable Feedstocks and Biocatalysis: Furoin Condensation

SUSTAINABLE FEEDSTOCKS

Vast quantities of biomass by-products are produced in agriculture, and these materials can be converted into feedstocks for organic chemistry.

Natalia Kuzmina/Alamy Stock Photo.

The use of plant materials as precursors for organic synthesis offers significant advantages over petroleum-based feedstocks. For example, plant material is a renewable resource rather than a depleting resource like petroleum, a fossil fuel which cannot be replaced on the human timescale. Moreover, compounds produced in bulk by plants, such as starches or vegetable oils, generally contain multiple functional groups that serve as handles for synthetic transformations. They may also contribute stereogenic centers that might otherwise be costly to introduce. In contrast, petroleum-based feedstocks are mostly saturated hydrocarbons that lack functional groups or stereogenic centers (**Figure 22.1**). Renewable resource feedstocks require a different way of thinking about chemical transformations of feedstocks—namely, you must selectively change or remove functionality, rather than add it in the first place. In recent years there has been a rapid growth in research in this area,[1] and there are now well-established pathways to access high-volume industrial feedstocks like ethanol and propylene glycol from renewable plant sources. There are also ways to convert biomass-derived feedstocks to high-value specialty chemicals that can be used in pharmaceuticals, polymers, and other products.

FIGURE 22.1

(a) Naturally occurring feedstock chemicals isolated from petroleum, a depleting resource, and (b) plant biomass, a renewable resource. The compounds isolated from plants contain more functional groups and stereochemical complexity.

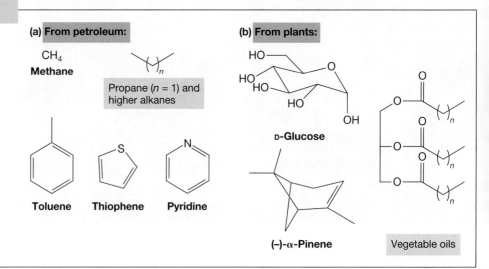

(a) From petroleum:

CH$_4$
Methane

Propane ($n = 1$) and higher alkanes

Toluene **Thiophene** **Pyridine**

(b) From plants:

D-Glucose

(–)-α-Pinene

Vegetable oils

biomimetic catalyst >>
A compound that mimics a catalytic function found in nature.

carbanion >>
A negatively charged reactive intermediate bearing anionic character at a carbon.

An attractive means of converting biomass into other useful chemicals is to take advantage of the catalysts found in nature and to mimic their biological role—that is, to use a biomimetic catalysis process to transform abundant compounds into more valuable ones. In this experiment, you obtain furfural from corn cobs. You then convert the furfural into a more complex material using naturally occurring thiamine (vitamin B$_1$) as a **biomimetic catalyst** for nucleophilic addition of a **carbanion** to an aldehyde.

Pre-Lab Reading Assignment

- Experiment: Chapter 22 (this chapter)
- Technique: Extraction (Chapter 5, section 5.2)
- Technique: Distillation (Chapter 5, section 5.3)

[1]Corma, A.; Iborra, S.; Velty, A. Chemical Routes for the Transformation of Biomass into Chemicals. *Chem. Rev.* **2007**, *107*, 2411–2502.

- Technique: Recrystallization (Chapter 5, section 5.4)
- Technique: Infrared (IR) spectroscopy (Chapter 6, section 6.2)
- Technique: Melting point measurement (Chapter 5, section 5.6B)
- Technique: Nuclear magnetic resonance (NMR) spectroscopy (Chapter 7, section 7.1)

22.1 BACKGROUND

22.1A Preparation of Furfural from Corn Cobs

Furfural (**Figure 22.2**) is a naturally occurring compound found in fruits, vegetables, and derived products such as beverages and bread products. It is also a large-scale commodity chemical; reduction of its aldehyde group yields furfuryl alcohol, which is widely used in furan resins for the metal casting industry, with smaller markets for lubrication oils and intermediates in the synthesis of pharmaceuticals and flavorings.[2] Furfural can be produced from corn cobs, oat hulls, and other agricultural waste by-products by heating in the presence of acid.

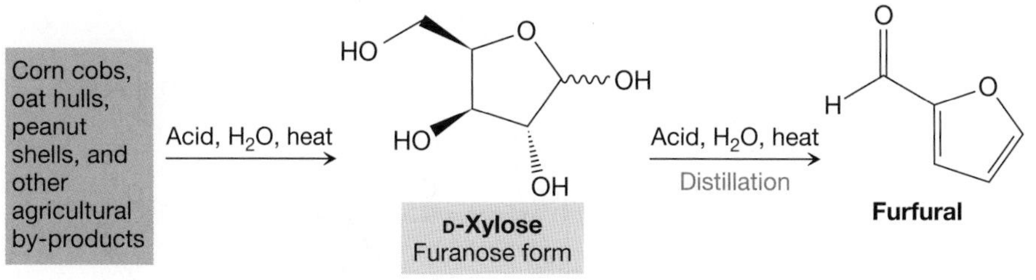

FIGURE 22.2

Heating with aqueous acid converts various agricultural waste materials into D-xylose, and the mixture can then be distilled to obtain furfural, an important industrial feedstock.

Corn is produced in huge quantities in many parts of the world as a source of food and fuels. Its main chemical constituents are polysaccharides, including starch, cellulose, and xylan. Polysaccharides consist of numerous individual sugars (monosaccharides) attached together via acetal linkages. The starch in corn kernels is digestible by humans, producing D-glucose. Corn cobs, on the other hand, consist of cellulose, which cannot be digested by humans, and xylan, a polysaccharide composed of D-xylose monosaccharides (**Figure 22.3a**).[3] The acetal linkages connecting the monosaccharide units in xylan can be hydrolyzed in aqueous acid, so heating corn cobs with dilute aqueous acid breaks apart the polysaccharide into its monosaccharide units, producing a water-soluble fraction containing D-xylose.

The acetal linkages connecting the monosaccharide units of xylan and other polysaccharides are also known as glycosidic bonds. They have the same characteristics of other acetals—namely, they are derivatives of aldehydes that are stable under basic conditions but react with aqueous acid or alcohols. When exposed to acid (**Figure 22.3b**), they become protonated on oxygen, creating a good leaving group and setting off the equilibrium reaction that interconverts acetals with hemiacetals and aldehydes. It is this reaction, catalyzed by aqueous acid, which breaks xylan into its D-xylose monosaccharide units.

---

[2]IHS Markit. Furfural. *Chemical Economics Handbook* [Online] **2020**. https://ihsmarkit.com/products/furfural-chemical-economics-handbook.html (accessed April 2022).
[3]Makishima, S.; Mizuno, M.; Sato, N.; Shinji, K.; Suzuki, M.; Nozaki, K.; Takahashi, F.; Kanda, T.; Amano, Y. Development of Continuous Flow Type Hydrothermal Reactor for Hemicellulose Fraction Recovery from Corncob. *Bioresource Technology* **2009**, *100*, 2842–2848.

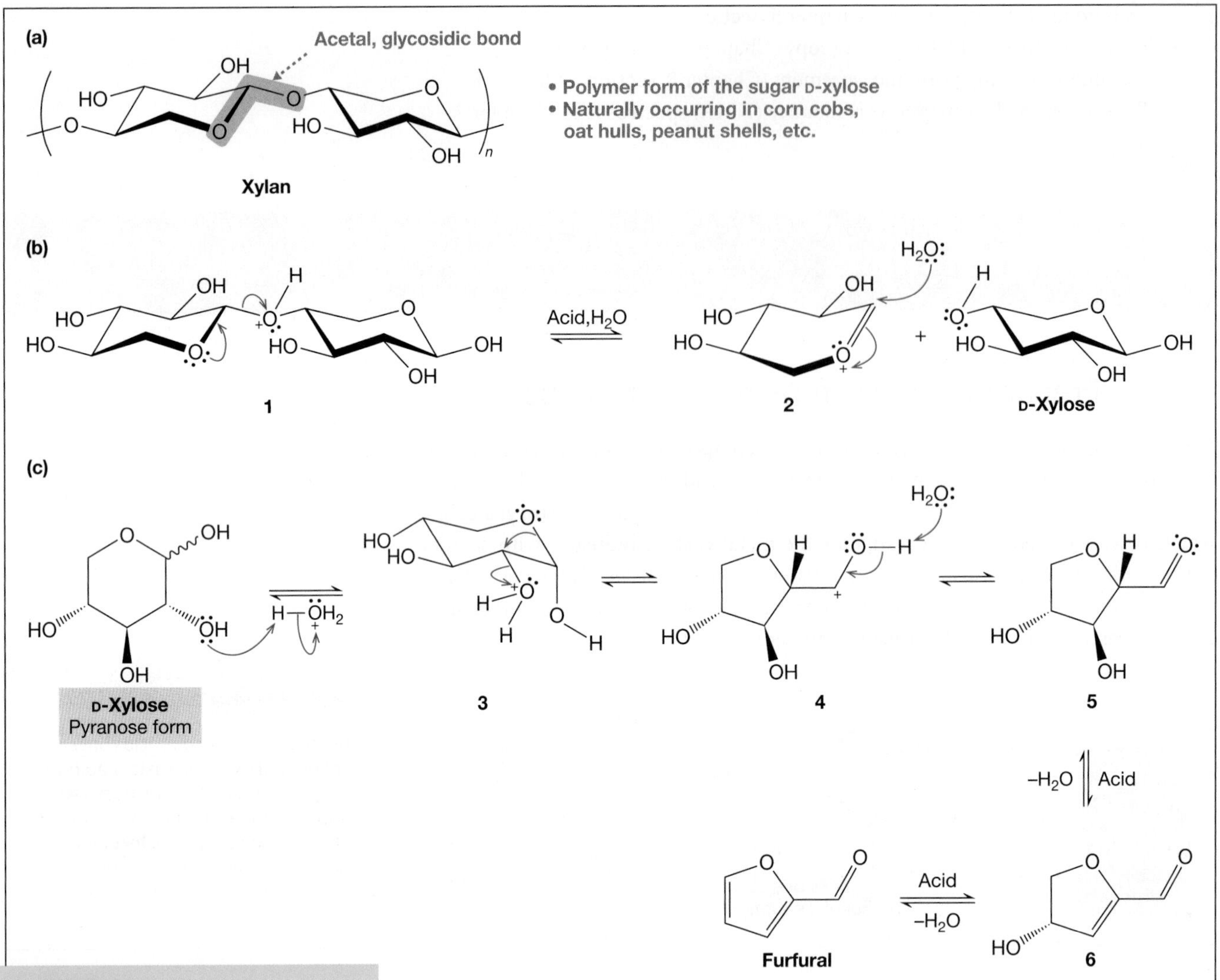

FIGURE 22.3

The conversion of corn cobs into furfural. (a) The structure of xylan, a carbohydrate polymer consisting of D-xylose units. (b) Heating with aqueous acid hydrolyzes xylan back into its monosaccharide (D-xylose) units. (c) Mechanistic proposal for the formation of furfural: ring-contraction isomerization of the D-xylose forms a five-membered ring, followed by two acid-catalyzed dehydrations.

Distillation of the acidic mixture of xylan leads to dehydration of the D-xylose, and an aromatic furan ring is formed. The mechanism by which furfural forms under these conditions is not completely understood, and alternative pathways may operate under different conditions.[4] One possibility (**Figure 22.3c**) involves an isomerization of D-xylose from a six-membered ring (**3**) to a five-membered ring (**4**), resulting in the aldehyde (**5**) after loss of a proton. Two successive acid-catalyzed dehydrations of **5** then provide furfural. The resonance stabilization upon forming the aromatic furan ring presumably makes the dehydration occur more readily.

22.1B Thiamine-Catalyzed Benzoin Condensation of Furfural

Acidic conditions are used to catalyze the hydrolysis and dehydration reactions to obtain furfural. After making furfural, carbanion intermediates are used to make a carbon–carbon bond. Carbanion intermediates possess a carbon bearing a negative charge, and are an important class of reactive intermediates in organic chemistry. They react with electrophiles, such as the polarized C=O bond of carbonyl groups,

---

[4]Binder, J. B.; Blank, J. J.; Cefali, A. V.; Raines, R. T. Synthesis of Furfural from Xylose and Xylan. *ChemSusChem* **2010**, *3*, 1268–1272.

Organolithium addition:

Grignard addition:

Benzoin condensation:

New carbon–carbon bond

Benzaldehyde

Benzoin
Racemic

to give new carbon–carbon bonds. Examples include organolithium and Grignard additions (**Figure 22.4**). Carbanion reagents can also be formed from aldehydes in the presence of cyanide ion. When these carbanions react with another aldehyde, the result is called the benzoin condensation.

The benzoin condensation was discovered in the 1830s by two chemistry pioneers, Friedrich Wöhler and Justus von Liebig (**Figure 22.5**), who were studying the chemistry of bitter almond oil. This oil, now known as benzaldehyde, reacted with cyanide ion to produce benzoin by carbon–carbon bond formation.

FIGURE 22.4

Aldehydes react with nucleophilic carbanion reagents, such as alkyllithium reagents and organomagnesium reagents, to give secondary alcohols, forming a new carbon–carbon bond in the process. The benzoin condensation is an example of this type of reaction, with a few extra twists to the mechanism.

(a) Friedrich Wöhler (1800–1882)   **(b)** Justus von Liebig (1803–1873)

FIGURE 22.5

(a) Friedrich Wöhler and (b) Justus von Liebig, the co-discoverers of the benzoin condensation. Wöhler is also credited with discoveries of beryllium and silicon compounds, along with the first synthesis of an organic compound using non-living sources of starting material. The latter was a landmark achievement in chemistry, because it refuted a hypothesis of vitalism—that organic compounds could only be made by living things through some special "vital force." Von Liebig discovered that nitrogen is an essential plant nutrient, and became known as the "father of the fertilizer industry." Among his students were August Kekulé, who first proposed the structure of benzene, and Emil Erlenmeyer, who first recognized the triple bond in acetylene (ethyne), and described the flasks which bear his name.

SPCOLLECTION/Alamy Stock Photo.
Chronicle/Alamy Stock Photo.

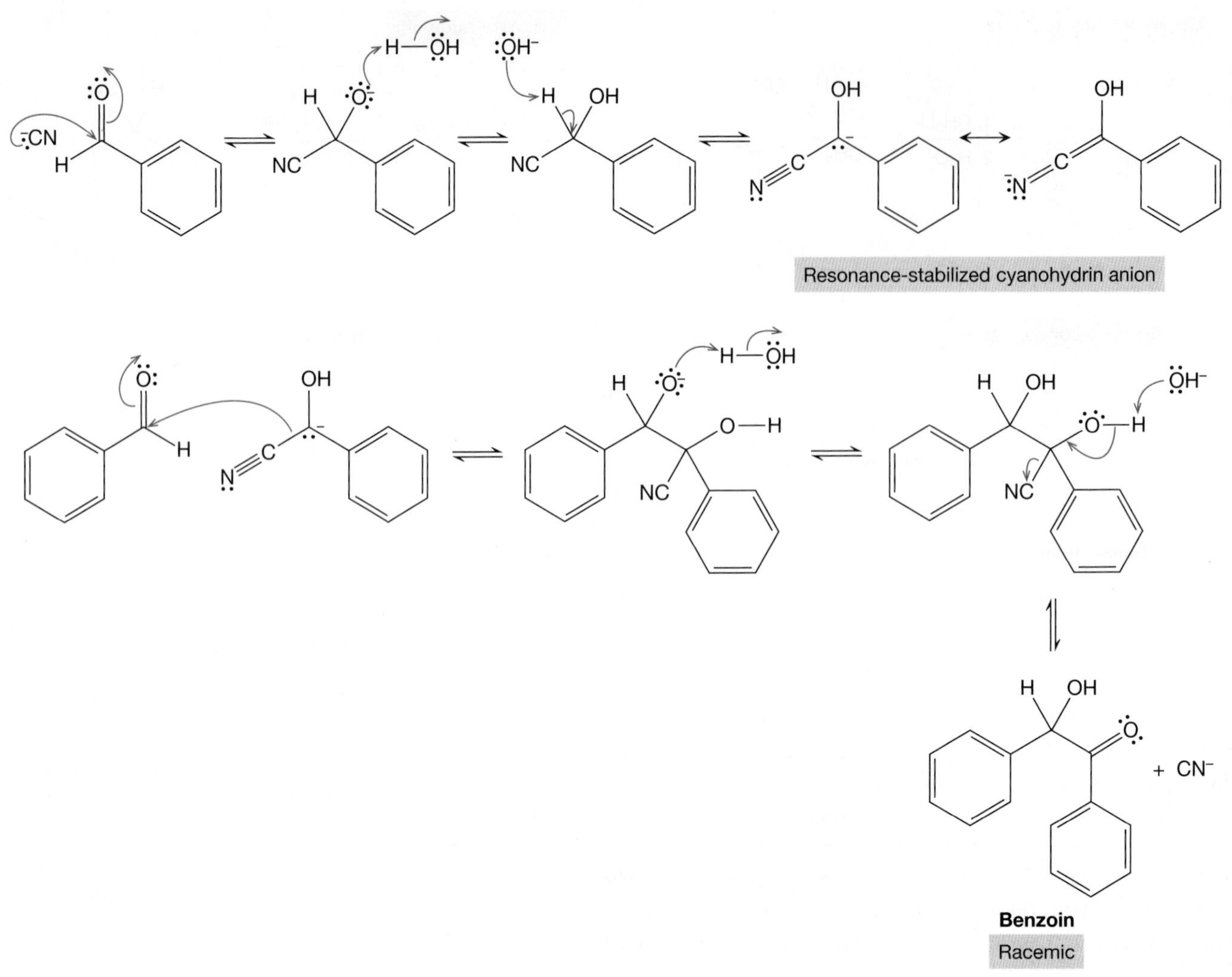

FIGURE 22.6

The mechanism of the cyanide-catalyzed benzoin condensation. The cyanide is not consumed in the reaction; instead, it is released at the end, and can carry out another catalytic cycle in the reaction.

umpolung >>
Reversed polarity in comparison with the usual polarization of a functional group or bond. For example, a C=O bond is usually polarized so that the carbon is positively charged; with umpolung, a structural modification of the C=O allows it to behave as if it has a negatively charged carbon.

The mechanism of this reaction (**Figure 22.6**) involves two types of nucleophilic additions to the carbonyl group of benzaldehyde. In the first step, cyanide adds to benzaldehyde to form the cyanohydrin. The nitrile group dramatically lowers the pK_a of the neighboring C—H bond, so that it may be deprotonated by the aqueous base (OH⁻) to make a resonance-stabilized carbanion. The original aldehyde C=O bond bears a partial positive charge on the same carbon that later becomes the carbanion! The carbanion intermediate then adds to another equivalent of benzaldehyde, creating the new carbon–carbon bond in benzoin. The resulting product loses cyanide ion to form the product. This is an example of **umpolung**, or reversal of polarity. The carbanion intermediate bears some similarities to alkyllithium (RLi), Grignard compounds (RMgX), and enolate ions: All are strong nucleophiles that add to carbonyl groups. Unlike alkyllithium or Grignard reactions, which are destroyed by water, this reaction can actually be conducted in aqueous solution.

In fact, similar chemistry occurs under aqueous conditions in human biology, where thiamine-dependent enzymes catalyze carbon–carbon bond formation or cleavage reactions using thiamine (vitamin B₁). The mechanism of these reactions was proposed by Ronald Breslow in 1958,[5] and involves the loss of a

---

[5](a) Breslow, R. On the Mechanism of Thiamine Action. IV. Evidence from Studies on Model Systems. *J. Am. Chem. Soc.* **1958**, *80*, 3719–3726. (b) Breslow later offered an entertaining retrospective essay on the discovery of biomimetic reactions of thiamine: Breslow, R. *This Week's Citation Classic* **1993**, *26*, 8. http://garfield.library.upenn.edu/classics1993/A1993LF83700001.pdf (accessed April 2022).

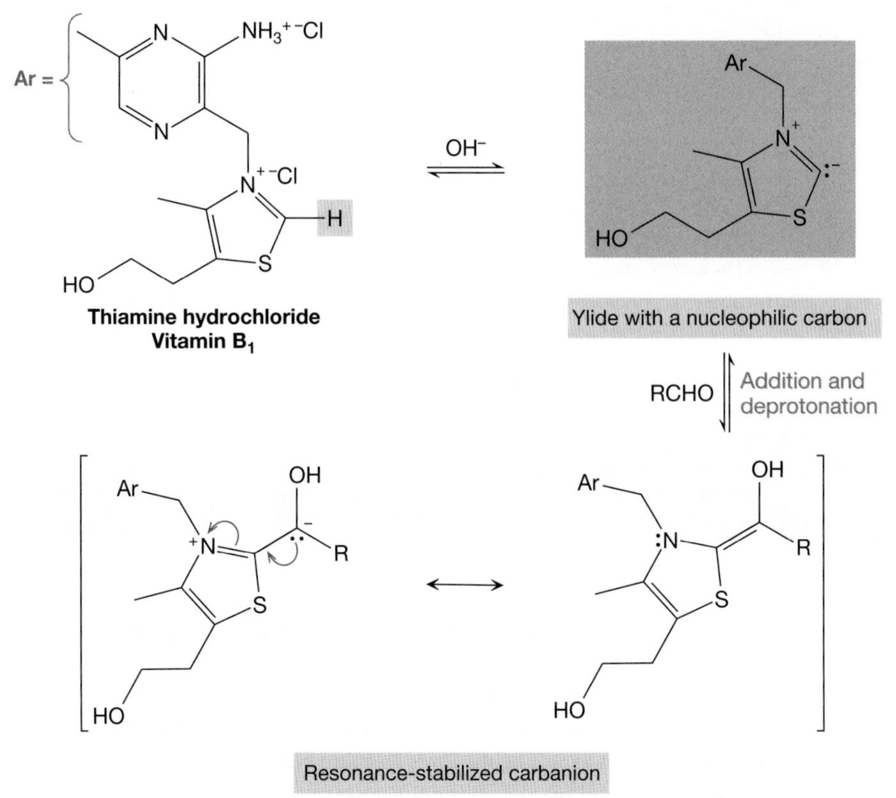

FIGURE 22.7

**Thiamine hydrochloride
Vitamin B₁**

Ylide with a nucleophilic carbon

Addition and
deprotonation

Resonance-stabilized carbanion

When thiamine is deprotonated, it forms an ylide that functions in the same way as CN⁻ in the mechanism of the benzoin condensation (Figure 22.6). Like CN⁻, it reversibly adds to the C=O of an aldehyde, and then lowers the pK_a of the adduct, where deprotonation gives a stabilized carbanion.

proton from thiamine, addition to an aldehyde, then deprotonation to generate a resonance-stabilized carbanion. Breslow's discovery inspired a whole new field of biomimetic chemistry, where synthetic chemists use the reactions of nature as inspiration for new laboratory processes. In these reactions, then, the thiamine performs the same functions as the cyanide ion by adding to an aldehyde and then stabilizing a carbanion structure (**Figure 22.7**).

Partly because of the less hazardous aqueous conditions, there have been increasing efforts to take advantage of thiamine-dependent enzymes as biocatalysts for C—C bond formation in organic synthesis.[6] In our experiment, we use thiamine as a biocatalyst to carry out a benzoin condensation of furfural, forming a new carbon–carbon bond and producing furoin.

Applying the thiamine-catalyzed benzoin condensation with the furfural you obtained from corn cobs, you will carry out the reaction in **Figure 22.8**, isolating pure furoin as a crystalline solid.

You will analyze the identity and purity of your furoin by infrared (IR) and ¹H NMR spectroscopy, as well as by its melting point.

[6]Brovetto, M.; Gamenara, D.; Saenz Mendez, P.; Seoane, G. A. C–C Bond-Forming Lyases in Organic Synthesis. *Chem. Rev.* **2011**, *111*, 4346–4403.

FIGURE 22.8

Thiamine-catalyzed furoin
condensation.

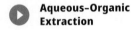

Furfural

Furoin

Racemic

22.2 EXPERIMENTAL PROCEDURE

22.2A Preparation of Furfural from Corn Cobs[7]

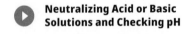

Simple Distillation

1. Assemble an apparatus for simple distillation. Carefully turn on the cooling water, checking the outlet hose to ensure the rate is set at the slowest continuous flow. Place a heat source (e.g., heating mantle or sandbath) on a laboratory jack and preheat at medium heat.

2. To a 200-mL round-bottomed flask, add 15 g corn cobs, ground into kernel-sized pieces[8] (**Figure 22.9a**), 15 g sodium chloride, and a boiling chip.

3. Clamp the flask securely, and add 75 mL of aqueous 3 M HCl solution.

4. Attach the flask to the distillation apparatus and heat on a high heat setting until boiling begins (**Figure 22.9b**). Then lower the heat to a setting that maintains a consistent distillation. If the material in the boiling flask "bumps" during distillation (i.e., suddenly splatters up into the condenser), quickly reduce the heat by moving the heat source away from the flask.

5. Distill until approximately 20 mL of distillate is collected. The distillate will form two layers, and may have a faint pink color (**Figure 22.9c**).

6. Lower the heat source away from the flask and turn the power off.

Neutralizing Acid or Basic Solutions and Checking pH

7. Remove the collection flask from the distillation apparatus. To the distillate, add a few drops of 2 M aqueous NaOH dropwise while swirling the flask. Check the acidity with pH paper, and continue dropwise addition of 2 M aqueous NaOH until the mixture is around pH 4–6. If the pH rises above 6, add a drop of aqueous 3 M HCl solution to bring it back down to pH 4–6.

Aqueous–Organic Extraction

8. Pour the mixture into a separatory funnel. Rinse the flask with two 1-mL portions of CH_2Cl_2, adding these rinse solutions to the separatory funnel. To avoid troublesome emulsions, do not shake vigorously; swirl the contents of the separatory funnel, then drain off the bottom organic layer into

---

[7]Adams, R.; Voorhees, V. *Organic Syntheses Collective Volume 1*, 2nd ed.; Wiley: New York, 1967; pp 280–283.
[8]Ground corn cobs can be purchased from animal feed manufacturers such as Best Cob (Independence, IA, 1-800-237-8262) or Waterloo Mills (Waterloo, IA, 319-234-7756).

(a)

(b)

(c)

(d)

FIGURE 22.9

Corn cob digestion. (a) This is what the corn cob pieces should look like before the digestion process. (b) A blackened mass forms during the digestion process. (c) Furfural collects as a two-phase liquid that can gradually turn a pinkish color. (d) After separating the organic phase, CH_2Cl_2 is removed by rotary evaporator.

Courtesy of Gregory K. Friestad.

a clean 25-mL Erlenmeyer flask. Extract the aqueous phase with two 1-mL portions of CH_2Cl_2, adding these extracts to the same flask.

   CAUTION: Use CH_2Cl_2 in the fume hood. Avoid inhalation or skin contact.

9. Dry the organic phase over Na_2SO_4. Decant into a tared 50-mL round-bottomed flask and remove the solvent using the rotary evaporator (**Figure 22.9d**). The product is a liquid, and it may be difficult to tell if all the solvent is removed. Continue evaporating it until the volume of liquid appears to reach a constant level. Record the mass of furfural that remains.

10. Obtain an infrared spectrum of your crude furfural.

 Rotary Evaporation

22.2B Thiamine-Catalyzed Benzoin Condensation of Furfural[9]

11. Into a 15-mL screw-cap vial, place 0.30 g thiamine hydrochloride, 0.45 mL water, and 4.0 mL 95% ethanol. Add a magnetic stir bar, and stir until a clear, colorless solution is obtained.

[9]Doxsee, K. M.; Hutchison, J. E. *Green Organic Chemistry: Strategies, Tools, and Laboratory Experiments*; Brooks/Cole: Pacific Grove, CA, 2004.

12. In dropwise fashion, add 0.90 mL of a 2 M aqueous solution of NaOH. A yellow color should develop, and then persist as a pale yellow when the addition is complete. If it is still colorless at this point, add drops of 2 M aqueous NaOH until the pale yellow color persists.

13. Add 0.73 mL of your crude furfural and mix thoroughly. Screw the cap onto the vial tightly, label it clearly, and store it at room temperature (ca. 20°C) until the next lab period (a range of 2–7 days is suitable).

14. Locate your vial containing your reaction mixture, stored from the prior lab period. Add an equal volume of water to the vial, cool in an ice bath, and isolate the solid product by vacuum filtration. If solid remains in the vial, use a second portion of cold water to aid in transferring to the filter. Allow the vacuum to pull air through the filtration apparatus for a few minutes until the crude product is dry. Record the mass of crude furoin.

15. Recrystallize the crude furoin (**Figure 22.10**) from 95% ethanol. In a 10-mL Erlenmeyer flask, mix the crude product with 1 mL 95% ethanol, and bring it to boiling on a hotplate. If the solid is not dissolved, add more 95% ethanol dropwise, using the minimum amount needed to dissolve the solid. Allow the flask to cool slowly on the benchtop, then cool it further in an ice bath, and recover the crystalline product by vacuum filtration. After the solid has dried, record the mass of recrystallized furoin.

16. Obtain an infrared spectrum and determine the melting point of your recrystallized furoin.

17. Obtain the ^{1}H NMR spectrum of your recrystallized furoin. No more than 10 mg should be used in preparing the sample for NMR acquisition.

▶ **Vacuum Filtration**

▶ **Recrystallization**

▶ **Melting Point Measurement**

▶ **NMR Sample Preparation**

(a)

(b)

(c)

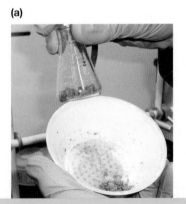

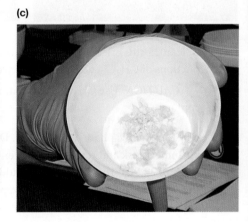

FIGURE 22.10

(a) The crude furoin product before recrystallization.
(b) The recrystallization in progress.
(c) The recrystallized furoin.
Courtesy of Gregory K. Friestad.

From all the evidence that you have at your disposal (IR, ^{1}H NMR, melting point, yield), discuss whether or not the reaction was successful. This discussion should refer specifically to the evidence of both the identity and the purity of the product. Do not forget to include literature values where appropriate. Attach IR and NMR spectra to your report. Include tables with assignments of all product peaks in the ^{1}H NMR spectrum and key peaks in the IR spectrum that are diagnostic for the structural changes in the reaction.

22.3A Post-Lab Questions

1. Reversal of polarity, or umpolung, is a key concept of the benzoin condensation. Can you think of other examples of umpolung? *Hint:* Review the Markovnikov and anti-Markovnikov hydration of alkenes from your organic chemistry lecture course, and consider the partial charges that develop during their mechanisms.

2. Draw a stepwise mechanism for the transformation shown in **Figure 22.11** that occurs in the presence of aqueous HCl, using curved arrows to show the movement of electrons at each step.

Attach the post-lab questions to your report.

FIGURE 22.11

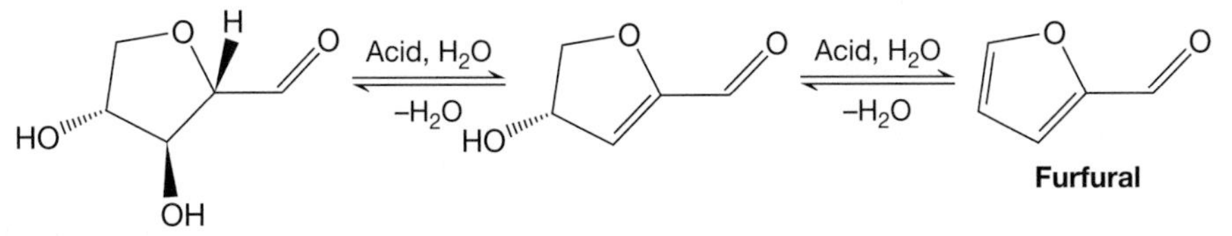

22.3B Data to Include in the Laboratory Report

A. PREPARATION OF FURFURAL FROM CORN COBS

1. Amount of corn cobs used (g):

2. Amount of furfural obtained (g):

B. THIAMINE-CATALYZED FUROIN CONDENSATION

1. Amount of furfural used (g):

2. Amount of crude furoin obtained (g):

3. Amount of furoin after recrystallization (g):

4. Yield of furoin from furfural (%):

Biocatalytic Reduction of a Prochiral Ketone

FROM GARDEN TO LAB

Enzymes found in carrots can accomplish a useful organic reaction: reducing ketones to afford secondary alcohols. These enzymes are chiral reagents that furnish products with a specific configuration.

5 second Studio/Shutterstock.

Naturally occurring chiral compounds are usually found in enantiomerically pure form because they have been synthesized by chiral catalysts that are enantiomerically pure. In nature, these catalysts are enzymes (**Figure 23.1**)—chiral protein molecules that not only catalyze a reaction, but also provide a chiral environment in which reactions can occur with **enantioselectivity**. Enantioselectivity arises when one enantiomeric product is preferentially formed from an achiral or racemic precursor. A synthesis that employs enantioselective reactions is called **asymmetric synthesis**. In the case of a racemic precursor reacting with an enantioselective enzyme, only one of the enantiomeric pair matches the shape of the pocket in which reaction occurs. One way to visualize this is by analogy with your hand in a glove; you have two mirror image hands, but only one of them will fit comfortably in a particular glove. In enzymes, this kind of selection of the proper fit also allows for one mirror image structure to be formed from an achiral precursor with exquisite enantioselectivity. That's extremely important, because the required biological properties of chiral reaction products often are associated with only one of the enantiomers. If the enzyme made racemic products (i.e., equal amounts of both enantiomers), then half of the material would be unusable, or possibly harmful. The same is true of chiral drug molecules; the desired properties are generally found in one of the enantiomers, so development of a chiral drug requires asymmetric synthesis.

Chiral drugs are often synthesized from small chiral organic compounds that can be purchased with one or more stereogenic centers already intact, or that can be extracted from natural sources in enantiomerically pure form. But what if a compound isn't readily available as one enantiomer? One way to obtain it is to prepare it as a racemic mixture, then separate the two enantiomers using a process known

enantioselectivity >>
When forming a chiral compound, the amounts of each of the enantiomers (nonsuperimposable mirror images) that are formed. The enantioselectivity is usually expressed as the enantiomer ratio (er) or percent enantiomeric excess (% ee). If a racemic product, or equal amounts of both enantiomers, is formed, the enantioselectivity would be a 50:50 ratio or 0% ee.

asymmetric synthesis >>
A reaction or sequence of reactions that produces a single enantiomer of a product, or an excess of one enantiomer of a chiral product.

FIGURE 23.1

The structure of an enzyme, aldose reductase. Here the chain of amino acids making up the protein structure of the enzyme is abbreviated with ribbons to illustrate structural features such as helices and sheets that form as the protein folds.

Figure 2 from Barski, O. A.; Tipparaju, S. M.; Bhatnagar, A. The Aldo-Keto Reductase Superfamily and Its Role in Drug Metabolism and Detoxification. *Drug Metab. Rev.* **2008**, *40*, 553–624. DOI: 10.1080/03602530802431439; PDB file 2F2K.

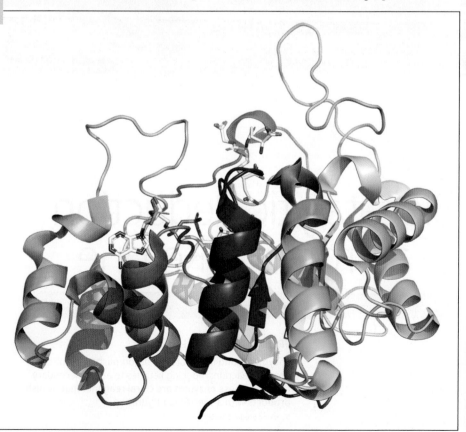

as **resolution**. Even when this is successful, it's an inherently wasteful approach because it discards half of the mass of the material as the undesired enantiomer. Therefore it is preferable to synthesize one enantiomer rather than a racemate, using asymmetric synthesis. In this experiment, we will use the latter approach, carrying out an enantioselective reduction of a ketone. Using an enzyme to catalyze the reduction, the chiral alcohol can be formed predominantly as one enantiomer.

<< **resolution**
A process that separates, or resolves, two enantiomers from a racemic mixture. A chiral reactant that reacts more rapidly with one enantiomer can resolve the enantiomers if the reaction is stopped near 50% conversion. Alternatively, if the compound is basic or acidic, a salt prepared from another chiral reactant in enantiomerically pure form can be recrystallized to separate the two diastereomeric salts.

Pre-Lab Reading Assignment

- Experiment: Chapter 23 (this experiment)
- Technique: Thin-layer chromatography (TLC) (Chapter 5, section 5.5)
- Technique: Column chromatography (Chapter 5, section 5.5)
- Technique: Polarimetry (Chapter 5, section 5.7)
- Technique: Nuclear magnetic resonance (NMR) spectroscopy (Chapter 7, section 7.1)

23.1 BACKGROUND

Over the last couple of decades, chemists and biologists have been increasingly working to harness the power of naturally occurring enzymes and their reactions, known as biocatalysis, to carry out organic synthesis processes in the lab.[1] One of the most widely used enzyme-catalyzed reactions in organic chemistry is reduction of ketones to secondary alcohols, which is catalyzed by a class of enzymes known as aldo- and ketoreductases (**Figure 23.2**).[2] These enzymes are involved in biosynthesis, steroid

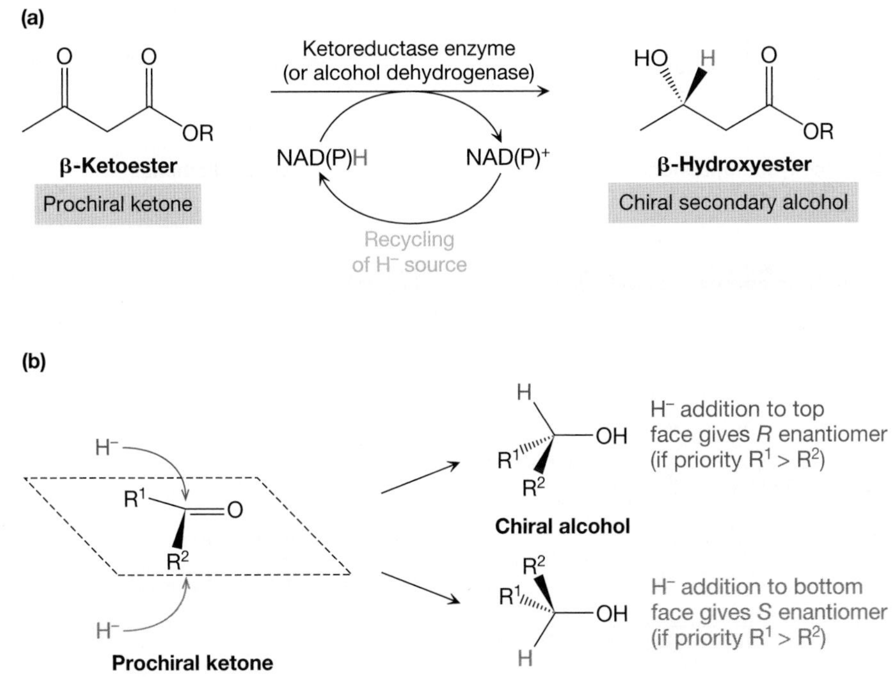

FIGURE 23.2

(a) The reduction of ketones to alcohols is catalyzed by a family of enzymes known as ketoreductases (or alcohol dehydrogenases) that transfer hydride (H^-) from NAD(P)H to the carbonyl carbon. (b) A planar sp^2-hybridized carbonyl carbon of a ketone is prochiral if its two carbon substituents are different. In this scenario, selective addition of H^- to each face results in a different enantiomer. The enzyme-catalyzed reduction is usually enantioselective—that is, one enantiomer of a racemic mixture is formed to a greater extent, or even exclusively.

[1]Kisukuri, C. M.; Andrade, L. H. Production of Chiral Compounds Using Immobilized Cells as a Source of Biocatalysts. *Org. Biomol. Chem.* **2015**, *13*, 10086–10107.
[2](a) Hall, M.; Bommarius, A. S. Enantioenriched Compounds via Enzyme-Catalyzed Redox Reactions. *Chem. Rev.* **2011**, *111*, 4088–4110. (b) Monti, D.; Ottolina, G.; Carrea, G.; Riva, S. Redox Reactions Catalyzed by Isolated Enzymes. *Chem. Rev.* **2011**, *111*, 4111–4140.

processing, drug detoxification, and numerous other biochemical reactions across all domains of life.[3] They catalyze the transfer of hydride (H⁻) from the reduced form of nicotinamide adenine dinucleotide phosphate (NAD(P)H) to the carbonyl carbon of the ketone. When these act upon a ketone with two different groups attached at the carbonyl, called a *prochiral ketone*, the addition of hydride creates a stereogenic carbon, and results in a chiral compound. Ketoreductase reactions can be highly enantioselective because the chiral environment provided by the structure of the enzyme catalyst causes the hydride to add selectively to one face of the planar carbonyl carbon. Under certain conditions, these enzymes can also catalyze the reverse reaction to oxidize an alcohol to a carbonyl group; such transformations are referred to as alcohol dehydrogenase reactions.

The effectiveness of an asymmetric synthesis can be determined by measuring the enantiomeric ratio of the product. There are a few ways to make this measurement, and the most convenient way is to measure the optical rotation. There are more precise methods involving analytical chromatography, but these methods can be time-consuming and are unsuitable for handling large quantities of samples.

This experiment uses biological material, carrot peelings, to reduce 3-methoxyacetophenone (**Figure 23.3**) in enantioselective fashion. This reaction will be monitored by thin-layer chromatography, which will separate reactant and product in order to see if the reaction is complete. You will also use a racemic control experiment to generate an authentic sample of the product as a standard for thin-layer chromatography. After purifying the biocatalytic reduction product, you will determine which enantiomer is formed, and how selectively, by measuring optical rotation on a polarimeter.

FIGURE 23.3

(a) The enantioselective reduction of 3-methoxyacetophenone; this reaction takes place with ketoreductase enzymes found in carrots. Both the enzyme and NAD(P)H, the hydride source, are present in very small amounts because the hydride source is recycled by the naturally occurring pathway in the carrots. (b) The racemic reaction using sodium borohydride as the hydride source affords a racemate (equal amounts of both enantiomers).

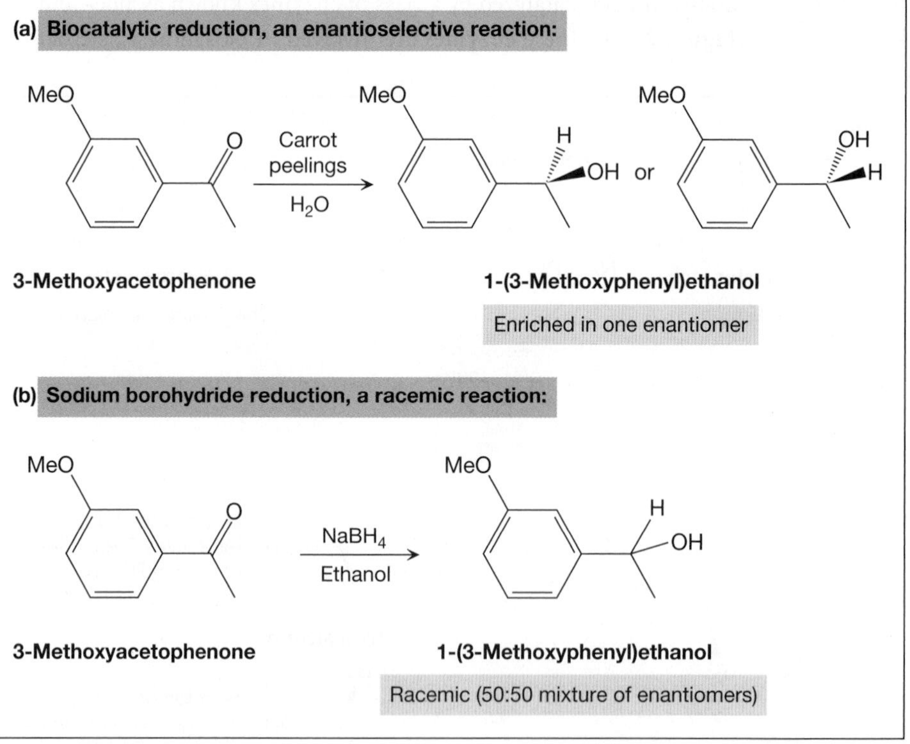

(a) Biocatalytic reduction, an enantioselective reaction:

3-Methoxyacetophenone → 1-(3-Methoxyphenyl)ethanol
Enriched in one enantiomer

(b) Sodium borohydride reduction, a racemic reaction:

3-Methoxyacetophenone → 1-(3-Methoxyphenyl)ethanol
Racemic (50:50 mixture of enantiomers)

[3](a) Barski, O. A.; Tipparaju, S. M.; Bhatnagar, A. The Aldo-Keto Reductase Superfamily and Its Role in Drug Metabolism and Detoxification. *Drug Metab. Rev.* **2008**, *40*, 553–624. (b) Sengupta, D.; Naik, D.; Reddy, A. R. Plant Aldo-Keto Reductases (AKRs) as Multi-Tasking Soldiers Involved in Diverse Plant Metabolic Processes and Stress Defense: A Structure-Function Update. *J. Plant Physiol.* **2015**, *179*, 40–55.

23.2A Biocatalytic Reduction: Enantioselective Reaction (Day 1)

1. Using an ordinary kitchen vegetable peeler, peel a washed carrot to give 25 g of thin slices.

2. To a 125-mL Erlenmeyer flask, add 3-methoxyacetophenone (about 90–100 mg) and record the exact mass that you measure. Add 75 mL of water and swirl the contents to mix thoroughly. Add the carrot peelings and swirl again. Let the mixture stand for about 1 hour, briefly swirling the flask at 15-minute intervals. During this time, proceed to part B below.

3. After approximately 1 hour, transfer 0.5 mL of the reaction mixture to a test tube. Add 0.5 mL ethyl acetate. Cover the opening of the tube with a piece of Parafilm and shake gently. After the layers separate, use the upper ethyl acetate layer for TLC analysis, comparing it to the product obtained in part B, the control reaction.

4. Label the flask and store it until the next lab period. (*Note to instructor:* This ketone reacts more rapidly than most other alkyl aryl ketones, but completion of the reaction may take 2–7 days. If column chromatography is used to separate the product, the reaction does not need to be complete. The enzyme-catalyzed reduction can be started in only a few minutes' time, so it is easy to set up at the end of a prior lab period, and it will be ready for purification and analysis at the next lab meeting. Occasional swirling or agitating the reaction mixture with magnetic stirring or a laboratory shaker can help ensure adequate conversion.)

23.2B Borohydride Reduction: Racemic Control Reaction (Day 1)

5. To a screw-cap vial, transfer one drop of 3-methoxyacetophenone, and dissolve it in 95% ethanol (1 mL). Add 5 mg sodium borohydride and swirl the contents of the vial. Allow the reaction to proceed for 10 minutes with occasional swirling.

6. Analyze the racemic control reaction mixture from step 5 by TLC on silica gel plates. Spot the 3-methoxyacetophenone starting material and the racemic control reaction mixture on the same plate, marking the origin spots with a pencil. Elute the plate with 85:15 hexanes/ethyl acetate and mark the solvent front with a pencil. Visualize the spots on the eluted plate with UV light, and draw around each spot with a pencil to indicate their size and location. Note whether any ketone reactant remains.

 Thin-Layer Chromatography

7. Store the remaining sample of the racemic control reaction mixture as a standard for TLC of the biocatalytic reduction of part A.

[4](a) The reduction reaction procedures are adapted from: Ravía, S.; Gamenara, D.; Schapiro, V.; Bellomo, A.; Adum, J.; Seoane, G.; Gonzalez, D. Enantioselective Reduction by Crude Plant Parts: Reduction of Benzofuran-2-yl Methyl Ketone with Carrot (*Daucus carota*) Bits. *J. Chem. Educ.* **2006**, *83*, 1049–1051. (b) Mazczka, W. K.; Mironowicz, A. Enantioselective Reduction of Bromo- and Methoxy-Acetophenone Derivatives Using Carrot and Celeriac Enzymatic System. *Tetrahedron: Asymmetry* **2004**, *15*, 1965–1967.

23.2C | ## Purification of the Biocatalytic Reduction Product (Day 2)

8. Repeat the TLC analysis of the biocatalytic reduction of part A, again comparing it to the racemic control sample, and note whether any ketone reactant remains.

Aqueous–Organic Extraction and Drying Organic Solutions

9. Decant the liquid from the carrot peelings, transferring it into a separatory funnel. Wash the carrot peelings with another 10 mL water, adding this wash to the separatory funnel. Extract with ethyl acetate (2 × 20 mL). Dry the combined organic extracts over anhydrous Na_2SO_4 and concentrate using the rotary evaporator.

10. Purify the biocatalysis product mixture using column chromatography. *CAUTION: Finely divided particles of silica gel are hazardous if inhaled. Keep silica gel in the hood.*

Column Chromatography

   a. *Pack the column.* Securely clamp a column of about 2 cm in diameter to the bars in the fume hood, close the stopcock, and loosely plug the inside end of the stopcock tube with cotton. Add about 1 cm of sand in a flat layer, followed by about 10 mL 85:15 hexanes/ethyl acetate. Gently tap the column with your finger to flatten the top of the sand layer if needed. Swirl together about 10 g of silica gel and 70 mL 85:15 hexanes/ethyl acetate until a smooth consistent slurry is obtained, and gently pour it in the column without disturbing the sand layer at the bottom. Open the stopcock and allow the solvent to drain into a clean Erlenmeyer flask as the silica gel settles. When the level of the silica gel stops settling, the height of the column should be 12–15 cm. If not, add more of the slurry and allow it to settle again. Gently add about 1 cm of sand, allowing it to fall evenly through the solvent and settle in a flat layer on top of the silica gel. Allow the solvent to drain until its level falls just below the top of the sand. Close the stopcock. (*Note:* The solvent that has drained out to this point is clean and should be reused.)

   b. *Load the sample.* Using a pipet, add the product mixture to the column. If the product mixture is difficult to draw up into the pipet, add 0.2 mL 85:15 hexanes/ethyl acetate, and turn the flask to move the liquid around and dissolve residues from the walls. Rinse the source flask (or vial) with two 0.2-mL portions of 85:15 hexanes/ethyl acetate, each time adding the rinse to the column and rinsing down any product mixture that may be on the inner walls of the column. Open the stopcock and allow the solvent to drain until its level falls just below the top of the sand. Close the stopcock and repeat the rinse with another 0.5-mL portion of 85:15 hexanes/ethyl acetate. Open the stopcock and allow the solvent to drain until its level falls just below the top of the sand. Close the stopcock.

   c. *Elute the column.* Using a pipet, gently add about 10 mL 85:15 hexanes/ethyl acetate without disturbing the top of the sand. Then pour in 85:15 hexanes/ethyl acetate until the column is full. Open the stopcock and allow the column to drip into a 125-mL Erlenmeyer flask until about 10 mL of eluent have collected. Then begin collecting fractions in 10-mL test tubes, filling each tube to about 75% capacity. Do not allow the column to run dry; replenish the solvent before it drains down to the top of the sand layer. Continue adding solvent and collecting fractions until you have 15 fractions. Label the fractions with numbers 1–15, taking care to keep them in sequence.

d. *Analyze by TLC.* Spot each of the even-numbered fractions, all on the same TLC plate. The fractions are very dilute solutions, so to make them visible, each spot should be repeated three times, allowing the solvent to evaporate between each repetition. As standards, include one spot of the 3-methoxyacetophenone starting material and one spot of the racemic control reaction mixture from part B. Elute the plate with 85:15 hexanes/ethyl acetate and visualize with a UV lamp. Circle all the spots. Identify the spots by comparing the R_f values with the standards.

e. *Pool fractions and evaporate solvent.* Label a round-bottom flask, weigh it, and write down its tare weight in your notebook. Those fractions whose TLC shows a single spot of the product alcohol should be pooled by pouring them together into the tared round-bottom flask, then place the flask on the rotary evaporator to remove the solvent.

f. *Remove the last traces of solvent.* Traces of solvent will remain after rotary evaporation. To get an accurate mass of the sample, all solvent must be removed. Connect the round-bottom flask to the vacuum available in your hood, using the simple assembly shown in **Figure 23.4**. Clamp the flask securely, plug the internal ground glass joint of a distillation adapter with a stopper, and attach the other end to the round-bottom flask. Attach a vacuum hose to the distillation adapter and apply vacuum for at least 10 minutes. Weigh the flask.

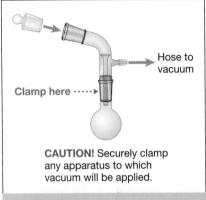

Clamp here ·····▶

Hose to vacuum

CAUTION! Securely clamp any apparatus to which vacuum will be applied.

FIGURE 23.4

The apparatus for the removal of the last traces of solvent after rotary evaporation.

23.2D Characterization of the Biocatalytic Reduction Product (Day 2)

11. Optional (check with your instructor): Prepare a sample for ^{1}H NMR spectroscopy. Transfer a small portion of the biocatalytic reduction product (from step 10g) to an NMR tube. Use $CDCl_3$ as the solvent for your sample. *If the amount in the flask is too small to remove a drop, add two drops of $CDCl_3$ to the flask first, and then transfer the solution to the NMR tube.*

 Preparing NMR Samples

12. Determine the optical rotation of the biocatalytic reduction product, 1-(3-methoxyphenyl)ethanol. After ensuring that solvents have been removed as in step 10g, measure the mass of the remaining biocatalytic reduction sample in the round-bottom flask. Add 1 mL chloroform ($CHCl_3$) to the flask containing the product. Transfer the solution to a 5-mL volumetric flask. Repeat twice with two additional 1-mL portions of $CHCl_3$, rinsing it down the inner walls of the flask, so that all of the reduction product is removed from the walls, and add each portion to the volumetric flask. Fill the volumetric flask up to the 5-mL volume line with $CHCl_3$. Stopper the volumetric flask and invert several times to ensure a homogeneous solution. Measure the optical rotation using a polarimeter. Record both the numerical value and its sign (+ or −).

 Polarimetry

Your results and discussion should include the following:

1. Determine the percent yield of 1-(3-methoxyphenyl)ethanol, using the amount of pure material after column chromatography.

2. Calculate the specific rotation of your 1-(3-methoxyphenyl)ethanol sample, and compare this to data reported in the literature for enantiomerically pure 1-(3-methoxyphenyl)ethanol.[5]

3. Determine the configuration and calculate the percent enantiomeric excess (% ee) of your 1-(3-methoxyphenyl)ethanol.

Useful equations:

$$[\alpha] = \frac{\alpha}{c \cdot l} \qquad \% \text{ ee} = \frac{[\alpha]_{calc}}{[\alpha]_{lit}}(100)$$

c = concentration in g/mL
l = path length of the cell (1 dm)
α = observed optical rotation reading from the polarimeter
$[\alpha]_{calc}$ = calculated specific optical rotation for laboratory sample
$[\alpha]_{lit}$ = literature specific optical rotation for the pure enantiomer

Some notes about polarimetry and % ee:

Besides concentration and the path length of the cell, the specific rotation also depends on temperature and the wavelength of light (the "sodium D-line" at 589 nm is typically used). In the literature, data are reported as $[\alpha]_D{}^{25}$, where the subscript is the wavelength (here the sodium D-line), and the superscripted number is the temperature (in this case, 25°C).

The equation for % ee is actually an approximation; it depends on the assumption that specific rotations are unaffected by concentration. A more accurate calculation of % ee is achieved by measuring the amount of R and S enantiomers with an analytical chromatography instrument, then using the following equation:

$$\% \text{ ee} = [(R - S)/(R + S)] \times 100$$

 ## Additional Questions

(Answers should be included in your lab report.)

1. Your 1-(3-methoxyphenyl)ethanol was enriched in one enantiomer. Assuming that enzymes generally cannot be made to produce either enantiomer as desired, use your knowledge of organic reactions to design a reaction sequence to convert your sample to the other enantiomer. Show reaction(s) and appropriate reagents.

[5]Literature data for (S)-1-(3-methoxyphenyl)ethanol are $[\alpha]_D{}^{27}$ –42.1 (c 1.0, CHCl₃). Xie, J.-H.; Liu, X.-Y.; Xie, J.-B.; Wang, L.-X.; Zhou, Q.-L. An Additional Coordination Group Leads to Extremely Efficient Chiral Iridium Catalysts for Asymmetric Hydrogenation of Ketones. *Angew. Chem., Int. Ed.* **2011**, *50*, 7329–7332.

2. Consider applying biocatalytic reduction with carrots to the following two ketones: *tert*-butyl methyl ketone, and 3-hexanone. Which would you expect to give higher enantioselectivity in biocatalytic reduction? Explain your answer.

23.3B Data to Include in the Laboratory Report

A. BIOCATALYTIC REDUCTION

Amount of 3-methoxyacetophenone used (g):

R_f data from TLC of reaction mixture:

R_f of 3-methoxyacetophenone:

R_f of 1-(3-methoxyphenyl)ethanol:

B. YIELD AFTER PURIFICATION

Mass of pure 1-(3-methoxyphenyl)ethanol (g):

C. POLARIMETRY OF 1-(3-METHOXYPHENYL)ETHANOL

Mass of 1-(3-methoxyphenyl)ethanol (g):

Observed optical rotation, α:

Calculated specific rotation, $[\alpha]_{calc}$:

Calculated percent enantiomeric excess, % ee:

24

LEARNING OBJECTIVES

- Recognize how polarization and resonance allow α deprotonation to form carbonyl enolate ions.

- Use the aldol reaction to form new carbon–carbon bonds.

- Employ solvent-free reaction conditions to minimize environmental impact.

- Implement asymmetric catalysis with (*S*)-proline for an enantioselective aldol reaction.

- Measure specific optical rotation of an aldol product and use it to calculate enantiomeric excess.

Aldol Reactions

AN AROMATIC ALDEHYDE SOURCE

Benzaldehyde derivatives such as veratraldehyde and vanillin are among the flavor and aroma compounds produced in the vanilla plant, mainly in the bean-like seed pods. Veratraldehyde is important in the flavorings and fragrances industry, and is also a suitable electrophilic reactant for aldol reactions.

Design Pics Inc/Alamy Stock Photo.

The carbonyl group has an extraordinarily powerful ability to alter reactivity at neighboring carbons through its polarizing effects. These effects, when harnessed, allow carbonyl groups to be versatile tools for carbon–carbon bond formation reactions. In most cases, the carbonyl carbon atom is electrophilic (**Figure 24.1**), and certain types of carbon nucleophiles can form a new carbon–carbon bond by nucleophilic addition to the carbonyl group. Also, the polarization of the C=O bond causes C—H bonds at the neighboring carbon (i.e., the alpha, or α, carbon) to become more acidic, enabling deprotonation to create a carbanion with nucleophilic behavior at the α carbon. The deprotonation is dramatically facilitated by the resonance stabilization of the conjugate base (Figure 24.1b), lowering the pK_a from about 50 (alkane) to about 20 for the position α to a carbonyl group.

FIGURE 24.1

(a) Polarization of the carbonyl group makes the C=O carbon electrophilic, while (b) deprotonation at the α carbon gives an enolate ion that can behave as a nucleophile.

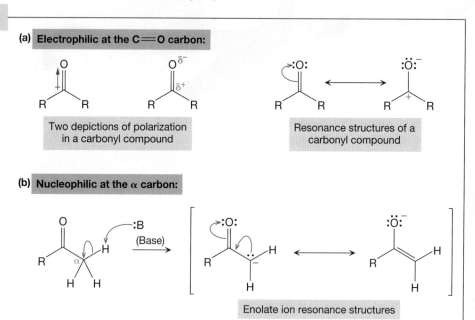

In this experiment, both types of reactivity are seen in an aldol reaction—one carbonyl compound serves as an electrophile, while the other serves as a nucleophile (at its α carbon). This pairing of nucleophile and electrophile results in carbon–carbon bond formation, linking the two carbonyl compounds together to make a more complicated molecular structure. The term aldol comes from the combination of <u>ald</u>ehyde and alcoh<u>ol</u> functional groups in the product, a β-hydroxycarbonyl compound (ketones may also serve as the carbonyl portion).

In this chapter, we examine two variations of this aldol reaction, both of which will employ conditions that address green chemistry principles. One variation carries out an aldol condensation under green conditions that avoid the use of volatile organic solvents (section 24.2). The other employs an environmentally benign amino acid as an organic catalyst to achieve an enantioselective aldol addition (section 24.3), with assessment of selectivity via polarimetry.

Pre-Lab Reading Assignment

- Experiment: Chapter 24 (this experiment)
- Technique: Filtration (Chapter 5, section 5.1)

- Technique: Recrystallization (Chapter 5, section 5.4)
- Technique: Melting point measurement (Chapter 5, section 5.6)
- Technique: Polarimetry (Chapter 5, section 5.7)
- Technique: Infrared (IR) spectroscopy (Chapter 6, section 6.2)
- Technique: Nuclear magnetic resonance (NMR) spectroscopy (Chapter 7, section 7.1)

24.1 BACKGROUND

In the aldol reaction (**Figure 24.2**), both types of reactivity illustrated in Figure 24.1 are combined into a single process. That is, two carbonyl compounds are linked together by forming a bond between the nucleophilic α carbon of one with the electrophilic carbonyl carbon of the other. Under base-catalyzed conditions,[1] the steps in this reaction include deprotonation at the α carbon to form an enolate ion, nucleophilic addition to the other carbonyl compound, and proton transfer to form the aldol addition product. Under some conditions, the reaction can be stopped at this stage, but dehydration occurs if the resulting alkene is conjugated with other unsaturated groups, or if the aldol addition reaction mixture is heated.

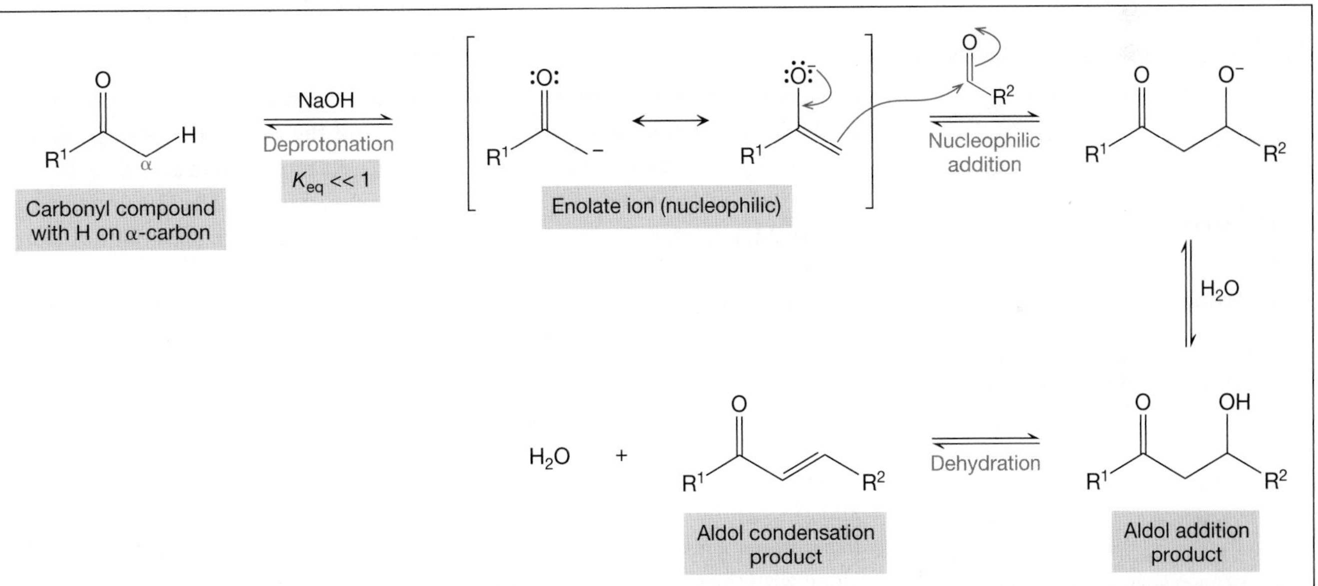

FIGURE 24.2

Steps in the aldol condensation reaction. Movements of electrons in the key C—C bond formation step are shown with curved arrows.

Consider the simplest case, involving only one carbonyl compound. Half of the reactant serves as the nucleophile and the other half serves as the electrophile, and because both are the same compound, there's only one way to couple them. If the two carbonyl reactants are different (a so-called "crossed aldol" reaction), however, then there can be a complex mixture of at least four products formed (**Figure 24.3**), because both of the carbonyl compounds can serve as either electrophile or nucleophile. In this scenario, only a small portion of the product mixture is the desired one, and the rest is undesirable waste, which may not be easy to separate.

There are ways to control the aldol reaction to avoid complex mixtures. If one of the aldol reaction components has no hydrogens on the α carbon, then it cannot be

---

[1]There are many variations on the powerful and versatile aldol reaction. For example, the aldol reaction can be carried out under kinetic control to avoid dehydration, or under acidic conditions (neutral enol as nucleophile and protonated carbonyl as electrophile). Reversible reaction conditions under thermodynamic control, as in the solventless aldol, will favor dehydration to the aldol condensation product. Catalysis with proline proceeds via enamine addition, which can avoid dehydration.

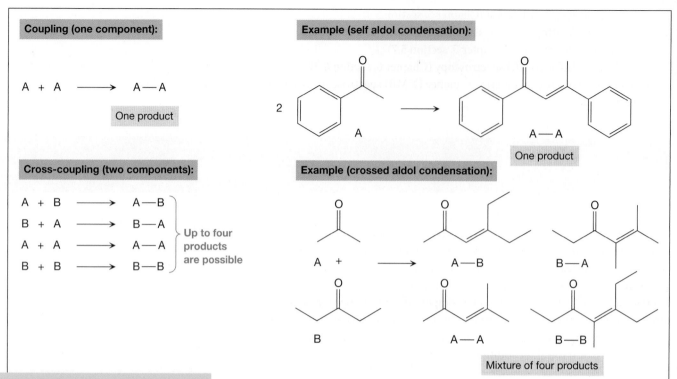

FIGURE 24.3

Selectivity issues emerge when atttempting a cross-coupling of two different components. The aldol reaction must be controlled to avoid a complex mixture of products from all possible combinations.

enamine >>
A combination of alkene and amine functional groups in which the nitrogen is directly attached to a π-bonded carbon of the alkene.

hydrolysis >>
Reaction with water, in which a reactant is effectively split into two products (lysed) by water.

converted to an enolate ion and therefore cannot serve as the nucleophilic partner in the aldol reaction. Also, steric hindrance and the electron-donating effects of alkyl groups both make a ketone a less reactive electrophile than an aldehyde in nucleophilic addition reactions. In a crossed aldol reaction such as the one shown in **Figure 24.4**, then, only one component can serve as the nucleophile and the other serves as a better electrophile, thus minimizing undesired products.

Enamines are a variation on the reactivity of enolate ions and serve as nucleophilic intermediates in aldol reactions; an example is the addition of acetone to butanal (**Figure 24.5**). Enamines may be generated from ketones under very mild conditions using a secondary amine as the only reagent, without any strong base required. These reactions start by condensation of the amine with the ketone with loss of water to yield an iminium ion; this is even more acidic than the starting ketone and loses a proton readily to afford the enamine. Next, the enamine adds to butanal, in a manner similar to addition of an enolate ion, regenerating another iminium ion along with the carbon–carbon bond. Since iminium ion formation is reversible, the reverse reaction (**hydrolysis**) can occur to regenerate the ketone functional group. The mild conditions of these reactions can avoid dehydration of the aldol addition product, enabling isolation of the β-hydroxyketone.

The aldol addition product in Figure 24.5 contains a new stereogenic center that was produced by addition to the planar carbonyl group of butanal. A typical secondary

FIGURE 24.4

An example of a good choice of reactants for the crossed aldol reaction. Only one carbonyl reactant can serve as a nucleophile by forming an enolate.

Nucleophile

Electrophile

Steric hindrance; poor electrophile

No H's at the α-carbon, so it cannot form the enolate

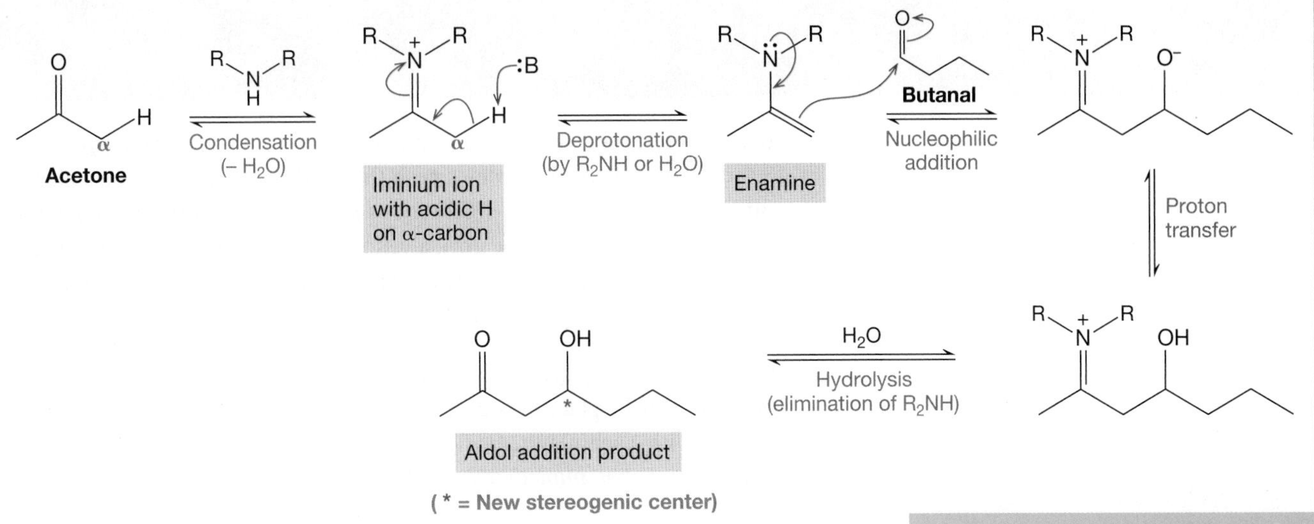

FIGURE 24.5

An example of the use of an enamine nucleophile for the crossed aldol reaction. The secondary amine is regenerated after the reaction, and is therefore a catalyst that may be used in small quantities.

amine will produce this in racemic form (equal parts of both R and S configurations), because addition can occur with equal probability from both faces of the planar C=O bond. However, if the secondary amine is a chiral compound, the enamine will also be chiral, and the new stereogenic center is then produced in a chiral environment. This means that there will be a nonracemic product; one enantiomer will be produced in greater proportion. Such an enantioselective reaction is discussed in section 24.3.

24.1A | Solventless Aldol Reaction

In this experiment, you use solvent-free conditions, avoiding some of the volatile organic compounds that are often employed in organic synthesis. A mixture of two solid carbonyl compounds would be unlikely to undergo an aldol reaction because the molecules diffuse too slowly through the solid matrix of their respective crystals. If the solid reactants are finely ground as a mixture, however, their melting points may become depressed. If the melting point depression is significant, the mixed melting point may be below room temperature. This causes the solid mixture to form a melt, wherein the reaction can occur more readily within the liquid state.

The reaction you carry out in this experiment (**Figure 24.6**) illustrates not only the use of solvent-free conditions, but also the characteristics of an effective crossed aldol reaction.

1-Indanone
(2,3-Dihydroinden-1-one)

Veratraldehyde
(3,4-Dimethoxybenzaldehyde)

(E)-2-(3,4-Dimethoxybenzylidene)-
2,3-dihydroinden-1-one

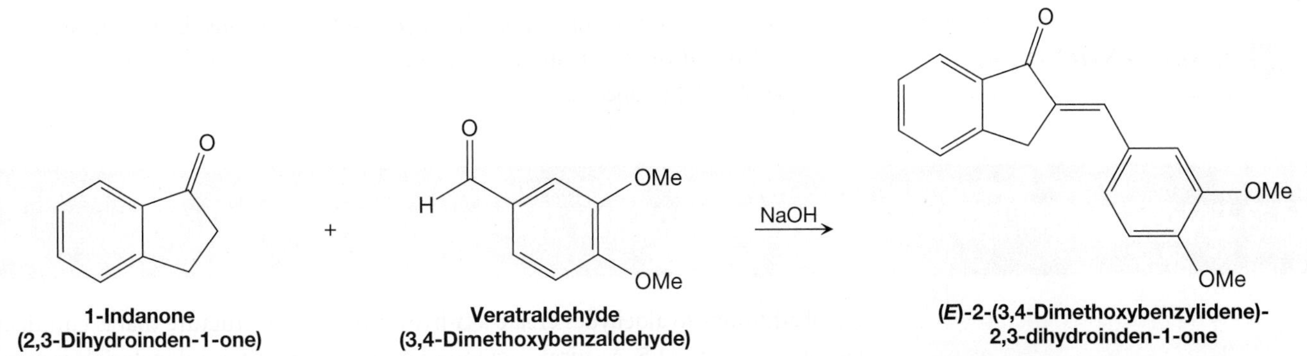

FIGURE 24.6

The aldol reaction of 1-indanone and veratraldehyde. Any small amount of the Z isomer that may form is removed during recrystallization.

1. In a small test tube, mix 0.20 g 1-indanone and 0.25 g veratraldehyde (3,4-dimethoxybenzaldehyde). Crush the solids completely, using a glass rod or metal spatula. (*CAUTION: Avoid pressing hard on the bottom of the test tube.*) Continue stirring until a yellowish oil forms. *Note to instructors: A mortar and pestle may be used in place of the test tube to allow for more efficient mixing and grinding of the solids.*

2. Add 50 mg of finely ground NaOH. If NaOH is furnished in pellet form, the pellet may be placed inside a piece of waxed paper or filter paper and crushed with a spatula. (*CAUTION: Avoid contact with NaOH! If you do get some on your skin, immediately wash thoroughly with water.*)

3. Stir the mixture thoroughly and continuously, scraping and crushing any solids that are present, until the whole mixture becomes solid. Allow another 15 min for the reaction to complete.

▶ **Neutralizing Acidic or Basic Solutions and Checking pH**

4. Quench the reaction by adding 10% aqueous HCl solution (2 mL). Scrape the solid off the walls of the test tube and break up any chunks. Using pH paper, check that the mixture is acidic (pH < 3).

▶ **Vacuum Filtration**

5. Isolate the solid product, (*E*)-2-(3,4-dimethoxybenzylidene)-2,3-dihydroinden-1-one, by vacuum filtration with a small Hirsch funnel and 1-cm filter paper, using small portions of water to rinse any remaining solid from the test tube.

▶ **Recrystallization**

6. Recrystallize from aqueous ethanol. Place the solid in a 50-mL Erlenmeyer flask, and place 20 mL 9:1 ethanol/water in another 50-mL Erlenmeyer flask. Heat the ethanol/water on a hotplate. Add small portions of hot ethanol/water to dissolve the solid product, while warming both flasks on the hotplate. Use only the minimum amount of solvent needed to dissolve the solid (less than 20 mL). Allow the mixture to cool to room temperature on the benchtop, then cool the product in an ice bath for a few minutes. Isolate the crystalline solid by vacuum filtration with a small Hirsch funnel and 1-cm filter paper, allowing vacuum to pull air through the product for several minutes to ensure complete dryness.

▶ **Melting Point Measurement**

7. Measure the mass of recrystallized (*E*)-2-(3,4-dimethoxybenzylidene)-2,3-dihydroinden-1-one and determine its melting point.[3] Acquire infrared and ^{1}H NMR spectra.

▶ **Preparing NMR Samples**

24.3 ENANTIOSELECTIVE PROLINE-CATALYZED ALDOL ADDITION

Aldol additions to aldehydes create a β-hydroxycarbonyl structure that is chiral, as we saw in Figure 24.5. A chiral compound produced from achiral materials is racemic; it consists of equal amounts of two enantiomers. Could we selectively produce

---

[2]Doxsee, K. M.; Hutchison, J. E. *Green Organic Chemistry: Strategies, Tools, and Laboratory Experiments*, 1st ed.; Thomson-Brooks/Cole: Boston, 2004.
[3]The melting point reported in the literature is 183–185°C. Rothenberg, G.; Downie, A. P.; Raston, C. L.; Scott, J. L. Understanding Solid/Solid Organic Reactions. *J. Am. Chem. Soc.* **2001**, *123*, 8701–8708.

one enantiomer? To achieve this from achiral reactants requires an enantioselective reaction, also known as an asymmetric synthesis. One way to do this is to include a chiral reagent or catalyst, causing the reaction to take place in a chiral environment.

Naturally occurring amino acids are generally found in enantiomerically pure form. One of these, (S)-proline, is a secondary amine, and serves as a chiral catalyst in aldol reactions.[4] It does this by forming a chiral enamine upon reaction with acetone. This enamine can add to an aldehyde, and the reaction occurs in the chiral environment of the (S)-proline. In this way, the new stereogenic center is created with selectivity for one enantiomer over the other. The approach of the enamine occurs selectively from one face because of the stereogenic center in the proline, with assistance from a hydrogen bond to the carboxylic acid group (**Figure 24.7**). This transition state is useful to predict the configuration of the product.

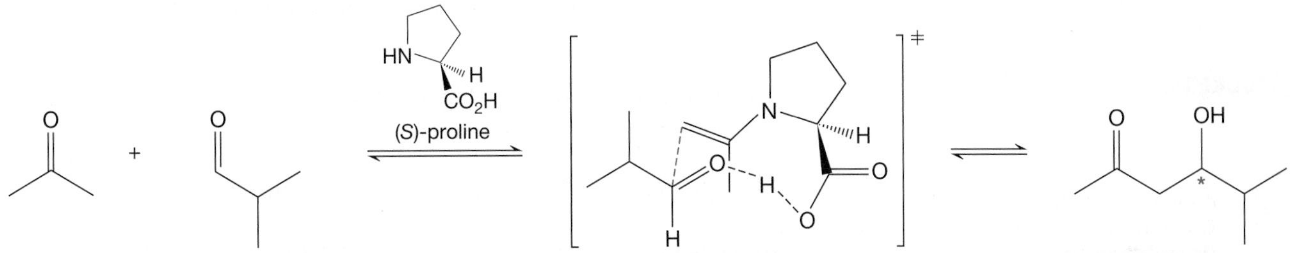

Acetone	**Isobutyraldehyde (2-Methylpropanal)**

Transition state

4-Hydroxy-5-methyl-2-hexanone

FIGURE 24.7

The aldol reaction of acetone and isobutyraldehyde, catalyzed by (S)-proline, showing the transition state for C—C bond construction. The forming C—C bond is highlighted in red (dashed line), and is being formed from behind the aldehyde carbonyl carbon in this perspective. The isobutyraldehyde and its hydrogen-bonding link to the (S)-proline (denoted by O---H) are in the foreground.

In this experiment, we will carry out the (S)-proline-catalyzed aldol addition reaction between acetone and isobutyraldehyde, and assess the enantiomeric purity of the product using polarimetry. One enantiomer rotates plane-polarized light clockwise (+), while the other rotates it counterclockwise (−). By comparison of the sign of rotation, we can determine which enantiomer was produced in the larger amount. Frequently, this is expressed in terms of percent enantiomeric excess (% ee), which is a measure of the amount of a single enantiomer that is present beyond the amount found in a racemic mixture. A racemic mixture is 0% ee, and a pure enantiomer is 100% ee. By dividing the observed specific rotation value by the value known in the literature for a pure single enantiomer, we can measure optical purity, which gives a close approximation of the % ee of the sample.

24.4 EXPERIMENTAL PROCEDURE[5]

24.4A Day 1

1. Place 0.45 mL isobutyraldehyde into a screw-cap vial.

2. Add 7.3 mL acetone and 0.11 g (S)-proline (L-proline).

[4]Schneider, J. F.; Ladd, C. L.; Bräse, S. *Sustainable Catalysis: Without Metals or Other Endangered Elements, Part 1*; Royal Society of Chemistry: London, 2015; pp 79–119.
[5]This reaction was originally reported in 2000. List, B.; Lerner, R. A.; Barbas, C. F., III. Proline-Catalyzed Direct Asymmetric Aldol Reactions. *J. Am. Chem. Soc.* **2000**, *122*, 2395–2396. The procedure in this chapter is adapted from Martínez, A.; Zumbansen, K.; Döhring, A.; van Gemmeren, M.; List, B. Improved Conditions for the Proline-Catalyzed Aldol Reaction of Acetone with Aliphatic Aldehydes. *Synlett*, **2014**, *25*, 932–934.

3. Cap the vial and swirl the mixture for about 5 minutes.

4. Store the reaction mixture for 1 week.

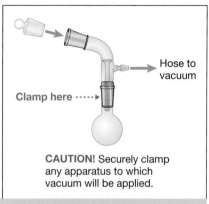

FIGURE 24.8

Clamp here ·····▶

Hose to vacuum

CAUTION! Securely clamp any apparatus to which vacuum will be applied.

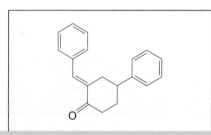

▶ **Aqueous–Organic Extractions and Drying Organic Solutions**

▶ **Rotary Evaporation**

▶ **Polarimetry**

24.4B **Day 2**

5. Transfer the mixture to a separatory funnel. Rinse the vial with 15 mL *tert*-butyl methyl ether and add that to the separatory funnel.

6. Wash the organic phase with 30 mL brine (saturated aqueous sodium chloride).

7. Dry the organic phase over anhydrous Na_2SO_4.

8. Transfer to a tared round-bottom flask and concentrate on the rotary evaporator.

9. Using the apparatus shown (**Figure 24.8**), apply vacuum for 15–20 minutes to remove the last traces of solvent.

10. Measure the mass of product, 4-hydroxy-5-methyl-2-hexanone.

11. Measure the optical rotation (section 5.7): Place about 50 mg of the sample in a volumetric flask (2 mL or 5 mL), recording the exact mass. Dilute to the volume line with $CHCl_3$ and invert the flask several times to ensure homogeneity. Transfer the solution to a clean polarimetry cell and measure the optical rotation on a polarimeter. Record both the sign and magnitude of the rotation.

12. Obtain an infrared spectrum, and submit a sample for ^1H NMR spectroscopy.

24.5 THE LABORATORY REPORT

FIGURE 24.9

In your laboratory report, include answers to the following questions:

1. In a solventless aldol condensation (section 24.2), what is the role of the melting point depression in ensuring the success of the reaction?

2. What are the structures of the precursors you would use to prepare the compound shown in **Figure 24.9** via an aldol condensation reaction?

3. In preparing the preceding compound from the precursors you've identified, would mixtures of different crossed aldol products be likely to form (as shown in Figure 24.3)?

4. What factors are likely involved in ensuring that the crossed aldol product is formed in the (*S*)-proline-catalyzed aldol addition reaction of acetone and isobutyraldehyde?

5. In the transition state for proline-catalyzed aldol addition (Figure 24.7), a hydrogen bond is shown between the carboxylic acid of proline and isobutyraldehyde. Explain how this could lower the energy barrier for the reaction.

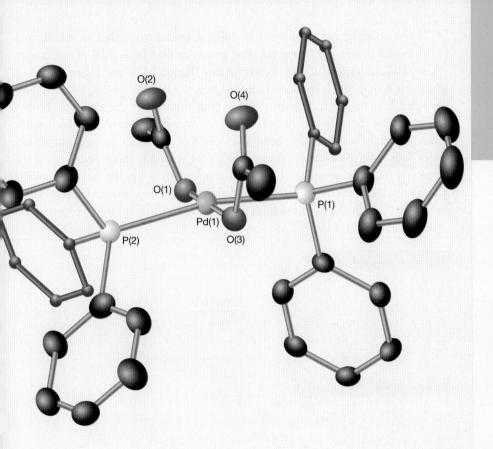

LEARNING OBJECTIVES

- Recognize the importance of transition metal–catalyzed reactions in forming carbon–carbon bonds.

- Execute a Pd-catalyzed Suzuki cross-coupling reaction under aqueous conditions.

- Calculate turnover number of a catalyst using stoichiometric data.

Suzuki Coupling: Pd-Catalyzed C—C Bond Construction in the Aqueous Phase

A TRANSITION METAL CATALYST

A palladium catalyst precursor, bis(triphenylphosphine)palladium(II) acetate, as seen by x-ray crystallography, containing two triphenylphosphine (PPh$_3$) ligands. In this experiment the triphenylphosphines are replaced by a water-soluble alternative so that the reaction may occur in aqueous media.

Republished with permission of The Royal Society of Chemistry. Scott, N. W. J.; Ford, M. J.; Schotes, C.; Parker, R. R.; Whitwood, A. C.; Fairlamb, I. J. S. The Ubiquitous Cross-Coupling Catalyst System 'Pd(OAc)$_2$'/2PPh$_3$ Forms a Unique Dinuclear Pd Complex: An Important Entry Point into Catalytically Competent Cyclic Pd$_3$ Clusters. *Chem. Sci.* **2019**, *10*, 7898–7906. DOI: 10.1039/C9SC01847F. ©2019; permission conveyed through Copyright Clearance Center, Inc.

C ross-coupling reactions join two different reactants together, increasing molecular complexity. These coupling processes may be carried out in crossed aldol and crossed Claisen condensations (**Figure 25.1a** and **Figure 25.1b**), in which one component is made into a nucleophilic enolate ion species. Not all desired cross-couplings involve components that can be deprotonated to form enolates, however, as shown for the cross-coupling of two aryl compounds in **Figure 25.1c**. Fortunately, different types of reactions have been developed that use transition metals to activate one or both of the precursors, so enolate ions are not needed. In this experiment you will perform a transition metal–catalyzed cross-coupling known as the Suzuki coupling.[1] This is a reaction that directly links two sp^2-hybridized carbons, an outcome that cannot be accomplished by the other reactions typically covered in an organic chemistry text.

FIGURE 25.1

Three different types of cross-coupling reactions. (a) Crossed aldol condensation, in which only one component is enolizable. (b) Crossed Claisen condensation, in which only one component is enolizable. (c) A cross-coupling reaction of two different aromatic rings using the Suzuki coupling. A variety of other substituents may be present on either or both rings. *Note:* The ambiguous positioning of the bond to the R substituent means that it could be ortho, meta, or para.

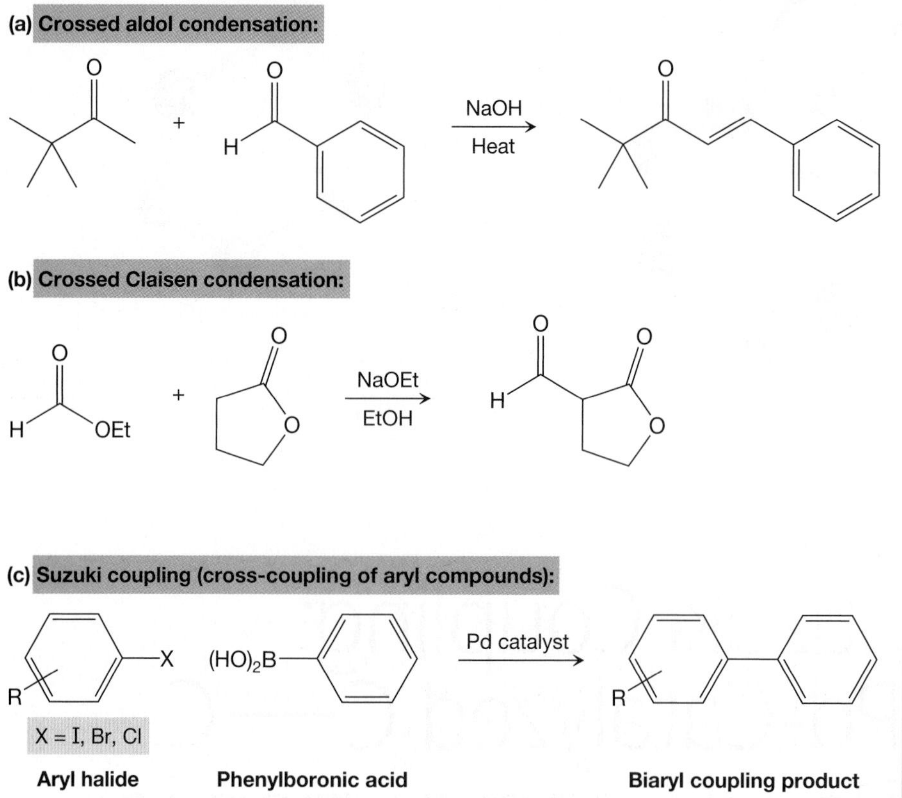

(a) **Crossed aldol condensation:**

(b) **Crossed Claisen condensation:**

(c) **Suzuki coupling (cross-coupling of aryl compounds):**

X = I, Br, Cl

Aryl halide   **Phenylboronic acid**   **Biaryl coupling product**

Pre-Lab Reading Assignment

- Experiment: Chapter 25 (this experiment)
- Technique: Extraction (Chapter 5, section 5.2)
- Technique: Recrystallization (Chapter 5, section 5.4)
- Technique: Melting point measurement (Chapter 5, section 5.6)
- Technique: Infrared (IR) spectroscopy (Chapter 6, section 6.2)
- Technique: Nuclear magnetic resonance (NMR) spectroscopy (Chapter 7, section 7.1)

---

[1]Miyaura, N.; Suzuki, A. Palladium-Catalyzed Cross-Coupling Reactions of Organoboron Compounds. *Chem. Rev.* **1995**, *95* (7), 2457–2483.

The Suzuki coupling (also known as the Suzuki–Miyaura reaction) can link two aromatic rings together in the presence of a wide range of other functional groups, such as carboxylic acids, which remain unchanged. In some cases, alkenyl or alkynyl reactants may also be coupled using variations of this reaction. The reaction works even in aqueous conditions that satisfy green chemistry principles! The versatility of the Suzuki coupling gives it special importance in medicinal chemistry, because chemists who work in drug discovery commonly need to forge libraries of many related compounds with linkages between aromatic rings. In other subdisciplines as well, synthetic chemists have come to rely on the Suzuki coupling for introducing new C—C bonds in a variety of contexts. The inventor of this reaction, Japanese chemist Akira Suzuki, shared the Nobel Prize in Chemistry in 2010 with Richard Heck and Ei-ichi Negishi, who invented very similar palladium-catalyzed coupling reactions. All three types are illustrated in a generalized way in **Figure 25.2**.

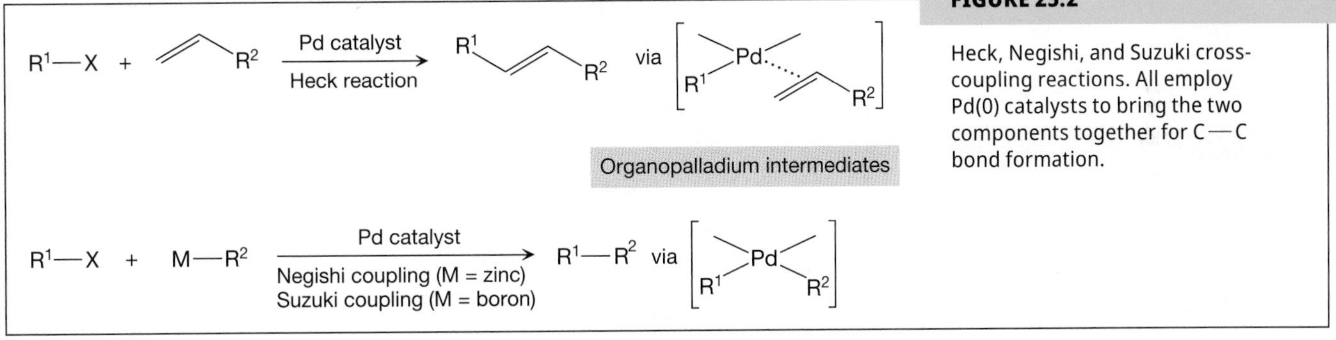

FIGURE 25.2

Heck, Negishi, and Suzuki cross-coupling reactions. All employ Pd(0) catalysts to bring the two components together for C—C bond formation.

The Suzuki coupling reaction enables sp^2–sp^2 cross-coupling, a type of reaction that is important in the synthesis of naturally occurring compounds and potential drug candidates, as well as various other applications. An example is the synthesis of the cardiovascular drug valsartan, which is an angiotensin II inhibitor that blocks metabolic processes that would otherwise narrow and tighten blood vessels; a key aryl–aryl bond was made using the Suzuki coupling (**Figure 25.3**).

In this experiment the Suzuki coupling is used to link two aromatic rings together. One component must have an aromatic ring bearing a halogen (usually Br or I) and behaves as an electrophile, while the other must have a boronic acid so that it behaves as a carbon nucleophile. The nucleophilicity of organoboronic acids derives from the polarized C—B bond that places a partial negative charge on carbon. Organomagnesium and organozinc reagents can also serve this purpose, although they lack compatibility with aqueous solvents. Palladium catalyzes the formation of a C—C bond from the nucleophilic and electrophilic reactants. The catalyst is recycled many times during the reaction, thus minimizing the amount of transition metal waste.

The mechanism of the Suzuki coupling proceeds through three key steps that are involved in many transition metal–catalyzed coupling reactions—namely, oxidative addition, transmetallation, and reductive elimination (**Figure 25.4**). The oxidative addition attaches the electrophilic partner (in our case, an aryl halide) to the palladium, the transmetallation attaches the nucleophilic partner (the boronic acid) to the palladium, and reductive elimination connects the two carbons, releasing the product from the palladium, which then can repeat the cycle.

These reactions are usually performed in organic solvents such as tetrahydrofuran, which present flammability hazards and waste disposal complications. Organic solvents are needed because the ligands attached to the Pd to keep the catalyst in solution are usually hydrophobic phosphines such as PPh$_3$. When these are attached

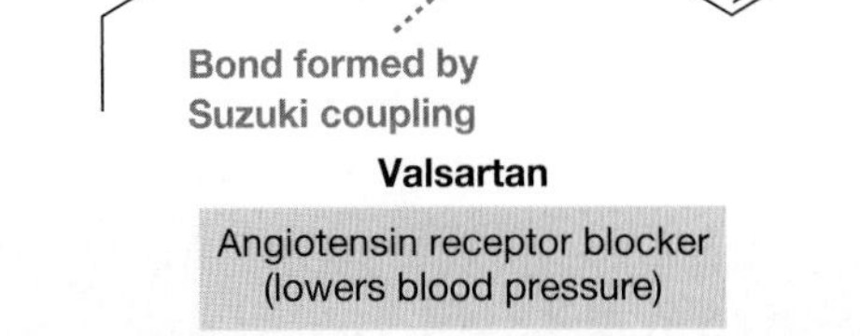

Bond formed by Suzuki coupling

Valsartan

Angiotensin receptor blocker
(lowers blood pressure)

FIGURE 25.3

The synthesis of valsartan, a cardiovascular drug, was accomplished using a Suzuki coupling to connect two aromatic rings.

FIGURE 25.4

The catalytic cycle for Suzuki coupling, showing the coupling of two different aryl groups, Ar^1 and Ar^2. Ligands are denoted by L, and are usually hydrophobic phosphine compounds. In some cases, alkyl or alkenyl groups may be used in place of the aryl groups in this reaction.

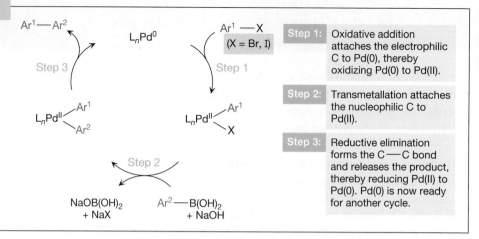

Step 1: Oxidative addition attaches the electrophilic C to Pd(0), thereby oxidizing Pd(0) to Pd(II).

Step 2: Transmetallation attaches the nucleophilic C to Pd(II).

Step 3: Reductive elimination forms the C—C bond and releases the product, thereby reducing Pd(II) to Pd(0). Pd(0) is now ready for another cycle.

to the Pd, the catalyst becomes hydrophobic as well. A recent innovation is to use a ligand called 2-amino-4,6-dihydroxypyrimidine (**Figure 25.5**), which is soluble in aqueous base. This ligand has two OH groups attached to an aromatic ring, and these can be deprotonated to make a water-soluble salt, much like the OH group of phenol can be deprotonated by NaOH to make a water-soluble phenolate ion. When these ligands are attached to Pd, the catalyst is also water soluble. As a result, Suzuki coupling reactions can now be carried out in environmentally benign aqueous solvents.[2]

FIGURE 25.5

The ligand 2-amino-4,6-dihydroxypyrimidine may be deprotonated by a base, making it into a salt that is water-soluble. This ligand is abbreviated **L**.

2-Amino-4,6-dihydroxypyrimidine $\xrightarrow[H_2O]{NaOH}$ **L** (Water soluble) + 2 Na⁺

In this experiment we will use Suzuki coupling in a reaction of *p*-bromobenzoic acid with phenylboronic acid, and the carboxylic acid group will be unchanged in the product (**Figure 25.6**, procedure A). However, many different aryl halides and boronic acids may be used in Suzuki coupling reactions. Your lab instructor will tell you if alternative aryl halides are to be used. Procedures B and C of Figure 25.6 use aryl iodides produced by electrophilic aromatic substitution (Chapter 19).

25.2 EXPERIMENTAL PROCEDURES

25.2A Suzuki Coupling with *p*-Bromobenzoic Acid

1. Begin heating a sand bath (or other heat source, such as heated metal block or heating mantle) to 70–80°C.

[2](a) This catalyst system has been employed for aqueous Suzuki–Miyaura couplings to a protein containing a halogenated aromatic amino acid. For details, see Chalker, J. M.; Wood, C. S. C.; Davis, B. G. A Convenient Catalyst for Aqueous and Protein Suzuki–Miyaura Cross-Coupling. *J. Am. Chem. Soc.* **2009**, *131*, 16346–16347. (b) Details on preparation of the catalyst are found in the supporting information of this paper: Hamilton, A. E.; Buxton, A. M.; Peeples, C. J.; Chalker, J. M. An Operationally Simple Aqueous Suzuki–Miyaura Cross-Coupling Reaction for an Undergraduate Organic Chemistry Laboratory. *J. Chem. Educ.* **2013**, *90*, 1509–1513.

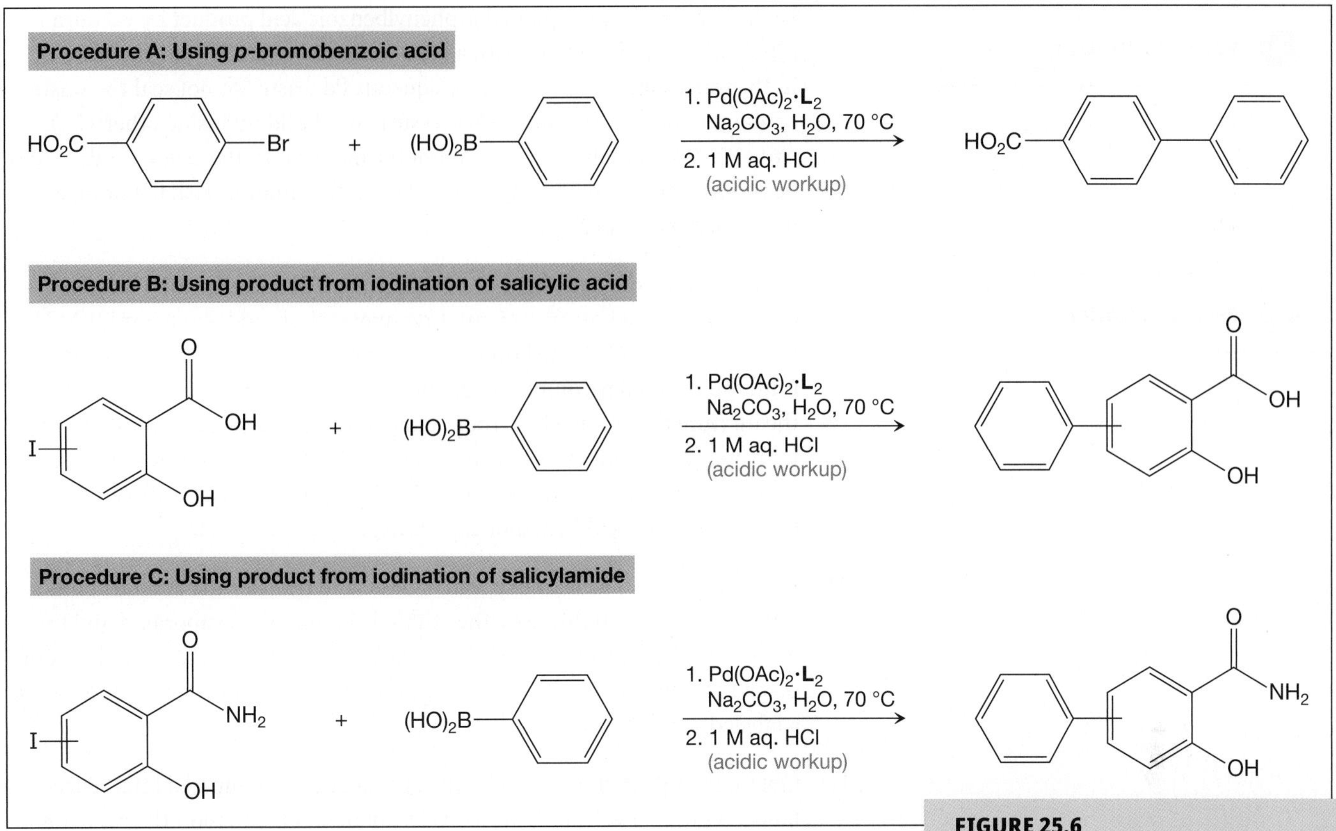

Procedure A: Using p-bromobenzoic acid

HO₂C–⟨benzene⟩–Br + (HO)₂B–⟨benzene⟩
→ (1. Pd(OAc)₂·L₂ / Na₂CO₃, H₂O, 70 °C; 2. 1 M aq. HCl (acidic workup))
→ HO₂C–⟨biphenyl⟩

Procedure B: Using product from iodination of salicylic acid

I–⟨salicylic acid⟩ (with CO₂H/OH, OH) + (HO)₂B–⟨benzene⟩
→ (1. Pd(OAc)₂·L₂ / Na₂CO₃, H₂O, 70 °C; 2. 1 M aq. HCl (acidic workup))
→ product

Procedure C: Using product from iodination of salicylamide

I–⟨salicylamide⟩ (with CONH₂, OH) + (HO)₂B–⟨benzene⟩
→ (1. Pd(OAc)₂·L₂ / Na₂CO₃, H₂O, 70 °C; 2. 1 M aq. HCl (acidic workup))
→ product

FIGURE 25.6

Suzuki coupling reactions that can be performed using the procedures outlined in this chapter. The ligand **L** comes from the deprotonation of 2-amino-4,6-dihydroxypyrimidine, as shown previously in Figure 25.5.

2. Prepare a solution of sodium carbonate (0.32 g, 3.0 mmol) in 6 mL deionized water in a 25-mL Erlenmeyer flask. Stir it with a magnetic stir bar until homogeneous.

3. Place 0.20 g of p-bromobenzoic acid (1.0 mmol) into a 50-mL round-bottom flask.

4. Add the solution of sodium carbonate (and the magnetic stir bar) to the flask containing the p-bromobenzoic acid. Then add 0.15 g of phenylboronic acid (1.2 mmol). Stir for about 10 minutes or until homogeneous.

5. Using a graduated Pasteur pipet or graduated plastic syringe (needle not necessary), transfer 0.4 mL of pre-prepared palladium catalyst (as a 0.025 M aqueous solution) into the reaction mixture.

6. Attach a reflux condenser to the flask and begin heating. Continue stirring the reaction mixture at 70°C for 60 minutes.

 Heating at Reflux

7. While the reaction continues, prepare an ice bath and and measure 1 M aqueous HCl (10 mL) into a small Erlenmeyer flask.

8. After 60 minutes at 70°C, allow the reaction to cool to room temperature. *Note:* If a black precipitate is present, it may be removed by pipet filtration through Celite (diatomaceous earth). Add a spatula-tip of Celite to the mixture, and a similar amount onto a cotton plug in a Pasteur pipet. Transfer the mixture to the pipet and allow it to drain through into a clean flask.

9. Using an ice bath, cool the flask containing the crude product mixture. While stirring, *slowly* add 10 mL of 1 M aqueous HCl. A precipitate should form at this time. Using pH paper, check to ensure that the mixture is acidic (pH < 4) and if not, adjust it by adding a few more drops of 1 M aqueous

 Neutralizing Acidic or Basic Solutions and Checking pH

Vacuum Filtration

HCl. Separate the precipitated p-phenylbenzoic acid product by vacuum filtration, rinsing the flask with a little water (2 × 2 mL). Pour the filtrate in the waste container marked for aqueous Pd waste (do not seal the waste container, though, because CO_2 pressure may build up inside otherwise). *Note:* This is a suitable stopping point if time is short; the crude product may be transferred to a screw-cap vial or 25-mL Erlenmeyer flask for storage until the next lab period.

Recrystallization

10. Recrystallize from a mixture of 1 M aqueous HCl and ethanol. Place the crude solid precipitate in a 25-mL Erlenmeyer flask, add about 2 mL 1 M HCl, heat to about 70°C on the hotplate or sand bath, then add ethanol dropwise until the material is completely dissolved. Remove from the heat and allow the mixture to cool slowly to room temperature, then place the flask into an ice bath for 15 minutes. Isolate the recrystallized p-phenylbenzoic acid using vacuum filtration. If necessary, any remaining crystals may be rinsed out of the flask with ice-cold ethanol (2 × 2 mL).

11. Preweigh ("tare") a clean dry watch glass, and transfer the crystals onto it to dry. After 5–10 minutes, the ethanol should have evaporated and the mass of p-phenylbenzoic acid should be constant. Calculate the amount of product by subtracting the tare weight of the paper. Transfer the product to a labeled screw-cap vial.

12. Obtain the infrared spectrum of the product using attenuated total reflectance (ATR), if available. Alternatively, obtain a transmission IR spectrum using a $CHCl_3$ solution on a salt plate. The product may be sparingly soluble, so place a few crystals in a test tube and add a few drops of $CHCl_3$, then warm it for a few seconds on the sand bath. Transfer a drop of the supernatant liquid to the salt plate and cover with a second salt plate. Acquire the spectrum.

Melting Point Measurement

13. Determine the melting point of the product (lit. 222–225°C).

Preparing NMR Samples

14. *Optional:* Submit a sample for ^{1}H NMR acquisition.

25.3 THE LABORATORY REPORT

In your laboratory report, provide evidence of both the identity and the purity of your product. Include answers to the following post-lab questions.

1. With respect to the aqueous Suzuki coupling you have just performed, would you expect this reaction to have similar success with both p-bromobenzoic acid and p-bromobenzyl alcohol? Why or why not?

2. The average number of times that catalyst molecules are recycled in the reaction is called the turnover number. It can also be described as the number of moles of product that are produced by a mole of catalyst before it loses its activity. Assuming the catalyst was no longer active at the end of your reaction, what was the turnover number for your experiment?

3. Calculate the mass (in grams) of elemental palladium that is present in the waste from your reaction. If the palladium reagent was used in a 1:1 mole ratio with the reactants, how many grams of elemental palladium would be present in the waste material?

LEARNING OBJECTIVES

- Determine the chemical, physical, and spectroscopic properties of an unknown compound.

- Choose appropriate laboratory procedures that distinguish between compounds of similar physical properties.

- Identify the structure of an unknown compound using logic and critical thinking to evaluate experimental results, some of which may conflict.

Identification of an Unknown Organic Compound

FLAME ANALYSIS

When a halogen-containing organic compound is present on a copper wire, a bright green color will appear in the flame when the wire is placed into a Bunsen burner. This reaction is known as the Beilstein test.

Organic chemists often need to identify unknown compounds in the course of organic chemistry research. They may also be consulted by specialists in other areas. For example, organic chemists may be needed to analyze a substance found at a crime scene or a pollutant found on the site of an environmental cleanup. Or, organic chemists may be asked to determine the identity of a metabolic by-product from an experimental drug that is causing liver damage in clinical trials. How do organic chemists approach problems like these? In this experiment you get to develop some logical reasoning and critical thinking skills that can be applied to these types of problems, as well as a wide range of future careers in science.

Pre-Lab Reading Assignment

- Experiment: Chapter 26 (this experiment)
- Technique: Recrystallization (Chapter 5, section 5.4)
- Technique: Determination of physical properties (Chapter 5, section 5.6)
- Technique: Infrared (IR) spectroscopy (Chapter 6, section 6.2)
- Technique: Nuclear magnetic resonance (NMR) spectroscopy (Chapter 7, section 7.1)

Pre-Lab Writing Assignment

To prepare for efficient and accurate interpretation of your infrared spectrum, fill out a table listing the diagnostic peaks for various functional groups. An example of a blank table is found at the end of the chapter.

26.1 BACKGROUND

You have already studied some of the ways to prove the identity of a product you obtain from a well-understood reaction—namely, physical constants and spectroscopic analysis may be compared with literature data. When presented with an unknown compound, however, there may be no knowledge of the history of the sample, and no knowledge of the reactants that were used to make it. The physical constants and spectroscopic data may be gathered as usual, but there are several million organic compounds in the literature with which to compare your data. Finding which one of these is your unknown becomes an entertaining problem of logical deduction—a tangled web of a puzzle that is best navigated with a systematic approach.

In this experiment, you determine the identity of an unknown compound using a combination of physical properties, chemical tests, and spectroscopic data. Your compound will be chosen at random, and each student will have a different compound. With deductions gathered from the analytical procedures you choose, you will apply a logical process to gradually eliminate functional group classes, then rule out specific compounds until there is only one left that is consistent with the data you've collected.

Beware! There is no universally applicable set of instructions for solving the structure of an unknown. The first steps may be the same for everyone, but then you will need to think for yourself and choose the tests and procedures that are best suited to your unknown. Be prepared to review your logic if you reach a dead end. You may need to continually modify your process of elimination as new information becomes available. Think very carefully about what you will do each day *before* you arrive in the lab, and have a backup plan in case of inconsistencies in the experimental results.

Keep a good written account of each step, and record your results and observations clearly. You'll need these observations in order to evaluate the relative importance of contradictory pieces of information. "False positives" and "false negatives" are commonly observed, so the results from two tests may contradict each other. If you recorded in your notebook that one of these tests gave an ambiguous result (perhaps a color change was not as dramatic as you expected), then you may be able to use this to decide which data should be given more weight in your analysis. Keeping good notes is also important because the quality of your instructor's advice will depend on the completeness of the observations in your notebook.

26.2 EXPERIMENTAL PROCEDURE, OVERVIEW

The following discussion outlines the general steps to be followed in the determination of your unknown. Every case is different, though, so be prepared to modify the steps as needed once you begin to accumulate information.

1. *Get a sample.* Obtain the randomly assigned unknown sample from your lab instructor. Note the unknown number in your notebook.

2. *Physical state.* Note the physical state (e.g., solid or liquid) and its appearance and color.

3. *Melting point or boiling point.* Obtain either a melting point (mp) or a micro-boiling point (bp) for your sample. These values, particularly micro-boiling points, can be very uncertain. When considering possible structures for your unknown, add ±5–10°C to the melting point range and ±10–20°C to the micro-boiling point. The micro-boiling points become increasingly unreliable above about 200°C. More reliable data can be obtained if the mp/bp measurement is repeated with a fresh capillary tube. Liquid samples can be saved after this procedure to provide a backup in the event that you run out of sample. However, these recovered samples should be used only as a last resort because they may contain impurities from any decomposition that occurred during the micro-boiling point determination.

Melting Point Measurement

Boiling Point Measurement

4. *Solubility tests.* Conduct solubility tests with water, aqueous NaOH, aqueous $NaHCO_3$, and aqueous HCl (see section 26.3) to help identify the major functional group(s) present.

5. *Infrared spectrum.* Obtain a *good, clean* IR spectrum of the unknown to identify the major functional groups present. **Table 26.1** lists the frequencies of the characteristic functional group absorbances in IR spectra. Depending on the equipment available, you may obtain a transmission spectrum by passing IR through the sample, or attenuated total reflectance (ATR) spectrum, in which solid or liquid samples are placed directly on a small window and analyzed by reflectance. ATR has lower peak intensities above 2000 cm^{-1}. For a liquid, placing a drop of the liquid on a salt plate is simple and effective. If the sample is a solid, dissolve your compound in the minimum amount of $CHCl_3$ or CH_2Cl_2 necessary and place a drop of

TABLE 26.1

Characteristic Functional Group Absorbances in Infrared Spectra

FREQUENCY (cm^{-1})	BOND	FUNCTIONAL GROUP
3640–3610 (s, sh)	O—H stretch, free hydroxyl	Alcohols, phenols
3500–3200 (s, br)	O—H stretch, H–bonded	Alcohols, phenols
3400–3250 (m)	N—H stretch	Primary, secondary amines, amides
3300–2500 (m, br)	O—H stretch	Carboxylic acids
3330–3270 (m, s)	—C≡C—H: C—H stretch	Alkynes (terminal)
3100–3000 (m)	C—H stretch	Aromatics
3100–3000 (m)	=C—H stretch	Alkenes
3000–2850 (m)	C—H stretch	Alkanes
2830–2695 (m)	H—C=O: C—H stretch	Aldehydes
2260–2210 (v)	C≡N stretch	Nitriles
2260–2100 (w)	—C≡C— stretch	Alkynes, non-symmetrical
1760–1665 (s)	C=O stretch	Carbonyls (general)^a
1760–1690 (s)	C=O stretch	Carboxylic acids
1750–1735 (s)	C=O stretch	Esters, saturated aliphatic
1740–1720 (s)	C=O stretch	Aldehydes, saturated aliphatic
1730–1715 (s)	C=O stretch	α,β-Unsaturated esters
1715 (s)	C=O stretch	Ketones, saturated aliphatic
1710–1665 (s)	C=O stretch	α,β-Unsaturated aldehydes, ketones
1680–1630 (s)	C=O stretch	Amides
1680–1640 (m)	—C=C— stretch	Alkenes
1600–1585 (m)	C—C stretch (in–ring)	Aromatics
1550–1475 (s)	N—O asymmetric stretch	Nitro compounds
1500–1400 (m)	C—C stretch (in–ring)	Aromatics
1470–1450 (m)	C—H bend	Alkanes
1370–1350 (m)	C—H rock	Alkanes
1360–1290 (m)	N—O symmetric stretch	Nitro compounds
1335–1250 (s)	C—N stretch	Aromatic amines
1320–1000 (s)	C—O stretch	Alcohols, carboxylic acids, esters, ethers
1300–1150 (m)	C—H wag (—CH$_2$X)	Alkyl halides
1250–1020 (m)	C—N stretch	Aliphatic amines

(continued)

TABLE 26.1

Characteristic Functional Group Absorbances in Infrared Spectra *(continued)*

FREQUENCY (cm^{-1})	BOND	FUNCTIONAL GROUP
1000–650 (s)	$=$C—H bend	Alkenes
950–910 (m)	O—H bend	Carboxylic acids
910–665 (s, br)	N—H wag	Primary, secondary amines
900–675 (s)	C—H "oop"	Aromatics
850–550 (m)	C—Cl stretch	Alkyl halides
725–720 (m)	C—H rock	Alkanes
700–610 (br, s)	—C$\equiv$C—H: C—H bend	Alkynes
690–515 (m)	C—Br stretch	Alkyl halides

(s) = strong, (m) = medium, (w) = weak, (br) = broad, (sh) = sharp, (v) = variable
^{a}Conjugation of other unsaturated groups with the carbonyl will lower the C$=$O frequency by 20–40 cm^{-1}.

the solution on the salt plate. Note that $CHCl_3$ or CH_2Cl_2 have their own peaks in the IR. If these interfere in your analysis, run a background with the $CHCl_3$ or CH_2Cl_2 alone so that those peaks are subtracted. If the solid is insoluble in $CHCl_3$ or CH_2Cl_2, prepare a very finely ground solid and use either a KBr pellet (refer to section 6.2B for details on the preparation of a pellet) or ATR probe to obtain the IR spectrum. In any of these methods, if more than one peak reaches 0% transmittance, then you have too much sample absorbing the infrared light. For a liquid or solution sample on salt plates, simply take the plates apart, wipe one off, put them back together, and obtain the spectrum again. For a KBr pellet, prepare a new pellet using less of the sample.

6. *Functional group tests.* Conduct chemical tests for functional group properties (see section 26.4) that are *necessary* to confirm or determine the nature of the functional groups present. To minimize false positives and/or false negatives, each selected functional group test should be run with two samples side-by-side. One should be a positive control, a known compound that contains the functional group you are testing, so that you'll know exactly what the positive test looks like. Negative controls or "blank" experiments may also be useful in some cases, and can be done by simply leaving out the unknown. You probably won't have to run all of the possible tests on your sample, but you must run at least three.

Confirming the functional group narrows your list of compounds from several hundred down to probably 50–100. Your melting or boiling point should then restrict your unknown's identity to a list of about 5–10 compounds. Quick, simple tests for aromaticity and halogen can rule out a few more possibilities.

7. *Flame test (section 26.4B).* Burn a small sample of the unknown in the hood to determine if aromatic rings are present. Check with your instructor before using any flames in the lab.

Preparing NMR Samples

Recrystallization

8. *Beilstein test (section 26.4C).* Conduct the Beilstein test (copper wire) to determine if halogen is present. Check with your instructor before using any flames in the lab.

9. *Short list.* Using your experimental data (mp/bp, functional group tests), develop a list of 3–5 possible structures from the tables available in your laboratory and on the web. If these are not all the same functional group, you may need to go back to solubility and IR data to identify your functional group. Submit this list of 3–5 possibilities to the person designated by the instructor to distribute NMR data. A sample submission form for such a list is found at the end of the chapter.

10. *Feedback.* If your list of 3–5 compounds contained your actual unknown, you may receive an NMR spectrum. If your list did *not* contain your unknown, you will receive feedback on whether you are on the "right track" or not, and you will be able to submit a second list later after further consideration of the results, including additional experimentation if necessary.

11. *Derivatives.* The melting point of an appropriate derivative can distinguish between alternative structures. Synthesize a derivative of your unknown (see section 26.5), recrystallize it, and obtain its melting point. Impure compounds exhibit melting point depression as well as a broader melting range, so a melting range of 2.0°C or less is indicative of a reasonably pure compound. If the range is greater than 2.0°C, or if the mp doesn't match any of the literature values from the tables, recrystallize it again, and get a more accurate melting point.

12. *Propose structure.* Using your accumulated information, propose a structure for your unknown.

13. *Write your report.* Discuss the logical process of elimination and explain how inconsistencies in the data (if any) were resolved.

CAUTION: If you run out of your unknown and require more, this may affect your grade. Work carefully, don't waste the unknown on unnecessary tests, and keep your sample safe and tightly sealed! If you need to repeat some procedures, it's worth noting that most of the tests and derivatives can be done on smaller scale, as long as all the other reactants and solvents are scaled accordingly.

26.3 SOLUBILITY TESTS

The solubility characteristics of an organic compound in water, aqueous base, and aqueous acid can often be enough to identify acidic or basic functional groups in the compound. Additionally, solubility may provide some information about the molecular weight or the presence or absence of other functional groups. There are four standard solubility tests that are most useful—namely, the solubility in water, 5% aqueous sodium hydroxide, 5% aqueous sodium bicarbonate, and 5% aqueous hydrochloric acid.

26.3A Procedure

Prepare a hot-water bath with about 2 cm of water in a beaker. A fresh sample of the unknown should be used for each solubility test. Place 0.1 g of a solid or 0.2 mL of a liquid in a test tube with 3 mL of the chosen solvent. It is best to test water first; if it is soluble the other tubes are not needed. The amounts of sample and solvent do not need to be exact. They may be measured for the first test and estimated thereafter. Mix the sample well and set it aside for a few minutes while you prepare tubes of solvent for the other solubility tests. If the unknown is not dissolved after a few minutes at room temperature, dip it briefly (less than 10 seconds) in the hot-water bath. Check the solubility after the tube reaches room temperature. Do not heat the acid or base solubility tests for more than a few seconds; false positives can occur upon heating, especially with functional groups that can be hydrolyzed.

Smaller-scale tests: Smaller amounts of unknown may be used, but to avoid false positives the solvent must be scaled down accordingly (e.g., use 10 mg unknown in 0.3 mL solvent).

26.3B | Interpretation

For the purpose of functional group classification, a substance is said to be soluble in H_2O if 30 mg or more of the solute dissolves completely in 1 mL of water. If the unknown is *soluble* in water, then its solubility in aqueous solutions is *not* meaningful for functional group identification, and there is no need to prepare all four tests. However, the pH of the aqueous solution can be tested using pH paper to reveal the presence of acidic or basic functional groups.

▶ **Neutralizing Acidic or Basic Solutions and Checking pH**

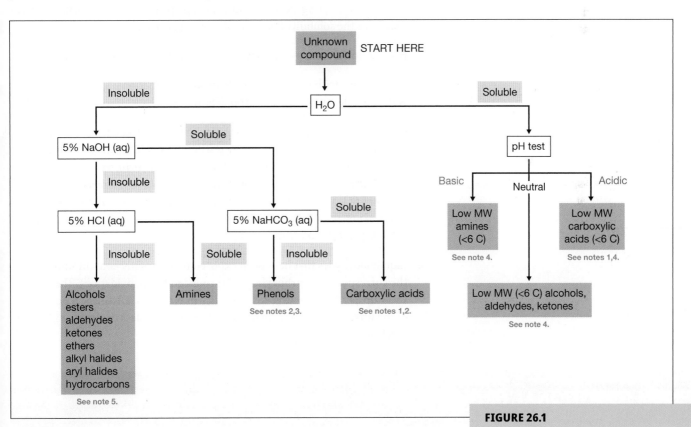

FIGURE 26.1

Solubility flowchart for determining the functional group of an unknown compound. The "notes" are listed in section 26.3C.

The solubility in aqueous NaOH, $NaHCO_3$, and HCl solutions should be determined by comparison to the water solubility. If the compound is noticeably more soluble in acid (or base) than in water alone, it is then considered to be soluble, even if it doesn't completely dissolve. Once you have determined whether your unknown is soluble or insoluble in each of the four solutions (H_2O, aqueous NaOH, aqueous $NaHCO_3$, and aqueous HCl), you may use the flowchart in **Figure 26.1** to identify the functional group(s) that may be present. Although it may not seem important to test the solubility in HCl if the compound is soluble in NaOH, experience suggests that (a) comparisons of positive and negative solubility results strengthen the conclusion, and (b) false positives and erroneous conclusions can waste greater quantities of time and sample than doing an extra solubility test. So, it is worthwhile to do all four simultaneously.

Occasionally, compounds may have inconclusive or misleading solubility data because of the electronic influences of a second functional group. For example, phenols bearing nitro groups may be soluble in $NaHCO_3$ because the electron-withdrawing nitro group decreases the pK_a of the phenolic OH group sufficiently to make it behave as a carboxylic acid. Therefore, the solubility data should be accompanied by additional evidence, such as the IR spectrum or functional group tests, before making a final decision on the identity of the functional group.

26.3C Important Notes for Figure 26.1 about the Interpretation of Solubility Data

1. If the compound is water soluble, most of the other solubility tests are useless because they are aqueous and may therefore yield false positives. One important exception is water-soluble acids, which vigorously evolve gas (CO_2) when treated with 5% $NaHCO_3$(aq).

2. Phenols with strong electron-withdrawing groups (especially ortho or para) may have the solubility behavior of carboxylic acids.

3. β-Dicarbonyl compounds may have the solubility behavior of phenols.

4. Compounds with more than one hydrophilic functional group may be water soluble, even with six carbons or more.

5. Organic halides and hydrocarbons can be distinguished from neutral oxygen-containing compounds by testing the solubility in concentrated H_2SO_4. (*CAUTION: Concentrated H_2SO_4 is highly corrosive and can cause severe burns. Obtain permission from your instructor before using concentrated H_2SO_4.*) (*Note to instructors: For this experiment, it is recommended that unknowns be selected so that students may omit the concentrated H_2SO_4 solubility test for safety reasons.*)

6. Liquid aldehydes that have been exposed to air usually contain some of the corresponding carboxylic acid (produced by air-oxidation). This may show up as a false positive for solubility in $NaHCO_3$. Distilling a portion of the unknown can resolve this problem, which is less common with solid aldehydes.

26.4 FUNCTIONAL GROUP TESTS

Solubility data and infrared spectroscopy are generally the best indicators of functional groups in your unknown. The following tests can confirm the identification of functional groups in your unknown, or in cases where solubility or IR are unclear, they can provide critical data to distinguish between different possibilities. There are many, many functional group tests, and this list gives the most important ones you'll need.

Bromine/CCl$_4$ Test for Alkenes

Reference: Harwood and Moody, p 245.

Procedure: In a small test tube, dissolve 10–20 mg (2–4 drops of a liquid) of your unknown in ~2 mL THF or CH$_2$Cl$_2$. Add a solution of 5% bromine in CCl$_4$ dropwise, with shaking, and observe the results.

Interpretation: A positive test for unsaturation results in a discharge of the bromine color (decolorization of the solution). Some alkenes react very slowly with this reagent, if at all. Also, the results of the test must be observed immediately because the color may fade over time as the bromine evaporates.

Flame Test for Aromatic Compounds

Reference: Harwood and Moody, p 232.

Procedure: Check with your instructor before using any flames in the lab. Place 10–20 mg (2–4 drops of a liquid) of your unknown on a spatula, and briefly place it in the flame of a Bunsen burner (in the hood). Pull the spatula away from the flame and observe the nature of combustion, the color of the flame, and the presence or absence of smoke. Try toluene, hexane, and 2-propanol as control experiments for comparison with your unknown.

Interpretation (smoke): A black, sooty smoke is indicative of an aromatic species. Some unsaturated molecules (i.e., alkenes and alkynes) and very long-chain alkanes may also give positive tests. The absence of sooty smoke suggests the presence of an aliphatic compound. Compounds with multiple functional groups or combined aromatic/aliphatic structures may give ambiguous results. The results are not strong evidence on their own, but may be useful in conjunction with other observations.

Interpretation (nature of combustion): Rapid and instantaneous combustion indicates high oxygen content, whereas combustion with sparks indicates a high nitrogen content or nitro groups.

Interpretation (color of the flame): A yellow *smoky* flame indicates aromatic, unsaturated, or high molecular weight aliphatic compounds, whereas a yellow *nonsmoky* flame indicates lower molecular weight aliphatic organic compounds. A clear bluish flame indicates oxygen-rich compounds (e.g., ethanol).

This flame test may provide additional evidence for the presence of some functional groups, but it should not be given more weight than the data from solubility tests or infrared spectroscopy.

Tests for Alkyl or Aryl Halides

1. BEILSTEIN TEST

Reference: Ault, p 242.

Procedure: Check with your instructor before using any flames in the lab. Form a small coil in the end of a copper wire by making a couple of turns around a nail or glass rod or similar object. Heat the tip of the copper wire in a Bunsen burner flame until no further coloration of the flame is noticed. Allow the wire to cool, then dip it into a small sample of the unknown (solid or liquid), and place the wire into the flame again.

Interpretation: A transient green color forms in the flame when traces of copper are made volatile in the presence of the halogenated organic compounds. The green flash, which may be very brief, constitutes a positive test for halogen, indicating the presence of chlorine, bromine, or iodine. Fluorine is not detected. Alkyl and aryl halides are indistinguishable because both give positive tests.

2. SILVER NITRATE TEST

Reference: Harwood and Moody, pp 247–248.

Procedure: Add 1 drop (about 10 mg if solid) of the unknown to 2 mL of a 0.1 M solution of silver nitrate in ethanol. Let it stand for 5 minutes at room temperature. If no precipitate is observed after this time, heat the solution in a water bath and observe any change. Note the color of any precipitate that is formed.

Interpretation: This reaction exhibits relative reactivities typical of S_N1 reactions. Thus, benzylic, allylic, and tertiary alkyl halides give an immediate precipitate at room temperature, whereas primary and secondary alkyl halides give a precipitate upon heating. Aryl and alkenyl halides do not react, even with heating. The color of the precipitate may suggest which halogen is present: White = Cl, pale yellow = Br, and yellow = I. Fluorine is not detected. Carboxylic acids may give a false positive—namely, a precipitate that is the silver carboxylate salt (RCO_2Ag). To detect this false positive, add two drops of 5% nitric acid; silver carboxylates will dissolve but silver halides will not.

26.4D Tests for Alcohols

1. CHROMIC ACID TEST

Reference: Shriner et al., pp 149–150.

Procedure: Dissolve 1 drop of a liquid unknown or ~10 mg of a solid unknown in 1 mL of reagent-grade acetone. In a second tube, place only the acetone as a blank or negative control. Add 1 drop of the chromic acid reagent to each tube and immediately examine the colors of both tubes.

Interpretation: A positive test for a primary or secondary alcohol is the immediate appearance of a blue-green color within 2 s. Tertiary alcohols do *not* react within that time period, so the color remains orange. Aldehydes can also be oxidized with chromic acid, so they give a positive in this test. Acetone may contain trace amounts of isopropanol, leading to a false positive, which can be detected in the blank test tube. The yellow-orange color should persist in the blank for at least 3 s. If the acetone blank (negative control) gives a color change, inform your TA or instructor, and request pure acetone.

2. LUCAS TEST

Reference: Harwood and Moody, p 245.

Procedure: Lucas reagent is an equimolar solution of anhydrous $ZnCl_2$ and concentrated HCl prepared by the instructor or other laboratory staff. Combine 0.5 mL (or about 0.4 g of solid) unknown with 3 mL Lucas reagent in a test tube. (*CAUTION: Strong acid!*) Stopper the tube, shake for 15 s, then allow the mixture to stand. After 5 minutes, observe whether there is a precipitate (usually the precipitate is a liquid).

Interpretation: This test depends on how easily carbocation intermediates form from the alcohol. In the Lucas test, these carbocation intermediates are converted to the corresponding alkyl chloride. A positive test usually appears as a liquid precipitate or a separate liquid layer. How rapidly this precipitate forms depends on the structure of the alcohol. Because you're looking for a precipitate, the test works well for those alcohols that are initially soluble in the reagent (generally liquid alcohols of low molecular weight).

Formation of a precipitate within 2 minutes indicates a benzylic, allylic, or tertiary alcohol (rapid S_N1). Formation of a precipitate after 10 minutes indicates a secondary alcohol (slow S_N1). And *no* precipitate indicates a primary alcohol (no S_N1 under these conditions).

26.4E | Tests for Phenols

1. FERRIC CHLORIDE TEST

Reference: Harwood and Moody, p 251.

Procedure: Add 1 drop of liquid unknown or ~10 mg of solid unknown to 2 mL of water. Add several drops of ferric chloride solution and observe the color immediately.

Interpretation: Most phenols produce an intense red, blue, purple, or green color. Some colors are transient and must be viewed immediately upon mixing. Some phenols, especially sterically hindered ones, do not give a positive test, so a negative test is unreliable evidence. Esters and 1,3-dicarbonyl compounds sometimes also show intense coloration (false positive).

2. BROMINE-WATER TEST

Reference: Harwood and Moody, p 269.

Procedure: Dissolve 1 drop of liquid unknown or ~10 mg of solid unknown in 1 mL of EtOH. Add the bromine reagent dropwise until the yellow color persists, shaking the reaction mixture after each addition. Moisten a piece of pH test strip or litmus paper and hold it at the mouth of the test tube, watching for color change.

Interpretation: The disappearance of the yellow bromine coloration indicates that the bromine has been consumed, and this suggests that the electron-rich aromatic ring of a phenol has been brominated. The color of the pH test strip or litmus paper turns pink in a positive test due to the evolution of HBr. A precipitate may form in some cases, and on a larger scale this could be a useful derivative. Other highly activated aromatic compounds (e.g., aromatic ethers) and β-dicarbonyl compounds could also give a positive result.

26.4F | Tests for Aldehydes and Ketones

1. 2,4-DINITROPHENYLHYDRAZINE (DNP) TEST

Reference: Harwood and Moody, pp 241–242.

Procedure: The DNP reagent is a solution of 2,4-dinitrophenylhydrazine and sulfuric acid in aqueous ethanol, prepared by the instructor or other laboratory staff. Dissolve 2–3 drops or ~50 mg unknown in a few drops of methanol. Add 1 mL of the DNP reagent and shake. If no precipitate is formed, heat the mixture for 3 minutes in a hot water bath (about 60°C) and cool in ice.

Interpretation: The appearance of a red, orange, or yellow precipitate (ppt) indicates the presence of an aldehyde or ketone. The color of the ppt can distinguish conjugated carbonyls (directly attached to an alkene or aromatic ring) from nonconjugated carbonyls. A yellow ppt indicates the carbonyl is nonconjugated, whereas an orange or red ppt indicates conjugation. Whether or not a precipitate forms is the strongest evidence from this test; the color can sometimes be misleading. The precipitate is the DNP derivative, but the amount may be too small to recrystallize conveniently. The derivative procedure (section 26.5C) should be used to make the product on a larger scale.

2. CHROMIC ACID TEST

Reference: Shriner et al., pp 149–150.

Procedure: (listed under Tests for Alcohols)

Interpretation: A positive test for an aldehyde is the appearance of a blue-green color within 2 s. Ketones are negative in this test. This test also is positive for primary and secondary alcohols. Chromic acid oxidizes both primary alcohols and aldehydes to carboxylic acids.

3. TOLLENS TEST

Do not use the Tollens test unless the DNP test confirms an aldehyde or ketone.

Reference: Harwood and Moody, p 242.

Reagent: The Tollens reagent must be freshly prepared. To a *new* test tube, add 2 mL of 5% $AgNO_3$ solution and one drop of 10% NaOH solution. Slowly add 2% NH_4OH solution dropwise, with thorough mixing between each drop added, until the precipitate of silver oxide has almost disappeared, but is not completely gone (too much ammonia will decrease the sensitivity of the reagent). Use only the supernatant solution in the following test procedure; transfer it by pipet and leave the solid behind.

Procedure: Have a 70°C water bath ready. Place 5–6 drops of the unknown liquid (or ~0.1 g of unknown solid) in a *new* test tube, then add 1 mL of freshly prepared Tollens reagent. Look for the formation of a silver mirror on the walls of the test tube, or a black precipitate. If there is no black precipitate or silver mirror forming after 10 minutes, warm the tube by immersing in the hot-water bath (>70°C) for at least 10 minutes, then check again.

Interpretation: The formation of Ag metal coating (silver mirror) inside the test tube, or a black ppt of finely divided Ag metal, constitutes a positive test, indicating the presence of an aldehyde. The Ag(I) in the reagent produces Ag(0) as it oxidizes the aldehyde to a carboxylate salt. Ketones are not oxidized under these conditions and give a negative test. Used or dirty test tubes may result in a black precipitate instead of the silver mirror. Some aromatic amines and phenols give a false positive in this test. Liquid aldehydes are likely to be contaminated by carboxylic acids due to air oxidation, so a distilled sample of these aldehydes may give more reliable results in the Tollens test.

4. IODOFORM TEST

Reference: Shriner et al., p 167.

Reagent: The iodoform test reagent is a solution of KI and I_2 in water, prepared by the instructor or other laboratory staff. It is a deep brown solution.

Procedure: Have a 60°C water bath ready. In a test tube, dissolve 2–3 drops or ~50 mg unknown in 2 mL tetrahydrofuran (for water-insoluble unknowns) or 1 mL water (for water-soluble unknowns), then add 1 mL 10% NaOH solution. Add the iodoform test reagent dropwise, taking note of the color and the volume of the reagent added. As each drop is added, the color should disappear upon mixing (if not, immerse the test tube in the 60°C water bath). Continue adding and shaking the test tube until the dark color persists for more than 2 minutes at 60°C. Add a couple of drops of 10% NaOH to decolorize the excess reagent. Fill the tube with water and allow it to stand for 15 minutes, noting the presence or absence of a yellow precipitate.

Interpretation: Formation of a pale yellow ppt indicates that the unknown contains a methyl ketone, $RCOCH_3$. Disappearance of the brown coloration as the reagent is added is consistent with the α-iodination of the ketone via a ketone enolate. With a methyl ketone, this reaction happens three times in succession to afford $RCOCI_3$, then CHI_3 (iodoform) is released by nucleophilic acyl substitution of $RCOCI_3$ by OH^-. Iodoform is a pale yellow solid (mp 119°C) with a foul odor. Ethanol and other alcohols of the type $CH_3CH(OH)R$ give a false positive, because they are oxidized to methyl ketones under the reaction conditions. Acetaldehyde also gives a positive result.

26.4G | Sodium Bicarbonate Test for Carboxylic Acids

Reference: Harwood and Moody, p 250.

Procedure: This is simply the solubility test using $NaHCO_3$. Observe whether bubbles form. If no bubbles form, immerse the tube in a hot-water bath (about 60–70°C) for several minutes and examine it again. Some solid carboxylic acids react quite slowly, and may not dissolve entirely.

Interpretation: The formation of bubbles suggests the evolution of CO_2 gas. This is expected when $NaHCO_3$ is mixed with an acid stronger than H_2CO_3. In such cases the HCO_3^- ion is protonated to make H_2CO_3, which then decomposes to H_2O and CO_2. Carboxylic acids ($pK_a \approx 4$) protonate HCO_3^- and produce CO_2, but most phenols ($pK_a \approx 10$) do not. However, phenols bearing strongly electron-withdrawing groups may give a false positive. Liquid aldehydes are likely to be contaminated by carboxylic acids due to air oxidation, and may give a false positive. If other evidence suggests an aldehyde, then a false positive here can be ruled out by repeating the test on a distilled sample of the unknown.

26.4H | Hinsberg Test for Amines

Reference: Shriner et al., pp 230–232.

Procedure: Combine 0.2 mL of liquid unknown (or 0.2 g of solid unknown) with 5 mL 10% NaOH solution in a test tube. Add 0.4 mL benzenesulfonyl chloride. Stopper the tube and shake vigorously for 5–10 minutes. Cool the test tube in a water bath if it becomes hot. Test the solution with pH paper to make sure it is still basic. If not, add another 1 mL 10% NaOH and shake for 5 minutes. Separate any insoluble material ("fraction A," which may be liquid or solid) by decanting or filtering and test its solubility in 5% HCl. Acidify the filtrate ("fraction B") by dropwise addition of concentrated HCl (check with pH paper) and promote crystallization by cooling and scratching the inside of the test tube.

Interpretation: Primary amines afford a secondary sulfonamide (RNHSO$_2$Ph) with a relatively acidic N——H bond; it is therefore soluble in NaOH solution and precipitates or crystallizes only after acidifying "fraction B." Secondary amines afford a tertiary sulfonamide (R$_2$NSO$_2$Ph), which has no N——H to deprotonate, so it is insoluble in NaOH solution and is detected as a "fraction A" that is *insoluble* in 5% HCl. Tertiary amines do not react with PhSO$_2$Cl, and the unreacted amine should be insoluble in NaOH, so it is detected as a "fraction A" that is *soluble* in 5% HCl. Some secondary amines react slowly and may require warming of the reaction mixture. Use the Hinsberg test only after establishing with some certainty that the unknown is an amine.

26.4I | Hydroxylamine/Ferric Chloride Test for Esters

Reference: Harwood and Moody, pp 243–244.

Procedure: Place 2–3 drops or ~50 mg unknown in a test tube, then add 10 drops of a saturated ethanolic solution of hydroxylamine hydrochloride and 10 drops of 20% ethanolic KOH. Heat the mixture to boiling, acidify with 5% HCl, then add a 5% solution of FeCl$_3$ dropwise. Examine the color.

Interpretation: Formation of a deep red or purple color (a positive result) suggests the presence of an ester. Do this test only if the compound is insoluble in NaOH, because phenols and carboxylic acids give a false positive. If a positive result is observed, repeat the test without using hydroxylamine hydrochloride—a positive test in this case indicates a phenol, not an ester.

26.4J | Iron(II) Hydroxide Test for Nitro Compounds

Reference: Harwood and Moody, pp 248–249.

Procedure: Add 2–3 drops or ~50 mg unknown to 2 mL of freshly prepared aqueous solution of 5% iron(II) ammonium sulfate. Add 3 drops of 1 M H$_2$SO$_4$, followed by 1 mL of 2 M ethanolic KOH. Stopper the tube and shake well. Check for a precipitate and examine its color immediately and after 1 minute.

Interpretation: The presence of a red-brown or brown precipitate within 1 minute constitutes a positive test for nitro groups. The precipitate may initially appear blue, then turn brown within 1 minute. A slight darkening of the solution or the appearance of a greenish color does *not* constitute a positive test.

26.5 DERIVATIVES

Materials for the following derivative preparations are commonly available in a typical organic chemistry laboratory. The amounts of compounds used in making these derivatives can be scaled to fit the amount of material you would like to use. The melting point should be determined when the product is thoroughly dry. In all cases, a melting point range greater than 2.0°C is unsatisfactory, in which case you should recrystallize your product and redo the melting point.

26.5A | Derivatives from Alcohols

1. PHENYLURETHANE OR 1-NAPHTHYLURETHANE

Reference: Shriner et al., p 156.

Procedure: Place 0.6 g of the anhydrous alcohol or phenol in a dry test tube and add 0.3 mL of phenyl isocyanate (alcohols) or α-naphthylisocyanate (phenols). If the compound is a phenol, add 2 or 3 drops of pyridine to catalyze the reaction. If the reaction is not spontaneous, heat the mixture in a hot-water bath (about 60–70°C) for 30 minutes, taking care to keep moisture out of the tube. While the mixture is still hot, remove insoluble by-products by gravity filtration (moisture leads to a diarylurea by-product). Cool the filtrate and scratch the inside of the tube with a glass rod to induce crystallization, then collect the product by vacuum filtration. A second recrystallization from hot petroleum ether, including another hot filtration, may be necessary to remove all of the diarylurea by-product.

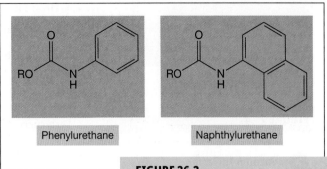

Phenylurethane   Naphthylurethane

FIGURE 26.2

2. 3,5-DINITROBENZOATE

Reference: Harwood and Moody, p 263.

Procedure: Combine ~0.5 mL (0.5 g) of the unknown alcohol, 1 g 3,5-dinitrobenzoyl chloride, and 2 mL pyridine in a dry test tube. Heat the mixture in a hot-water bath (about 60–70°C) for 15 minutes (30 minutes if the unknown is believed to be a 3° alcohol), taking care to keep moisture out of the tube. Pour the mixture into 10 mL ice-water while stirring, then acidify (pH < 4) by cautious addition of concentrated HCl (check with pH paper). Decant the water and thoroughly triturate the residue (solid or oil) twice with 5 mL of 5% sodium carbonate solution. Collect the solid by vacuum filtration and recrystallize from petroleum ether or aqueous ethanol.

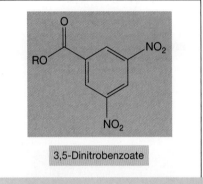

3,5-Dinitrobenzoate

FIGURE 26.3

26.5B | Derivatives from Phenols

1. PHENYLURETHANE OR 1-NAPHTHYLURETHANE

Reference: Shriner et al., p 156.

Procedure: Same as for alcohols (see section 26.5A).

2. 3,5-DINITROBENZOATE

Reference: Harwood and Moody, p 263.

Procedure: Same as for alcohols (see section 26.5A).

3. BROMINATED DERIVATIVES

Reference: Harwood and Moody, p 269.

Procedure: Dissolve 0.5 g of the unknown phenol in 5 mL ethanol and add the aqueous bromine reagent dropwise, shaking the reaction mixture until a yellow color persists. Add 20 mL ice-water and isolate the precipitate via vacuum filtration. If the precipitate is yellow, rinse with 5 mL saturated aqueous NaHSO₃ and then water (3 × 15 mL). Recrystallize the crude product from ethanol or aqueous ethanol. *Note: If an unknown gives no precipitate in the bromine water test for phenols, this is not likely to be a useful derivative procedure.*

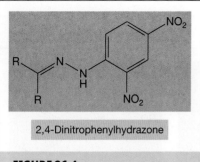

2,4-Dinitrophenylhydrazone

FIGURE 26.4

 Vacuum Filtration

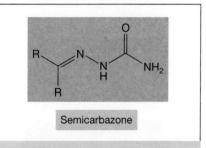

Semicarbazone

FIGURE 26.5

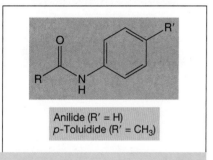

Anilide (R′ = H)
p-Toluidide (R′ = CH₃)

FIGURE 26.6

 Heating at Reflux

 Aqueous–Organic Extractions and Drying Organic Solutions

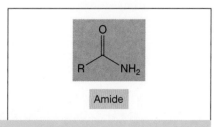

Amide

FIGURE 26.7

26.5C Derivatives from Aldehydes and Ketones

1. 2,4-DINITROPHENYLHYDRAZONE

Reference: Harwood and Moody, p 260.

Procedure: Combine ~0.20 g or 0.3 mL unknown with 5 mL 2,4-DNP reagent. Swirl to mix. If a solid does not form immediately, warm briefly in a water bath, then allow to stand for 10 minutes. If precipitation still does not occur, add water dropwise until a precipitate forms. Isolate the solid by vacuum filtration. Wash successively with 5% $NaHCO_3$ and aqueous methanol (1:1). Recrystallize from ethanol or ethyl acetate.

2. SEMICARBAZONE

Reference: Harwood and Moody, p 261.

Procedure: Dissolve 1.0 g semicarbazide hydrochloride and 2.0 g sodium acetate in 5 mL distilled water, then add ~0.5 g or 0.6 mL of the unknown. If the unknown is insoluble, add ethanol dropwise until it dissolves. Swirl the mixture with warming in a water bath (60°C) for 10 minutes, then cool in ice. Collect the solid by vacuum filtration and recrystallize from ethanol, ethyl acetate, or aqueous ethanol.

26.5D Derivatives from Carboxylic Acids

Carboxylic acids are first converted to the acid chloride (step 1), with subsequent reaction leading to the amide derivative (step 2).

1. ANILIDES AND *p*-TOLUIDIDES

Reference: Ault, pp 286–287.

Procedure (step 1—preparation of acid chloride): Place 0.3 g unknown acid, 1.8 mL thionyl chloride, and 1 drop of dimethylformamide (DMF) in a 50-mL round-bottom flask. (*CAUTION: The reaction may be exothermic.*) Fit the flask with a reflux condenser with a drying tube containing fresh anhydrous calcium chloride. Heat under reflux for 30 minutes, taking care to exclude moisture. Allow the mixture to cool before proceeding to the second step.

Procedure (step 2): Dissolve 5 mmol of the aromatic amine (aniline or *p*-toluidine) in 20 mL toluene. Slowly add this solution to the acid chloride prepared in step 1. (*CAUTION: There may be considerable fuming and a vigorous reaction.*) Heat under reflux for 15 minutes. Cool to room temperature, transfer to a separatory funnel, and extract with 2 mL water, 5 mL 5% HCl, 5 mL 5% NaOH, and 2 mL water. Save the organic layer and discard the aqueous layers. Concentrate the organic layer on the rotary evaporator, then recrystallize from ethanol or aqueous ethanol.

2. AMIDES

Reference: Ault, pp 285–286.

Procedure (step 1—preparation of the acid chloride): Same as for anilides and *p*-toluidides.

Procedure (step 2): Pour the acid chloride cautiously into 5 mL ice-cold concentrated ammonium hydroxide. (*CAUTION: There may be considerable fuming and a vigorous reaction.*) Collect the product by vacuum filtration and recrystallize from water or aqueous ethanol.

Derivatives from Esters

Esters can be hydrolyzed, or saponified, into alcohol and carboxylic acid components; either of these components, if solid, can serve as the derivative. If the component of interest is a liquid, then it will have to be made into an appropriate solid derivative, as described under alcohols and carboxylic acids. The component of interest depends on whether it distinguishes between the possible structures of your unknown. For example, if you are trying to distinguish methyl *o*-chlorobenzoate or methyl *m*-chlorobenzoate, the alcohol component (methanol) is the same for both. If the acid is the component of interest and is water soluble (i.e., < 6 carbons, such as acetic acid), then saponification is *not* recommended because of the difficulty in recovering the product.

1. SAPONIFICATION OF ESTERS

Reference: Harwood and Moody, pp 272–273.

This procedure can produce liquid alcohols and/or carboxylic acids that are not water soluble.

Procedure: In a 25-mL round-bottom flask equipped with a reflux condenser, dissolve 1 g KOH in 3 mL diethylene glycol. Add ~1 mL (1 g) unknown ester and a boiling stone, then warm to reflux. Heat until a single liquid phase is visible. Cool to room temperature, then distill the alcohol from the reaction mixture. Do not collect diethylene glycol (bp 245°C). Note the temperature at which the alcohol distills, because that value may assist you in a preliminary identification. When distillation is complete, cool the residue, add 10 mL water, and acidify (pH < 4) with 20% H_2SO_4. If at this point the carboxylic acid precipitates, collect by vacuum filtration, wash with cold water (3 × 5 mL), and recrystallize. If a solid does *not* form, extract with ethyl acetate, dry the organic phase over Na_2SO_4, and concentrate on the rotary evaporator to obtain the carboxylic acid. Recrystallize or prepare a solid derivative.

2. TRANSESTERIFICATION—3,5-DINITROBENZOATES

Reference: Ault, pp 290–291.

This is a method to obtain a derivative of the alcohol component directly, without the saponification. It is ineffective for esters of alcohols that are unstable to strong acid, such as tertiary alcohols or unsaturated alcohols. Higher molecular weight esters may react slowly.

Procedure: In a 25-mL round-bottom flask equipped with a reflux condenser, mix 0.5 g of the ester with 0.5 g of powdered 3,5-dinitrobenzoic acid. Add a drop of concentrated H_2SO_4, and heat under reflux (or to 150°C if the ester boils above 150°C) until the 3,5-dinitrobenzoic acid dissolves, and then for an additional 30 minutes. Pour the mixture into a beaker containing ice and water (10 mL). Add 5% $NaHCO_3$ solution (*CAUTION: foaming will occur*) until the evolution of CO_2 no longer occurs. Transfer the mixture to a separatory funnel, using 5 mL ethyl acetate to rinse product residues from the beaker into the separatory funnel. Extract with ethyl acetate (10 mL), concentrate the organic phase on the rotary evaporator, and recrystallize the residue from ethanol or aqueous ethanol.

$$RCO_2R' \longrightarrow RCO_2H \ + \ R'OH$$

FIGURE 26.8

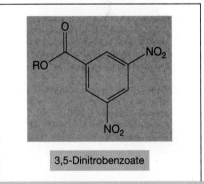

3,5-Dinitrobenzoate

FIGURE 26.9

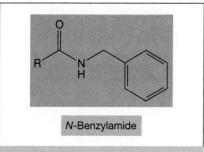

FIGURE 26.10

N-Benzylamide

3. *N*-BENZYLAMIDES

Reference: Ault, p 290.

This is a method to obtain a derivative of the acid component directly, without saponification. The procedure is effective for methyl and ethyl esters. Higher esters may react very slowly.

Procedure: In a 25-mL round-bottom flask equipped with a reflux condenser, mix 0.5 g of the ester with 1.5 mL benzylamine ($PhCH_2NH_2$) and 50 mg ammonium chloride. Heat under reflux for 1 hour. Pour, while stirring, into a mixture of 5% HCl (5 mL) and ice-water (10 mL). Collect the solid by suction filtration and recrystallize from aqueous ethanol.

26.5F Derivatives from Primary and Secondary Amines

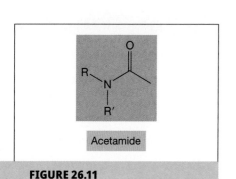

FIGURE 26.11

Acetamide

1. ACETAMIDE

Reference: Ault, pp 280–281.

Procedure: Prepare a solution of 5 g sodium acetate trihydrate in 5 mL water and set this aside for later. In a separate flask, dissolve about 0.5 g of the amine in 5% HCl solution (25 mL). Add 5% NaOH solution dropwise until the mixture just begins to become cloudy from precipitation of the amine. Add a couple of drops of 5% HCl—only as much as required to remove the cloudiness. Add 10 g ice and 5 mL acetic anhydride. With stirring, add the previously prepared sodium acetate solution in one portion. Cool the mixture in an ice bath. If no crystallization occurs after 30 minutes, cover the flask and allow it to stand undisturbed on the lab bench until the next lab period. Collect the solid by vacuum filtration, wash with 5 mL water, and recrystallize from aqueous ethanol.

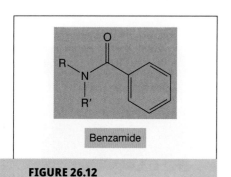

FIGURE 26.12

Benzamide

2. BENZAMIDE

Reference: Harwood and Moody, pp 279–280.

Procedure: Combine 0.2 mL (0.2 g) amine with 3 mL 10% NaOH in a small flask. Add ~0.4 mL benzoyl chloride in 4 portions, stoppering the flask securely (use a glass stopper) and shaking vigorously for 2 minutes between each addition. After all the benzoyl chloride has been added, let the mixture stand for 10 minutes, then destroy any reagent residue by adding ammonium hydroxide. Check to see if the reaction mixture is still alkaline (pH > 10) (add more ammonium hydroxide if necessary), and collect the precipitate by vacuum filtration. Wash the residue with 10 mL water, then recrystallize from aqueous ethanol.

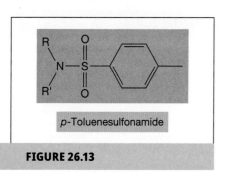

FIGURE 26.13

p-Toluenesulfonamide

3. *p*-TOLUENESULFONAMIDE

Reference: Ault, p 282.

Procedure: In a 25-mL round-bottom flask equipped with a reflux condenser, mix 0.5 g of the amine, 1.5 g *p*-toluenesulfonyl chloride, and 3 mL pyridine. Heat under reflux for 30 minutes. Pour the reaction mixture into 5 mL cold water and stir until the product crystallizes. Collect the precipitate by vacuum filtration, wash with water, and recrystallize from ethanol or aqueous ethanol.

4. PHENYLTHIOUREA

Reference: Shriner et al., p 234.

Procedure: To a 10-mL test tube, add 0.5 g of the amine and 0.5 g of phenylisothio-cyanate. With a glass rod or spatula, stir the mixture for 2 minutes. If no reaction is obvious, cover the tube with a piece of Parafilm and heat in a hot-water bath for 20 minutes, stirring occasionally. Place the tube in an ice bath until the mixture solidifies. Break up the solid and wash it with hexane (or petroleum ether), followed by aqueous ethanol (50%). Recrystallize from ethanol.

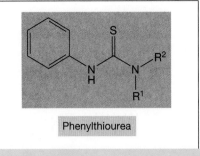

Phenylthiourea

FIGURE 26.14

26.5G Quaternary Ammonium Salt—Methiodide Derivative from Tertiary Amines

Reference: Ault, p 285.

Procedure: Prepare solutions of 0.5 g of the unknown amine in 1 mL acetonitrile, and 1 g methyl iodide in 1 mL acetonitrile. (*CAUTION: Methyl iodide—iodomethane—must be handled and used in the hood!*) In a round-bottom flask, combine the two solutions and allow the mixture to stand for 1 hour, then carefully heat the mixture in a boiling-water bath for 30 minutes. If the methiodide salt crystallizes on cooling, collect the product by vacuum filtration. If crystallization does *not* occur, remove the solvent on the rotary evaporator. Recrystallize from ethanol or ethyl acetate.

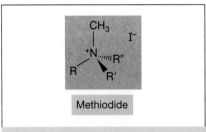

Methiodide

FIGURE 26.15

26.6 THE LABORATORY REPORT

Once you have gathered convincing evidence for the identity of your unknown, you'll need to make the case to your instructor by way of a laboratory report. An example of a report form is found at the end of the chapter. Check with your instructor, who may provide a version of this form for your use, or may require a different style of report.

Ault, A. *Techniques and Experiments for Organic Chemistry*, 6th ed., University Science Books: Sausalito, CA, 1998.

Harwood, L. M.; Moody, C. J. *Experimental Organic Chemistry: Principles and Practice*, Blackwell Scientific Publications: Oxford, 1989.

Shriner, R. L.; Fuson, R. C.; Curtin, D. Y.; Morrill T. C.; et al. *The Systematic Identification of Organic Compounds*, 6th ed.; Wiley: New York, 1980.

Submission Form—First Attempt

Name:

Section:

ID number:

Date:

In order to receive a ^{1}H NMR spectrum of your unknown, submit a list of at least three, but no more than five, possibilities. These are hypotheses, not guesses. In other words, they should be informed by evidence. In addition to the compound names, include the structure of each compound and its literature mp or bp.

UNKNOWN NUMBER:

Observed boiling point:

Observed melting point:

POSSIBLE COMPOUNDS (INCLUDE NAME, STRUCTURE, AND LITERATURE mp/bp)

1. Name: Structure:

    mp/bp:

2. Name: Structure:

    mp/bp:

3. Name: Structure:

    mp/bp:

4. Name: Structure:

    mp/bp:

5. Name: Structure:

    mp/bp:

Submission Form—Second Attempt

Name:

Section:

ID number:

Date:

In order to receive a ^{1}H NMR spectrum of your unknown, submit a list of at least three, but no more than five, possibilities. These are hypotheses, not guesses. In other words, they should be informed by evidence. In addition to the compound names, include the structure of each compound and its literature mp or bp. If you have made a derivative, include its name and mp.

UNKNOWN NUMBER:

Observed boiling point:

Observed melting point:

POSSIBLE COMPOUNDS (INCLUDE NAME, STRUCTURE, AND LITERATURE mp/bp)

1. Name: Structure:

    mp/bp:

2. Name: Structure:

    mp/bp:

3. Name: Structure:

    mp/bp:

4. Name: Structure:

    mp/bp:

5. Name: Structure:

    mp/bp:

DERIVATIVE PREPARED:

Observed mp (range):

Pre-Lab Writing Assignment: Identification of an Unknown

Name:

Section:

ID number:

Date:

Indicate with an **X** the infrared (IR) absorbances you would expect to see for each of the major functional groups listed below and write the frequency range in the comments section. Record any comments that you think might be helpful in identifying each group. Also, indicate in the comments column the IR absorbances you would expect to see if a nitro group were present in your unknown compound.

FUNCTIONAL GROUP	IR ABSORBANCE(S) EXPECTED			
	— NH	— OH	C＝O	COMMENTS
Carboxylic acid				
Ester				
Aldehyde				
Ketone				
Alcohol				
Phenol				
1° Amine				
2° Amine				
3° Amine				
Nitro group				

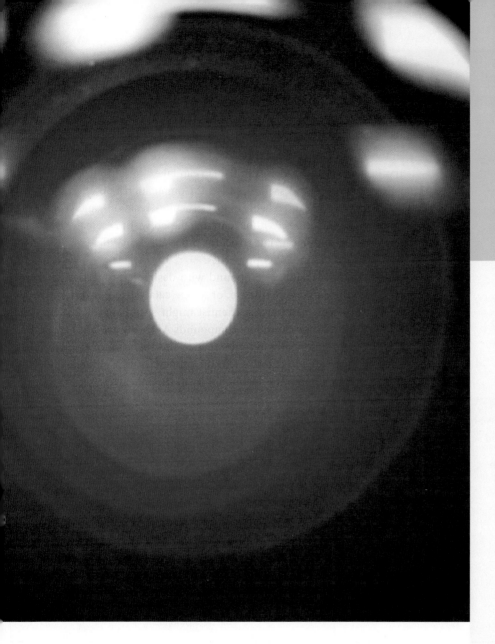

Molecular Modeling of Ketones and Hydrogen Bonds

AI IN CHEMISTRY?

The infamous eye lens of HAL 9000, an artificial intelligence character in *2001: A Space Odyssey*. Modern computational tools can provide valuable chemical insights, yet are bound by the limits of their programming, as was HAL in the classic Arthur C. Clarke–Stanley Kubrick sci-fi film.

Allstar Picture Library Ltd/Alamy Stock Photo.

O wing to the development of computational methods and exponential increase in the speed of computers, they are playing more significant roles in chemistry. Performing experiments in the lab can take a lot of time and resources. Can computers make this process more efficient? Computational power has increased enough that we can quickly and reliably model structures, reactions, and properties of increasingly complex molecules. Without entering the lab, we can evaluate the stability of a molecular target, study electronic structure, probe effects of different substituents, and predict spectroscopic and other properties.

Imagine trying to develop a compound with desirable properties, such as a new drug to treat an infectious disease, or a new catalyst to convert waste carbon dioxide into useful materials. A chemist might spend significant time and resources to synthesize four candidate compounds **A–D** (**Figure 27.1**), or perhaps many more, from commercially available materials. Then, all of the candidates would have to be evaluated to determine if the properties were suitable. This adds up to a lot of work, impacting the sustainability of the endeavor.

FIGURE 27.1

Molecular modeling can be combined with experimental work to more efficiently discover a compound with desirable properties.

Focusing Resources Through Use of Computational Chemistry

• Synthesis and measurement are costly (labor, materials, time, waste disposal). Computational chemistry can be done first, focusing the discovery efforts upon a few compounds that are more likely to succeed.

• Without computation: Many syntheses, many measurements

• With computation: Fewer syntheses, fewer measurements, which results in greater efficiency, greater sustainability

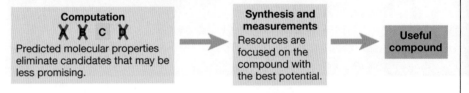

computational chemistry >>
A discipline of chemistry in which computer calculations and atomic/ molecular theory are used to predict the structures, properties, and functions of compounds.

On the other hand, calculating the properties of molecules or compounds with a computer is simple and fast, providing useful predictions. These calculations are referred to as *molecular modeling*, or **computational chemistry**. Incorporating these calculations into the discovery process in the hypothetical example depicted in Figure 27.1, led to predicted properties that eliminated three of the four candidates *before* synthesis, saving considerable resources.

Computational chemistry doesn't eliminate the need for experimental chemistry— it just makes the experiments more focused and more efficient, with less environmental impact.

This experiment[1] consists of two parts. The first part uses computational chemistry to predict the molecular properties of ketones and how they are impacted by electronic effects. Bond order, aromaticity, and substituents strongly affect bond strengths, bond energies, and reactivities of molecules. Molecular modeling of the carbonyl group of various compounds will illustrate these features. In the second

---

[1]This experiment was designed by Professor Leonard R. MacGillivray, University of Iowa.

part, computational chemistry is used to probe how electronic effects impact hydrogen bonding in molecular complexes. For all of these calculations, we will use a computer program known as *Spartan* (Wavefunction, Inc.).

Pre-Lab Reading Assignment

- Experiment: Chapter 27 (this experiment)

27.1 BACKGROUND

27.1A Molecular Modeling of Ketones

In the first part of this lab, the vibrational spectral frequencies for a variety of cyclic ketones (**Figure 27.2**) are predicted using molecular orbital (MO) methods, and trends in the bond length and C$=$O stretching frequency are compared to expectations based on resonance theory and aromaticity models.

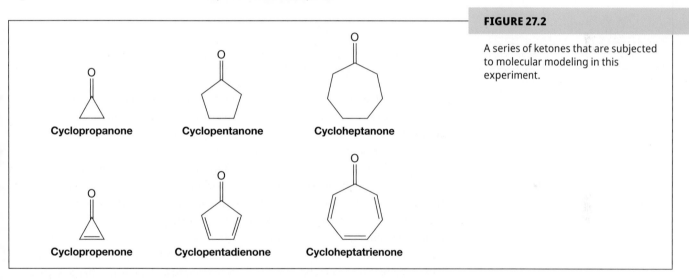

FIGURE 27.2

A series of ketones that are subjected to molecular modeling in this experiment.

1. PRINCIPLES OF RESONANCE

Lewis structures are essential tools for understanding organic chemistry, but the structures of some compounds and ions cannot be represented by a single Lewis structure. For example, sulfur dioxide (SO$_2$; **Figure 27.3a**) and nitric acid (HNO$_3$; **Figure 27.3b**) may each be described by two equivalent Lewis structures. They differ in the locations of electrons, but not in the locations of atoms.

If only one Lewis structure for sulfur dioxide were correct and accurate, then the double bond to oxygen would be shorter and stronger than the single bond. Experimental evidence indicates, however, that this molecule is bent (bond angle 120°) and the two sulfur–oxygen bonds have equal lengths (1.432 Å). Thus, a single Lewis structure is inadequate, and the actual structure resembles an average of the two. The averaging of electron distribution over two or more hypothetical contributing structures to produce a hybrid electronic structure is called resonance. Likewise, the structure of nitric acid is best described as a resonance hybrid of two structures, where the double-headed arrows in Figure 27.3 indicate resonance.

FIGURE 27.3

The resonance structures of (a) sulfur dioxide, (b) nitric acid, and (d) acetone. (c) An addition reaction of acetone.

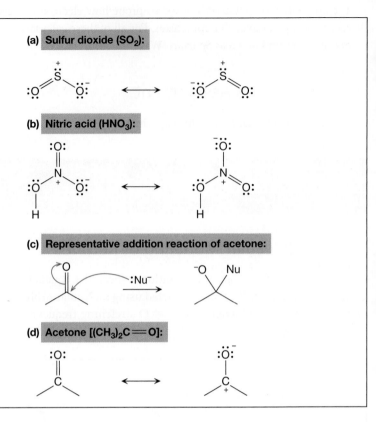

(a) Sulfur dioxide (SO₂):

(b) Nitric acid (HNO₃):

(c) Representative addition reaction of acetone:

(d) Acetone [(CH₃)₂C=O]:

SO_2 and HNO_3 represent one extreme in the application of resonance—namely, two structurally and energetically equivalent electronic structures for a stable compound can be written, but no single structure provides an accurate or even an adequate representation of the true molecule.

Principles of resonance are very useful in rationalizing the chemical behavior of many such compounds. For example, the carbonyl group of aldehydes and ketones reacts readily with many nucleophiles to give addition products (**Figure 27.3c**). The course of these reactions can be explained by a small contribution of a dipolar resonance contributor, where the carbonyl carbon is positively charged and the carbonyl oxygen is negatively charged. For acetone (**Figure 27.3d**), the first contributor (on the left) is the best representation of this molecule because there is no charge separation and both the carbon and oxygen atoms have achieved an octet of electrons. If the π bond is broken heterolytically, formal charge pairs result, as shown in the other structure. The preferred charge distribution has the *positive* charge on the *less* electronegative atom (carbon) and the *negative* charge on the *more* electronegative atom (oxygen). Because it has more bonding, and less charge separation, the double-bonded Lewis structure represents a more reasonable and stable structure for acetone. The double-bonded structure is the major contributor and the charge-separated structure is a minor contributor. The charge-separated contributor has an electron-deficient carbon atom, which helps explain why nucleophiles tend to bond at this site.

The following factors are important in evaluating the contribution each resonance structure makes to the actual molecule:

The amount of bonding in a structure: The more covalent bonds a structure has, the more stable it will be, and the more important its contribution will be to the resonance hybrid.

Formal charge separation: Other factors aside, charge separation makes a structure less stable, decreasing its contribution to the resonance hybrid.

Electronegativity of charge-bearing atoms and charge density: High charge density is destabilizing. Positive charge is best accommodated on atoms of low electronegativity and negative charge on atoms of high electronegativity.

The stability of a resonance hybrid is always greater than the stability of any canonical contributor: Consequently, if one form has a much greater stability than all others, the hybrid will closely resemble it electronically and energetically. This is the case for the carbonyl group (Figure 27.3d). The double-bonded structure has much greater total bonding than the charge-separated structure, so the C=O structure describes this functional group rather well.

2. AROMATICITY AND THE HÜCKEL RULE

Unsaturated compounds contain π bonds, as in alkenes or alkynes, which react readily with electrophiles to give addition products (**Figure 27.4a**). However, certain cyclic unsaturated compounds behave very differently, undergoing substitution instead of addition (**Figure 27.4b**). The substitution occurs in order to preserve a special form of stabilization known as aromaticity, and these compounds are called aromatic compounds.

FIGURE 27.4

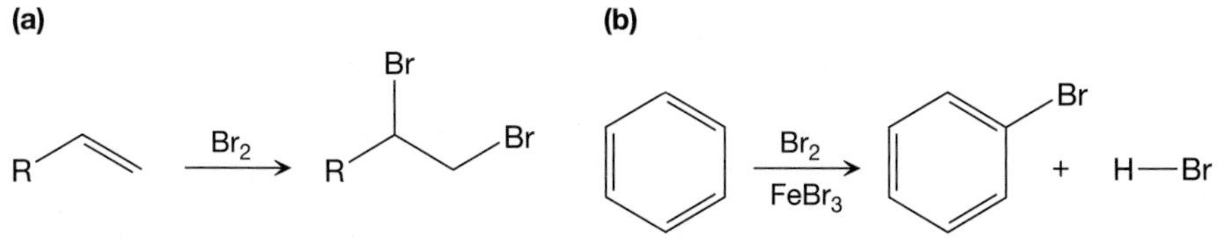

Reactions of unsaturated compounds with bromine, an electrophile, showing (a) addition to alkenes, and (b) substitution of aromatic compounds.

An aromatic compound has the following characteristics:

A planar (or near planar) cycle of sp^2-hybridized atoms, the p orbitals of which are oriented parallel to each other. These overlapping p orbitals generate an array of π molecular orbitals.

A π orbital system occupied by $4n + 2$ electrons (where $n = 0, 1, 2, \ldots$). This requirement is known as the **Hückel rule**.

Several typical unsaturated cyclic compounds are illustrated in **Figure 27.5**. The properties of the first three compounds (cyclic polyenes) are most like alkenes in general. That is, each reacts readily with bromine to give addition products. The thermodynamic change on introducing double bonds to the carbon atom ring is also typical of alkenes (a destabilization of ~26 kcal/mol for each double bond). Conjugation offsets the increase in energy by 4–6 kcal/mol. 1,3-Cyclopentadiene and 1,3,5-cycloheptatriene both fail to meet the first requirement of aromaticity, because one carbon atom of each ring is sp^3-hybridized and has no p orbital. 1,3,5,7-Cyclooctatetraene fails both requirements, even though each carbon in the ring is sp^2-hybridized. This molecule is not planar and it has eight π electrons, a number inconsistent with the Hückel rule. If it were planar, its C—C—C bond angles would be 135°. The resulting angle strain is relieved by adopting a tub-shaped conformation; consequently, the p orbitals can only overlap as isolated pairs, not over the entire ring.

The remaining four compounds (i.e., benzene, pyridine, furan, and pyrrole) exhibit very different properties, and are considered to be aromatic. Benzene and pyridine are relatively unreactive with bromine, requiring heat and/or catalysts to force the reaction. And, when they do react, they undergo substitution rather than addition. Furan and pyrrole react more rapidly with bromine, but they also

<< **Hückel rule**
A rule applied to predict aromaticity in compounds in which all atoms of a cyclic system are unsaturated and planar (sp or sp^2-hybridized). If such a compound has a π-electron count of $4n + 2$ (where n is an integer) in the ring, it will be aromatic, conferring ring exceptional stability.

FIGURE 27.5

COMPOUND	STRUCTURE	ALL RING ATOMS sp^2 HYBRIDIZED?	NUMBER OF π ELECTRONS	THERMODYNAMIC STABILIZATION	REACTION WITH Br_2
1,3-Cyclopentadiene	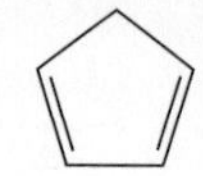	No	4	Slight	Addition
1,3,5-Cycloheptatriene		No	6	Slight	Addition
1,3,5,7-Cyclooctatetraene	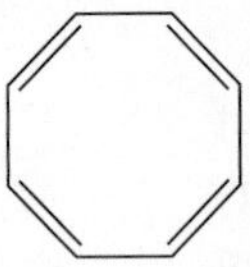	Yes	8	Slight	Addition
Benzene	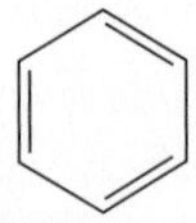	Yes	6	Large	Substitution
Pyridine	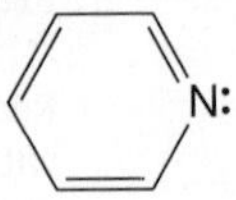	Yes	6	Large	Substitution
Furan	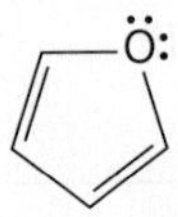	Yes	6	Moderate	Substitution
Pyrrole	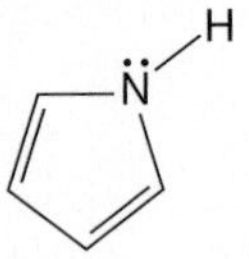	Yes	6	Moderate	Substitution

Cyclic unsaturated compounds that illustrate the characteristics of aromatic or nonaromatic compounds.

give substitution products. The tendency to favor substitution rather than addition suggests that the parent unsaturated ring system has exceptional stability. Thermodynamic measurements support this conclusion. The enhanced stability, often referred to as aromatic stabilization, ranges (for the compounds in Figure 27.5) from a low of 16 kcal/mol for furan to a high of 36 kcal/mol for benzene.

Benzene is the archetypical aromatic compound (**Figure 27.6**). It is planar (with bond angles of 120°), all carbon atoms in the ring are sp^2-hybridized, and the π orbitals are occupied by six electrons (one from each of the carbons).

Pyridine is similar to benzene, but the nonbonding electron pair on nitrogen occupies an sp^2 orbital and is *not* part of the π electron system. Thus, pyridine

FIGURE 27.6

A closer look at benzene, showing its resonance structures. The six sp^2-hybridized carbons provide six p orbitals that combine to make the π bonding system of benzene. Each carbon also contributes three sp^2-hybridized orbitals that are used to make the C—C and C—H σ bonds.

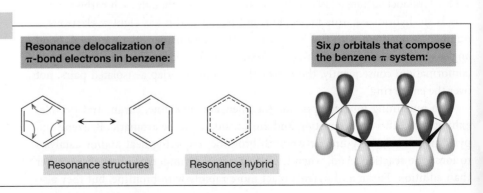

Resonance delocalization of π-bond electrons in benzene:

Resonance structures   Resonance hybrid

Six p orbitals that compose the benzene π system:

has six π electrons and two nonbonding sp^2 electrons. Furan and pyrrole have five-membered rings (Figure 27.5), in which one atom is a heteroatom instead of carbon, and which has at least one pair of nonbonding valence shell electrons. By hybridizing this heteroatom to an sp^2 state, a p orbital occupied by a pair of electrons and oriented parallel to the carbon p orbitals is created. The resulting planar ring meets the first requirement for aromaticity, and the π system is occupied by six electrons, four from the two double bonds and two from the heteroatom, thus satisfying the Hückel rule.

Carbanions and carbocations, such as in **Figure 27.7**, may also show aromatic stabilization.

FIGURE 27.7

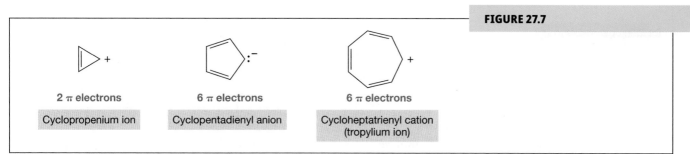

2 π electrons	6 π electrons	6 π electrons
Cyclopropenium ion	Cyclopentadienyl anion	Cycloheptatrienyl cation (tropylium ion)

The three-membered ring cation has two π electrons and is surprisingly stable, considering its ring strain. Cyclopentadiene is as acidic as ethanol, reflecting the stability of its six π-electron conjugate base. Salts of the cycloheptatrienyl cation (the tropylium ion) are stable in aqueous solution, again reflecting the stability of this six π-electron cation. In all of these examples, the π electrons and charge are delocalized so that the true structure is more accurately represented by a resonance hybrid (as shown for benzene in Figure 27.6).

3. INFRARED SPECTRA OF ALDEHYDES AND KETONES

For simple aldehydes and ketones, the stretching vibration of the carbonyl group gives rise to a strong and distinctive infrared absorption at 1710 to 1740 cm^{-1}. The **dipole moment** is increased on stretching (the single-bond character is greater), and this results in a strong absorption. Through hyperconjugation, alkyl substituents stabilize the carbocation character of the charge-separated resonance contributor, so ketone carbonyls have slightly lower stretching frequencies (1715 ± 7 cm^{-1}) compared to aldehydes (1730 ± 7 cm^{-1}). The values cited here are for pure liquid or CCl_4 solution spectra. Hydrogen-bonding solvents lower these frequencies by 15–20 cm^{-1}.

Three factors are known to perturb the carbonyl stretching frequency:

a. *Conjugation with a double bond or benzene ring lowers the stretching frequency.* A 30- to 40-cm^{-1} decrease in carbonyl stretching frequency is observed for conjugated carbonyl compounds compared to nonconjugated carbonyl compounds (**Figure 27.8**). The stretching frequency of the conjugated double bond is also lowered and may be enhanced in intensity. The cinnamaldehyde example shows that extended conjugation further lowers the absorption frequency, although not to the same degree.

b. *Incorporation of the carbonyl group in a small ring (5-, 4-, or 3-membered) raises the stretching frequency.* The increase in frequency ranges from 30–45 cm^{-1} for a five-membered ring, to 50–60 cm^{-1} for a four-membered ring, and to nearly 130 cm^{-1} for a three-membered ring (Figure 27.8c). Typical bond angles are 120° at the carbonyl carbon because it is sp^2-hybridized and its bonds are of 33% s-character. Smaller rings enforce smaller bond angles within the ring (108° for five-membered, 90° for

<< **dipole moment**
A measurement of the net polarization of a molecule, related to the combination of all internal dipoles associated with the bonds within the molecule.

(a) Carbonyl:

   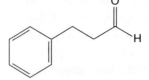 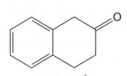 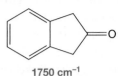

1716 cm⁻¹	1748 cm⁻¹	1731 cm⁻¹	1718 cm⁻¹	1750 cm⁻¹
Cyclohexanone	Cyclopentanone	Hydrocinnamaldehyde (3-Phenylpropanal)	2-Tetralone	2-Indanone

1716 cm⁻¹   1748 cm⁻¹   1731 cm⁻¹   1718 cm⁻¹   1750 cm⁻¹

Cyclohexanone   Cyclopentanone   Hydrocinnamaldehyde   2-Tetralone   2-Indanone
(3-Phenylpropanal)

(b) Conjugated carbonyl:

   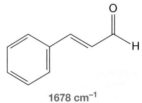 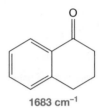

1685 cm⁻¹   1710 cm⁻¹   1678 cm⁻¹   1683 cm⁻¹   1713 cm⁻¹
(C=C 1617 cm⁻¹)   (C=C 1627 cm⁻¹)

2-Cycloalken-1-ones   Cinnamaldehyde   1-Tetralone   1-Indanone

(c) Carbonyls of varying ring size:

1716 cm⁻¹   1748 cm⁻¹   1783 cm⁻¹   1813 cm⁻¹

Cyclohexanone   Cyclopentanone   Cyclobutanone   Cyclopropanone

FIGURE 27.8

Carbonyl stretching frequencies (a) are lowered when conjugation is present (b), whereas reducing ring size raises it (c). These data are from experimental infrared spectra.

four-membered, and 60° for three-membered). Such ring bonds have increasing p-character, leaving more s-character in the exocyclic C=O bond. This causes the carbonyl bond to be stronger and shorter, with higher stretching frequency. This shift also occurs in the presence of conjugation.

c. *Replacing an alkyl substituent on a ketone with an electron-donating group raises the stretching frequency, whereas an electron-withdrawing group lowers it.* The effect is particularly important when an alkyl substituent is replaced by a heteroatom such as N, O, or halogen. The examples in **Figure 27.9** show that a strongly electron-withdrawing group (—CCl₃) raises the carbonyl stretching frequency almost 40 cm⁻¹, whereas a conjugatively electron-donating group (—OCH₃) lowers it almost 10 cm⁻¹.

FIGURE 27.9

Electron-withdrawing groups raise, whereas electron-donating groups lower, the carbonyl stretching frequency. These data are from experimental infrared spectra.

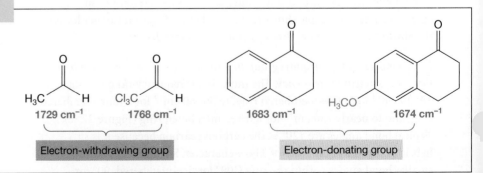

1729 cm⁻¹   1768 cm⁻¹   1683 cm⁻¹   1674 cm⁻¹

Electron-withdrawing group   Electron-donating group

4. DIPOLE MOMENT

When two different atoms are covalently bonded, the shared electrons are attracted to the more electronegative atom of the bond, resulting in a shift of electron density toward the more electronegative atom. This can be illustrated with partial charges, or with arrows pointing from the more positively charged atom to the one bearing more negative charge (**Figure 27.10**). This kind of covalent bond is *polar* and will have a **dipole** (one end is positive and the other end negative). The degree of polarity and the magnitude of the bond dipole (which is measured by the *dipole moment*) is proportional to the difference in electronegativity (or charge separation) of the bonded atoms and the distance between these atoms. That is, the dipole moment *increases* as the charge separation *increases*, whereas it *decreases* as the distance between the atoms *increases*. For example, the C—H bond will have a small dipole moment because the electronegativities of carbon and hydrogen are similar (2.55 and 2.20, respectively). Oxygen (3.44) is much more electronegative than hydrogen, so the O—H bond has a larger dipole moment than the C—H bond.

<< **dipole**
An unequal electron density distribution across a bond, causing a polarization of the bond.

FIGURE 27.10

Polar covalent bonds have dipole moments that can be predicted by electronegativity differences, as illustrated with partial charges.

$$\overset{\delta^-}{C}\!-\!\overset{\delta^+}{H} \qquad \overset{\delta^-}{O}\!-\!\overset{\delta^+}{H} \qquad \overset{\delta^+}{C}\!-\!\overset{\delta^-}{Cl} \qquad \overset{\delta^-}{C}\!-\!\overset{\delta^+}{Li}$$

27.1B	## Molecular Modeling of Hydrogen-Bonded Molecular Complexes

In the second part of this lab, the stabilities of hydrogen-bonded molecular complexes and the strengths and lengths of hydrogen bonds are predicted. Trends are rationalized based on substituent effects.

For our purpose, a molecular complex (**Figure 27.11**) is an assembly of two or more neutral molecules held together by noncovalent interactions (e.g., hydrogen bonds).

FIGURE 27.11

A series of molecular complexes of pyridine with substituted phenols that are subjected to molecular modeling in this experiment.

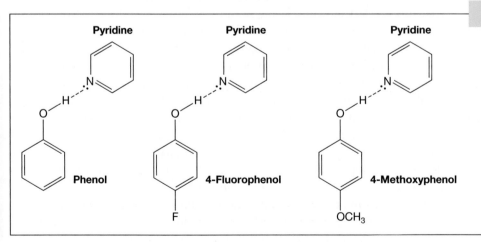

1. HYDROGEN BOND

The **hydrogen bond** is a noncovalent interaction (it may be intermolecular or intramolecular) and is therefore much weaker than a covalent bond. It is relatively strong, however, compared to other noncovalent interactions, such as van der Waals forces and dipole attractions. A hydrogen bond is best described as an attractive electrostatic interaction between polar molecules where at least one of the molecules has a hydrogen atom covalently bonded to an electronegative atom (typically

<< **hydrogen bond**
A noncovalent attractive force between a hydrogen bearing partial positive charge (e.g., one that is part of an O—H or N—H bond) and a different electronegative atom (generally O or N) bearing partial negative charge.

FIGURE 27.12

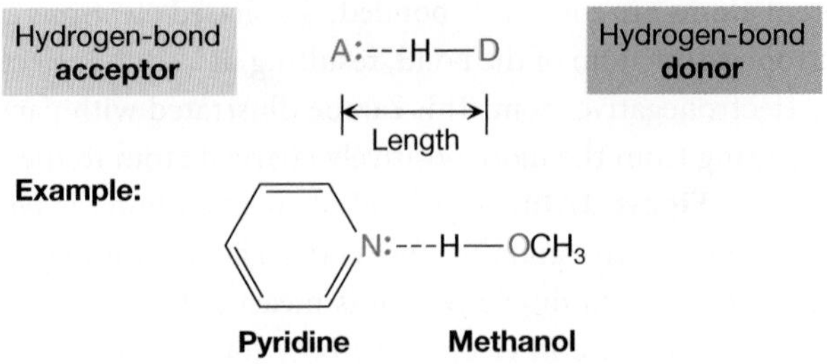

Definitions of the hydrogen bond terminology, and an example.

O, N, or F). The hydrogen bond is formed between the *hydrogen-bond donor (D)*, which is the atom that is covalently bonded to hydrogen, and the *hydrogen-bond acceptor (A)*, which has at least one lone pair of electrons (**Figure 27.12**). Because the hydrogen-bond donor atom is highly electronegative, it pulls electron density away from hydrogen and forms a partial positive charge on hydrogen. The partial positive charge on the hydrogen attracts the lone pair of electrons on the hydrogen-bond acceptor. Thus, the hydrogen is "shared" between the hydrogen-bond donor and the hydrogen-bond acceptor.

The hydrogen bond is defined by the distance between the hydrogen-bond donor (D) and the hydrogen-bond acceptor (A). Specifically, if we use the notation D—H---A, the length of the hydrogen bond is actually the distance between D and A, d(D—A). Similar to covalent bonds, the strength of the hydrogen bond is inversely proportional to the length of the bond, so a *shorter bond is stronger*. Additionally, the covalent bond between the hydrogen-bond donor and the hydrogen atom (D—H) *gets longer* when the hydrogen bond is formed because the hydrogen atom moves towards the acceptor in the process of being "shared" between donor and acceptor.

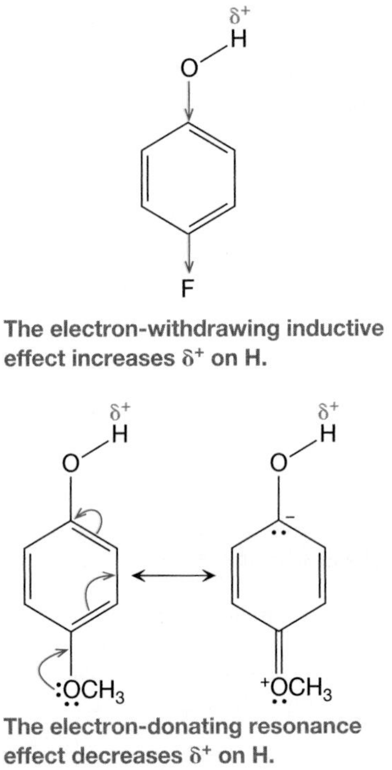

2. HAMMETT PLOT

To increase the strength of a hydrogen bond, the hydrogen-bond donor must more readily share hydrogen, resulting in a weaker D—H bond. If the hydrogen-bond donor is attached to an aromatic ring such as benzene, this can be achieved by attaching an electron-withdrawing substituent. Strongly electronegative substituents, such as fluorine (—F), pull electrons from the benzene ring, in turn impacting an O—H group on the ring (**Figure 27.13**). The pull of electrons is transmitted to O—H primarily through polarization of the bonding electrons from one atom to the next, which is called an *inductive effect*. Thus, an electronegative substituent stabilizes the negative charge on the donor and build-up of positive charge on the hydrogen atom. Overall, the O—H group becomes *more acidic*, so it more readily shares the hydrogen atom with a hydrogen-bond acceptor. The opposite is true if a substituent (e.g., a methoxy group, —OCH$_3$) donates electrons to the ring. An electron-donating substituent makes the O—H group *less acidic*, which weakens the hydrogen bond. This is called a *resonance effect*.

For a series of phenols with different substituents, acidity (pK_a) can be plotted versus σ, a substituent parameter for electron-donating or -withdrawing groups (**Figure 27.14**). This type of graph is called a *Hammett plot* and it depicts how different substituents impact acidity of an O—H group, and by extension, the strength of its hydrogen bonds.

The electron-withdrawing inductive effect increases δ⁺ on H.

The electron-donating resonance effect decreases δ⁺ on H.

FIGURE 27.13

The inductive and resonance effects of fluorine and methoxy groups as aromatic ring substituents.

3. STABILITY OF MOLECULAR COMPLEXES AND BONDING ENERGY (BE)

The relative stability of a molecule or a molecular complex is generally expressed in terms of its energy, where the more stable complex is lower in energy. In the molecular complexes that you will study, the most important interaction is hydrogen bonding. Hence, the stability of the complex is directly affected by the strength of the hydrogen

FIGURE 27.14

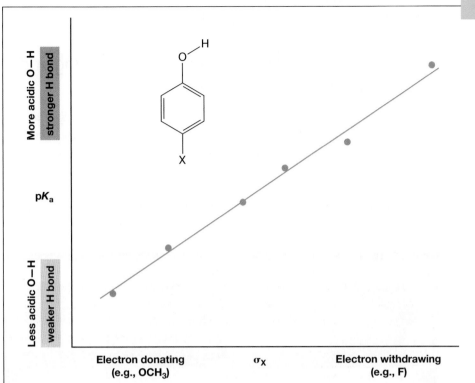

Example of a Hammett plot, which correlates electron-donating and electron-withdrawing effects of different substituents (X) to various other properties, which in this experiment is the acidity of OH groups in substituted phenols. On the x axis, σ_X is a quantity that reflects the extent of electron-donating and -withdrawing behavior of various substituents.

bond. The complex with the strongest hydrogen bond will be stabilized the most and will therefore be the lowest in energy.

Understanding the stability of a complex depends on understanding how the energy of the complex relates to the energy of the individual molecules that make up the complex. Moreover, the energy of a favorable complex will be lower than the energies of the individual molecules combined. The difference in energy is called the **bonding energy** (BE), which can be expressed by the following equation:

$$BE = energy_{complex} - (energy_{molecule\ 1} + energy_{molecule\ 2})$$

<< **bonding energy**
The amount of enthalpy or free energy needed to break a bond.

4. IR SPECTRA OF HYDROGEN-BONDED COMPLEXES

The stretching vibration of the hydroxyl group (—OH) in phenols and alcohols gives an intense, broad peak at 3550–3200 cm^{-1}. Based on the discussions so far, —DH is actually weaker and longer for a stronger hydrogen bond, while it is stronger and shorter for a weaker hydrogen bond. In an IR spectrum, stronger hydrogen bonds result in weaker D—H stretching frequencies and, thus, are shifted to lower frequencies (i.e., to the right in an IR spectrum). A weaker hydrogen bond has a stronger D—H stretching frequency and is shifted to higher frequencies (i.e., to the left in the spectrum).

27.2 EXPERIMENTAL PROCEDURE

The *Spartan* computer program used in this lab is a commonly used molecular modeling and computational chemistry application that provides a variety of computational approaches based on different theoretical models.

You will need to begin with the calculations on the hydrogen-bonded complexes because they require more CPU time to complete. While these calculations are running, you will then proceed to the second part of the procedure, which deals with resonance forms and the aromaticity of cyclic ketones.

Stage I: Sample Procedures for Drawing and Calculations of Structures

The sample procedures for drawing and running calculations using the *Spartan* program are described with the pyridine-phenol complexes as an example. The same procedure should be repeated for all molecules and molecular complexes in this lab.

DRAWING THE FIRST STRUCTURE

1. Go to the *Start Menu* and under *All programs* find *Spartan'20*. A green window with a command tab similar to Word or Excel will show up (**Figure 27.15**).

FIGURE 27.15

Spartan window and some useful tools.

Courtesy of Gregory K. Friestad.

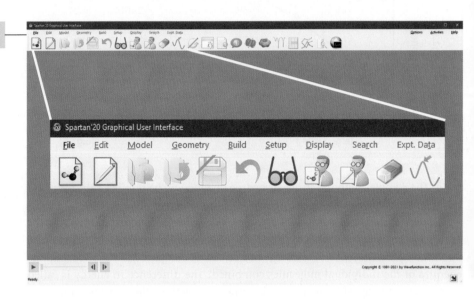

2. Under *File* (**Figure 27.16**), click on the second option, *New Sketch*.

(a)

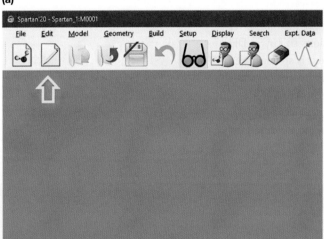

(b)

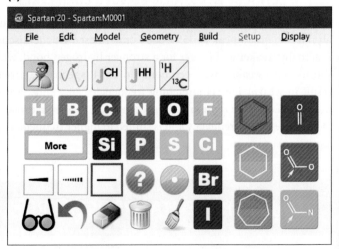

FIGURE 27.16

Screenshot from *Spartan'20* showing (a) how to open a new file and (b) some functions of the model kit.

Courtesy of Gregory K. Friestad.

3. Now, on the right-hand side, you have a *Model Kit* (Figure 27.16). The *Model Kit* is an interface for drawing molecules. It gives you an option to start drawing using an individual atom (with different hybridizations), a group (alkenyl, carbonyl, nitro, etc.), or a ring (cyclopropane, benzene, etc.).

4. Start by choosing *Benzene*, and then double click on the white space.

5. The pyridine molecule has an aromatic six-membered ring, similar to the benzene molecule, except that it has a nitrogen atom in place of one of the carbon atoms. To change one of the carbon atoms to the nitrogen atom, choose the nitrogen (N), then double click on one of the carbons in a benzene ring.

6. Next, draw the phenol molecule underneath the pyridine molecule. First draw another benzene, and then click on the oxygen (O). Starting from the carbon that is closest to the pyridine nitrogen, click and drag outwards, creating a new bond to an oxygen. The hydrogen atom of the —OH group should point towards the nitrogen atom in the pyridine molecule.

Important: To ensure that your calculations run smoothly, the initial locations of the atoms involved in the hydrogen bond need to be reasonable. This means that the nitrogen atom in the pyridine molecule and the —OH group in the phenol molecule should be placed close to each other.

If a mistake is made in drawing the structures, use the eraser tool in the *Model Kit* to remove any atoms, bonds, or structures that need correction.

RUNNING THE FIRST CALCULATION

1. After you have drawn a reasonable model of the complex, click on the glasses icon at the lower left corner of the *Model Kit*.

2. Under *Build*, click on *Minimize*.

3. Under the *Model* tab, select *Hydrogen bonds*. Do you have a bright green dotted line connecting the —OH group and the nitrogen atom (**Figure 27.17**)? If not, you have to start with a better model.

4. Under the *Setup* tab, choose *Calculations*.

(a)

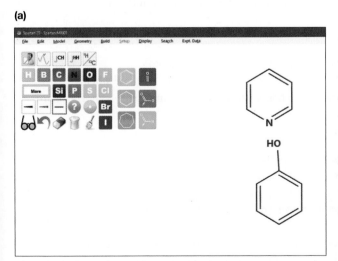

(b)

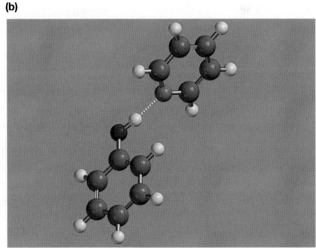

FIGURE 27.17

Screenshot from *Spartan* showing (a) a reasonable initial model of the phenol and pyridine structures, and (b) a minimized structure of the hydrogen-bonded complex.

Courtesy of Gregory K. Friestad.

FIGURE 27.18

Screenshot from *Spartan* showing
the setup of the calculation
commands.

Courtesy of Gregory K. Friestad.

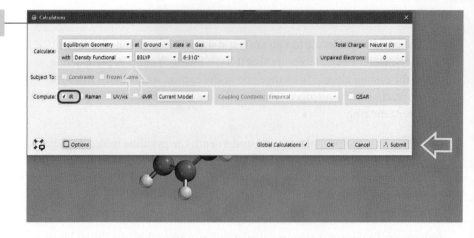

5. Select the following in the *Calculations* window (**Figure 27.18**): *Equilibrium geometry*, at *Ground* state in *Gas* with *Density Functional*, B3LYP, 6-31G*.

6. Under *Compute*, choose *IR*.

7. Click *Submit*.

8. The *Save As* window will appear. Rename the calculation as *pyr_phen*. When done, click *Save*.

9. The window that informs you that the calculation has started will appear. Click *OK*.

10. The calculations can last up to 45 minutes. While waiting for your calculations to finish, open a new file and begin working on the next complex.

DRAWING AND CALCULATIONS ON REMAINING COMPLEXES AND MOLECULES

You need to draw and run calculations for all of the following molecules and complexes.

Complexes:

- pyridine + phenol (suggested file name: pyr_phen)
- pyridine + 4-fluorophenol (suggested file name: pyr_4-F-phen)
- pyridine + 4-methoxyphenol (suggested file name: pyr_4-OCH3-phen)

Individual molecules:

- pyridine (suggested file name: pyr)
- phenol (suggested file name: phen)
- 4-fluorophenol (suggested file name: 4-F-phen)
- 4-methoxyphenol (suggested file name: 4-OCH3-phen)

After you draw and start calculations for all three complexes and four individual molecules, move to stages II and III. You will return to analyze the results of these calculations at stage IV.

Stage II: Drawings and Calculations on Cyclic Ketones

SAMPLE PROCEDURE FOR CYCLOPROPANONE MOLECULE

1. After you have drawn a reasonable model of the molecule, click on the *Minimize* icon.

2. After a few seconds, *Spartan* will display an improved version of your model.

3. Under *Setup*, choose *Calculations*.

4. Click on the boxes to choose *Equilibrium Geometry* at *Ground* state in *Gas* with *Density Functional, B3LYP, 6-31G**.

5. Under *Compute*, choose *IR*.

6. Click *Submit*.

7. The *Save As* window will appear. Rename the calculation as "cyclopropanone." When done, click *Save*.

8. The window that informs you that the calculation has started will appear. Click *OK*.

9. When your calculation has finished a smaller window will appear informing you that the *Spartan* calculation is completed. Click *OK*.

Draw and run calculations for all of the following cyclic ketones.

- cyclopropanone (suggested file name: cyclopropanone)
- cyclopentanone (suggested file name: cyclopentanone)
- cycloheptanone (suggested file name: cycloheptanone)
- cyclopropenone (suggested file name: cyclopropenone)
- cyclopentadienone (suggested file name: cyclopentadienone)
- cycloheptatrienone (suggested file name: cycloheptatrienone)

Stage III: Analysis of Calculation Results on Cyclic Ketones

After all the cyclic ketones have been drawn and the calculations are completed, the calculated properties need to be collected and analyzed.

1. Under the *Display* menu, choose *Properties* to read the energy (in a.u.) and the dipole (in debyes).

2. In the same window, select *Display Dipole Vector* to visualize the direction of the dipole moment.

3. To find the atomic charges, right-click on an atom. The selected atom will turn yellow. Select *Properties*. Record Mulliken charges.

4. Under *Display*, click *Spectra* and click on the + sign, then choose the "IR calculated" and click on the different frequency values to view the bond vibrations. Find the $C{=}O$ stretching vibration and record its frequency. Frequencies are given in cm^{-1}.

5. Under *Geometry*, choose *Measure Distance*. This way you can measure a distance between any two selected atoms. In this case, select the C and O atoms of the carbonyl (C═O) group. The distance will appear in the bottom-right corner. Record the distance in angstroms (Å).

Use the information you gather in stage III to complete **Table 27.1** (see Data to Include in the Laboratory Report at the end of this experiment).

27.2D Stage IV: Analysis of Results for Phenol–Pyridine Complexes

Properties of the phenol–pyridine complexes need to be collected in **Table 27.2** and **Table 27.3** (see Data to Include in the Laboratory Report at the end of this experiment) and analyzed for your report.

1. Under the *Display* menu, choose *Properties* to read the energy in a.u. (**Figure 27.19**). Record the value. You will need these values later to calculate bonding energies.

   Note: For the pyridine molecule, you need to record the energy only, so you can stop after this step. For the phenols and the complexes, proceed to the following steps.

2. Under *Display*, click *Spectra* and click on the different frequency values to view the bond vibrations. Find the O—H stretching vibration and record its frequency. Frequencies are given in cm^{-1}.

(a)

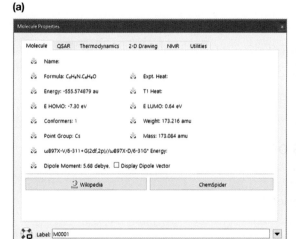

(b)

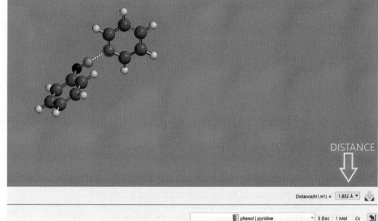

FIGURE 27.19

Screenshot from *Spartan* showing (a) the properties window and (b) the measured distance.

Courtesy of Gregory K. Friestad.

3. Under the *Geometry* tab, choose *Measure Distance*. In this case, you need to record two distances—namely, the O—H distance between the O and H atoms in the hydroxyl group and the O—N distance between the O atom of the hydroxyl group and the N atom of the pyridine. Click on the O and H atoms of the hydroxyl group. Record the distance in angstroms (Å). Now, click on the O and N atoms. Record the distance in angstroms (Å). *Tip:* A more detailed output of completed calculations can be found if you choose *Output* under the *Display* tab. The file has information such as energy after each step, details about theory level, calculation time, and so on.

Use the information you gather in stage IV to complete Table 27.2 and Table 27.3 (see Data to Include in the Laboratory Report at the end of this experiment).

27.3 THE LABORATORY REPORT

Address each of the following points and questions in your results and discussion section.

1. Resonance forms and aromaticity: Draw the resonance forms (neutral and charge separated) for each cyclic ketone studied in this experiment. Also, identify the number of π electrons that are present in each ring and indicate whether a resonance form is aromatic.

2. Calculated IR frequencies for C=O: Comment on the trend observed in the C=O stretching frequencies for the saturated cyclic ketones that you calculated. Likewise, comment on the trend observed in the frequencies for the unsaturated cyclic ketones. Arrange the saturated and unsaturated molecules according to increasing C=O stretching frequency.

3. Rationalization of trends for IR frequencies for C=O: Discuss the trends observed in the C=O stretching frequencies for the saturated and unsaturated cyclic ketones (neutral and charge separated) that you calculated in terms of ring size and aromaticity. Explain why the various trends are observed.

4. Literature IR frequencies for C=O: The following are literature C=O stretching frequencies for the molecules studied in this experiment.

   cyclopropanone: 1813 cm^{-1}
   cyclopentanone: 1748 cm^{-1}
   cycloheptanone: 1702 cm^{-1}
   cyclopropenone: 1840 cm^{-1}
   cyclopentadienone: 1709 cm^{-1}
   cycloheptatrienone: 1651 cm^{-1}

   Is the trend observed in the literature for the saturated and unsaturated cyclic ketones consistent with the trends observed in your calculated structures? Explain.

5. Bond lengths and dipole moments for C=O: Compare the calculated bond lengths of your calculated molecules for those saturated and unsaturated ring systems: (1) of the same size and (2) of increasing size. Discuss whether the trends are consistent with your rationalization of your calculated C=O stretching frequencies.

   Compare the calculated dipole moments of your calculated molecules. What trend, if any, can you recognize?

6. Calculated IR frequencies for O—H: Comment on the trend observed in the O—H stretching frequencies for the phenols in the hydrogen-bonded complexes in terms of substituent effects. Compare the values to the O—H stretching frequencies in the individual phenols.

7. Literature IR frequencies for O—H: The following are literature O—H stretching frequencies for the substituted phenols.

phenol: 3386 cm^{-1}
4-fluorophenol: 3367 cm^{-1}
4-methoxyphenol: 3401 cm^{-1}

Is the trend reported in the literature consistent with the trend observed in your calculated complexes?

8. Hydrogen bond distance O---N: Comment on the trend observed in the hydrogen bond distance for all three molecular complexes in terms of substituent effects. What do your observations suggest about hydrogen bond strength?

9. Bonding energy: Comment on the trend observed in the bonding energies of the three complexes. What does this trend suggest about hydrogen bond strength? Some useful formulas are as follows:

To convert kJ/mol to kcal/mol: 1 kJ/mol = 0.239 kcal/mol
To convert a.u. to kcal/mol: 1 a.u. = 627.5 kcal/mol
To calculate bonding energy (BE) of a complex:

$$BE_{complex} = energy_{complex} - (energy_{pyridine} + energy_{phenol})$$

Data to Include in the Laboratory Report

TABLE 27.1

Calculated Data for Cyclic Ketones

CYCLIC KETONE	ENERGY (kJ/mol)	DIPOLE (debyes)	MULLIKEN CHARGES	IR FREQUENCY $C{=}O$ (cm^{-1})	DISTANCE $C{=}O$ (Å)
Cyclopropanone					
Cyclopentanone					
Cycloheptanone					
Cyclopropenone					
Cyclopentadienone					
Cycloheptatrienone					

TABLE 27.2

Calculated Data for Individual Molecules

MOLECULE	ENERGY (a.u.)	ENERGY (kcal/mol)	IR FREQUENCY $O{-}H$ (cm^{-1})	DISTANCE $O{-}H$ (Å)
Pyridine			*Not required*	
Phenol				
4-F-phenol				
4-OCH$_3$-phenol				

TABLE 27.3

Calculated Data for Hydrogen-Bonded Complexes

COMPLEX	ENERGY (a.u.)	ENERGY (kcal/mol)	BONDING ENERGY (kcal/mol)	DISTANCE $O{-}{-}{-}N$ (Å)	IR FREQUENCY $O{-}H$ (cm^{-1})	DISTANCE $O{-}H$ (Å)
Pyridine + phenol						
Pyridine + 4-F-phenol						
Pyridine + 4-OCH$_3$-phenol						

Appendix 1

Infrared Spectroscopic Data

APPENDIX 1A

Typical Infrared Absorbances of Various Bonds

FREQUENCY (cm^{-1})	BOND	FUNCTIONAL GROUP
3500–3200 (s, b)	O—H	Alcohols, phenols
3400–3250 (m, sh)	N—H	Amines, amides
3300–2500 (m)	O—H	Carboxylic acids
3330–3270 (s, sh)	C≡C—H, C—H	Alkynes (terminal)
3100–3000 (m)	C—H, C (sp^2)	Aromatics, alkenes
3000–2850 (m)	C—H, C (sp^3)	Alkanes
2830–2695 (m)	H—C=O, C—H	Aldehydes
2260–2210 (v)	C≡N	Nitriles
2260–2100 (w)	C≡C—	Alkynes
1760–1665 (s)	C=O	Carboxylic acids, esters, amides, aldehydes, ketones
1680–1640 (m)	C=C	Alkenes
1600–1585 (m)	C—C (in-ring)	Aromatics
1320–1000 (s)	C—O	Alcohols, carboxylic acids, esters, ethers
850–515 (m)	C—Cl or C—Br	Alkyl halides

(s) = strong, (m) = medium, (w) = weak, (v) = variable intensity, (b) = broad, (sh) = sharp

Infrared Data for Representative Compounds with C=O Bonds

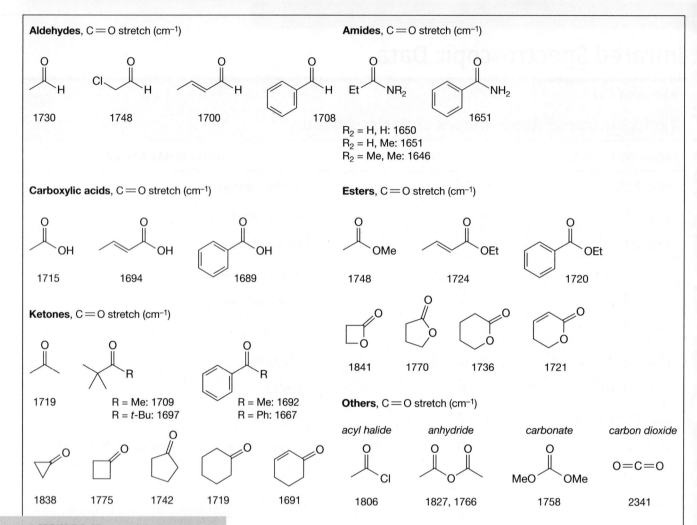

Aldehydes, C=O stretch (cm⁻¹)

1730   1748   1700   1708

Amides, C=O stretch (cm⁻¹)

1651

R_2 = H, H: 1650
R_2 = H, Me: 1651
R_2 = Me, Me: 1646

Carboxylic acids, C=O stretch (cm⁻¹)

1715   1694   1689

Esters, C=O stretch (cm⁻¹)

1748   1724   1720

1841   1770   1736   1721

Ketones, C=O stretch (cm⁻¹)

1719   R = Me: 1709   R = Me: 1692
        R = t-Bu: 1697   R = Ph: 1667

Others, C=O stretch (cm⁻¹)

acyl halide	anhydride	carbonate	carbon dioxide
1806	1827, 1766	1758	2341

O=C=O

1838   1775   1742   1719   1691

FIGURE A.1B

Silverstein, R. M.; Bassler, G. C.; Morrill, T. C. *Spectrometric Identification of Organic Compounds*, 4th ed.; Wiley: Hoboken, NJ, 1981. Crews, P.; Rodriguez, J.; Jaspars, M. *Organic Structure Analysis*; Oxford University Press: Oxford, 1998. Reich, H., University of Wisconsin, Chem 605 Handouts, personal communication, 2005. *AIST Spectral Database for Organic Compounds*. https://sdbs.db.aist.go.jp /sdbs/cgi-bin/cre_index.cgi (accessed March 2022).

Appendix 2

¹H NMR Spectroscopic Data

APPENDIX 2A

Typical Chemical Shift Ranges for Types of H

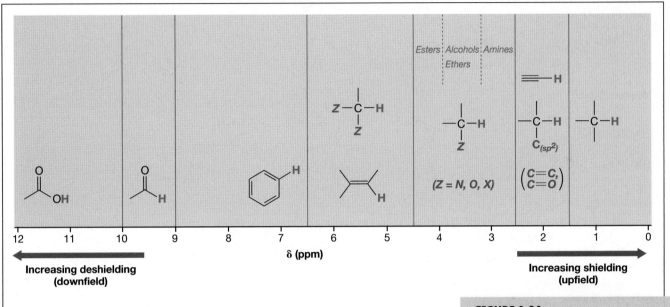

FIGURE A.2A

¹H NMR Chemical Shift Data for Representative Compounds

Acetals, δ (ppm)

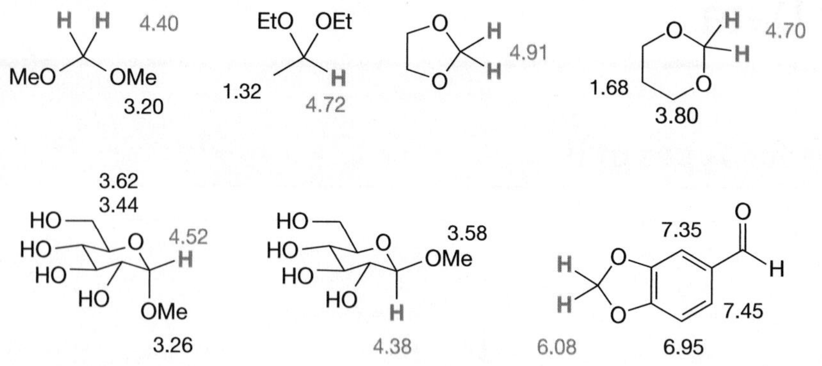

Alcohols, δ (ppm)

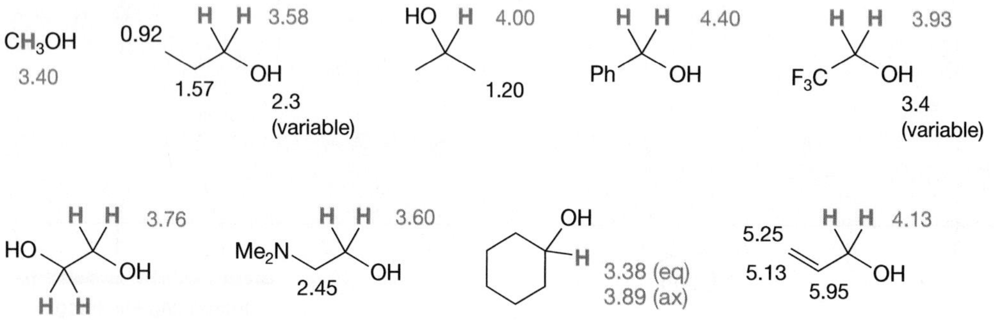

Aldehydes, δ (ppm)

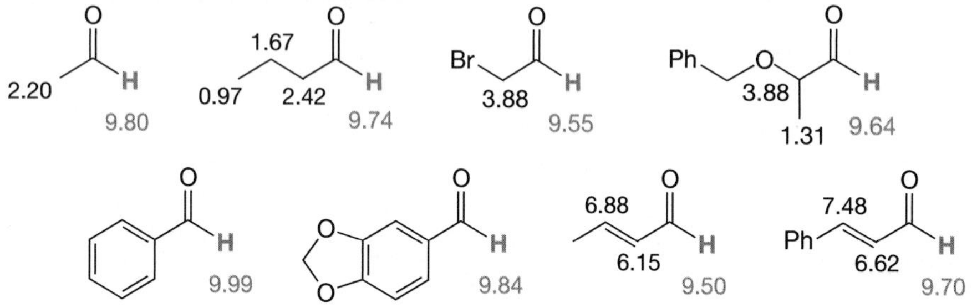

Alkenes, δ (ppm)

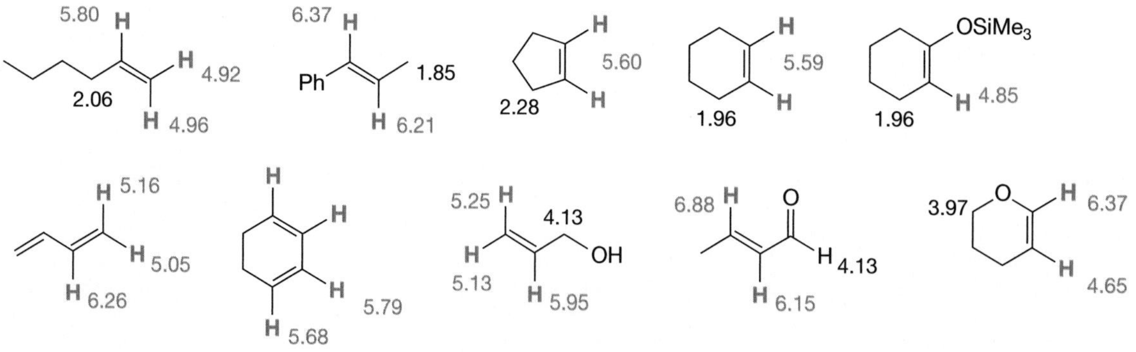

(continued)

^{1}H NMR Chemical Shift Data for Representative Compounds *(continued)*

Alkynes, δ (ppm)

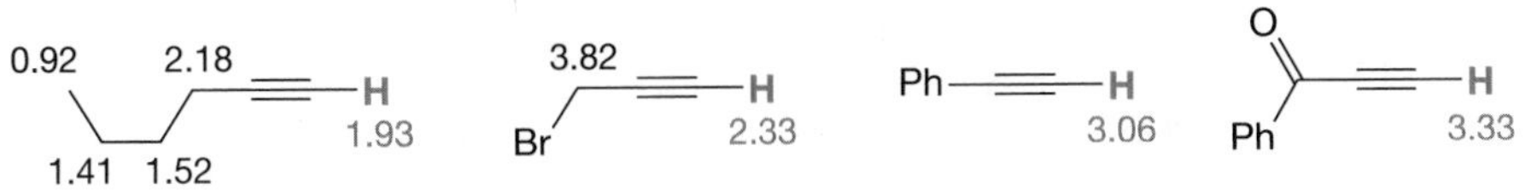

Amides and Lactams, δ (ppm)

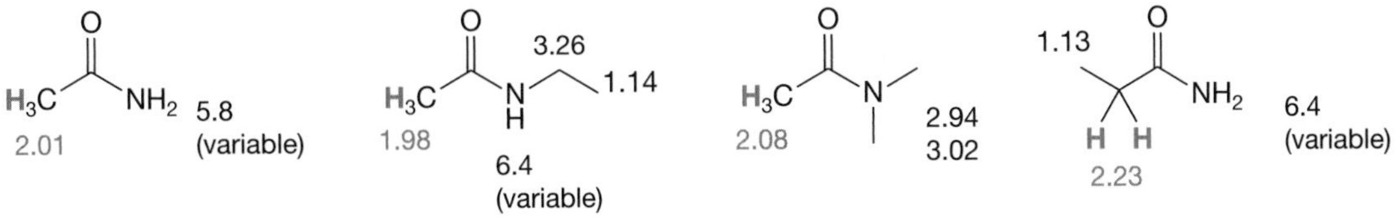

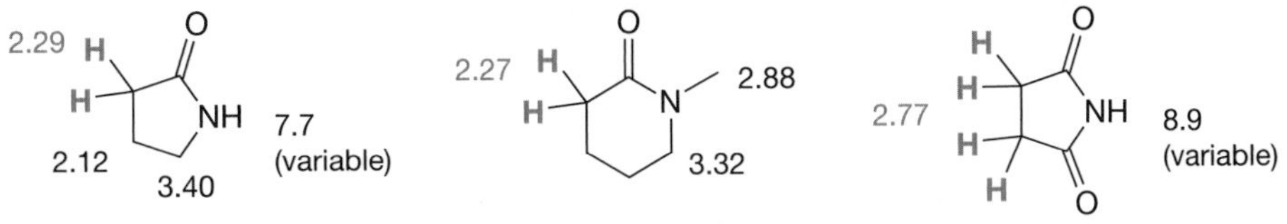

Amines, δ (ppm)

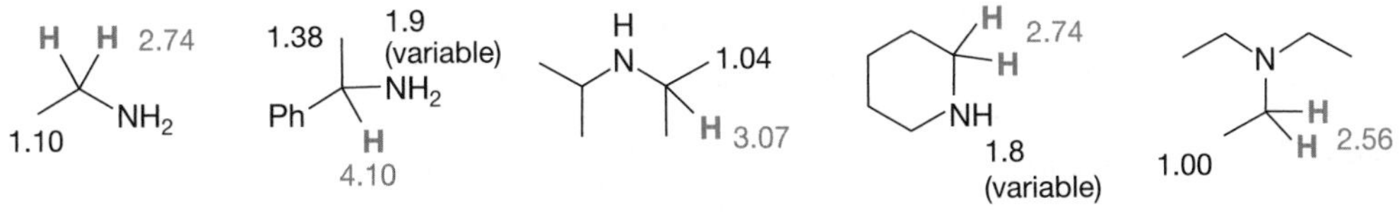

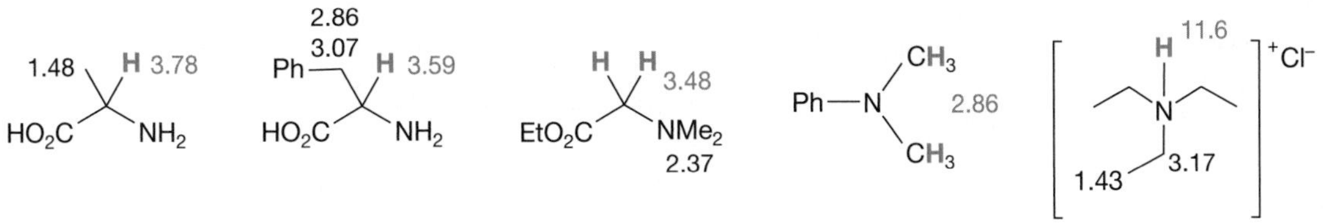

Aromatics, δ (ppm)

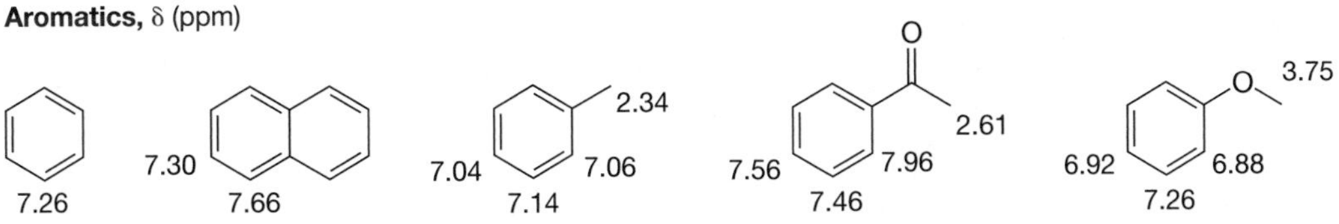

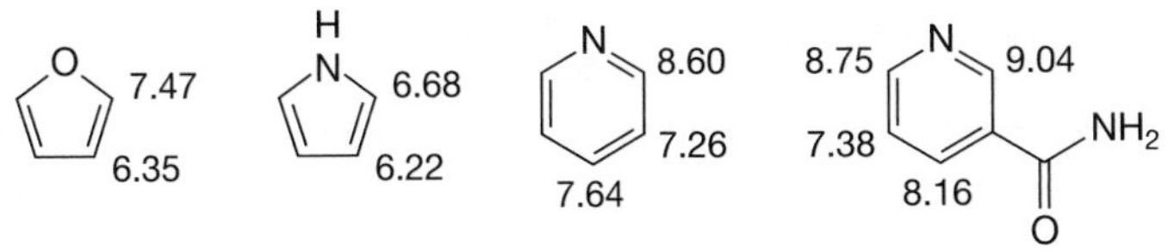

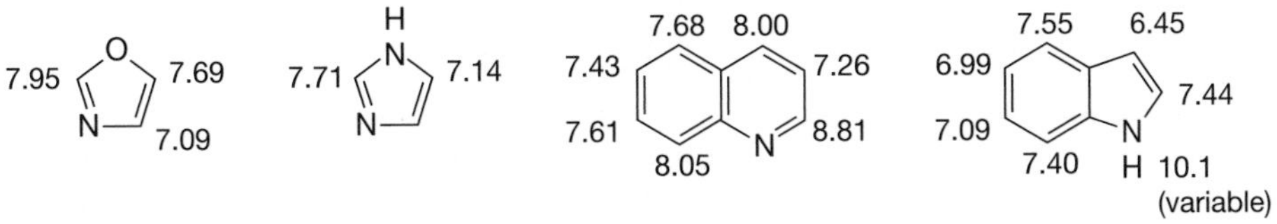

(continued)

^{1}H NMR Chemical Shift Data for Representative Compounds *(continued)*

Carboxylic acids, δ (ppm)

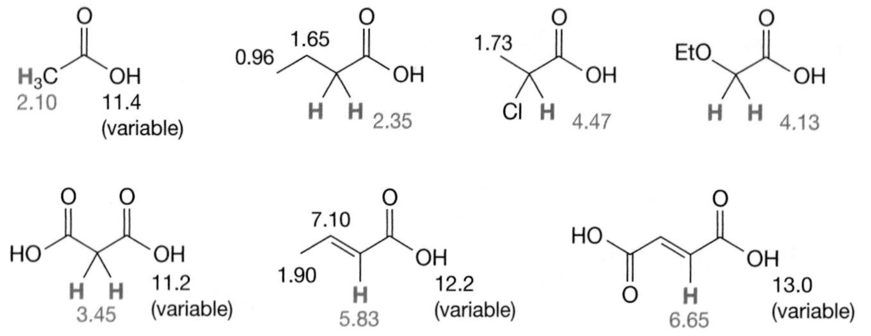

Esters and Lactones, δ (ppm)

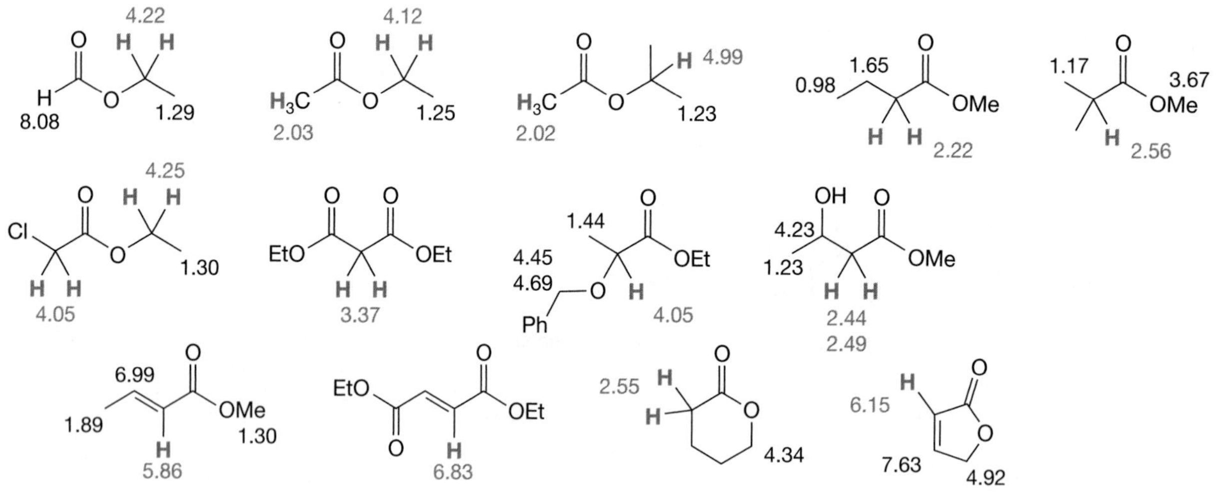

Ethers, δ (ppm)

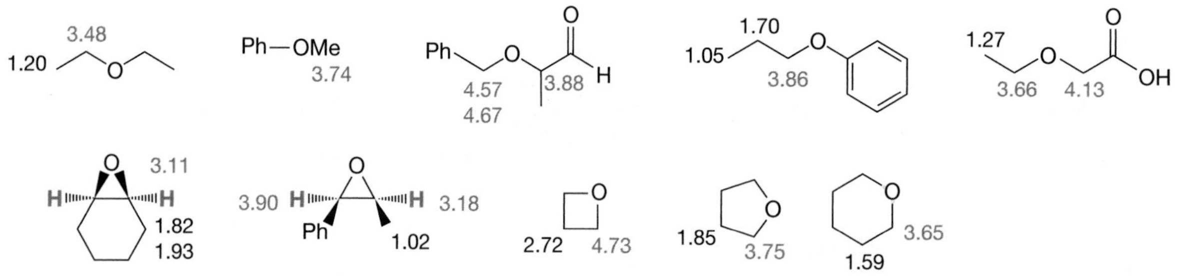

(continued)

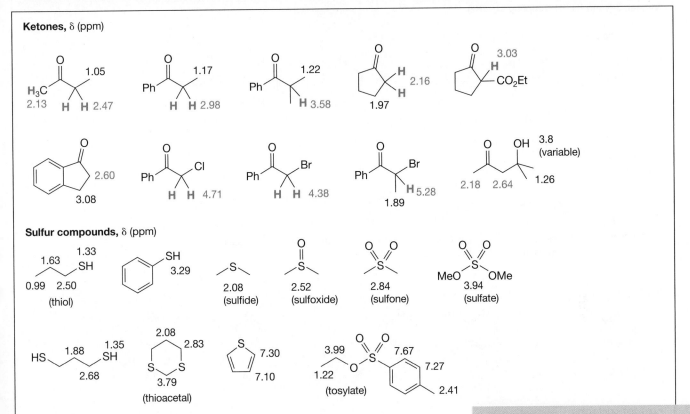

Ketones, δ (ppm)

Sulfur compounds, δ (ppm)

FIGURE A.2B

Data highlighted in blue are chemical shifts of hydrogens generally regarded as strongly diagnostic for the functional group. Numerical values should be regarded as approximations because solvent and concentration are not specified.

Silverstein, R. M.; Bassler, G. C.; Morrill, T. C. *Spectrometric Identification of Organic Compounds*, 4th ed.; Wiley: Hoboken, NJ, 1981. Crews, P.; Rodriguez, J.; Jaspars, M. *Organic Structure Analysis*; Oxford University Press: Oxford, 1998. Reich, H. University of Wisconsin, Chem 605 Handouts, personal communication, 2005. *AIST Spectral Database for Organic Compounds*. https://sdbs.db.aist.go.jp /sdbs/cgi-bin/cre_index.cgi (accessed March 2022).

Typical Ranges of Coupling Constants for Various ¹H–¹H Relationships

Vicinal, 3-Bond ¹H–¹H Coupling, J_{ab} **(Hz)**

• *alkenyl and dienyl*

| 17 | 10 | 7 | 5–7 | 9–11 | 10 |

• *alkyl*

| 6–8 | cis: 7 trans: 6 | cis: 6–9 trans: 5–7 | ax–ax: 12 | ax–eq: 4 | eq–eq: 4 |

• *aryl*

| 7.5 | 5.5 | 7.5 | 1.8 | 3.4 |

• *oxygen-containing*

| 5 | 2–3 | 6 | 2.5 | 4 |

Geminal, 2-Bond ¹H–¹H Coupling, J_{ab} **(Hz)**

| 12–15 | 0–2 |

6

4-Bond ¹H–¹H Coupling, J_{ab} **(Hz)**

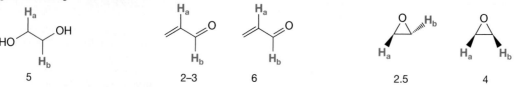

| 1–3 | 2 | 1.5 |

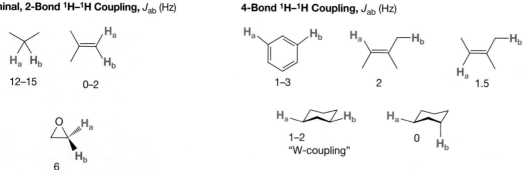

| 1–2 "W-coupling" | 0 |

FIGURE A.2C

These are abbreviated and generalized structures; observation of signal splitting according to these coupling constants presumes that other substituents are present, rendering H_a and H_b nonequivalent.

Silverstein, R. M.; Bassler, G. C.; Morrill, T. C. *Spectrometric Identification of Organic Compounds*, 4th ed.; Wiley: Hoboken, NJ, 1981. Crews, P.; Rodriguez, J.; Jaspars, M. *Organic Structure Analysis*; Oxford University Press: Oxford, 1998.

Appendix 3

13C NMR Spectroscopic Data

APPENDIX 3A

Typical Chemical Shift Ranges for Types of C

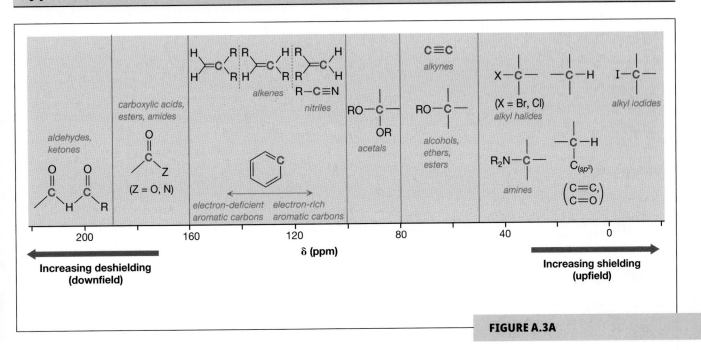

FIGURE A.3A

¹³C NMR Data for Representative Compounds

Acetals, δ (ppm)

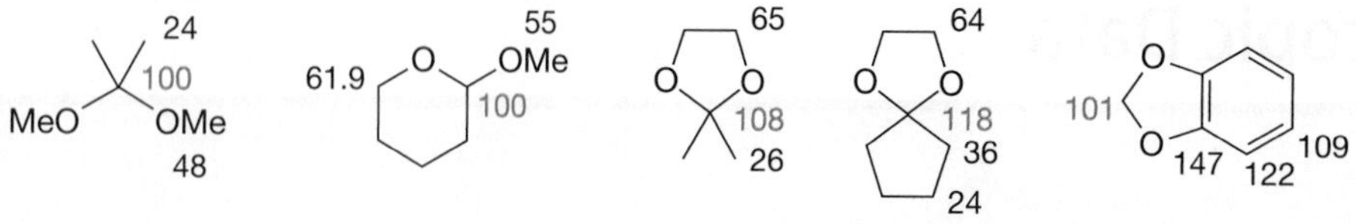

Alcohols, δ (ppm)

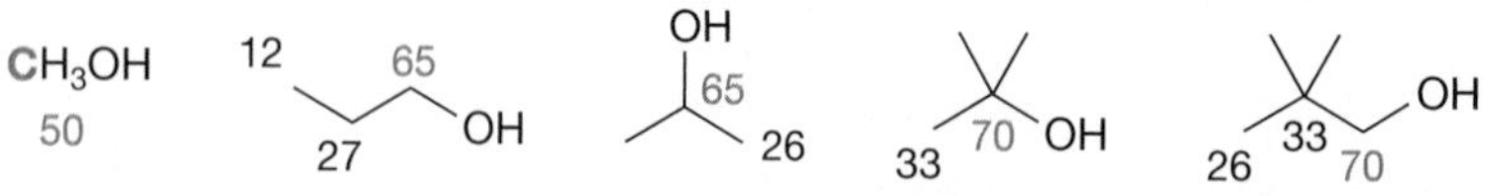

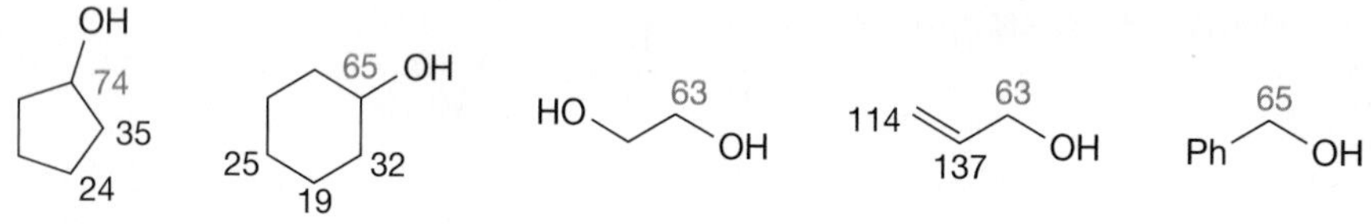

CH₃OH

Aldehydes, δ (ppm)

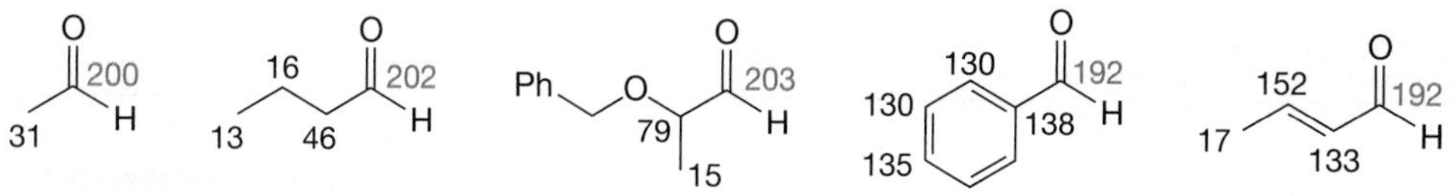

Alkenes, δ (ppm)

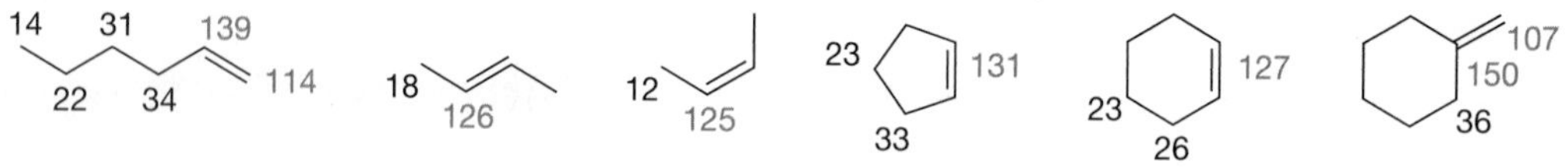

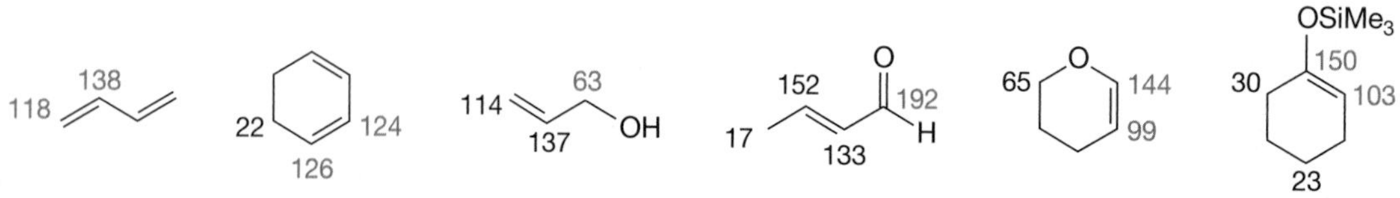

Alkynes, δ (ppm)

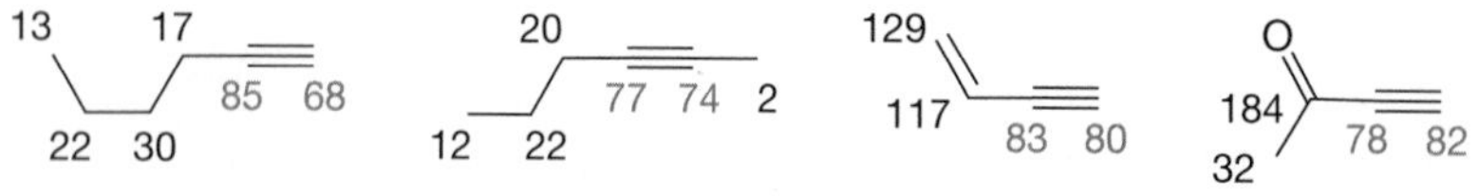

Amines, δ (ppm)

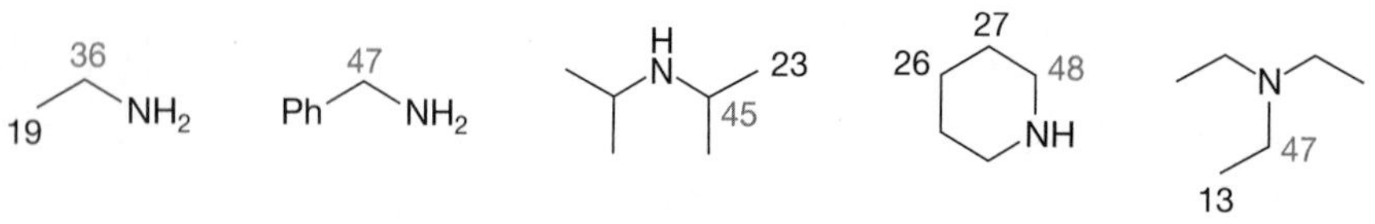

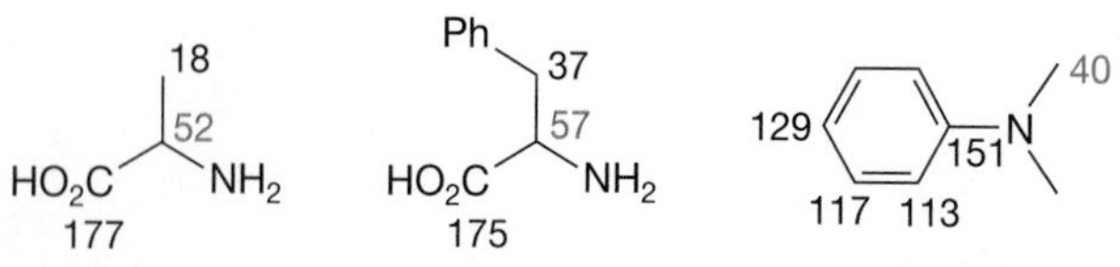

(continued)

^{13}C NMR Data for Representative Compounds *(continued)*

Amides and Lactams, δ (ppm)

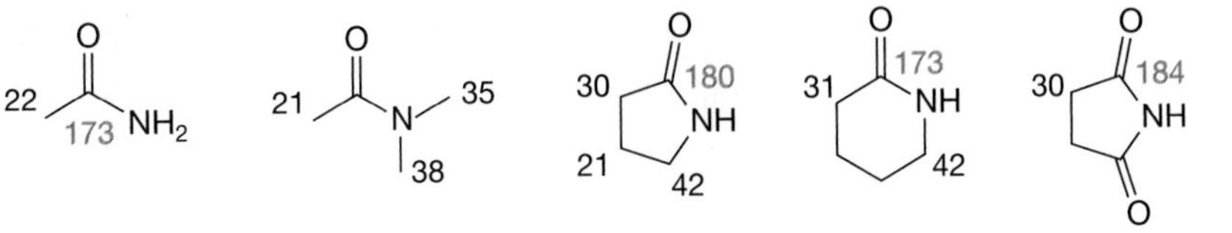

Aromatics, δ (ppm)

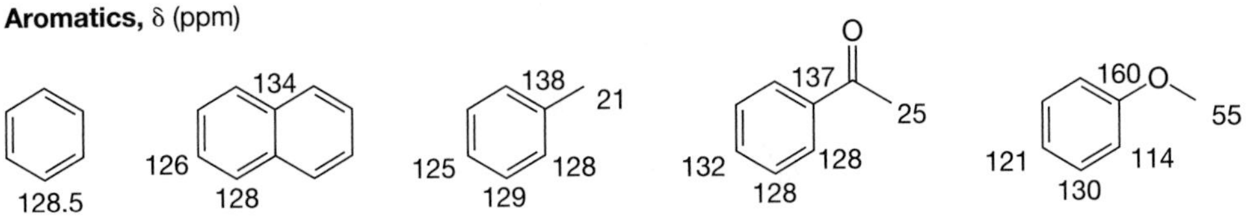

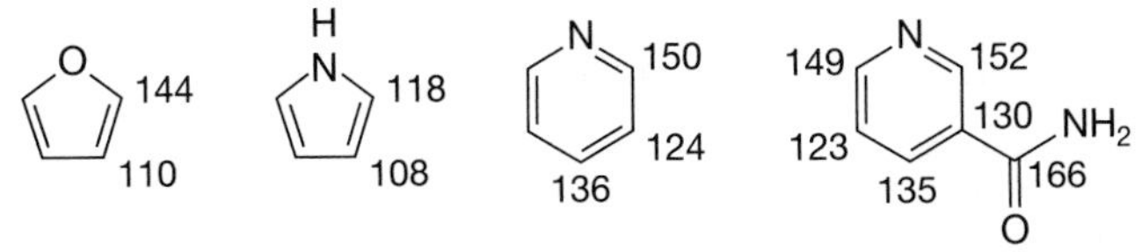

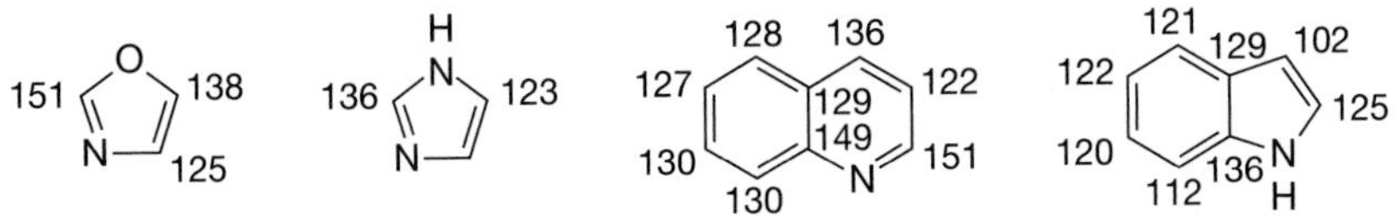

Carboxylic acids, δ (ppm)

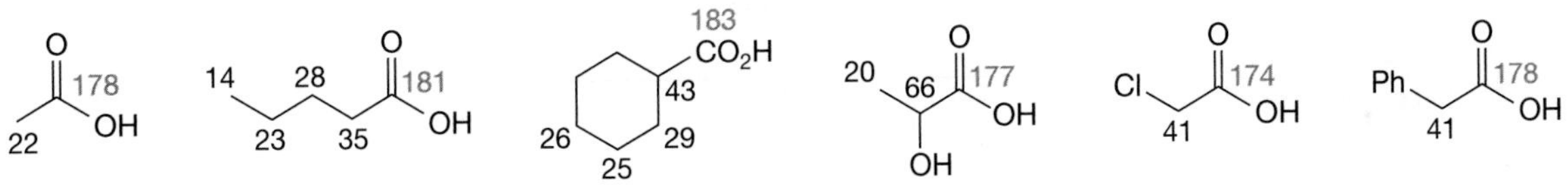

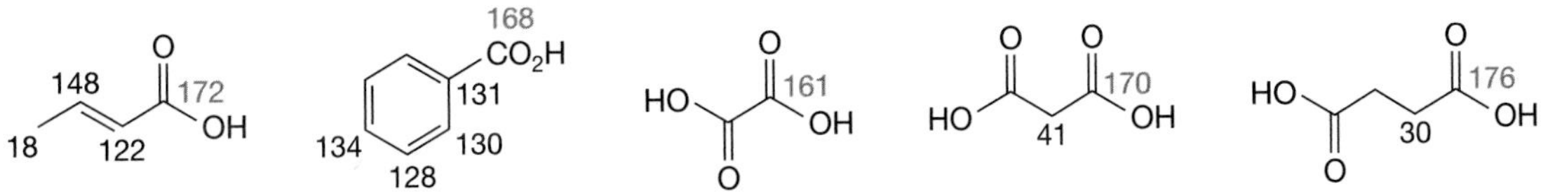

Esters and Lactones, δ (ppm)

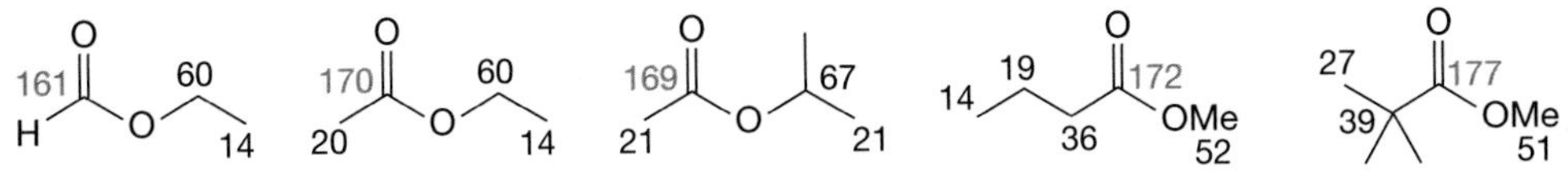

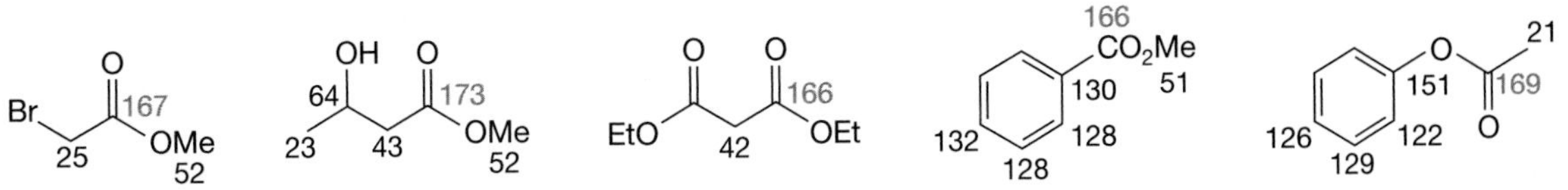

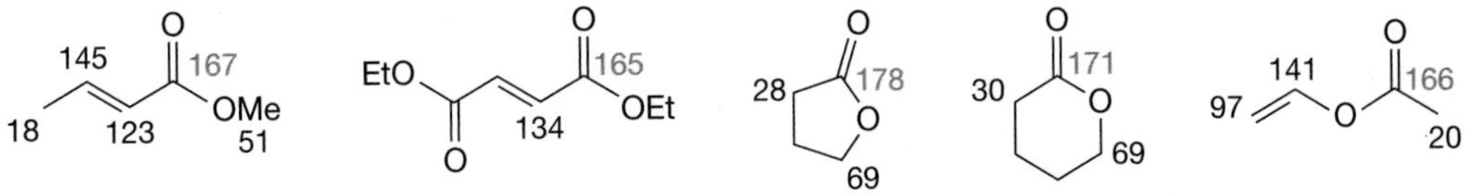

(continued)

^{13}C NMR Data for Representative Compounds *(continued)*

Ethers, δ (ppm)

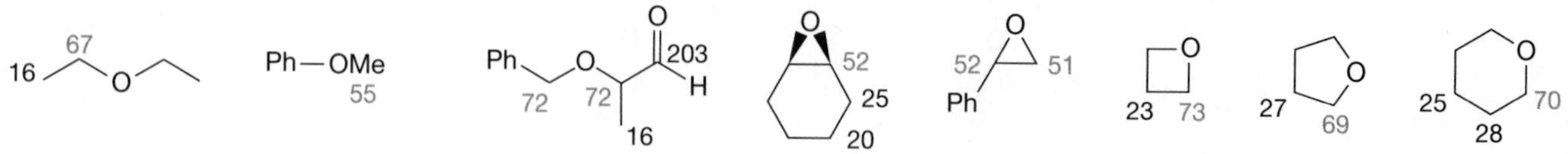

Ketones, δ (ppm)

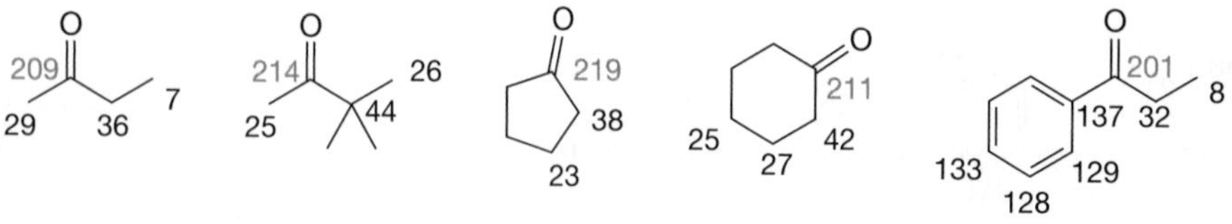

Sulfur compounds, δ (ppm)

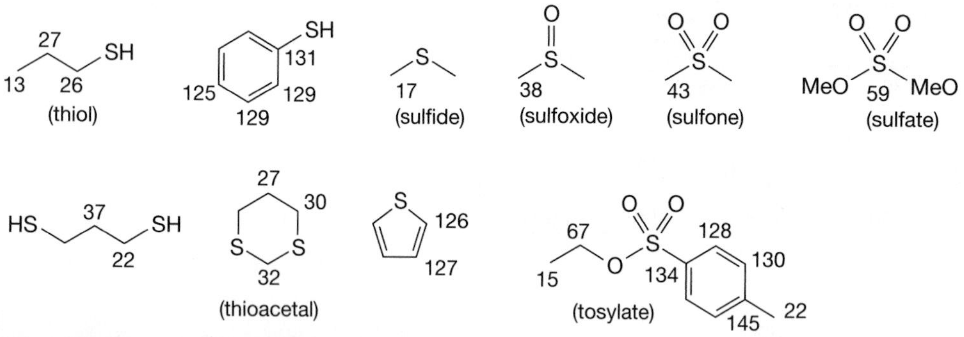

FIGURE A.3B

Reich, H. University of Wisconsin, Chem 605 Handouts, personal communication, 2005. *AIST Spectral Database for Organic Compounds*. https://sdbs.db.aist.go.jp/sdbs/cgi-bin/cre_index.cgi (accessed March 2022).

Appendix 4

¹H NMR Data for Identification of Common Impurities

Observed ¹H NMR Chemical Shifts (δ, ppm) of Residual Solvent Impurity Peaks in Three Common NMR Solvents

	NMR SOLVENT		
IMPURITY	CHLOROFORM-d, $CDCl_3$	ACETONE-d_6, $(CD_3)_2C{=}O$	DIMETHYLSULFOXIDE-d_6, $(CD_3)_2S{=}O$
Water	1.56	2.84	3.33
Acetic acid	2.10	1.96	1.91
Acetone	2.17	2.09	2.09
tert-Butyl methyl ether	3.22, 1.19	3.13, 1.13	3.09, 1.11
Chloroform	7.26	8.02	8.32
Dichloromethane	5.30	5.63	5.76
Diethyl ether	3.48 (q), 1.21 (t)	3.41 (q), 1.11 (t)	3.38 (q), 1.09 (t)
Dimethylsulfoxide	2.62	2.52	2.54
Ethanol	3.72 (q), 1.32 (s, OH), 1.25 (t)	3.57 (q), 3.39 (s, OH), 1.12 (t)	4.63 (s, OH), 3.44 (q), 1.06 (t)
Ethyl acetate	4.12 (q), 2.05 (s), 1.26 (t)	4.05 (q), 1.97 (s), 1.20 (t)	4.03 (q), 1.99 (s), 1.17 (t)
n-Hexane	1.26 (m), 0.88 (t)	1.28 (m), 0.88 (t)	1.25 (m), 0.86 (t)
Methanol	3.49, 1.09 (s, OH)	3.31, 3.12 (s, OH)	4.01 (s, OH), 3.16
Pyridine	8.62 (m), 7.68 (m), 7.29 (m)	8.58 (m), 7.76 (m), 7.35 (m)	8.58 (m), 7.79 (m), 7.39 (m)
Tetrahydrofuran	3.76 (m), 1.85 (m)	3.63 (m), 1.79 (m)	3.60 (m), 1.76 (m)
Triethylamine	2.53 (q), 1.03 (t)	2.45 (q), 0.96 (t)	2.43 (q), 0.93 (t)

Unless noted, peak is observed as singlet (s). Other multiplicities are noted in parentheses: m = multiplet, q = quartet, t = triplet.

Gottlieb, H. E.; Kotlyar, V.; Nudelman, A. NMR Chemical Shifts of Common Laboratory Solvents as Trace Impurities. *J. Org. Chem.* **1997**, *62*, 7512–7515.

Appendix 5

Tables of Unknowns (Chapter 26): mp/bp and Derivative mp

APPENDIX 5A

Alcohols: Physical Properties and Solid Derivatives

COMPOUND	SYNONYM	mp (°C)	bp (°C)	3,5-DNB (°C)	PU (°C)
Methanol			65	108	47
Ethanol			78	93	52
2-Propanol	Isopropyl alcohol		82	123	88
2-Methyl-2-propanol	*tert*-Butyl alcohol	26	83	142	136
2-Propen-1-ol	Allyl alcohol		97	49	70
1-Propanol	*n*-Propyl alcohol		97	74	51
2-Butanol	*sec*-Butyl alcohol		98	76	65
2-Methyl-2-butanol	*tert*-Pentyl alcohol		102	116	42
2-Methyl-1-propanol	Isobutyl alcohol		108	87	86
3-Pentanol			115	101	48
1-Butanol	*n*-Butyl alcohol		118	64	63
2-Pentanol			119	62	
3-Methyl-3-pentanol			123	96	43
2-Methoxyethanol			124		113[a]
2-Methyl-3-pentanol			128	85	50
2-Chloroethanol			129	95	51
4-Methyl-2-pentanol			132	65	143
3-Hexanol			135	97	
1-Pentanol	*n*-Amyl alcohol		137	46	46
Cyclopentanol			140	115	132
2-Ethyl-1-butanol			146	51	
2,2,2-Trichloroethanol			151	142	87
1-Hexanol	*n*-Hexyl alcohol		156	58	42

(continued)

Alcohols: Physical Properties and Solid Derivatives *(continued)*

COMPOUND	SYNONYM	mp (°C)	bp (°C)	3,5-DNB (°C)	PU (°C)
Cyclohexanol			160	113	82
2-Hydroxymethylfuran	Furfuryl alcohol		170	80	45
1-Heptanol	*n*-Heptyl alcohol		176	47	65
2-Octanol			179	32	114
2-Ethyl-1-hexanol			185		61[a]
1-Octanol	*n*-Octyl alcohol		195	61	74
2-Nonanol			198	43	56[a]
3,7-Dimethyl-1,6-octadien-3-ol	Linalool		199		66
Benzyl alcohol			205	113	77
1-Nonanol	*n*-Nonyl alcohol		214	52	62
2-(4-Methyl-3-cyclohexenyl)isopropanol	α-Terpineol	36	221	78	112
1-Tetradecanol	Myristyl alcohol	39		67	74
2-Isopropyl-5-methylcyclohexanol	Menthol	41	212	158	111
1-Hexadecanol	Cetyl alcohol	49		66	73
2,2-Dimethyl-1-propanol	Neopentyl alcohol	56	113		144
4-Methylbenzyl alcohol		59	217	117	79
1-Octadecanol	Stearyl alcohol	59		77	79
Diphenylmethanol	Benzhydrol	68		141	139
4-Nitrobenzyl alcohol		93		157	
2-Hydroxy-2-phenylacetophenone	Benzoin	136			165
Cholesterol		148			168
2-Hydroxy-2,2-diphenylacetic acid	Benzilic acid	150		b	b

3,5-DNB = mp of 3,5-dinitrobenzoate derivative
PU = mp of phenylurethane derivative (except those with [a])
[a] Naphthylurethane derivative
[b] For derivatives, see carboxylic acids, Appendix 5F.

Aldehydes: Physical Properties and Solid Derivatives

COMPOUND	SYNONYM	mp (°C)	bp (°C)	DNP (°C)	SC (°C)
Propanal	Propionaldehyde		48	150	89
Propenal	Acrolein		52	165	171
2-Methylpropanal	Isobutyraldehyde		64	187	125
Butanal	Butyraldehyde		75	123	95
3-Methylbutanal	Isovaleraldehyde		92	123	107
Pentanal	Valeraldehyde		102	106	
2-Butenal	Crotonaldehyde		104	190	199
2-Ethylbutanal			117	134	99
Hexanal	Caproaldehyde		130	104	106
Heptanal	Heptaldehyde		153	108	109
Cyclohexanecarboxaldehyde			162		173
Furan-2-carboxaldehyde	Furfural		162	229	202
2-Ethylhexanal			163	114	254
Octanal			171	105	101
Benzaldehyde			179	237	222
Nonanal			185	100	100
Phenylethanal	Phenylacetaldehyde	33	195	121	153
2-Hydroxybenzaldehyde	Salicylaldehyde		197	248	231
4-Methylbenzaldehyde	Tolualdehyde		204	234	234
Decanal			207	104	102
3,7-Dimethyl-6-octenal	Citronellal		207	77	82
2-Chlorobenzaldehyde		11	213	213	225
3-Chlorobenzaldehyde		18	214	248	228
3-Bromobenzaldehyde			235		205
4-Methoxybenzaldehyde	*p*-Anisaldehyde		248	253	210
3,4-Methylenedioxybenzaldehyde	Piperonal	37		266	230
3,4-Dimethoxybenzaldehyde	Veratraldehyde	43		261	177
2-Nitrobenzaldehyde		44		265	256

(continued)

Aldehydes: Physical Properties and Solid Derivatives *(continued)*

COMPOUND	SYNONYM	mp (°C)	bp (°C)	DNP (°C)	SC (°C)
4-Chlorobenzaldehyde		48		254	230
4-Bromobenzaldehyde		57		257	228
3-Nitrobenzaldehyde		58		293	246
4-(*N*,*N*-Dimethylamino)benzaldehyde		74		325	222
4-Hydroxy-3-methoxybenzaldehyde	Vanillin	82		271	230
3-Hydroxybenzaldehyde		104		259	198
4-Nitrobenzaldehyde		106		320	221
4-Hydroxybenzaldehyde		116		280	224

DNP = mp of 2,4-dinitrophenylhydrazone derivative
SC = mp of semicarbazone derivative

Primary Amines: Physical Properties and Solid Derivatives

COMPOUND	SYNONYM	mp (°C)	bp (°C)	AA (°C)	BA (°C)	BSA (°C)	TSA (°C)	PT (°C)
tert-Butylamine			46		134			120
n-Propylamine			49		84	36	52	63
Allylamine			56			39	64	98
sec-Butylamine			63		76	70	55	101
Isobutylamine			69		57	53	78	82
n-Butylamine			77		42			65
Isopentylamine			95				65	102
1,2-Diaminoethane	Ethylenediamine		116	172	249	168	160	102
1,2-Diaminopropane			120	139	192		103	
n-Hexylamine			128		40	96		77
Cyclophexylamine			134	104	149	89		148
2-Aminomethylfuran	Furfurylamine		145	31	103			
n-Heptylamine			155					75
1,4-Diaminobutane		27	160					168
2-Hydroxyethylamine			171					135
Aniline			183	114	160	112	103	154
Benzylamine			184	60	105	88	116	156
1-Phenylethylamine	α-Phenethylamine		185	57	120			
2-Phenylethylamine	β-Phenethylamine		198	114	116	69		135
2-Methylaniline	*o*-Toluidine		199	112	143	124	108	136
3-Methylaniline	*m*-Toluidine		203	65	125	95	114	94
2-Chloroaniline			207	87	99	129	105	156
2-Ethylaniline			210	111	147			
4-Ethylaniline			216	94	151			104
2-Bromoaniline		31	229	99	116			146
1,6-Diaminohexane		42			155 (di)	154 (di)		
4-Methylaniline	*p*-Toluidine	45	200	153	158	120	117	141
2,5-Dichloroaniline		50		132	120			

(continued)

Primary Amines: Physical Properties and Solid Derivatives *(continued)*

COMPOUND	SYNONYM	mp (°C)	bp (°C)	AA (°C)	BA (°C)	BSA (°C)	TSA (°C)	PT (°C)
1-Naphthylamine		50		159	160	167	157	165
4-Methoxyaniline	*p*-Anisidine	58		127	154	95	114	154
3-Aminoaniline	*m*-Phenylenediamine	63		191	240	194	172	
2,4-Dichloroaniline		63		145	117	128	126	
4-Bromoaniline		66		167	204	134	101	148
4-Chloroaniline		70		179	192	121	95	152
2-Nitroaniline		71		92	94	104		142
4-Methyl-3-nitroaniline		77		148	172	160	164	145
2,4-Dibromoaniline		79		146	134		134	
2,6-Dibromoaniline		83		210				
Ethyl 4-aminobenzoate		89		110	148			
2-Methyl-3-nitroaniline		95		158				
4-Aminoacetophenone		106		167	205	128	203	
2-Methyl-5-nitroaniline		107		150		172		
4-Chloro-2-nitroaniline		118		104				
3-Aminophenol		122		101 (di)			157	156
4-Aminoaniline	*p*-Phenylenediamine	140		304	300	247	266	
4-Nitroaniline		147		210	199	139	191	
4-Aminoquinoline		154		178				
4-(Acetylamino)aniline	*p*-Aminoacetanilide	162		304				
2-Aminophenol		174		201	182	141	139	146
2,4-Dinitroaniline		180		120	202		219	
4-Aminophenol		184		150 (di)	234	125		150

AA = mp of acetamide derivative
BA = mp of benzamide derivative
BSA = mp of benzenesulfonamide derivative (Hinsberg product)
TSA = mp of *p*-toluenesulfonamide derivative
PT = mp of phenylthiourea derivative
di = derivatized at two sites

APPENDIX 5D

Secondary Amines: Physical Properties and Solid Derivatives

COMPOUND	SYNONYM	mp (°C)	bp (°C)	AA (°C)	BA (°C)	BSA (°C)	TSA (°C)	PT (°C)
Diethylamine			55		42	42	60	34
Diisopropylamine			86		67		81	
Pyrrolidine			88		48		133	
Piperidine			106		48	93	96	101
Di-n-propylamine			110			51		69
2-Methylpiperidine			116		45		55	
Morpholine			129		75	118	147	136
Diisobutylamine			139	86		55		113
Di-n-butylamine			160		49			86
N-Methylbenzylamine			185				95	
N-Methylaniline			192	102	63	79	94	87
N-Ethylaniline			205	54	60		87	89
Di-n-pentylamine			205					72
N-Methyl-m-toluidine			206	66				
N-Methyl-o-toluidine			207	55	66		120	
N-Methyl-p-toluidine			208	83	53	64	60	89
N-Benzylaniline		37		58	107	119		103
2-Nitro-N-methylaniline		37		70				
Indole		52			68			
Diphenylamine		54		101	180	124	141	152
3-Nitro-N-methylaniline		66		95	156	83		
Di-p-tolylamine		79		85	125			
4-Nitro-N-ethylaniline		96		118	98			
Piperazine		104	140	134	191		173	

AA = mp of acetamide derivative
BA = mp of benzamide derivative
BSA = mp of benzenesulfonamide derivative (Hinsberg product)
TSA = mp of p-toluenesulfonamide derivative
PT = mp of phenylthiourea derivative

Tertiary Amines: Physical Properties and Solid Derivatives

COMPOUND	SYNONYM	mp (°C)	bp (°C)	QAM (°C)
Triethylamine			89	280
Pyridine			115	117
2-Methylpyridine	2-Picoline		129	230
2,6-Dimethylpyridine	2,6-Lutidine		143	233
3-Methylpyridine	3-Picoline		144	92
Tri-*n*-propylamine			157	207
N,N-Dimethylbenzylamine			183	179
N,N-Dimethylaniline			193	228
Tri-*n*-butylamine			216	186
N,N-Diethylaniline			217	102
Quinoline			237	133
4-Bromo-*N,N*-dimethylaniline		55	264	185
N,N-Dibenzylaniline		70		135
4-Dimethylaminobenzaldehyde		74		161
Tribenzylamine		91		184

QAM = mp of quaternary ammonium methiodide derivative

Carboxylic Acids: Physical Properties and Solid Derivatives

COMPOUND	SYNONYM	mp (°C)	bp (°C)	AMIDE (°C)	ANILIDE (°C)	*p*-TOLUIDIDE (°C)
Formic acid	Methanoic acid	8	101	43	47	53
Acetic acid	Ethanoic acid	17	118	82	114	148
Propenoic acid	Acrylic acid	13	139	85	104	141
Propanoic acid	Propionic acid		141	81	103	124
2-Methylpropanoic acid	Isobutyric acid		154	128	105	104
Butanoic acid	Butyric acid		162	115	95	72
2-Methylpropenoic acid	Methacrylic acid	16	163	102	87	
2,2-Dimethylpropanoic acid	Pivalic acid	35	164	178	127	
3-Methylbutanoic acid	Isovaleric acid		176	135	109	109
Pentanoic acid	Valeric acid		186	106	63	70
2-Methylpentanoic acid			186	79	95	80
2-Chloropropanoic acid			186	80	92	124
Dichloroacetic acid			194	98	118	153
Hexanoic Acid	Caproic acid		205	101	95	75
2-Bromopropanoic acid		24	205	123	99	125
Octanoic acid	Caprylic acid	16	237	107	57	70
4-Oxopentanoic acid	Levulinic acid	33	246	108	102	108
Nonanoic acid		12	254	99	57	84
Decanoic acid		32	268	108	70	78
Lauric acid	Dodecanoic acid	43		98	76	
3-Phenylpropanoic acid	Hydrocinnamic acid	48		105	98	135
Bromoacetic acid		50		91	131	
Tetradecanoic acid	Myristic acid	54		103	84	93
Trichloroacetic acid		57	198	141	97	113
Hexadecanoic acid	Palmitic acid	62		106	90	98
Chloroacetic acid		63	189	121	137	162
Octadecanoic acid	Stearic acid	69		109	95	102

(continued)

Carboxylic Acids: Physical Properties and Solid Derivatives *(continued)*

COMPOUND	SYNONYM	mp (°C)	bp (°C)	AMIDE (°C)	ANILIDE (°C)	*p*-TOLUIDIDE (°C)
trans-2-Butenoic acid	Crotonic acid	72		158	118	132
Phenylacetic acid		77		156	118	136
4-Methoxyphenylacetic acid		87		189		
3,4-Dimethoxyphenylacetic acid		97		147		
Pentanedioic acid	Glutaric acid	98		176 (di)	224 (di)	218 (di)
2-Methoxybenzoic acid	*o*-Anisic acid	100		129	131	
2-Methylbenzoic acid	*o*-Toluic acid	104		142	125	144
3-Methoxybenzoic acid	*m*-Anisic acid	107		136		
3-Methylbenzoic acid	*m*-Toluic acid	111		94	126	118
4-Bromophenylacetic acid		117		194		
Benzoic acid		122		130	163	158
2-Benzoylbenzoic acid		127		165	195	
cis-Butenedioic acid	Maleic acid	130		260 (di)	187 (di)	142 (di)
2-Furoic acid		133		143	124	170
(*E*)-3-Phenylprop-2-enoic acid	*trans*-Cinnamic acid	133		147	153	168
2-Acetylsalicylic acid	Aspirin	138		138	136	
2-Chlorobenzoic acid		140		139	118	131
3-Nitrobenzoic acid		140		143	155	162
4-Chloro-2-nitrobenzoic acid		142		172		
2-Nitrobenzoic acid		146		176	155	
Diphenylacetic acid		148		167	180	172
2-Hydroxy-2,2-diphenylacetic acid	Benzilic acid	150		154	175	190
Adipic acid	Hexanedioic acid	152		220	235	
3-Chlorobenzoic acid		156		134	123	
2,4-Dichlorobenzoic acid		158		194		
2-Hydroxybenzoic acid	Salicylic acid	158		142	136	156
3,4-Dimethylbenzoic acid		165		130	104	

(continued)

Carboxylic Acids: Physical Properties and Solid Derivatives *(continued)*

COMPOUND	SYNONYM	mp (°C)	bp (°C)	AMIDE (°C)	ANILIDE (°C)	*p*-TOLUIDIDE (°C)
2-Chloro-5-nitrobenzoic acid		166		178		
Tartaric acid		169		196 (di)	264 (di)	
4-Methylbenzoic acid	*p*-Toluic acid	180		160	145	160
4-Methoxybenzoic acid	*p*-Anisic acid	184		167	169	186
Butanedioic acid	Succinic acid	188		260 (di)	230 (di)	255 (di)
4-Ethoxybenzoic acid		198		202	170	
3-Hydroxybenzoic acid		201		170	157	163
3,5-Dinitrobenzoic acid		202		183	234	
3,4-Dichlorobenzoic acid		209		133		
Benzene-1,2-dicarboxylic acid	Phthalic acid	210		220 (di)	253 (di)	201 (di)
4-Hydroxybenzoic acid		214		162	197	204
4-Nitrobenzoic acid		240		201	211	204
4-Chlorobenzoic acid		242		179	194	
4-Bromobenzoic acid		251		190	197	

Amide = mp of primary amide derivative (via acyl chloride with ammonia)
Anilide = mp of *N*-phenylamide derivative (via acyl chloride with aniline)
p-Toluidide = mp of *N*-(*p*-tolyl)amide derivative (via acyl chloride with *p*-toluidine)
di = derivatized at two sites

Esters: Physical Properties and Solid Derivatives

COMPOUND	SYNONYM	mp (°C)	bp (°C)	NBA (°C)	3,5-DNB (°C)	ACID (°C)
Methyl formate	Methyl methanoate		32	60	108	
Ethyl formate	Ethyl methanoate		54	60	93	
Methyl acetate	Methyl ethanoate		57	61	108	
Ethyl acetate	Ethyl ethanoate		77	61	93	
Methyl propanoate	Methyl propionate		80	47	108	
Isopropyl acetate	Propan-2-yl ethanoate		88	61	123	
Methyl 2-methylpropanoate	Methyl isobutyrate		93	88	108	
Ethyl propanoate	Ethyl propionate		99	47	93	
Propyl acetate	Propyl ethanoate		102	61	74	
Methyl butanoate	Methyl butyrate		102	38	108	
Ethyl 2-methylpropanoate	Ethyl isobutyrate		110	88	93	
2-Butyl acetate	sec-Butyl acetate		112	61	76	
2-Methyl-1-propyl acetate	Isobutyl acetate		118	61	87	
Ethyl butanoate	Ethyl butyrate		122	38	93	
Propyl propanoate	n-Propyl propionate		123	47	74	
Butyl acetate			126	61	64	
Methyl chloroacetate			132		108	63
3-Methylbutan-1-yl acetate	Isoamyl acetate		142	61	61	
Ethyl chloroacetate			145		93	63
Pentyl acetate	n-Amyl acetate		149	61	46	
Ethyl 2-hydroxypropanoate	Ethyl lactate		154	48	93	
Ethyl hexanoate	Ethyl caproate		168	55	93	
Hexyl acetate			169	61	58	
Methyl 3-oxobutanoate	Methyl acetoacetate		170	100	108	
Dimethyl malonate			180	142 (di)	108	
Ethyl 3-oxobutanoate	Ethyl acetoacetate		181	100	93	
Diethyl oxalate			185	221 (di)	93	
Heptyl acetate			192	61	47	

(continued)

Esters: Physical Properties and Solid Derivatives *(continued)*

COMPOUND	SYNONYM	mp (°C)	bp (°C)	NBA (°C)	3,5-DNB (°C)	ACID (°C)
Phenyl acetate			197	61	146	
Methyl benzoate			199	106	108	122
Ethyl octanoate	Ethyl caprylate		207	66	93	
2-Methylphenyl acetate	*o*-Tolyl acetate		208	61	135	
3-Methylphenyl acetate	*m*-Tolyl acetate		212	61	165	
Ethyl benzoate			212	106	93	122
4-Methylphenyl acetate	*p*-Tolyl acetate		213	61	189	
Methyl 2-methylbenzoate	Methyl *o*-toluate		215		108	104
Methyl 3-methylbenzoate	Methyl *m*-toluate		215	76	108	111
Benzyl acetate			217	61	113	
Methyl phenylacetate			220	122	108	77
Methyl 4-methylbenzoate	Methyl *p*-toluate	33	223	133	108	180
Methyl 2-hydroxybenzoate	Methyl salicylate		224		108	158
Ethyl phenylacetate			228	122	93	77
Methyl (2*E*)-3-phenylprop-2-enoate	Methyl cinnamate	36		226	108	133
Phenyl 2-hydroxybenzoate	Phenyl salicylate	42			146	158
Methyl 4-chlorobenzoate		44		163	108	242
Ethyl 4-nitrobenzoate		56		142	93	240
Phenyl benzoate		69		106	146	122
Methyl 3-nitrobenzoate		78		101	108	140
Methyl 4-bromobenzoate		81		168	108	251
Methyl 4-nitrobenzoate		94		142	108	240

NBA = mp of *N*-benzylamide derivative of the acyl component
3,5-DNB = mp of 3,5-dinitrobenzoate derivative of the alcohol component
Acid = mp of carboxylic acid obtained via saponification

Ketones: Physical Properties and Solid Derivatives

COMPOUND	SYNONYM	mp (°C)	bp (°C)	DNP (°C)	SC (°C)
2-Propanone	Acetone		56	126	187
2-Butanone	Methyl ethyl ketone		80	117	146
3-Methyl-2-butanone	Isopropyl methyl ketone		94	120	112
2-Pentanone			101	143	112
3-Pentanone			102	156	138
3,3-Dimethyl-2-butanone	Pinacolone, *tert*-butyl methyl ketone		106	125	157
4-Methyl-2-pentanone			117	95	132
2,4-Dimethyl-3-pentanone	Diisopropyl ketone		124	95	160
3-Hexanone			125	130	113
2-Hexanone			128	106	121
Cyclopentanone			130	146	210
2,4-Pentanedione	Acetylacetone		139	209	209 (di)
4-Heptanone			144	75	132
3-Heptanone			148		101
2-Heptanone			151	89	123
Cyclohexanone			156	162	166
3-Octanone			167		117
2-Octanone			173	58	122
Cycloheptanone			181	148	163
Ethyl 3-oxobutanoate	Ethyl acetoacetate		181	93	129
5-Nonanone			186		90
3-Nonanone			187		112
2-Nonanone			195		118
Acetophenone	Methyl phenyl ketone	20	202	238	198
1-Phenyl-2-propanone	Phenylacetone	27	216	156	198
1-Phenyl-1-propanone	Propiophenone	21	218	191	173
Carvone			225	193	142
1-Phenyl-2-butanone			226		135

(continued)

Ketones: Physical Properties and Solid Derivatives *(continued)*

COMPOUND	SYNONYM	mp (°C)	bp (°C)	DNP (°C)	SC (°C)
4-Methylacetophenone	Methyl *p*-tolyl ketone	28	226	258	205
2-Undecanone			231	63	122
4-Chlorophenyl-1-propanone	4-Chloropropiophenone	36			176
4-Methoxyphenyl-1-ethanone	4-Methoxyacetophenone	38		220	198
1-Indanone		41		258	233
Benzophenone	Diphenyl ketone	48		238	164
4-Bromophenyl-1-ethanone	4-Bromoacetophenone	51		230	208
1,2-Diphenylethanone	Benzyl phenyl ketone	60		204	148
1,1-Diphenyl-2-propanone	1,1-Diphenylacetone	61			170
4-Chlorobenzophenone		76			185
4-Bromobenzophenone		82		230	
Fluorenone		83		234	283
4-Hydroxyphenyl-1-ethanone	4-Hydroxyacetophenone	109		210	199
2-Hydroxy-2-phenylacetophenone	Benzoin	136		245	206
Camphor, (+)		179	205	177	237

DNP = mp of 2,4-dinitrophenylhydrazone derivative
SC = mp of semicarbazone derivative
di = derivatized at two sites

Phenols: Physical Properties and Solid Derivatives

COMPOUND	SYNONYM	mp (°C)	bp (°C)	NU (°C)	BROMINATION (°C)	OTHER (°C)
2-Chlorophenol			176	120	Mono 48, di 76	
2-Methylphenol	o-Cresol	32	191	142	Di 56	
3-Methylphenol	m-Cresol		203	128	Tri 84	
2-Methoxyphenol	Guaiacol	32	204	118	Tri 116	
2,4-Dimethylphenol		23	212	135		
Phenyl 2-hydroxybenzoate	Phenyl salicylate	42			Mono 113, di 128	Phenylurethane 112
4-Chlorophenol		43	217	166	Mono 33, di 90	
2,4-Dichlorophenol		45	210		Mono 68	
4-Ethylphenol		45	219	128		
2-Isopropyl-5-methylphenol	Thymol	51	234	160	Mono 55	
4-Bromophenol		64	238	169	Tri 95	
3,5-Dimethylphenol		68	220	109	Tri 166	
2,5-Dimethylphenol		75	212	173	Tri 178	
4-Hydroxy-3-methoxybenzaldehyde	Vanillin	82				(see Aldehydes)
1-Naphthol		96		152	Di 105	
2-Hydroxyphenol	Catechol	105		175	Tetra 192	
3-Hydroxyphenol	Resorcinol	109			Tri 112	
4-Hydroxyphenyl-1-ethanone	4-Hydroxyacetophenone	109				Phenylurethane 154
4-Nitrophenol		112		150	Di 142	
4-Hydroxybenzaldehyde		116			Mono 125	Phenylurethane 136
2-Naphthol		121		157	Mono 84	
1,2,3-Trihydroxybenzene	Pyrogallol	133			Di 158	
2-Hydroxybenzoic acid	Salicylic acid	158				Amide 142
3-Hydroxybenzoic acid		201				Amide 170
4-Hydroxybenzoic acid		214				Amide 162

NU = mp of naphthylurethane derivative
Bromination = mp of arene bromination products, with number of Br substitutions

Appendix 6

Bond Strengths of Representative Organic Compounds

C—H BONDS, HYDROCARBONS		C—H BONDS, OTHER FUNCTIONAL GROUPS		C—C BONDS		C—X BONDS	
COMPOUND, BOND	STRENGTH (kJ/mol)	COMPOUND, BOND	STRENGTH (kJ/mol)	COMPOUND, BOND	STRENGTH (kJ/mol)	COMPOUND, BOND	STRENGTH (kJ/mol)
H₃C—H	439	HOCH₂—H	401	H₃C—CH₃	377	H₃C—Cl	350
CH₃CH₂CH₂CH₂—H	421	CH₃CH(OH)—H	384	(CH₃)₂CH—CH₃	369	Cl₃C—Cl	289
(CH₃)₂CH—H	411	CH₂=CHCH(OH)—H	341	(CH₃)₃C—CH₃	364	(CH₃)₂CH—Cl	355
(CH₃)₃C—H	400	CH₃CH(OEt)—H	389	Ph—CH₃	434	(CH₃)₃C—Cl	352
HC≡C—H	558	(tetrahydrofuran C2)—H	385	Ph—CH₂CH₃	325	Ph—Cl	406
HC≡CCH₂—H	384	CH₃CH₂C(O)—H	375	Ph—Ph	493	PhCH₂—Cl	300
CH₂=CH—H	464	CH₃CH₂CH₂C(O)—H	384	CH₃CH₂CH₂CH₂—OH	357	CH₃CH₂C(O)—Cl	353
CH₂=CHCH₂—H	369	(ketone C(O))—H	404	H₃C—CHO	355	ClCH₂C(O)—OH	311
(allylic sec)—H	351	(ketone)—H	386	H₃C—C(O)CH₃	352	H₃C—Br	294
(allylic tert)—H	333	H—CH₂C(O)OH	399	Ph—C(O)CH₃	414	H₃C—I	232
cyclopropyl—H	445	H—CH₂C(O)OEt	402	H₃C—C(O)OH	385	H₃C—OH	385
cyclobutyl—H	409	H₂NCH₂—H	393	H₂NCH₂—C(O)OH	349	H₃C—OCH₃	352
cyclopentyl—H	400	H₂NCH(CH₃)—H	377			H₃C—NH₂	356
cyclohexyl—H	416	EtHNCH—H	371	**O—H BONDS**			
phenyl—H	466	Et₂NCH—H	380	HO—H	497	**O—O, N—N, S—S BONDS**	
benzyl—H	376			EtO—H	435	HO—OH	210
PhC(CH₃)₂—H	348			PhO—H	363	EtO—OEt	166
				HOO—H	366	t-BuO—OH	186
				ONO—H	330	H₂N—NH₂	277
						PhS—SPh	214

Strength = Bond dissociation enthalpy
To convert data to kcal/mol, divide by 4.184 kJ/kcal.

Appendix 7

Typical pK_a Values of Representative Organic Compounds

Values of pK_a for Various Acids[a]

ACID	CONJUGATE BASE	pK_a	ACID	CONJUGATE BASE	pK_a
F$_3$C—S(=O)$_2$—OH Trifluoromethanesulfonic acid (TfOH)	F$_3$C—S(=O)$_2$—O$^-$	−13	Cl$_3$C—C(=O)—OH Trichloroethanoic acid (Trichloroacetic acid)	Cl$_3$C—C(=O)—O$^-$	0.77
HI Hydroiodic acid	I$^-$	−10	H$_2$ClC—C(=O)—OH Chloroethanoic acid (Chloroacetic acid)	H$_2$ClC—C(=O)—O$^-$	2.87
HO—S(=O)$_2$—OH Sulfuric acid	HO—S(=O)$_2$—O$^-$	−9	HF Hydrofluoric acid	F$^-$	3.2
HBr Hydrobromic acid	Br$^-$	−9	H—C(=O)—OH Methanoic acid (Formic acid)	H—C(=O)—O$^-$	3.75
HCl Hydrochloric acid	Cl$^-$	−7	C$_6$H$_5$—C(=O)—OH Benzoic acid	C$_6$H$_5$—C(=O)—O$^-$	4.2
p-CH$_3$C$_6$H$_4$—S(=O)$_2$—OH p-Toluenesulfonic acid (TsOH)	p-CH$_3$C$_6$H$_4$—S(=O)$_2$—O$^-$	−2.8	CH$_3$—C(=O)—OH Ethanoic acid (Acetic acid)	CH$_3$—C(=O)—O$^-$	4.75
H$_3$C—S(=O)$_2$—OH Methanesulfonic acid (MsOH)	H$_3$C—S(=O)$_2$—O$^-$	−2	Pyridinium ion	Pyridine	5.2
H$_3$O$^+$ Hydronium ion	H$_2$O	0.0[b]	HO—C(=O)—OH Carbonic acid	$^-$O—C(=O)—O$^-$	6.3
F$_3$C—C(=O)—OH Trifluoroethanoic acid (Trifluoroacetic acid)	F$_3$C—C(=O)—O$^-$	0.0			

(continued)

ACID	CONJUGATE BASE	pK_a	ACID	CONJUGATE BASE	pK_a
Thiophenol (C6H5–SH)	C6H5–S⁻	6.6	2,2,2-Trifluoroethanol (CF3CH2OH)	CF3CH2O⁻	12.4
4-Nitrophenol	(conjugate base)	7.2	Diethyl propanedioate (Diethyl malonate)	(conjugate base)	13.5
H_2S Hydrogen sulfide	HS⁻	7.2	H_2O Water	HO⁻	14.0[b]
2,4-Pentanedione	(conjugate base)	8.9	2-Chloroethanol	(conjugate base)	14.3
N≡C–H Hydrocyanic acid	N≡C⁻	9.2	Pyrrole	(conjugate base)	15
H_4N^+ Ammonium ion	NH_3	9.4	CH_3OH Methanol	CH_3O^-	15.5
$(CH_3)_3\overset{+}{N}H$ Trimethylammonium ion	$(CH_3)_3N$	9.8	Ethanol	(conjugate base)	16
Phenol	(conjugate base)	10.0	Cyclopentadiene	(conjugate base)	16
$O_2N–CH_3$ Nitromethane	$O_2N–\overset{..}{C}H_2$	10.2	Propan-2-ol (Isopropyl alcohol)	(conjugate base)	16.5
Ethanethiol	(conjugate base)	10.6			
$H_3C–\overset{+}{N}H_3$ Methylammonium ion	$H_3C–NH_2$	10.63	Ethanamide (Acetamide)	(conjugate base)	17
Ethyl 3-oxobutanoate (Acetoacetic ester)	(conjugate base)	11	Methylpropan-2-ol (*tert*-Butyl alcohol)	(conjugate base)	19

(continued)

Values of pK_a for Various Acids[a] *(continued)*

ACID	CONJUGATE BASE	pK_a	ACID	CONJUGATE BASE	pK_a
Propanone (Acetone) $H_3C-CO-CH_3$	$H_3C-CO-\bar{C}H_2$	20	NH_3 Ammonia	H_2N^-	36
Ethyl ethanoate (Ethyl acetate) $EtO-CO-CH_3$	$EtO-CO-\bar{C}H_2$	25	N-Methylmethanamine (Dimethylamine) $H_3C-NH-CH_3$	$H_3C-N^--CH_3$	38
Ethanenitrile (Acetonitrile) $N\equiv C-CH_3$	$N\equiv C-\bar{C}H_2$	25	Toluene (Methylbenzene)		40
Ethyne (Acetylene) $HC\equiv CH$	$HC\equiv C^-$	25	Benzene		43
Aniline (Phenylamine)		27	Ethene (Ethylene) $H_2C=CH_2$	$H_2C=\bar{C}H$	44
Hydrogen gas H_2	H^-	35	Ethoxyethane (Diethyl ether)		45
Dimethyl sulfoxide		35	Methane CH_4	H_3C^-	48
			Ethane CH_3CH_3	$CH_3\bar{C}H_2$	50

[a] p$K_a = -\log K_a$. The less positive (or more negative) the pK_a value, the stronger the acid relative to another acid.
[b] In older textbooks, the pK_a values of H_3O^+ and H_2O are reported to be −1.7 and 15.7, respectively, but by definition they are 0 and 14.

Karty, J. *Organic Chemistry: Principles and Mechanisms*, 3rd ed.; W. W. Norton: New York, 2022.

Index

D

laboratory report, 283–284
pre-lab reading assignment, 274
gum, 17, 24

H

H_2O hydrogen bonds, 243, *243*
HAL 9000, 343
halichondrin B, 218
halogen, 321
halohydrin formation, 221, *222*
Hammett plot, 352, *353*
Handbook of Chemistry and Physics, 150
Handbook of Fine Chemicals, 150
hazard communication standard (HCS)
　pictogram, 17, *18*
hazard data, 17
HBr (hydrogen bromide), 255, *255*
Heck, Richard, 317, *317*
Heck reaction, 317, *317*
heme B, 264–265, *265*
heterogeneous catalysis, 194, *195*
heterolysis, 202–203, *203*
(E,E)-2,4-hexadienal, 236–237, *236*
hexanedioic acid. *See* adipic acid (hexanedi-
　oic acid) synthesis lab
Hickman still, 282–283, *282*
highest occupied molecular orbital
　(HOMO), 210–212
high-performance liquid chromatography
　(HPLC), 74–75, 130
high-resolution mass spectrometry
　(HRMS), 134–135
Hinsberg test, 333–334
HNO_3 (nitric acid), 222–223, 345–346,
　345–346
[1]H nuclear magnetic resonance (NMR)
　spectroscopy, 94–120
　acquiring NMR spectra, *111*, 114, *115*
　of *trans*-1,2-bis(4-pyridyl)ethene,
　　251–252
　calibration, 114, 176
　C—C rotation and, 109
　characterization of known compound,
　　115–116
　chemical shifts, 97, 99–100, *100–101*, 102,
　　103, *104–105*
　^{13}C NMR spectroscopy comparison,
　　120–121
　coupling and chemical exchange, 104, 109
　data interpretation, 97, 109–110, 115–116
　data reporting, 116–117, *116–117*
　examples of, *101–102*, 102–103
　integration, 97, 100, *101*, 102, 177
　mixed samples and, *110–111*, 112–113,
　　113
　molecular formula overview, 126
　multiplicity, 97, 103–109, *104–108*, 177
　number of signals, 97–99, *98*, 102, 177
　percent conversion determination,
　　251–252

product ration measurement, 205, *205*,
　207
real-life practical situations with,
　110–112, *110–112*
sample preparation, 114, *115*, 175, *175*
sample problems, 97, 102–103, *102*, 113,
　113, 118–120, *118–120*, 140–142,
　140–142
sample spinning, 114, *115*, 176
solvent choice, *111*, 114, 176
theory of, 94–97, *94–96*
Hodgkin, Dorothy Crowfoot, 265
HOI (hypoiodous acid), 256, *256*
HOMO (highest occupied molecular orbit-
　al), 210–212
homogeneous catalysis, 194, *195*
honeybee sex pheromone, *274*, 275
Hooke's law, 82–83, *83*
hormones, 275
hot filtration, 41, *41*, 55, 58, *58*
household bleach (NaOCl), 256, *256*, 277,
　278, *278*, 281
HPLC (high-performance liquid chroma-
　tography), 74–75, 130
HRMS (high-resolution mass spectrome-
　try), 134–135
Hückel rule, 347–349, *347–349*
hydride reduction of aldehydes and ketones
　background, 231, *231*
　experimental procedure, 236–237, *236*
hydrocarbon, saturated, 255, *255*
hydrochloric acid, 165, *166*
hydrogen and hydrogen bonds
　background, 243, *243*
　molecular modeling of, 351–353, *351–353*
　safety practices, 195
hydrogenation reaction, 192, *192*. *See also*
　catalytic transfer hydrogenation lab
hydrogen peroxide, 223, *223*, 277, 282
hydrolysis, 310
hydroxide ion, 203
hydroxylamine/ferric chloride test, 334
hyperconjugation, 203, *203*
hypochlorite, 277, *278*
hypoiodous acid (HOI), 256, *256*

I

ibuprofen, 184, *184*
imine(s), 193, *193*
1-indanone, *311*, 312
indole, 264, *264*
inductive effect, 352
industrial wastes, 12
informal laboratory reports, 34
infrared (IR) light, 78, *78*, 79
infrared (IR) spectroscopy, 82–89
　acquiring an infrared spectrum, 85–86,
　　85–86
　of aldehydes, 349–350
　for chalcones, 199

characterization of known compound,
　86–87
database for, 150
data reporting, 36, 88–89, *88–89*
of hydrogen-bonded molecular
　complexes, 353
identification of unknown compound,
　87–88
interpretation of data, 86–88
introduction, *78*, 82, *82*
of ketones, 349–350
of ketones and unsaturated ketones, 196,
　196, *197*
molecular formula overview, 126
molecular modeling and, 349–350, *350*
molecular transitions and, 82–83, *83–84*
NMR spectroscopy comparison, 94
preparation for liquids, 85–86
preparation for solids, 86
for substitution pattern of aromatic rings,
　257, *258*
injuries, responses to, 23
in-process monitoring, 12
insect pheromones, 273, 274, *274*, 275
integrated pest management, 274, 275
internet citation formats, 153
intrinsic atom economy, 8–9, *9*
inverted tree diagram, *107–108*
iodination of salicylic acid derivatives lab,
　253–261
　about, 253–254, *254*
　background, 255–257, *255–257*, *258*
　experimental procedure, 258–260
　laboratory report, 261
　pre-lab reading assignment, 255
iodine, in protein-drug interactions, 253
iodoform test, 332–333
iodosalicylamide, *257*
iodosalicylic acid, *257*
IR (infrared) light, 78, *78*, 79
iron(II) hydroxide test, 334
IR spectroscopy. *See* infrared (IR) spectros-
　copy
isobutyraldehyde, 313–314, *313*
isoprene units, 160, *161*
isopropyl bromide, 205, *205*, 207
isotope peaks, 131–132, *131–132*

J

J (coupling constant), 106, *107*
Johnson, W. S., 228
Jones oxidation, 277
Journal of the American Chemical Society,
　152
journals
　abstracts for, 147, 148–149
　citation formats, 153
　as primary sources, 146–147
　review journals, 151
　standard abbreviations, 146–147

K

K (partition coefficient), 43–44
Karplus curve, *106*
KBr pellet, 86
Kekulé, August, *289*
ketone(s)
    derivatives from, 336, *336*
    enamine generation from, 310
    functional group tests for, 331–333
    Grignard reagents and, 276, *276*
    hydrogenation of, 193–194, *193*, 196
    molecular modeling of, 345–351, *345–351*
    reduction of, 297. *See also* biocatalytic reduction lab
    Wittig reaction and, 245, 246, *247*
ketoreductase, 299–300, *299*
Kevlar, 217
KHSO₄ (potassium bisulfate), 224
KMnO₄ (potassium permanganate), 222, *222*
Kwolek, Stephanie, 217

L

laboratory notebook, 25–32
    about, 26
    general guidelines, 26
    informal laboratory reports and, 34
    pre-lab flowchart, 28, 29–32, *30–32*
    sections to be filled out pre-lab, 27–28
    section to be filled out during lab, 28, 29, 34, 37
laboratory reports. *See also specific labs*
    about, 34
    evaluation, 37
    format, 34
    sections, 34, 35–37
λ (wavelength), 78, 79
λ_max (wavelength maximum absorption), 79, 80–81, *80–81*
languages and article translations, 146
Le Châtelier's principle, 219
letters (communications), 146
Lewis acid catalysts, 256
Lewis structures, 345, *345*
Liebig, Justus von, 136–137, 289, *289*
life expectancy, 4–6, *5*
limiting reagent, 27
limonene, 160–161, *161*, 162
Lindahl, Tomas, 241
lipophilicity, 10, 43
liquid-liquid extraction lab
    about and objectives, 159–160
    lab report, 170–171
    for obtaining organic products, 160–164, *161–163*, 170–171
    for separation of organic compounds, 164–169, *166–169*, 170, 171

liquid-liquid extraction techniques, 43–48
    background, 43–44, *44*
    examples of, 47–48, *47–48*
    for obtaining organic products, 44, *45*
    practical aspects of, 46–47, *47*
    for separation of organic compounds, 44–45, *46*
liquid sample preparation, for IR spectroscopy, 85–86
liquid–vapor phase diagram, 50–51, *51*
literature of organic chemistry, 145–157
    about and objectives, 145, *146*
    assignment, 153–157
    assignment examples, 156–157
    citation formats, 152–153
    primary sources, 146–147
    secondary sources, 147–152
longer-range coupling, 111–112, *112*
Lucas test, 330–331
LUMO (lowest unoccupied molecular orbital), 210

M

magnetically active (spin-active) nucleus, 95, *95*
magnetic anisotropy, 100, *101*
magnetic dipole (μ), 79, 95, *95*, 103
magnetic dipole transitions, 94
magnetic resonance imaging (MRI), *173–174*, 174. *See also* nuclear magnetic resonance (NMR) spectroscopy
magnetic sector analyzer, 128, *128*
MALDI (matrix-assisted laser desorption ionization), 129
mass spectrometry, 125–142
    database for, 150
    fragmentation and structure observations, 134, *134*
    high-resolution mass spectrometry, 134–135, *135*
    introduction, 127–128, *127–128*
    ionization techniques, 129
    mass spectrum, 127–128, *127*, 130–132, *131–132*
    molecular formula determination, 132–134
    molecular formula overview, 126, *126*
    practical aspects of, 130
    sample preparation, 130
    sample problems, 133–134, 135, 140–142, *140–142*
mass spectrum, 127–128, *127*, 130–132, *131–132*
mass-to-charge ratio (*m/z*), 127, 128, *128*, 132–133, *132*
matrix-assisted laser desorption ionization (MALDI), 129
m-chloroperbenzoic acid (MCPBA), 221, *222*

medical conditions, safety practices regarding, 17
medicines, database of, 150
melting point determination, 70–72, *71*, 334
Merck Index, 150
mercury disposal guidelines, *21*
mercury vapor lamp, 244, *244*
methiodide, 339, *339*
3-methoxyacetophenone, 300, *300*
1-(3-Methoxyphenyl)ethanol, *300*
4-methyl-3-heptanol, 273, 274, *274*, 275, *277*, 279–281. *See also* Grignard synthesis lab
4-methyl-3-heptanone, 274, *274*, 275
(*R*)-(+)-3-methyladipic acid, synthesis of, 30–32, *30–32*
methylenecyclohexane, 245
methyltransferases, 201
methyltrioctylammonium (Aliquat 336), 223, 224
mixed melting point, 72
mobile phase
    of chromatography (general), 60–61, *60*
    in column chromatography, 67–68
    in gas chromatography, 61–62
    in thin-layer chromatography, 64
Modrich, Paul, 241
molar absorptivity or molar extinction coefficient (ε), 81–82
molar equivalent, 27
molecular bond vibrations, 78
molecular formula
    combustion analysis of, 136–138, *137*
    confirmation of, 138
    introduction, 126, *126*
    mass spectrometry and, 126, *126*, 132–134
    sample problems, 137–138, *137*, 139–142, *139–142*
    structural clues from, 138–140
molecular modeling, 343–361
    about, 343–345, *344*
    experimental procedure, 353–359, *354–356, 358, 361*
    of hydrogen-bonded molecular complexes, 351–353, *351–353*
    of ketones, 345–351, *345–351*
    laboratory report, 359–360, *361*
    pre-lab reading assignment, 345
    software for, 354–358, *354–356, 358*
mother liquor, 55, 57
MRI (magnetic resonance imaging), *173–174*, 174. *See also* nuclear magnetic resonance (NMR) spectroscopy
multiplicity (signal splitting)
    in ¹³C NMR spectroscopy, 120–121, 122
    data interpretation, 97, 177
    data reporting, 116–117, *116–117*
    introduction, 103–110, *104–108*, 110
m/z (mass-to-charge ratio), 127, 128, *128*, 132–133, *132*

vibration frequency (*v*; Hooke's law), 82–83, *83*

vicinal relationship, 103–106, *104–106*, 107, *108*

vicinal σ bonds, 203, 211

vitamin B$_{12}$, 264–265, *265*

W

Warner, John, 7

waste
    disposal of, 20, *21*
    handling of, 12
    prevention of, 7, *8*

water hydrogen bonds, 243, *243*

wavelength (λ), 78, 79

wavelength maximum absorption (λ$_{max}$), 79, 80–81, *80–81*

Web of Science, 150

willow trees, *183*

Wilstätter, R. M., 137

Wittig, Georg, 245

Wittig olefination lab, 239–252
    background, 245–246, *245–247*
    experimental procedure, 248–249
    laboratory report, 252

Wöhler, Friedrich, 289, *289*

workup (procedure), 43

X

xeroderma pigmentosum, 239

x-ray crystallography, 315

xylan, 287–288, *288*

D-Xylose, 287–288, *287*

Y

ylide, 245, *245–247*, 246

Z

zwitterionic intermediate, 246